An Introduction to Ansys® Fluent® 2023

John E. Matsson, Ph.D., P.E.

Initializing project ...

SDC Publications
P.O. Box 1334
Mission, KS 66222
913-262-2664
www.SDCpublications.com
Publisher: Stephen Schroff

ISBN-13: 978-1-63057-648-6
ISBN-10: 1-63057-648-4

Printed and bound in the United States of America.

Acknowledgements

I would like to thank Stephen Schroff of SDC Publications for his help in preparing this book for publication.

About the Author

Dr. John Matsson is a Professor of Engineering and Chair of the Engineering Department at Oral Roberts University in Tulsa, Oklahoma. He earned M.S. and Ph.D. degrees from the Royal Institute of Technology in Stockholm, Sweden, in 1988 and 1994, respectively, and completed postdoctoral work at the Norwegian University of Science and Technology in Trondheim, Norway. His teaching areas include Finite Element Methods, Fluid Mechanics, Manufacturing Processes, and Aerodynamics. He is a member of the American Society of Mechanical Engineers ASME Mid-Continent Section. Please contact the author jmatsson@oru.edu with any comments, questions, or suggestions on this book.

Notes:

TABLE OF CONTENTS

TABLE OF CONTENTS

TABLE OF CONTENTS

TABLE OF CONTENTS

TABLE OF CONTENTS

TABLE OF CONTENTS

Notes:

CHAPTER 1. INTRODUCTION

A. Ansys Workbench

Ansys Workbench is an integrated simulation platform that includes a wide range of systems in the toolbox including analysis systems, component systems, custom systems, and design exploration, see Table 1.1.

Analysis Systems	Component Systems	Custom Systems	Design Exploration
Design Assessment	ACP (Post)	FSI: Fluid Flow (CFX) > Static Structural	Direct Optimization
Eigenvalue Buckling	ACP (Pre)	FSI: Fluid Flow (FLUENT) > Static Structural	Parameters Correlation
Electric	Autodyn	Pre-Stress Modal	Response Surface
Explicit Dynamics	BladeGen	Random Vibration	Response Surface Optimization
Fluid Flow (CFX)	CFX	Response Spectrum	ROM Builder
Fluid Flow (Fluent)	Engineering Data	Thermal-Stress	Six Sigma Analysis
Harmonic Acoustics	External Data		
Harmonic Response	External Model		
IC Engine (Fluent)	Fluent		
IC Engine (Forte)	Fluent (with Fluent Meshing)		
Magnetostatic	Forte		
Modal	Geometry		
Modal Acoustics	Mechanical APDL		
Random Vibration	Mechanical Model		
Response Spectrum	Mesh		
Rigid Dynamics	Microsoft Office Excel		
Static Acoustics	Performance Map		
Static Structural	Results		
Steady-State Thermal	System Coupling		
Thermal-Electric	Turbo Setup		
Topology Optimization	Turbo Grid		
Transient Structural	Vista AFD		
Transient Thermal	Vista CCD		
TurbomachineryFluid Flow	Vista CPD		
	Vista RTD		

Table 1.1 Available systems in Ansys Workbench Toolbox

Ansys Workbench also has a Project Schematic where the details for the process of execution of the different analysis systems are available. These details of each analysis system in the project schematic include Engineering Data, Geometry, Model, Setup, Solution and Results. Furthermore, properties of schematics are available for user input.

B. Ansys DesignModeler

The geometry for the Fluid Flow (Fluent) flow study project can be created either in DesignModeler or alternatively using SpaceClaim. DesignModeler is a parametric modeler designed to draw 2D sketches and model 3D CAD parts. Ansys DesignModeler has nine different toolbars as listed in Table 1.2.

Toolbars
Active Plane/Sketch toolbar
Display toolbar
Feature toolbar
File toolbar
Graphics Options toolbar
Menus toolbar
Rotation Modes toolbar
Selection toolbar
Undo/Redo toolbar

Table 1.2 Toolbars in Ansys DesignModeler

Moreover, the Sketching Toolboxes in DesignModeler include Draw, Modify, Dimensions, Constraints and Settings as shown in Table 1.3.

Constraints	Dimensions	Draw	Modify	Settings
Auto Constraints	Angle	Arc by 3 Points	Chamfer	Grid
Coincident	Animate	Arc by Center	Copy	Major Grid Spacing
Concentric	Diameter	Arc by Tangent	Corner	Minor-Steps per Major
Equal Distance	Display	Circle	Cut	Snaps per Minor
Equal Length	Edit	Circle by 3 Tangents	Drag	
Equal Radius	General	Construction Point	Duplicate	
Fixed	Horizontal	Construction Point at Intersection	Extend	
Horizontal	Length/Distance	Ellipse	Fillet	
Midpoint	Move	Line	Move	
Parallel	Radius	Line by 2 Tangents	Offset	
Perpendicular	Semi-Automatic	Oval	Paste	
Symmetry	Vertical	Polygon	Replicate	
Tangent		Polyline	Spline Edit	
Vertical		Rectangle	Split	
		Rectangle by 3 Points	Trim	
		Spline		
		Tangent Line		

Table 1.3 Sketching Toolboxes in Ansys DesignModeler

C. Ansys Meshing

Ansys Meshing produces the mesh using Mesh Methods and Mesh Controls as shown in Tables 1.4 and Table 1.5.

Hexahedral Meshing	Tetrahedral Meshing	Surface Meshing
Cut cell Cartesian	ANSYS CFX-Mesh	Default quad, quad/tri or tri
General Sweep	Patch conforming	Uniform quad or quad tri
Hex-dominant	Patch independent	
MultiZone		
Thin Sweep		

Table 1.4 Mesh methods in Ansys Meshing

CHAPTER 1. INTRODUCTION

Global Mesh Control	Local Mesh Control
Curvature-based refinement settings	Automatic Contact Detection
Element midside node settings	Body, face, edge curvature-based refinement
Inflation settings	Body, face, edge, vertex sizings
Physics preference settings	Body, face, edge, vertex sphere of influence
Pinch (defeaturing) settings	Body mesh method controls
Proximity-based refinement settings	Body of Influence
Quality settings	Contact Sizing
Relevance settings	Gap tool
Rigid-body behavior settings	Inflation controls
Smoothing settings	Mapped-face meshing controls
Transition/growth settings	Match mesh controls
	Pinch controls
	Solver-based refinement controls
	Virtual topologies

Table 1.5 Mesh controls in Ansys Meshing

D. Ansys Fluent

Ansys Fluent is a computational fluid dynamics software package that is written in the C language. Fluent has numerous capabilities for simulations as shown in Table 1.6.

Fluent Capabilities
2D Planar flows, Axisymmetric flows, Axisymmetric flows with swirl, 3D Flows
Acoustics
Cavitation flows
Chemical species mixing and reaction
Compressible flows, Incompressible flows
Forced heat transfer, Mixed convection heat transfer, Natural heat transfer
Free surface flows
Ideal gases, Real gases
Inviscid flows, Laminar flows, Turbulent flows
Lumped Parameter Models
Melting and Solidification
Multiphase flows
Newtonian flows, Non-Newtonian flows
Porous media
Steady flows, Time-dependent flows

Table 1.6 Some of the capabilities for Ansys Fluent

There are eleven add on modules available in Ansys as shown in Table 0.7.

Fluent Addon Modules
Adjoint Solver
Dual-Potential MSMD Battery Model
Fiber Model
Fuel Cell and Electrolysis Model
Macroscopic Particle Model
MHD Model
PEM Fuel Cell Model
Population Balance Model
Single-Potential Battery Model
SOFC Model with Unresolved Electrolyte
Reduced Order Model

Table 1.7 Add on modules for Ansys Fluent

CHAPTER 1. INTRODUCTION

The user interface of Fluent includes the Outline View and the Task Page. The Outline View includes Setup, Solution, Results and Parameters & Customization, see Table 1.8. The different categories and variables available in setup, solution and results are shown in Tables 1.9-1.12.

Setup	Solution	Results	Parameters & Customization
Boundary Conditions	Calculation Activities	Animations	Custom Field Functions
Cell Zone Conditions	Cell Registers	Graphics	Parameters
Dynamic Mesh	Controls	Plots	User Defined Functions
General	Initialization	Reports	User Defined Memory
Materials	Methods	Scene	User Defined Scalars
Models	Monitors	Surfaces	
Named Expressions	Report Definitions		
Reference Frames	Run Calculation		
Reference Values			

Table 1.8 Outline view options in Ansys Fluent

Setup Models
Acoustics (Ffowcs-Williams & Hawkings, Wave Equation)
Discrete Phase (Numerics, Parallel, Physical Models, Tracking, UDF)
Electric Potential (Potential Equation)
Energy (Energy Equation)
Heat Exchanger (Dual Cell Model, Macro Model Group, Ungrouped Macro Model)
Multiphase (Eulerian, Mixture, Volume of Fluid, Wet Stream)
Radiation (Discrete Ordinates DO, Discrete Transfer DTRM, P1, Rosseland, Surface to Surface S2S)
Solidification and Melting (Back Diffusion, Solidification/Melting)
Species (Composition PDF Transport, Non-Premixed Combustion, Partially Premixed Combustion, Premixed Combustion, Species Transport)
Structure (Linear Elasticity)
Viscous (Detached Eddy Simulation DES, Inviscid, k-epsilon 2 eqn, k-omega 3 eqn, Laminar, Reynolds Stress 5 eqn, Scale-Adaptive Simulation SAS, Spalart-Allmaras 1 eqn, Transition k-kl-omega 3 eqn, Transition SST 4 eqn)

Table 1.9 Models options in Ansys Fluent

Setup Reference Values
Area (m2)
Density (kg/m3)
Enthalpy (j/kg)
Length (m)
Pressure (pascal)
Ratio of Specific Heats
Temperature (k)
Velocity (m/s)
Viscosity (kg/m-s)

Table 1.10 Reference values in Ansys Fluent

Solution	
Calculation Activities	Monitors
Autosave	Convergence Conditions
Execute Commands	Report Files
Cell Register Operations	Report Plots
Solution Animation	Residual

Table 1.11 Calculation activities and monitors options in Ansys Fluent

Results			
Animations	**Graphics**	**Plots**	**Reports**
Scene Animation	Contours	FFT	Discrete Phase
Solution Animation Playback	Mesh	File	Fluxes
Sweep Surface	Vectors	Histogram	Forces
	Particle Tracks	Interpolated Data	Surface Integrals
	Pathlines	Profile Data	Volume Integrals
	Vectors	XY Plot	

Table 1.12 Animation, graphics, plots and reports options in Ansys Fluent

The different options that are available for the General Setup are Gravity, Mesh and Solver with the details as shown in Table 1.13.

General		
Gravity	**Mesh**	**Solver**
X (m/s2)	Check	2D Space (Axisymmetric, Axisymmetric Swirl, Planar)
Y (m/s2)	Display...	Time (Steady, Transient)
Z (m/s2)	Report Quality	Type (Density-Based, Pressure-Based)
	Scale...	Velocity Formulation (Absolute, Relative)
	Units...	

Table 1.13 Gravity, mesh and solver options in Ansys Fluent

The different Solution Methods are shown in Table 1.14.

Solution Methods	
Pressure-Velocity Coupling	**Pressure-Based Solver , Spatial Discretization**
Coupled	Gradient (Green-Gauss Cell Based, Green-Gauss Node Based, Least Squares Cell Based)
PISO	Pressure (Body Force Weighted, Linear, PRESTO!, Second Order, Standard)
SIMPLE	Momentum (First Order Upwind, Power Law, QUICK, Second Order Upwind, Third-Order MUSCL)
SIMPLEC	Energy (First Order Upwind, Power Law, QUICK, Second Order Upwind, Third-Order MUSCL)
	Density-Based Solver , Spatial Discretization
	Gradient (Green-Gauss Cell Based, Green-Gauss Node Based, Least Squares Cell Based)
	Flow (First Order Upwind, Second Order Upwind, Third-Order MUSCL)

Table 1.14 Solution methods options in Ansys Fluent

E. References

1. ANSYS Fluent Getting Started Guide, Release 19.1, April 2018, ANSYS, Inc.
2. ANSYS FLUENT Theory Guide, Release 15.0, January 2013, ANSYS, Inc.
3. ANSYS Fluent Tutorial Guide, Release 18.0, January 2017, ANSYS, Inc.
4. ANSYS FLUENT User's Guide, Release 15.0, November 2013, ANSYS, Inc.
5. ANSYS Meshing User's Guide, Release 15.0, November 2013, ANSYS, Inc.
6. ANSYS Workbench User's Guide, Release 12.1, November 2009, ANSYS, Inc.
7. DesignModeler User Guide, Release 14.5, October 2012, ANSYS, Inc.

F. Notes

CHAPTER 2. FLAT PLATE BOUNDARY LAYER

A. Objectives

- Creating Geometry in Ansys Workbench for Ansys Fluent
- Setting up Ansys Fluent for Laminar Steady 2D Planar Flow
- Setting up Mesh
- Selecting Boundary Conditions
- Running Calculations
- Using Plots to Visualize Resulting Flow Field
- Compare with Theoretical Solution using Mathematica Code

B. Problem Description

In this chapter, we will use Ansys Fluent to study the two-dimensional laminar flow on a horizontal flat plate. The size of the plate is considered being infinite in the spanwise direction and therefore the flow is 2D instead of 3D. The inlet velocity for the 1 m long plate is 5 m/s and we will be using air as the fluid for laminar simulations. We will determine the velocity profiles and plot the profiles. We will start by creating the geometry needed for the simulation.

C. Launching Ansys Workbench and Selecting Fluent

1. Start by launching Ansys Workbench. Double click on Fluid Flow (Fluent) that is located under Analysis Systems in Toolbox.

Figure 2.1 Selecting Fluid Flow

D. Launching Ansys DesignModeler

2. Select Geometry under Project Schematic in Ansys Workbench. Right-click on Geometry and select Properties. Select 2D Analysis Type under Advanced Geometry Options in Properties of Schematic A2: Geometry. Right-click on Geometry in Project Schematic and select to launch New DesignModeler Geometry. Select Units>>Millimeter as the length unit from the menu in DesignModeler.

Figure 2.2a) Selecting Geometry and 2D Analysis Type

Figure 2.2b) Launching DesignModeler

Figure 2.2c) Selecting the length unit

3. Next, we will be creating the geometry in DesignModeler. Select XYPlane from the Tree Outline on the left-hand side in DesignModeler. Select Look at Sketch [icon]. Click on the Sketching tab in the Tree Outline and select the Line sketch tool.

Draw a horizontal line 1,000 mm long from the origin to the right. Make sure you have a *P* at the origin when you start drawing the line. Also, make sure you have an *H* along the line so that it is horizontal and a *C* at the end of the line.

Select Dimensions within the Sketching options. Click on the line and enter a length of 1000 mm. Draw a vertical line upward 100 mm long starting at the end point of the first horizontal line. Make sure you have a *P* when starting the line and a *V* indicating a vertical line. Continue with a horizontal line 100 mm long to the left from the origin followed by another vertical line 100 mm long.

The next line will be horizontal with a length 100 mm starting at the endpoint of the former vertical line and directed to the right. Finally, close the rectangle with a 1,000 mm long horizontal line starting 100 mm above the origin and directed to the right.

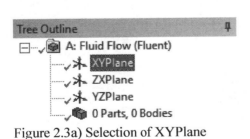

Figure 2.3a) Selection of XYPlane   Figure 2.3b) Selection of Line tool

Figure 2.3c) Rectangle with dimensions

Details View	
Details of Sketch1	
Sketch	Sketch1
Sketch Visibility	Show Sketch
Show Constraints?	No
Dimensions: 4	
☐ H1	1000 mm
☐ H3	100 mm
☐ V2	100 mm
☐ V4	100 mm

Figure 2.3d) Dimensions in Details View

4. Click on the Modeling tab under Sketching Toolboxes. Select Concept>>Surfaces from Sketches in the menu. Control select the six edges of the rectangle as Base Objects and select Apply in Details View. Click on Generate in the toolbar ⌐ Generate. The rectangle turns gray. Right click in the graphics window and select Zoom to Fit and close DesignModeler.

Figure 2.4a) Selecting Surfaces from Sketches

Details View		무
Details of SurfaceSk1		
Surface From Sketches	SurfaceSk1	
Base Objects	Apply	Cancel
Operation	Add Material	
Orient With Plane Normal?	Yes	
Thickness (>=0)	0 mm	

Figure 2.4b) Applying Base Objects

Figure 2.4c) Completed rectangle in DesignModeler

E. Launching Ansys Meshing

5. We are now going to double click on Mesh under Project Schematic in Ansys Workbench to open the Meshing window. Select Mesh in the Outline of the Meshing window. Right click and select Generate Mesh. A coarse mesh is created. Select Unit Systems>>Metric (mm, kg, N …) from the bottom of the graphics window.

Figure 2.5a) Launching mesh

Figure 2.5b) Generating mesh

Figure 2.5c) Coarse mesh

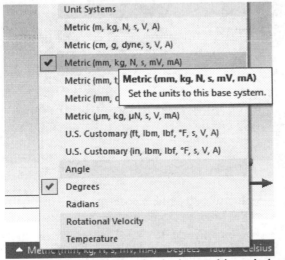

Figure 2.5d) Selection of units in graphics window

Select Mesh>> Controls>>Face Meshing from the menu. Click on the yellow region next to Geometry under Scope in Details of Face Meshing. Select the rectangle in the graphics window. Click on the Apply button for Geometry in Details of "Face Meshing". Select Mesh>>

Controls>>Sizing from the menu and select Edge above the graphics window.

Figure 2.5e) Selection of face meshing   Figure 2.5f) Sizing

Select the upper longer horizontal edge of the rectangle, control select the lower horizontal edges and the vertical edges for a total of 5 edges. Click on Apply for the Geometry in "Details of Edge Sizing". Under Definition in "Details of Edge Sizing", select Element Size as Type, 1.0 mm for Element Size, Capture Curvature as No, and Hard as Behavior. Select the second Bias Type - - — —— and enter 12.0 as the Bias Factor.

Figure 2.5g) Details of edge sizing for three of the horizontal edges and the vertical edges

Repeat the selection of Mesh>>Controls>>Sizing from the menu once again but this time select the upper horizontal edge to the left of the origin. Enter the same Element Size, Behavior, and Bias Factor but the first Bias Type —— — - - . Click on Home>>Generate Mesh in the menu and select Mesh in the Outline. The finished mesh is shown in the graphics window.

Details of "Edge Sizing 2" - Sizing ▾ ⊕ □ ✕	
⊟ **Scope**	
Scoping Method	Geometry Selection
Geometry	1 Edge
⊟ **Definition**	
Suppressed	No
Type	Element Size
☐ Element Size	1.0 mm
⊟ **Advanced**	
Behavior	Hard
Capture Curvature	No
Capture Proximity	No
Bias Type	___ __ _ _
Bias Option	Bias Factor
☐ Bias Factor	12.0
Reverse Bias	No Selection

Figure 2.5h) Details of edge sizing for the remaining horizontal edge

Figure 2.5i) Details of finished mesh

> **Why did we create a biased mesh?** The reason for using a biased mesh is that we need a finer mesh close to the wall where we have velocity gradients in the flow. We also included a finer mesh where the boundary layer starts to develop on the flat plate.

We are now going to rename the edges for the rectangle. Select the left edge of the rectangle, right click and select Create Named Selection. Enter *inlet* as the name and click on the OK button.

Repeat this step for the right vertical edge of the rectangle and enter the name *outlet*. Create a named selection for the lower longer horizontal right edge and call it *wall*. Finally, control-select the remaining three horizontal edges and name them *ideal wall*. An ideal wall is an adiabatic and frictionless wall.

Figure 2.5j) Named selections

Select File>>Export...>>Mesh>>FLUENT Input File>>Export from the menu. Select Save as type: FLUENT Input Files (*.msh). Enter *boundary-layer-mesh* as file name and click on the Save button. Select File>>Save Project from the menu. Name the project *Flat Plate Boundary Layer*. Close the Ansys Meshing window. Right click on Mesh in Project Schematic and select Update.

F. Launching Ansys Fluent

6. We are now going to double click on Setup under Project Schematic in Ansys Workbench to open Fluent. Launch the Dimension 2D and Double Precision solver of Fluent. Check Double Precision under Options. Set the number of Solver Processes equal to the number of computer cores. To check the number of physical cores, press the Ctrl + Shift + Esc keys simultaneously to open the Task Manager. Go to the Performance tab and select CPU from the left column. You'll see the number of physical cores on the bottom-right side. Ansys Student is limited to a maximum of 4 solver processes. Close the Task Manager window. Click on the Start button to launch Ansys Fluent. Click OK to close the Key Behavioral Changes window if it appears.

> **Why do we use double precision?** Double precision will give more accurate calculations than single precision.

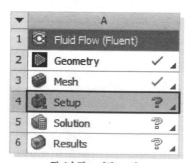

Fluid Flow (Fluent)

Figure 2.6a) Launching Setup

Figure 2.6b) Taskmanager

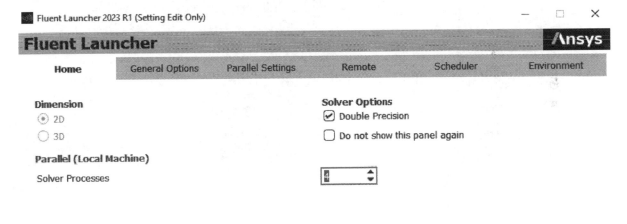

Figure 2.6c) Fluent Launcher

```
-----------------------------------------------------------------------
ID      Hostname     Core  O.S.         PID    Vendor
-----------------------------------------------------------------------
n3      1Jmatsson4   4/28  Windows-x64  31012  Intel(R) Xeon(R) W-2275
n2      1Jmatsson4   3/28  Windows-x64  32248  Intel(R) Xeon(R) W-2275
n1      1Jmatsson4   2/28  Windows-x64  28444  Intel(R) Xeon(R) W-2275
n0*     1Jmatsson4   1/28  Windows-x64  26128  Intel(R) Xeon(R) W-2275
host    1Jmatsson4         Windows-x64  16308  Intel(R) Xeon(R) W-2275

MPI Option Selected: intel
Selected system interconnect: default
-----------------------------------------------------------------------
```

Figure 2.6d) Example of console printout for four cores

7. Check the scale of the mesh by selecting the Scale button under Mesh in General on the Task Page. Make sure that the Domain Extent is correct and close the Scale Mesh window.

Figure 2.7a) Scale Check

Figure 2.7b) Scale Mesh

8. Double click on Models and Viscous (SST k-omega) under Setup in the Outline View. Select Laminar as Viscous Model. Click OK to close the window. Double click on Boundary Conditions under Setup in the Outline View. Double click on *inlet* under Zone on the Task Page. Choose Components as Velocity Specification Method and set the X-Velocity [m/s] to 5. Click on the Apply button followed by the Close button. Double click on *ideal_wall* under Zones. Check Specified Shear as Shear Condition and keep zero values for specified shear stress since an ideal wall is frictionless. Click on the Apply button followed by the Close button.

Why did we select Laminar as Viscous Model? For the chosen free stream velocity 5 m/s the Reynolds number is less than 500,000 along the plate and the flow is therefore laminar. Turbulent flow along a flat plate occurs at Reynolds numbers above 500,000.

Figure 2.8a) Viscous model

Figure 2.8b) Boundary Conditions for inlet

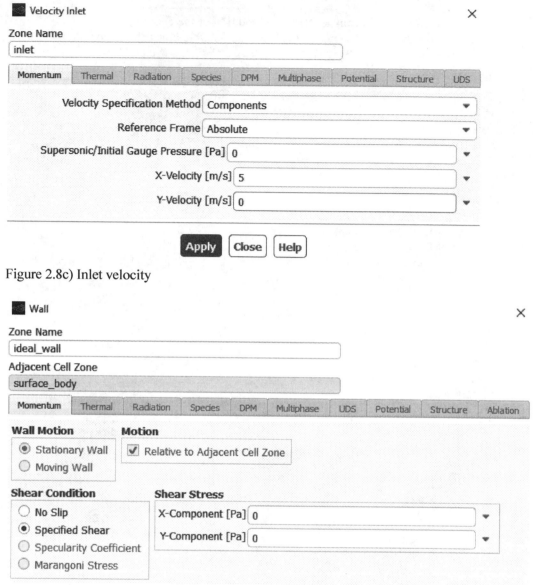

Figure 2.8c) Inlet velocity

Figure 2.8d) Specified shear for ideal wall

9. Double click on Methods under Solution in the Outline View. Select *Standard* for Pressure and *First Order Upwind* for Momentum. Double click on Reference Values under Setup in the Outline View. Select *Compute from inlet* on the Task Page.

Figure 2.9a) Solution Methods Task Page

Why do we use First Order Upwind method for Spatial Discretization of Momentum?
The First Order Upwind method is generally less accurate but converges better than the Second Order Upwind method. It is common practice to start with the First Order Upwind method at the beginning of calculations and continue with the Second Order Upwind method.

Figure 2.9b) Reference values

10. Double click on Initialization under Solution in the Outline View, select *Standard Initialization*, select *Compute from inlet*, and click on the *Initialize* button.

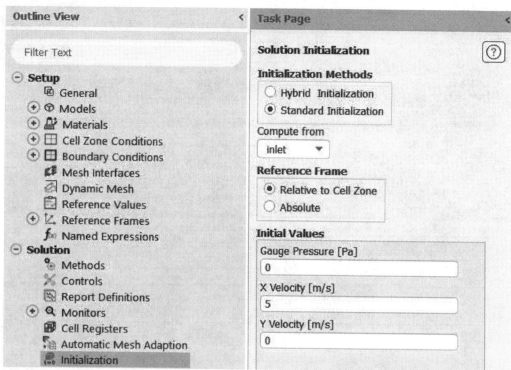

Figure 2.10 Solution Initialization

11. Double click on Monitors under Solution in the Outline View. Double click on Residual under Monitors in the Outline View and enter 1e-9 as *Absolute Criteria* for all Residuals. Click on the OK button to close the window. Select File>>Save Project from the menu. Select File>>Export>>Case… from the menu. Save the Case File with the name *Flat Plate Boundary Layer.cas.h5*

Figure 2.11a) Residual monitors

Figure 2.11b) Residual Monitors settings

Why did we set the Absolute Criteria to 1e-9? Generally, the lower the absolute criteria, the longer time the calculation will take and give a more exact solution. We see in Figure 2.12b) that the x-velocity and y-velocity equations have lower residuals than the continuity equation. The slopes of the residual curves for all three equations are about the same with a sharp downward trend.

12. Double click on Run Calculation under Solution and enter 5000 for *Number of Iterations*. Click on the *Calculate* button. The calculations will be complete after 196 iterations, see Figure 2.12b). Click on Copy Screenshot of Active Window to Clipboard, see Figure 2.12c). The Scaled Residuals can be pasted into a Word document.

Figure 2.12a) Running the calculations

Figure 2.12b) Scaled Residuals

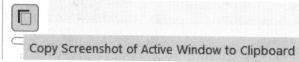

Figure 2.12c) Copying screenshot

G. Post-Processing

13. Select the Results tab in the menu and select Create>>Line/Rake… under Surface. Enter 0.2 for x0 (m), 0.2 for x1 (m), 0 for y0 (m), and 0.02 m for y1 (m). Enter *x=0.2m* for the New Surface Name and click on Create. Repeat this step three more times and create vertical lines at x = 0.4m with length 0.04 m, x=0.6m with length 0.06 m, and x=0.8m with length 0.08 m. Close the window.

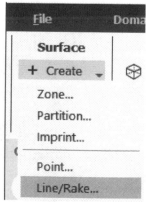

Figure 2.13a) Selecting Line/Rake from the Post-processing menu

38

Figure 2.13b) Line/Rake Surfaces at x = 0.2 m and x = 0.4m

14. Double click on Plots and XY Plot under Results in the Outline View. Uncheck Position on X Axis under Options and check Position on Y Axis. Set Plot Direction for X to 0 and 1 for Y. Select Velocity… and X Velocity as X Axis Function. Select the four lines x=0.2m, x=0.4m, x=0.6m, and x=0.8m under Surfaces.

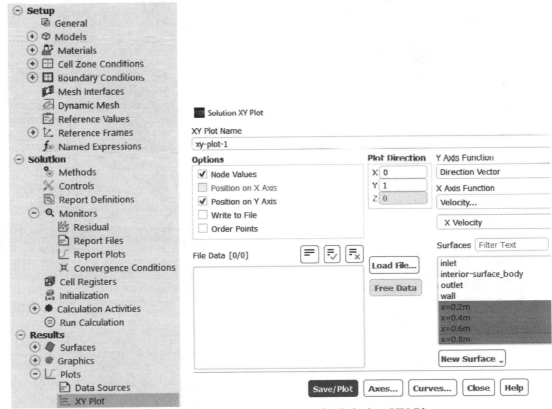

Figure 2.14a) XY Plot setup   Figure 2.14b) Settings for Solution XY Plot

15. Click on the *Axes...* button in the Solution XY Plot window. Select the X Axis, uncheck Auto Range under Options, enter 6 for Maximum Range, select general Type under Number Format and set Precision to 0. Click on the Apply button. Select the Y Axis, uncheck the Auto Range, enter 0.01 for Maximum Range, select general Type under Number Format, and click on the Apply button. Close the Axes window.

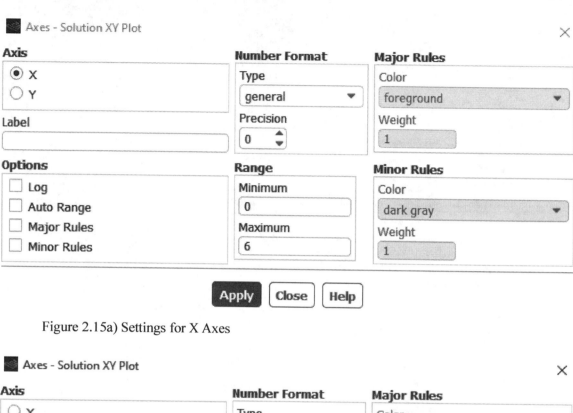

Figure 2.15a) Settings for X Axes

Figure 2.15b) Settings for Y Axes

16. Click on the Curves... button in the Solution XY Plot window. Select the first pattern under Line Style for Curve # 0. Select no Symbol for Marker Style and click on the Apply button. Next select Curve # 1, select the next available Pattern for Line Style, no Symbol for Marker Style, and click on the Apply button. Continue this pattern of selection with the next two curves # 2 and # 3. Close the Curves – Solution XY Plot window. Click on the Save/Plot button in the Solution XY Plot window and Close this window. Click on Copy Screenshot of Active Window to Clipboard, see Figure 2.16c). The XY Plot can be pasted into a Word document.

Figure 2.16a) Settings for curve # 0

Figure 2.16b) Velocity profiles for a laminar boundary layer on a flat plate

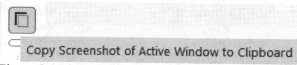

Figure 2.16c) Copying screenshot

Select the *User Defined* tab in the menu and *Custom* under *Field Functions*. Select a specific *Operand Field Functions* from the drop-down menu by selecting *Mesh...* and *Y-Coordinate*. Click on *Select* and enter the definition as shown in Figure 2.16f). You need to select *Mesh...* and *X-Coordinate* to include the x coordinate and complete the definition of the field function. Enter *eta* as *New Function Name*, click on *Define* and close the window.

Repeat this step to create another custom field function. This time, we select *Velocity...* and *X Velocity* as *Field Functions* and click on *Select*. Complete the *Definition* as shown in Figure 2.16g) and enter *u-divided-by-freestream-velocity* as *New Function Name*, click on *Define* and close the window.

Figure 2.16d) Custom Field Functions   Figure 2.16e) Operand Field Function

Figure 2.16f) Custom field function for self-similar coordinate

> **Why did we create a self-similar coordinate?** It turns out that by using a self-similar coordinate, the velocity profiles at different streamwise positions will collapse on one self-similar velocity profile that is independent on the streamwise location.

Figure 2.16g) Custom field function for non-dimensional velocity

17. Double click on Plots and XY Plot under Results in the Outline View. Set X to 0 and Y to 1 as Plot Direction. Uncheck *Position on X Axis* and uncheck *Position on Y Axis* under *Options*. Select *Custom Field Functions* and *eta* for *Y Axis Function* and select *Custom Field Functions* and *u-divided-by-freestream-velocity* for *X Axis Function*.

Place the file *blasius.dat* in your working directory. This file can be downloaded from *sdcpublications.com*. See Figure 2.19 for the Mathematica code that can be used to generate the theoretical Blasius velocity profile for laminar boundary layer flow over a flat plate. As an example, in this textbook the working directory is *C*:\Users\jmatsson. Click on *Load File*. Select Files of type: All Files (*) and select the file *blasius.dat* from your working directory. Select the four surfaces *x=0.2m, x=0.4m, x=0.6m, x=0.8m* and the loaded file *Theory*.

Figure 2.17a) Solution XY Plot for self-similar velocity profiles

43

Click on the Axes… button. Select *Y Axis* in Axes-Solution XY Plot window and uncheck *Auto Range*. Set the *Minimum Range* to 0 and *Maximum Range* to 10. Set the Type to float and Precision to 0 under Number Format. Enter the Label as *eta* and click on *Apply*.

Select *X Axis* in Axes-Solution XY Plot window and enter the Label to *u/U*. Check the box for *Auto Range*. Set the Type to float and Precision to 1 under Number Format. Click on Apply and Close the window.

Click on the Curves… button in the Solution XY Plot window. Select the first pattern under Line Style for Curve # 0, see Figure 2.16a). Select no Symbol for Marker Style and click on the Apply button. Next select Curve # 1, select the next available Pattern for Line Style, no Symbol for Marker Style, and click on the Apply button. Continue this pattern of selection with the next two curves # 2 and # 3. Close the Curves – Solution XY Plot window. Click on the Save/Plot button in the Solution XY Plot window and close this window.

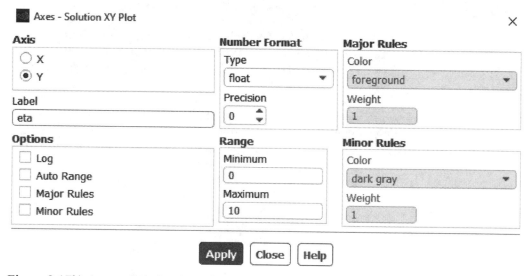

Figure 2.17b) Axes – Solution XY Plot window settings for Y Axis

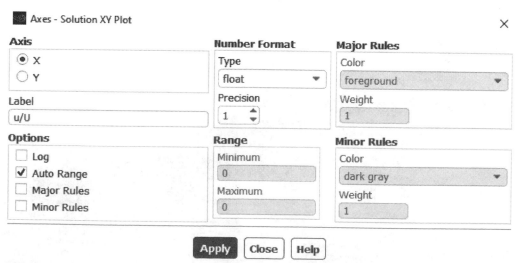

Figure 2.17c) Axes – Solution XY Plot window settings for X Axis

Figure 2.17d) Self-similar velocity profiles for a laminar boundary layer on a flat plate

Click on Copy Screenshot of Active Window to Clipboard, see Figure 2.16c). The XY Plot can be pasted into a Word document.

Select the User Defined tab in the menu and Custom. Select a specific Operand Field Functions from the drop-down menu by selecting Mesh… and X-Coordinate. Click on Select and enter the definition as shown in Figure 2.17e). Enter *re-x* as New Function Name, click on Define and Close the window.

Figure 2.17e) Custom field function for Reynolds number

45

18. Double click on Plots and XY Plot under Results in the Outline View. Set X to 0 and Y to 1 under Plot Direction. Uncheck Position on X Axis and uncheck Position on Y Axis under Options. Select *Wall Fluxes* and *Skin Friction Coefficient* for Y Axis Function and select Custom Field Functions and *re-x* for X Axis Function.

Place the file "*Theoretical Skin Friction Coefficient*" in your working directory. Click on Load File. Select Files of type: All Files (*) and select the file "*Theoretical Skin Friction Coefficient*". Select *wall* under Surfaces and the loaded file Skin Friction under File Data. Click on the Axes... button.

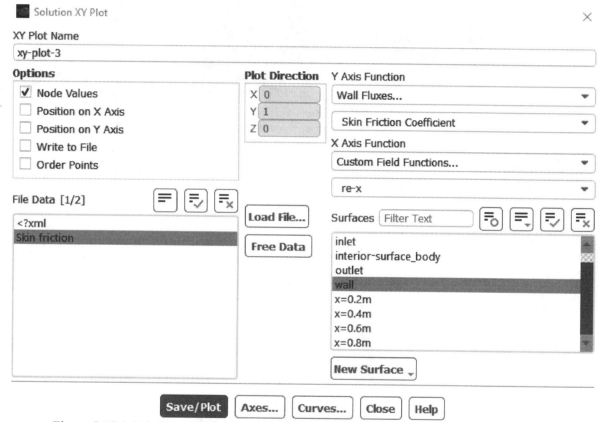

Figure 2.18a) Solution XY Plot for skin friction coefficient

Check the X Axis, check the box for Log under Options, enter Re-x as Label, uncheck Auto Range under Options, set Minimum to 100 and Maximum to 1000000. Set Type to float and Precision to 0 under Number Format and click on Apply.

Check the Y Axis, check the box for Log under Options, enter Cf-x as Label, uncheck Auto Range, set Minimum to 0.001 and Maximum to 0.1, set Type to float, Precision to 3 and click on Apply. Close the window. Click on Save/Plot in the Solution XY Plot window.

Click on the Curves... button in the Solution XY Plot window. Select the first pattern under Line Style for Curve # 0. Select no Symbol for Marker Style and click on the Apply button.

Next select Curve # 1, select the next available Pattern for Line Style, no Symbol for Marker Style, and click on the Apply button. Close the Curves – Solution XY Plot window. Click on the Save/Plot button in the Solution XY Plot window and close this window.

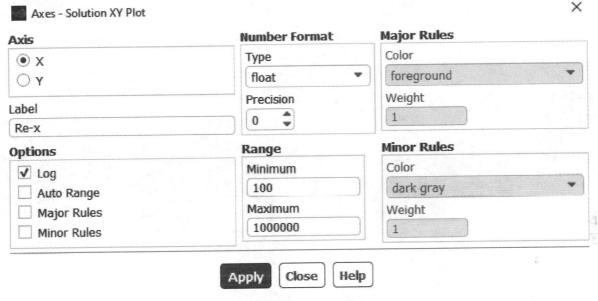

Figure 2.18b) Axes – Solution XY Plot window settings for X Axis

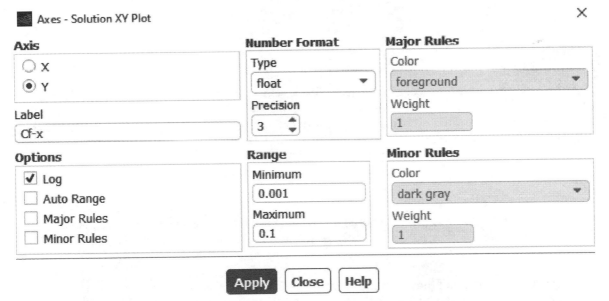

Figure 2.18c) Axes – Solution XY Plot window settings for Y Axis

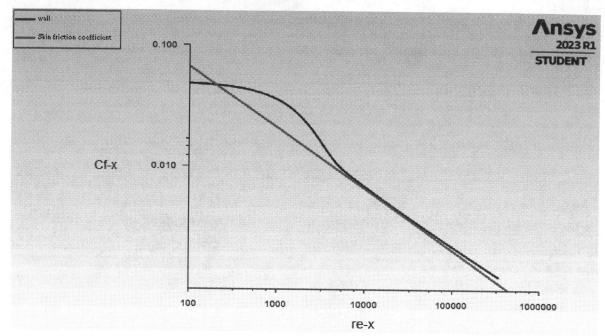

Figure 2.18d) Comparison between Ansys Fluent and theoretical skin friction coefficient for laminar boundary layer flow on a flat plate

Click on Copy Screenshot of Active Window to Clipboard, see Figure 2.16c). The XY Plot can be pasted into a Word document.

H. Theory

19. In this chapter we have compared Ansys Fluent velocity profiles with the theoretical Blasius velocity profile for laminar flow on a flat plate. We transformed the wall normal coordinate to a similarity coordinate for comparison of profiles at different streamwise locations. The similarity coordinate is defined by

$$\eta = y\sqrt{\frac{U}{\nu x}} \qquad (2.1)$$

where y (m) is the wall normal coordinate, U (m/s) is the free stream velocity, x (m) is the distance from the streamwise origin of the wall and ν (m²/s) is the kinematic viscosity of the fluid.

We also used the non-dimensional streamwise velocity u/U where u is the dimensional velocity profile. u/U was plotted versus η for Ansys Fluent velocity profiles in comparison with the Blasius theoretical profile and they all collapsed on the same curve as per definition of self-similarity. The Blasius boundary layer equation is given by

$$f'''(\eta) + \tfrac{1}{2}f(\eta)f''(\eta) = 0 \qquad (2.2)$$

with the following boundary conditions

$$f(0) = f'(0) = 0, f'(\infty) = 0 \tag{2.3}$$

The Reynolds number for the flow on a flat plat is defined as

$$Re_x = \frac{Ux}{\nu} \tag{2.4}$$

The boundary layer thickness δ is defined as the distance from the wall to the location where the velocity in the boundary layer has reached 99% of the free stream value. For a laminar boundary-layer we have the following theoretical expression for the variation of the boundary layer thickness with streamwise distance x and Reynolds number Re_x.

$$\delta = \frac{4.91x}{\sqrt{Re_x}} \tag{2.5}$$

The corresponding expression for the boundary layer thickness in a turbulent boundary layer is given by

$$\delta = \frac{0.16x}{Re_x^{1/7}} \tag{2.6}$$

The local skin friction coefficient is defined as the local wall shear stress divided by dynamic pressure.

$$C_{f,x} = \frac{\tau_w}{\frac{1}{2}\rho U^2} \tag{2.7}$$

The theoretical local friction coefficient for laminar flow is determined by

$$C_{f,x} = \frac{0.664}{\sqrt{Re_x}} \qquad Re_x < 5 \cdot 10^5 \tag{2.8}$$

and for turbulent flow we have the following relation

$$C_{f,x} = \frac{0.027}{Re_x^{1/7}} \qquad 5 \cdot 10^5 < Re_x < 10^7 \tag{2.9}$$

```
sol = NDSolve[{f'''[η] + 0.5 f[η] f''[η] == 0, f[0] == f'[0] == 0, f'[10] == 1}, f, η];
myplot = Plot[f'[η] /. First[sol], {η, 0, 10}, AxesLabel → {"u/U", "η"}];
axisFlip ≠ /. {x_Line | x_GraphicsComplex ↦ MapAt[♯~Reverse~2 &, x, 1],
    x : (PlotRange → _) ↦ x~Reverse~2} &;
myplot // axisFlip


SetDirectory["C:\\Users\\jmatsson"];
mylist = Table[{f'[η] /. First[sol], η}, {η, 0, 10, 0.5}];
TableOfValues1 = Prepend[mylist, {""}];
TableOfValues1 = Prepend[TableOfValues1, {"((xy/key/label \"Blasius profile\")"}];
TableOfValues1 = Prepend[TableOfValues1, {""}];
TableOfValues1 = Prepend[TableOfValues1, {"(labels \"X Velocity\" \"η-Coordinate\")"}];
TableOfValues1 = Prepend[TableOfValues1, {"(title \"Theory\")"}];
TableOfValues1 = Append[TableOfValues1, {")"}];
Grid[TableOfValues1]
Export["blasius.dat", TableOfValues1]
```

Figure 2.19 Mathematica code for theoretical Blasius laminar boundary layer

I. References

1. Çengel, Y. A., and Cimbala J.M., Fluid Mechanics Fundamentals and Applications, 1st Edition, McGraw-Hill, 2006.
2. Richards, S., Cimbala, J.M., Martin, K., ANSYS Workbench Tutorial – Boundary Layer on a Flat Plate, Penn State University, 18 May 2010 Revision.
3. Schlichting, H., and Gersten, K., Boundary Layer Theory, 8th Revised and Enlarged Edition, Springer, 2001.
4. White, F. M., Fluid Mechanics, 4th Edition, McGraw-Hill, 1999.

J. Exercises

2.1 Use the results from the Ansys Fluent simulation in this chapter to determine the boundary layer thickness at the streamwise positions as shown in the table below. Fill in the missing information in the table. U_δ is the velocity of the boundary layer at the distance from the wall equal to the boundary layer thickness and U is the free stream velocity.

x (m)	δ (mm) Fluent	δ (mm) Theory	Percent Difference	U_δ (m/s)	U (m/s)	ν (m²/s)	Re_x
0.2						.0000146	
0.4						.0000146	
0.6						.0000146	
0.8						.0000146	

Table 2.1 Comparison between Fluent and theory for boundary layer thickness

2.2 Change the element size to 2 mm for the mesh and compare the results in XY Plots of the skin friction coefficient versus Reynolds number with the element size 1 mm that was used in this chapter. Compare your results with theory.

2.3 Change the free stream velocity to 3 m/s and create an XY Plot including velocity profiles at $x = 0.1, 0.3, 0.5, 0.7$ and 0.9 m. Create another XY Plot with self-similar velocity profiles for this lower free stream velocity and create an XY Plot for the skin friction coefficient versus Reynolds number.

2.4 Use the results from the Ansys Fluent simulation in Exercise 2.3 to determine the boundary layer thickness at the streamwise positions as shown in the table below. Fill in the missing information in the table. U_δ is the velocity of the boundary layer at the distance from the wall equal to the boundary layer thickness and U is the free stream velocity.

x (m)	δ (mm) Fluent	δ (mm) Theory	Percent Difference	U_δ (m/s)	U (m/s)	ν (m²/s)	Re_x
0.1						.0000146	
0.2						.0000146	
0.5						.0000146	
0.7						.0000146	
0.9						.0000146	

Table 2.2 Comparison between Fluent and theory for boundary layer thickness

K. Notes

CHAPTER 3. FLOW PAST A CYLINDER

A. Objectives

- Using Ansys Workbench to Model the Cylinder and Mesh
- Inserting Boundary Conditions and Free Stream Velocity
- Running Ansys Fluent Simulation for Laminar Transient 2D Planar Flow
- Using FFT Analysis, Velocity, Pressure and Vorticity Plots for Visualizations

B. Problem Description

We will study the flow of air past a cylinder and we will analyze the problem using Ansys Fluent. The diameter of the cylinder is 50 mm and the free stream velocity is 0.06 m/s.

C. Launching Ansys Workbench and Selecting Fluent

1. Start by launching Ansys Workbench. Double click on Fluid Flow (Fluent) that is located under Analysis Systems in the Toolbox.

Figure 3.1 Selecting Fluid Flow (Fluent)

D. Launching Ansys DesignModeler

2. Select Geometry under Project Schematic in Ansys Workbench. Right-click on Geometry and select Properties. Select 2D Analysis Type under Advanced Geometry Options in Properties of Schematic A2: Geometry. Right-click on Geometry under Project Schematic and select to launch New DesignModeler Geometry. Select Units>>Millimeter as the length unit from the menu in DesignModeler.

Figure 3.2a) Selecting Geometry

	A	B
1	Property	Value
2	⊟ General	
3	Component ID	Geometry
4	Directory Name	FFF
5	⊟ Notes	
6	Notes	
7	⊟ Used Licenses	
8	Last Update Used Licenses	
9	⊟ Geometry Source	
10	CAD Plug-In	DesignModeler[11992]
11	⊟ Advanced Geometry Options	
12	Analysis Type	2D
13	Compare Parts On Update	No

Properties of Schematic A2: Geometry

Figure 3.2b) Selecting 2D Analysis Type

Figure 3.2c) Launching DesignModeler

Figure 3.2d) Selecting the length unit

3. Next, we will be creating the geometry for the simulation. Select XYPlane from the Tree Outline on the left-hand side. Select Look at Sketch ⎯. Click on the Sketching tab in the Tree Outline, select the Draw tab within the Sketching options, and select the Circle sketch tool.

Figure 3.3a) Selection of XYPlane   Figure 3.3b) Selection of Circle

Draw a circle centered at the origin of the coordinate system in the graphics window. Make sure that you move the cursor so that you get the letter *P* in the graphics window indicating that your circle will be centered at the origin. Select the Dimensions tab within the Sketching options and select Diameter. Click on the circle in the graphics window and enter 50 mm for the diameter D1 under Dimensions:1 in Details View.

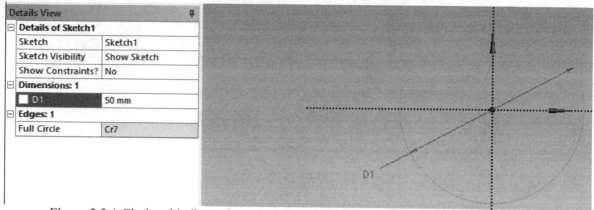

Figure 3.3c) Circle with dimension

Click on the Modeling tab in the Tree Outline and select Concepts>>Surfaces from Sketches from the menu. Click on the circle in the graphics window and select Apply as Base Objects in Details View. Click on ⚡ Generate . The circle turns gray.

Figure 3.3d) Surfaces from Sketches

Figure 3.3e) Applying Base Objects

Click on the Modeling tab in the Tree Outline and select the XYPlane. Click on New Sketch 🗏 . Select Rectangle by 3 Points from the Sketching tools. Click anywhere outside of the circle in the upper left corner (2nd quadrant) of the graphics window, followed by clicking anywhere outside of the circle in the lower left corner (3rd quadrant) of the graphics window (make sure you have a *V* before clicking), and finally click in the right-hand plane outside of the circle (4th quadrant).

Select the Dimensions tab under Sketching Toolboxes, select Horizontal and click on the vertical edges of the rectangle. Enter 1575 mm as the horizontal length of the rectangle. Select the vertical axis and the vertical edge of the rectangle on the left-hand side. Enter 575 mm as the length. Select the Vertical dimensioning tool and click on the horizontal edges of the rectangle. Enter 1250 mm as the vertical length of the rectangle. Click on the horizontal axis and the lower horizontal edge of the rectangle. Enter 625 mm as the length. Right-click in the graphics window and select Zoom to Fit.

Figure 3.3f) Applying Base Objects

Select Concept>>Surface from Sketches from the menu. Click on Sketch2 in the Tree Outline and select Apply for Base Objects in Details View. Select Operation>>Add Frozen in Details View followed by Generate.

Details of SurfaceSk2	
Surface From Sketches	SurfaceSk2
Base Objects	1 Sketch
Operation	Add Frozen
Orient With Plane Normal?	Yes
Thickness (>=0)	0 mm

Figure 3.3g) Add frozen operation

Select Create>>Boolean from the menu. Select Subtract Operation from the Details View. Click on the mesh region around the cylinder in the graphics window and click on Apply as Target Bodies. Select the Cylinder as the Tool Body, see Figure 3.3j), and click on Generate.

Figure 3.3h) Creating a Boolean

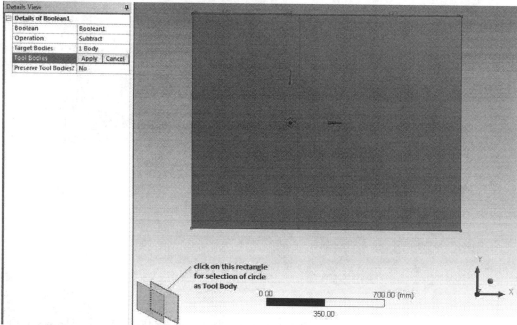

Figure 3.3i) Selecting a subtracting operation

Figure 3.3j) Selecting tool body

Figure 3.3k) Mesh region around cylinder

4. Select *XYPlane* in the Tree Outline and create a New Sketch . Select the Sketching tab and select Line under the Draw tab. Draw a vertical line in the mesh region from top to bottom to the left of the cylinder, see Figure 3.4a). Make sure you have a *C* when you start the line, a *V* indicating that the line is vertical and another *C* at the end of the line. The vertical line will intersect the entire mesh region. Position the line 75 mm from the vertical axis using the Horizontal dimension. Select Concepts>>Lines from Sketches from the menu. Select from the graphics window the newly created vertical line and Apply it as a Base Object in Details View. Click on Generate.

Create another sketch in the *XYPlane* with a vertical line 75 mm to the right of the vertical axis and repeat all the different steps. Finally, repeat these steps two more times with one line from left to right intersecting the mesh region and positioned 75 mm above the cylinder and another line positioned 75 mm below the cylinder.

Figure 3.4a) First vertical line located 75 mm from the vertical axis

Figure 3.4b) Lines from sketches

Figure 3.4c) Lines added to the mesh region

Select Tools>>Projection from the menu. Select Edges on Face as type in Details View. Control-select all 12 line-segments for the 4 lines that you have created and Apply as the Edges in the Details View. Select the mesh region and Apply it as the Target under Details View. Click Generate.

Figure 3.4d) Selection of the projection tool

You will now have 9 different regions for meshing. Select File>>Save Project from the menu and save the project with the name *Cylinder Flow Study*. Close DesignModeler.

E. Launching Ansys Meshing

5. Next, we will be creating the mesh around the cylinder. We are now going to double click on Mesh under Project Schematic in Ansys Workbench to open the Meshing window.

Figure 3.5a) Launching mesh

Select Mesh under Model (A3) in Project under Outline on the left-hand side. Select Mesh>> Controls>>Face Meshing from the menu. Control-select all 9 faces of the mesh. Apply the Geometry under Details of "Face Meshing".

Figure 3.5b) Selection of face meshing

Figure 3.5c) Details of face meshing

Select Mesh>>Controls>>Sizing from the menu, right-click in the graphics window and select Cursor Mode>>Edge. Control-select the three edges to the left and bottom of the mesh region as shown in Figure 3.5d). Apply the Geometry under Details of "Edge Sizing". Select Number of Divisions as Type under Details of "Edge Sizing". Set the Number of Divisions to 100. Set Capture Curvature to No and set the Behavior to Hard. Select the first Bias Type option from the drop-down menu, see Figure 3.5d). Enter a Bias Factor of 10.

Figure 3.5d) Details for first mesh sizing

Select Mesh>>Controls>>Sizing from the menu. Control-select the three edges to the right and bottom of the mesh region as shown in Figure 3.5e). Apply the Geometry under Details of "Edge Sizing 2". Select Number of Divisions as Type under Details of "Edge Sizing 2". Set the Number of Divisions to 150. Set Capture Curvature to No and set the Behavior to Hard. Select the second Bias Type from the drop-down menu, see Figure 3.5e). Enter a Bias Factor of 10.

Figure 3.5e) Details for second mesh sizing

Select Mesh>>Controls>>Sizing from the menu. Select the horizontal upper left edge. Apply the Geometry under Details of "Edge Sizing 3". Select Number of Divisions as Type under Details of "Edge Sizing 3". Set the Number of Divisions to 100. Set Capture Curvature to No and set the Behavior to Hard. Select the second Bias Type from the drop-down menu. Enter a Bias Factor of 10, see Figure 3.5f).

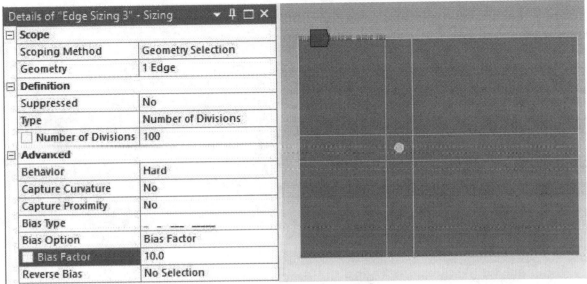

Figure 3.5f) Details for third mesh sizing

Select Mesh>>Controls>>Sizing from the menu. Select the upper horizontal right edge. Apply the Geometry under Details of "Edge Sizing 4". Select Number of Divisions as Type under Details of "Edge Sizing 4". Set the Number of Divisions to 150. Set Capture Curvature to No and set the Behavior to Hard. Select the first Bias Type from the drop-down menu. Enter a Bias Factor of 10, see Figure 3.5g).

Details of "Edge Sizing 4" - Sizing	
Scope	
Scoping Method	Geometry Selection
Geometry	1 Edge
Definition	
Suppressed	No
Type	Number of Divisions
Number of Divisions	150
Advanced	
Behavior	Hard
Capture Curvature	No
Capture Proximity	No
Bias Type	
Bias Option	Bias Factor
Bias Factor	10.0
Reverse Bias	No Selection

Figure 3.5g) Details for fourth mesh sizing

Select Mesh>>Controls>>Sizing from the menu. Control-select the 8 short edges around the cylinder and on the edges as shown in Figure 3.5h). Apply the Geometry under Details of "Edge Sizing 5". Select Number of Divisions as Type under Details of "Edge Sizing 5". Set the number of Divisions to 50. Set Capture Curvature to No and set the Behavior to Hard.

Figure 3.5h) Details for fifth mesh sizing

Select Mesh>>Controls>>Sizing from the menu. Control-select the four edges as shown in Figure 3.5i). Apply the Geometry under Details of "Edge Sizing 6". Select Number of Divisions as Type under Details of "Edge Sizing 6". Set the Number of Divisions to 100. Set Capture Curvature to No and set the Behavior to Hard. Select the first Bias Type from the drop-down menu. Enter a Bias Factor of 20.

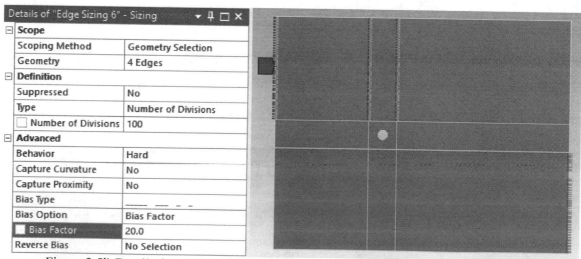

Figure 3.5i) Details for sixth mesh sizing

Select Mesh>>Control>>Sizing from the menu. Control-select the four edges as shown in Figure 3.5j). Apply the Geometry under Details of "Edge Sizing 7". Select Number of Divisions as Type under Details of "Edge Sizing 7". Set the Number of Divisions to 100. Set Capture Curvature to No and set the Behavior to Hard. Select the second Bias Type from the drop-down menu. Enter a Bias Factor of 20. Right click on Mesh in the tree outline and select Generate Mesh.

Details of "Edge Sizing 7" - Sizing	▼ ⊿ ⬜ ✕
Scope	
Scoping Method	Geometry Selection
Geometry	4 Edges
Definition	
Suppressed	No
Type	Number of Divisions
☐ Number of Divisions	100
Advanced	
Behavior	Hard
Capture Curvature	No
Capture Proximity	No
Bias Type	_ _ __ ___
Bias Option	Bias Factor
☐ Bias Factor	20.0
Reverse Bias	No Selection

Figure 3.5j) Details for seventh mesh sizing

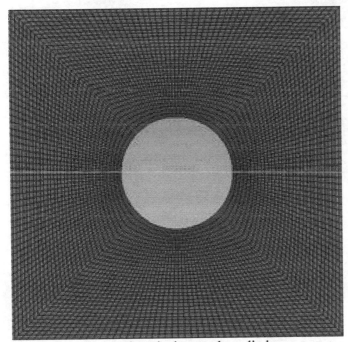

Figure 3.5k) Finished mesh close to the cylinder

Why do we use a biased mesh? A biased mesh enables us to create a finer mesh closer to the cylinder. In this way we create a fine mesh where it is needed the most and the calculation time will be much shorter compared with a finer mesh in the entire mesh region.

6. Select Geometry in the outline. Select the Edge Selection filter . Control-select the 3 vertical edges on the right-hand side, right click and select Create Named Selection. Name the edges "outlet" and click OK. Name the 3 vertical edges to the left "inlet" and the edge of the cylinder with the name "cylinder". Name the 6 horizontal edges (3 at the top and 3 at the bottom) as "symmetry".

Figure 3.6a) Named selections

Open Geometry in the outline, select Line Body, and select Thermal Fluid as Model Type in Details of "Line Body". In Geometry, right click the Line Body and select Suppress Body.

Details of "Line Body"	▼ 廿 □ ×
⊞ Graphics Properties	
⊟ Definition	
☐ Suppressed	No ▼
Model Type	Thermal Fluid
Coordinate System	Default Coordinate System
Offset Mode	Refresh on Update
Offset Type	Centroid
Treatment	None
☐ Fluid Cross Area	0. m²
Fluid Discretization	Upwind/Linear
⊟ Material	
Assignment	
Fluid/Solid	Defined By Geometry (Solid)
⊞ Bounding Box	
⊞ Properties	
⊞ Statistics	

Figure 3.6b) Details of Line Body

Select File>>Export>>Mesh >>FLUENT Input File>>Export from the menu. Enter the name *cylinder-flow-mesh* and select FLUENT Input Files (*,msh) as Save as Type. Select File>>Save Project… from the menu and close the meshing window. Right click on Mesh under Project Schematic in Ansys Workbench and select Update.

> **Why do we export the mesh?** When we save the mesh file, we can read the mesh using standalone mode directly in Ansys Fluent without having to design the mesh all over again each time that we run the simulation for various flow variables.

F. Launching Ansys Fluent

7. Double click Setup under Project Schematic in Ansys Workbench. Check the Options box for Double Precision. Set the number of Solver Processes equal to the number of computer cores. To check the number of physical cores, press the Ctrl + Shift + Esc keys simultaneously to open the Task Manager. Select the Performance tab and select CPU from the left column. You'll see the number of physical cores on the bottom-right side. Ansys Student is limited to 4 Solver Processes. Click OK to launch Fluent. Click OK when you get the Key Behavioral Changes window. Select Transient Time under General on the Task Page.

Figure 3.7 Fluent launcher window

> **Why do we use the Pressure-Based Solver?** The pressure-based solver is usually being chosen for incompressible flows such as the flow past a cylinder at low Reynolds numbers.

> **Why do we use a planar 2D Space study?** The plane 2D Space study is completed under the assumption that the cylinder is infinitely long such that there is no variation of the flow along the length of the cylinder.

8. Double click on Models and Viscous (SST k-omega) under Setup in the Outline View. Select Laminar as Viscous Model. Click OK to close the window. Double click on Boundary Conditions under Setup in the Outline View. Double click on *inlet* under Zone on the Task Page.

> **Why do we use a Transient study and the Laminar model?** The flow past a cylinder is time dependent at the Reynolds number that we are using in this chapter. Vortex shedding will occur behind the cylinder and we are interested in determining the frequency and Strouhal number for the vortex shedding process. The Reynolds number is not sufficiently high to have turbulent flow and therefore we choose Laminar as Viscous model in this case.

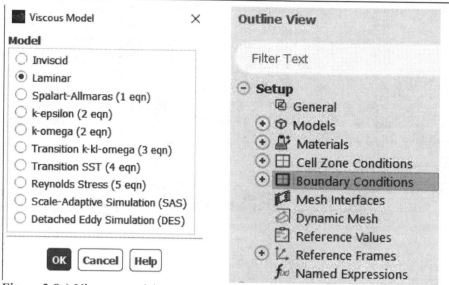

Figure 3.8a) Viscous model   Figure 3.8b) Defining boundary conditions

Select Components as Velocity Specification Method. Enter 0.06 m/s as X-Velocity. Click Apply and Close the Velocity Inlet window. Double-click on Methods in the Outline View under Solution. Set the Pressure-Velocity Coupling Scheme to PISO. Uncheck the box for Skewness-Neighbor Coupling. Select Green-Gauss Cell Based Gradient under Spatial Discretization.

Figure 3.8c) Velocity inlet boundary condition

> **Why do we choose PISO as Pressure-Velocity Coupling Scheme?** PISO is used for transient solutions.

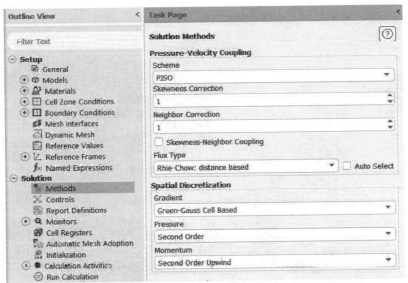

Figure 3.8d) Settings for solution methods

9. Double-click on Initialization in the Outline View under Solution. Select Standard Initialization under Solution Initialization on the Task Page and enter 0.06 for X-Velocity (m/s). Click on Initialize.

Double-click on Reference Values under Setup in the Outline View. Select Compute from inlet from the drop-down menu. Set the value for Area (m2) to 0.05 and Length (m) to the same value. Select *solid-surface_body* as Reference Zone.

Double-click on Report Definitions under Solution in the Outline View. Select New>>Force Report>>Lift... from the drop-down menu. Select Lift Coefficient as Report Output Type and set the Force Vector to X 0, Y 1 and Z 1. Select cylinder under Zones. Check the boxes for Report File, Report Plot and Print to Console under Create. Click on the OK button to close the Lift Report Definition window. Close the Report Definitions window.

Figure 3.9a) Solution initialization

Figure 3.9b) Reference values

Figure 3.9c) New Force Report

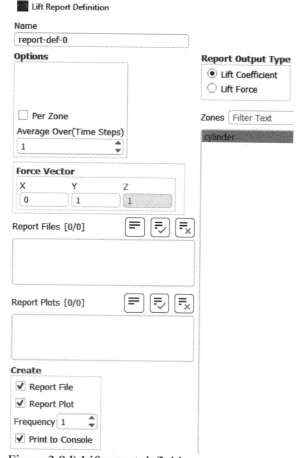

Figure 3.9d) Lift report definition

Select File>>Save Project from the menu. Select File>>Export>>Case & Data… from the menu. Save the Case/Data File with the name *Flow Past a Cylinder.cas.h5*.

Double-click on Run Calculation under Solution on the left-hand side in the Outline View. Enter 0.2 for the Time Step Size (s) and 600 for number of Time Steps. Enter 30 for Max Iterations/Time Step and click on Calculate.

> **Why do we choose 0.2 s as Time Step Size?** The Time Step Size must be sufficiently small to resolve the wave form of the time signal as shown in Figure 3.9g).

Figure 3.9e) Calculation settings

Figure 3.9f) Scaled residuals

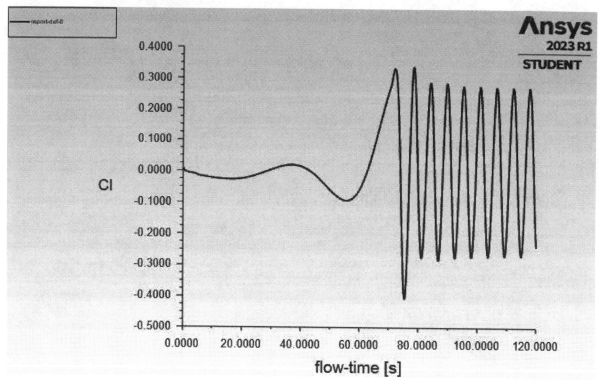

Figure 3.9g) Lift coefficient versus time

Click on Copy Screenshot of Active Window to Clipboard, see Figure 3.9h). The Scaled Residuals and lift coefficient versus flow-time graphs can be pasted into a Word document.

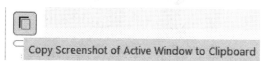

Figure 3.9h) Copying screenshot

G. Post-Processing

10. Open Plots under Results in the Outline View. Double-click on FFT. Select Load Input File in the Fourier Transform window. Select the file with the name *report-def-0-rfile.out* and click OK. Select Axes… at the bottom of the Fourier Transform window. Select Y axis, set Type to float and Precision to 1 as Number Format. Click on Apply and Close the Axes – Fourier Transform window. Click on Plot FFT in the Fourier Transform window. Click on Copy Screenshot of Active Window to Clipboard, see Figure 3.9h). The Power Spectral Density versus Frequency graph can be selected and pasted into a Word document.

Check the box for Write FFT to File under Options in the Fourier Transform window and click on the Write FFT button. Save the file with All Files (*) as file type and the name *fft-data* in the working directory. This file can be opened in Notepad, see Figure 3.10d). The power spectral density maximum is at 0.1747 Hz. The power spectral density shows the variation of power for the different frequencies that makes up the time signal and is based on the mathematical theory for Fourier analysis.

Fourier Transform ✕

Options
☐ Write FFT to File
☐ Write FFT to Simulation Report
☐ Acoustics Analysis

[Ansys Sound Analysis ...]

Reference Acoustic Pressure [Pa]
2e-05

Y Axis Label
Power Spectral Density

Y Axis Variable
report-def-0 ▼

Files
C:/Users/Instructor/Desktop/An Introduction to Ansys Fluent 2023/
◄ ▒▒▒▒▒▒▒▒▒▒▒▒▒▒▒▒▒▒ ►
1. 3/Cylinder Flow Study_files/dp0/FFF/Fluent/report-def-0-rfile.out

Y Axis Function
Power Spectral Density ▼

X Axis Function
Frequency [Hz] ▼

[Plot/Modify Input Signal...]

Plot Title
Spectral Analysis of report-def-0-rfile

X Axis Label
Frequency [Hz]

X Axis Variable
flow-time ▼

[Load Input File...]

[Free File Data]

[Plot FFT] [Axes...] [Curves...] [Close] [Help]

Figure 3.10a) Fourier transform window

Axes - Fourier Transform ✕

Axis
○ X
◉ Y

Label
[   ]

Options
☐ Log
☑ Auto Range
☐ Major Rules
☐ Minor Rules

Number Format
Type
float ▼

Precision
1 ▲▼

Range
Minimum
0

Maximum
0

Major Rules
Color
foreground ▼

Weight
1

Minor Rules
Color
dark gray ▼

Weight
1

[Apply] [Close] [Help]

Figure 3.10b) Axes settings

Figure 3.10c) Plot of power spectral density versus frequency

fft-data - Notepad

File   Edit   Format   View   Help

(title "Spectral Analysis of report-def-0-rfile")
(labels "Frequency [Hz]" "Power Spectral Density")

((xy/key/label "report-def-0")
8.3194673e-03   3.0495193e-02
1.6638935e-02   2.5178339e-02
2.4958402e-02   1.0613659e-01
3.3277869e-02   7.9150878e-02
4.1597337e-02   6.2252596e-02
4.9916804e-02   3.1356338e-02
5.8236271e-02   3.5616327e-02
6.6555738e-02   2.3006709e-02
7.4875206e-02   2.0606434e-02
8.3194673e-02   2.8362103e-02
9.1514140e-02   1.4312185e-02
9.9833608e-02   2.4942609e-02
1.0815307e-01   2.1785064e-02
1.1647254e-01   1.2043310e-02
1.2479201e-01   3.5544541e-02
1.3311148e-01   1.3886261e-02
1.4143094e-01   3.0343210e-02
1.4975041e-01   5.8982324e-02
1.5806988e-01   8.0928123e-03
1.6638935e-01   2.7683389e-01
1.7470881e-01   7.3251098e-01
1.8302828e-01   5.9891981e-01
1.9134775e-01   1.3208690e-01
1.9966722e-01   2.0020199e-03
2.0798668e-01   3.1679697e-02
2.1630615e-01   4.6473574e-03
2.2462562e-01   6.1070612e-03
2.3294508e-01   8.0552222e-03
2.4126455e-01   1.2382607e-04

Figure 3.10d) Data for plot in figure 3.10c)

Double-click on Contours in Graphics under Results in the Outline View. Enter *static-pressure* as Contour Name in the Contours window. Select Contours of Pressure… and Static Pressure. De-select all Surfaces and click on Save/Display.

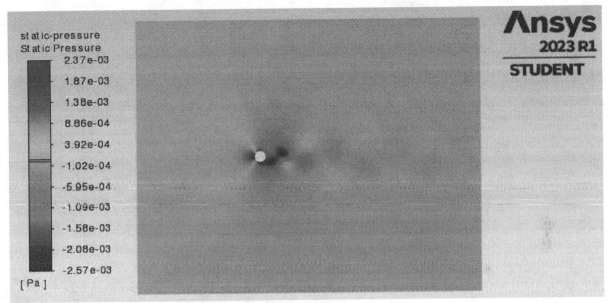

Figure 3.10e) Contours plot of pressure

Create another contour plot and select Contours of Velocity and Vorticity Magnitude. Enter *vorticity-magnitude* as Contour Name in the Contours window. Uncheck the box for Filled under Options. Select Colormap Options… in the Contours window and select Top as Colormap Alignment in the Colormap window. De-select all surfaces and click on Save/Display.

Figure 3.10f) Contour plot of vorticity

Click on Copy Screenshot of Active Window to Clipboard, see Figure 3.9h). The contour plots can be pasted into a Word document. Find the coefficient of drag by double-clicking on Reports under Results in the Outline View. Double-click on Forces under Reports. Select the cylinder under Wall Zones and select Print to obtain the value for the total drag coefficient that you have calculated: 1.1785.

```
Console
Forces - Direction Vector (1 0 0)
                         Forces [N]                               Coefficients
Zone                    Pressure        Viscous         Total       Pressure      Viscous         Total
cylinder                0.00010551857   2.1412398e 05   0.00012993097   0.95708457   0.22142764   1.1785122
---------------------   -------------   -------------   -------------   -----------  ----------   -----------
Net                     0.00010551857   2.4412398e-05   0.00012993097   0.95708457   0.22142764   1.1785122
```

Figure 3.10g) Forces and coefficients

H. Theory

11. We determine the Reynolds number for the flow around the cylinder as

$$Re = Ud/v = 0.06 * 0.050 / 1.5 * 10^{-5} = 200 \tag{3.1}$$

where U (m/s) is the magnitude of the free stream velocity, d (m) is the diameter of the cylinder, and v (m²/s) is the kinematic viscosity of air at room temperature. The drag coefficient from experiments for a smooth cylinder is found using the following curve-fit formula:

$$C_{D,Exp} = 1 + \frac{10}{Re^{2/3}} = 1.29 \qquad 0 \le Re \le 250,000 \tag{3.2}$$

The difference between Ansys Fluent results from Figure 3.10g) and experiments for the drag coefficient is 8.6%.

The Strouhal number S is one of the many non-dimensional numbers used in fluid mechanics and based on the oscillation frequency of fluid flows. In this case the Strouhal number is based on the vortex shedding frequency f as shown in Figures 3.10c), 3.10d) and can be defined as

$$S = \frac{fd}{U} = \frac{0.1747Hz*0.050m}{0.06\ m/s} = 0.1456 \tag{3.3}$$

This value of the Strouhal number from Ansys Fluent can be compared with DNS (Direct Numerical Simulation) results by Henderson[3], see also Williamson and Brown[6]. The difference is 26%.

$$S_{DNS} = 0.2698 - \frac{1.0272}{\sqrt{Re}} = 0.1968 \tag{3.4}$$

I. References

1. Cimbala, J.M., Fluent-Laminar Flow over a cylinder, Penn State University, 29 January 2008.
2. Flow Over a Cylinder, Fluent Inc., FlowLab 1.2, April 16, 2007.
3. Henderson, R., Nonlinear dynamics and pattern formation in turbulent wake transition, J. of Fluid Mech., 352, (65-112), 1997.
4. Tutorial 6. Flow Past a Circular Cylinder. Fluent Inc. January 17, 2007.
5. Validation 3. Laminar Flow Around a Circular Cylinder, Fluent Inc. September 13, 2002.
6. Williamson, C. H. K., and Brown G.L., A series in 1/√Re to represent the Strouhal-Reynolds number relationship of the cylinder wake, Journal of Fluids and Structures, 12, (1073 – 1085), 1998.

J. Exercises

3.1 Rerun the simulation with the same free stream velocity and Reynolds number and the same mesh but use a Time Step Size of 0.25 s and 500 as the Number of Time Steps. Set Max Iterations /Time Step to 30. Determine the drag coefficient and compare with equation (3.1). Determine the Strouhal number and compare with equation (3.4). Include plots of the lift coefficient over time and power spectral density versus frequency.

3.2 Rerun the simulation shown in this chapter for a lower free stream 0.05 m/s but with the same mesh and for the same diameter 0.05 m of the cylinder. Use the same Time Step Size, Number of Time Steps and Max Iterations/Time Step as in this chapter. Determine the Reynolds number. Determine the drag coefficient and compare with equation (3.1). Determine the Strouhal number and compare with equation (3.4). Include plots of the lift coefficient over time and power spectral density versus frequency. Also, include contour plots of static pressure, vorticity and velocity magnitude.

3.3 Rerun the simulation shown in this chapter for the free stream velocities as shown in Table 3.1 with the same mesh and for the same diameter 0.05 m of the cylinder. Use the same Time Step Size, Number of Time Steps and Max Iterations/Time Step as in this chapter. Determine the drag coefficients and compare with equation (3.1). Determine the Strouhal numbers and compare with equation (3.4).

U (m/s)	f (Hz)	Re	$C_{D,Exp}$	$C_{D,Fluent}$	Percent Diff.	S_{DNS}	S_{Fluent}	Percent Diff.
0.00006		0.2						
0.0006		2						
0.006		20						
0.06	0.1747	200	1.29	1.1785	8.6	0.1968	0.1456	26
0.6		2000						
6		20000						
60		200000						

Table 3.1 Comparison between Fluent and theory for boundary layer thickness

3.4 Rerun the simulation shown in this chapter for the flow around a square with a size of 50 mm x 50 mm. Use the same Time Step Size, Number of Time Steps and Max Iterations/Time Step as in this chapter. Determine the Reynolds number. Determine the drag coefficient and the Strouhal number. Include plots of the lift coefficient over time and power spectral density versus frequency. Also, include contour plots of static pressure, vorticity and velocity magnitude.

3.5 Rerun the simulation shown in this chapter for the flow around an equilateral triangle with a side length of 50 mm. Use the same Time Step Size, Number of Time Steps and Max Iterations/Time Step as in this chapter. Determine the Reynolds number. Determine the drag coefficient and the Strouhal number. Include plots of the lift coefficient over time and power spectral density versus frequency. Also, include contour plots of static pressure, vorticity and velocity magnitude.

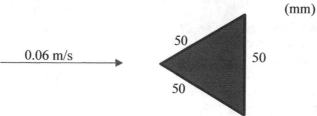

CHAPTER 4. FLOW PAST AN AIRFOIL

A. Objectives

- Using Ansys Workbench to Model the NACA 2412 Airfoil and Mesh
- Inserting Boundary Conditions and Free Stream Velocity
- Running Turbulent Steady 2D Planar Ansys Fluent Simulations
- Using Velocity and Pressure Plots for Visualizations
- Determine Lift and Drag Coefficients for Zero Angle of Attack

B. Problem Description

The purpose of this simulation is to become familiar with a few basic aerodynamic concepts and terms such as Reynold's number, coefficients of lift, drag, and pressure. We will therefore study the flow of air past the NACA 2412 airfoil and we will analyze the problem using Ansys Fluent. The chord length of the airfoil is 230 mm, the angle of attack is 0° and the free stream velocity is 12.7 m/s.

Another airfoil shape that has been studied extensively is the Clark-Y airfoil. One of the first measurements of pressure distributions on this airfoil was completed by Jacobs et al.[1], followed by Marchman and Werme[2] and Stern et al. [3] that made low Reynolds number measurements. Warner[4] and Matsson[5] have also determined results from tests of the Clark-Y airfoil. The NACA 2412 airfoil has been tested by Matsson et al.[6].

For the four digit series[7] of NACA airfoils, the first digit stands for the camber which is the distance between chord line and mean camber line. The second digit stands for the location of maximum camber along the chord line from the leading edge. The last two digits stands for the maximum thickness of the airfoil. For the NACA 2412 studied in this chapter, there is a 2% max camber at 40% chord length with 12% thickness.

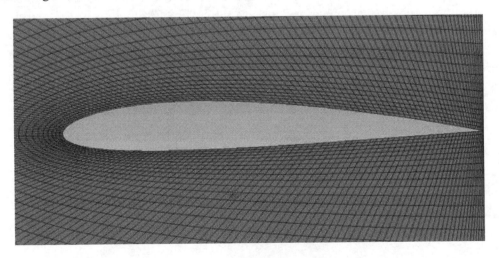

C. Launching Ansys Workbench and Selecting Fluent

1. Start by launching Ansys Workbench. Double click on Fluid Flow (Fluent) that is located under Analysis Systems in Toolbox.

Figure 4.1 Selecting Fluid Flow in ANSYS Workbench

D. Launching Ansys DesignModeler

2. Right click Geometry and select Properties. In Properties of Schematic A2: Geometry, select Analysis Type 2D under Advanced Geometry Options. Right-click on Geometry in the Project Schematic window and select New DesignModeler Geometry. Select **Units>>Millimeter** from the menu as the desired length unit in the new DesignModeler window.

Select Create>>Point from the menu. We will use the coordinates file *NACA_2412_Airfoil.txt* that is available for download at *sdcpublications.com*. Open the Coordinates File from Details View and click on Generate. Select Look At Face ⬚ . Select Concept>>Lines From Points in the menu. Right click in the graphics window and select Point Chain. Right click in the graphics window and select Select All. Apply the Point Segments in Details View. Click on Generate. Select Concept>>Surfaces From Edges from the menu. Right click in the graphics window and Select All. Apply Edges in Details View. Click on Generate.

Figure 4.2a) Selection of units in DesignModeler (DM)

Figure 4.2b) Creating points in DM

Figure 4.2c) Creating lines from points

Figure 4.2d) Creating surfaces from edges

Figure 4.2e) NACA 2412 Airfoil in DesignModeler

3. Next, we will be sketching the mesh region around the airfoil. Create a new coordinate system by first clicking on New Plane ![]. Select Type>>From Coordinates in the Details View. Enter 230 mm for FD11, Point X. Click on Generate. Select the new Plane4 in the tree outline and click on ![] to generate a new sketch. Click on the Sketching tab and select Arc by Center from the Draw tab. Click on Look At Face/Plane/Sketch ![]. Click on the origin of the coordinate system at the trailing edge of the airfoil. Make sure that you see the letter *P* when you move the cursor over the origin of the coordinate system. Next, click on the vertical axis above the origin (make sure you have a *C*) and finally on the vertical axis below the origin (make sure you have a *C* once again), see Figure 4.3b).

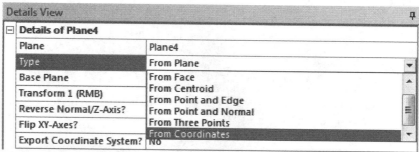

Figure 4.3a) A new plane from coordinates

Figure 4.3b) Half-circle around the NACA 2412 airfoil

4. Select Rectangle by 3 Points from the Draw tab. Click at the intersection of the arc and the positive vertical axis, at the intersection between the arc and the negative vertical axis, and finally click in the right-hand side plane. Select Modify and Trim from Sketching Toolboxes. Click on the lines of the rectangle that are aligned with and on top of the vertical axis.

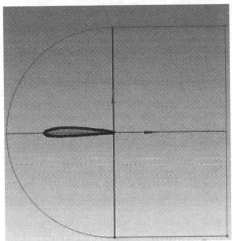

Figure 4.4a) Half-circle and rectangular mesh region around the NACA 2412 airfoil

Select Dimensions under Sketching Toolboxes, select Radius under Dimensions and select the arc in the graphics window. Set the radius of the arc to 2875 mm which gives a ratio of the radius of the arc to the chord length of 12.5. Next, select Horizontal under Dimensions from the Sketching Toolboxes and click on the vertical axis and the vertical edge of the rectangle on the right-hand side. Enter 5750 mm as the length.

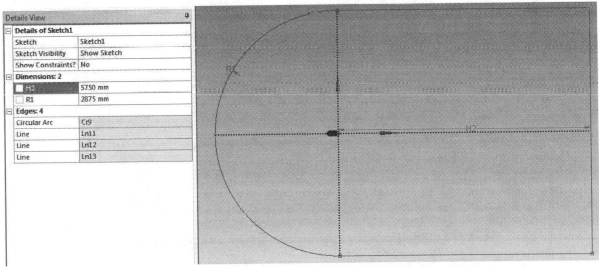

Figure 4.4b) Dimensions for mesh region

Select Concept>>Surface from Sketches. Select Sketch1 under Plane4 in the Tree Outline and Apply as Base Objects in Details View. Select Operation>>Add Frozen in Details View followed by Generate.

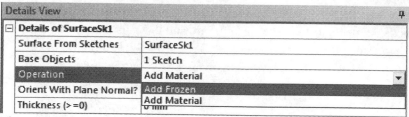

Figure 4.4c) Add frozen operation

5. Select Create>>Boolean from the menu. Select Subtract Operation from the Details View. Click on the mesh region in the graphics window and click on Apply as Target Bodies in Details View. Zoom in on the airfoil, select the Airfoil as Tool Bodies in Details View and click on Generate, see Figure 4.5c) and Figure 4.5d).

Figure 4.5a) Creating a Boolean

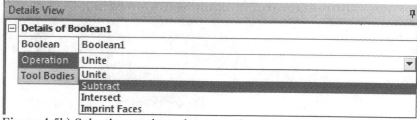

Figure 4.5b) Selecting a subtracting operation

click on this rectangle
for selection of airfoil
as Tool Body

Figure 4.5c) Selection of Tool Body

Figure 4.5d) Completed subtraction operation

6. Select Plane 4 in the tree outline and create a new sketch ⧉. Select the Sketching tab and the Line tool under Draw. Draw a line on the vertical axis from top to bottom of the mesh that intersects the entire mesh region. Select Concepts>>Lines from Sketches from the menu. Select the newly created vertical line and Apply it as a Base Object in the Details View. Click on Generate.

Repeat this step as described above and create a new sketch with a line on the horizontal axis in Plane4 that is drawn from left to right and goes through the whole mesh region including the leading and trailing edges of the airfoil. You will need to create a Coincident Constraint using the Sketching Toolbox and insert the constraint between the new line and the horizontal axis. This is completed by selecting the Coincident Constraint and clicking on both the new line and the horizontal axis. It will be helpful to draw the horizontal line a little bit below the horizontal axis and use the constraint to fix the line with the horizontal axis. Make sure you have the letter C show up when you start and end the horizontal line. Select Concepts>>Lines from Sketches from the menu. Select the newly created vertical line and Apply it as a Base Object in the Details View. Click on Generate.

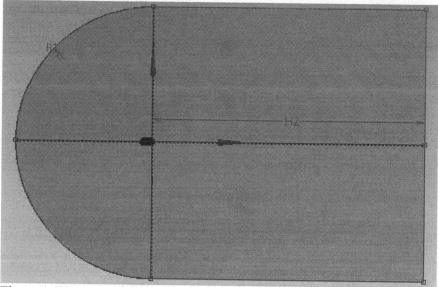

Figure 4.6 Completed vertical and horizontal lines through the mesh region

7. Select Tools>>Projection from the menu. Select Edges on Face as Type in Details View. Control-select both parts of the newly created vertical line and the two parts of the newly created horizontal line and Apply the four Edges in the Details View. Click in the yellow region next to Target in Details View. Select the mesh region around the airfoil and Apply it as the Target under Details View. Click Generate. You should now have four different mesh regions.

Figure 4.7a) Selection of the projection tool

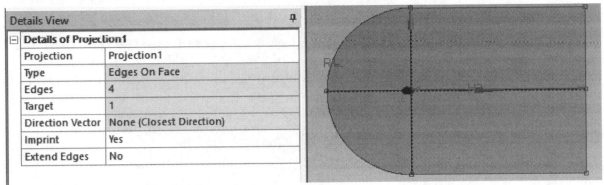

Figure 4.7b) Details view for projection

Click on the plus sign next to 2 Parts, 2 Bodies in the Tree Outline. Right click the Line Body and select Suppress Body. Select File>>Save Project from the menu and save the project with the name *NACA 2412 Airfoil Flow Study*. Close DesignModeler.

E. Launching Ansys Meshing

8. Double click on Mesh under Project Schematic in Ansys Workbench. Select Mesh in the Outline of the Meshing window and select Mesh>>Controls>>Face Meshing from the menu. Control-select all four faces of the mesh. Apply the Geometry under Details of "Face Meshing".

Figure 4.8a) Selection of face meshing

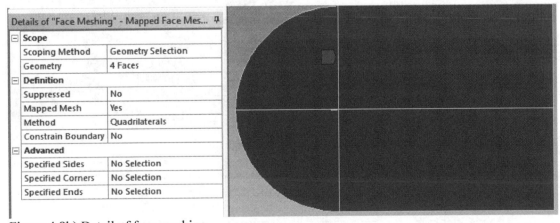

Figure 4.8b) Detail of face meshing

Select Mesh Control>>Sizing from the menu and click on the Edge Selection Filter. Control-select the four edges as shown in Figure 4.8c). Apply the Geometry under Details of "Edge Sizing". Select Number of Divisions as Type under Details of "Edge Sizing". Set the number of Divisions to 50. Set Capture Curvature to No and set the Behavior as Hard. Select the first Bias Type from the drop-down menu. Enter a Bias Factor of 150.

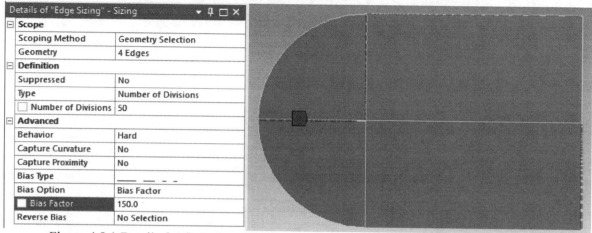

Figure 4.8c) Details for first edge sizing

Create another identical edge sizing but this time select the four remaining straight edges, see Figure 4.8d). Select the same number of divisions 50, Capture Curvature to No and behavior Hard. Select the second Bias Type ———— from the top from the drop-down menu and the same bias factor 150.

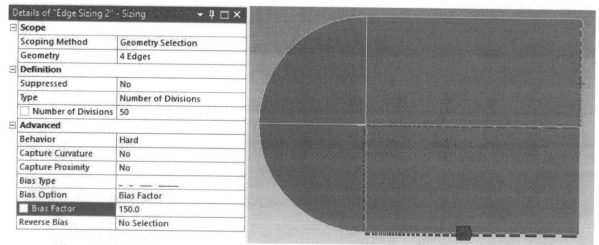

Figure 4.8d) Details for second edge sizing

Why do we use a biased mesh? A biased mesh enables us to create a finer mesh closer to the airfoil. We create a fine mesh where it is needed the most and the calculation time will be much shorter compared with a finer mesh over the entire mesh region.

The final edge sizing will be for the two round edges. Set the Number of Divisions to 100, Capture Curvature to No and Hard behavior. No bias will be used for the round edges. Right click on Mesh in the tree outline and select Generate Mesh.

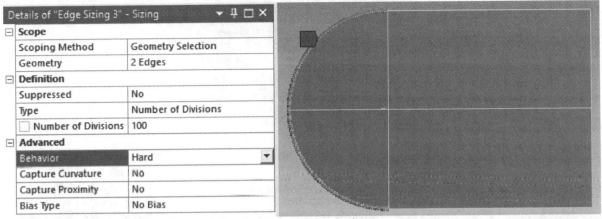

Figure 4.8e) Details of edge sizing for the two round edges

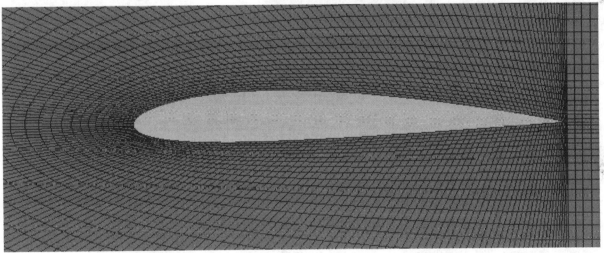

Figure 4.8f) Finished mesh close to the airfoil

9. Select Geometry in the Outline. Select the Edge Selection filter ⬚. Control-select the two vertical edges on the right-hand side of the mesh region, right click and select Create Named Selection. Name the edges "outlet" and click OK. Name the two round edges "inlet". Name the horizontal upper and lower edges "symmetry". Save the project. Select File>>Export>>Mesh>>FLUENT Input File>>Export from the menu. Enter *airfoil-flow-mesh* as file name and Save as type: FLUENT Input Files (*.msh). Close the meshing window. Right click on Mesh and select Update in Ansys Workbench.

> **Why do we export the mesh?** When we save the mesh file, we can read the mesh using standalone mode directly in Ansys Fluent without having to design the mesh all over again each time that we run the simulation for various flow variables.

Figure 4.9 Named selections for mesh

F. Launching Ansys Fluent

10. Double click on Setup under Project Schematic in Ansys Workbench. Check the Options box for Double Precision. Select the number of Solver Processes under Parallel (Local Machine) to the same number as the number of cores for your computer. To check the number of physical cores, press the Ctrl + Shift + Esc keys simultaneously to open the Task Manager. Go to the Performance tab and select CPU from the left column. You'll see the number of physical cores on the bottom-right side. Close the Task Manager window. Click Start to launch Fluent.

Figure 4.10a) Fluent launcher window

Select the Density-Based Solver under General on the Task Page. Double-click on Models under Setup in Outline View and double click on the Viscous (SST k-omega) on the Task Page. Select the Spalart-Allmaras (1 eqn) turbulence model. Select OK to close the Viscous Model window.

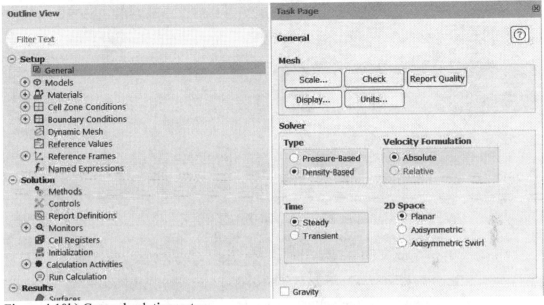

Figure 4.10b) General solution setup

Why did we select the Density-Based Solver? Both solver type options can be used for a variety of flow cases. Traditionally, the density-based solver has been used for compressible flow. For the flow past an airfoil that is studied in this chapter, by using the density-based solver we can choose to include the study of flows where compressibility of air is important.
Why do we use the Spalart-Allmaras Viscous Model? This viscous model was designed for boundary layers in aerospace applications.

Figure 4.10c) Details for the viscous model

11. Double click on Boundary Conditions under Setup in Outline View. Double click the inlet Zone on the Task Page. Select Components as Velocity Specification Method. Enter 12.7 as X-Velocity [m/s]. Click on Apply and Close the Velocity Inlet window.

Figure 4.11 Velocity inlet boundary condition

12. Double click on Monitors under Solution in the Tree. Double click on Residual under Monitor. Change the Convergence Criterion for all four residuals to 1e-8. Click on OK to exit the Residual Monitors window. Double click on Initialization under Solution under the Outline View. Select Standard Initialization as Initialization Method, select Compute from inlet and click on Initialize. Select File>>Save Project from the menu. Select File>>Export>>Case & Data... from the menu. Save the Case/Data File with the name *Flow Past an Airfoil.cas.h5*. Double click on Run Calculation under Solution in the Outline View. Set number of Iterations to 10000. Click on Calculate. Click OK when the calculations are complete.

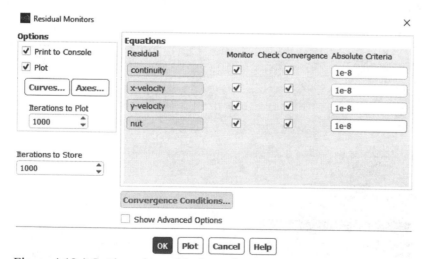

Figure 4.12a) Settings for residual monitors

Why did we select 1e-8 as the Absolute Criteria? We want to choose a sufficiently small value for the absolute criteria such that the drag and lift coefficients are not depending on the value of the absolute criteria. As an example, using 1e-12 instead of 1e-8 will not change the results for lift and drag coefficients. Therefore, the higher value was chosen as calculation times will be shorter.

Figure 4.12b) Solution initialization

Figure 4.12c) Running calculations

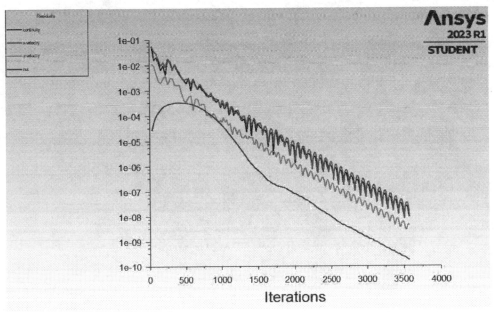

Figure 4.12d) Residuals window

Click on Copy Screenshot of Active Window to Clipboard, see Figure 4.12e). The Scaled Residuals can be pasted into a Word document.

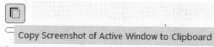

Copy Screenshot of Active Window to Clipboard

Figure 4.12e) Copying screenshot

G. Post-Processing

13. Double click on Graphics under Results in the Outline View. Display a plot of pressure contours by double clicking on Contours under Graphics, check the box for Filled under Options, select all Surfaces and select Pressure and Static Pressure under Contours of. Click on Colormap Options and set Colormap Alignment to Top. Set the Font Behavior to Fixed and Font Size under Font to 16. Set Type to general and Precision to 2 under Number Format. Select Apply followed by Close for the Colormap window. Click on Save/Display in the Contours window. Create another contour plot for Velocity and Velocity Magnitude.

Figure 4.13a) Pressure contours around NACA 2412 airfoil

Figure 4.13b) Velocity contours around the NACA 2412 airfoil

> **Why did we select all surface when we created the pressure contour plots?** When we select all surfaces, the mesh will be shown on top of the contour plot. Deselecting all surfaces will cause the mesh to disappear.

14. Double click on Reference Values under Setup in the Outline View. Select Compute from inlet. Enter the value 0.23 for Length [m] which is the chord length and enter the value 12.7 for Velocity [m/s]. Enter the value 0.3048 for Depth [m] and 0.070104 for Area [m²]. The Area is equal to the Depth times the Length.

Figure 4.14a) Reference values

Find the coefficient of drag by double clicking on Reports under Results in the Outline View. Double click on Forces under Reports. Set the Direction Vector in the Force Reports window to 1 for X and 0 for Y. Select *wall-surface_body* under Wall Zones and click on Print. Write down the value for the total drag coefficient that you have calculated $C_d = 0.017688831$. Repeat this step to find the value for the lift coefficient using 0 for X and 1 for Y for the Direction Vector in the Force Reports window. The value for the lift coefficient is $C_l = 0.19927405$.

Figure 4.14b) Force report for drag

Select the Domain tab in the menu and Adapt>>Manual…. Select Cell Registers>>New>>Region in the Manual Mesh Adaption window. Define the region with X Min [m] -0.1, X Max [m] 0.5, Y Min [m] -0.1 and Y Max [m] 0.1. Click on Save and Close the Region Register window. Set the Maximum Refinement Level to 9 and the Minimum Edge Length [m] to 0 in the General Adaption Controls. Click on OK in the General Adaption Controls to close the window. Select *region_0* as Refinement Criterion in the Manual Mesh Adaption window. Select Adapt in the Manual Mesh Adaption window. Select OK in the Information window.

Reinitialize the solution and run the calculations. Enter the reference values once again. Recalculate the drag and lift coefficients, see Table 4.1. Repeat this step, refine the mesh once again, initialize the solution and rerun the calculations.

	Original Mesh	*Refinement Level 1*	*Refinement Level 1*	*Refinement Level 3*
C_d	0.01769	0.01553	0.01560	0.01876
C_l	0.19927	0.19885	0.20348	0.19557
No. Cells	15000	29586	87765	319983
No. Faces	30300	59722	176582	642018
No. Nodes	15300	30136	88817	322035

Table 4.1 Drag and lift coefficient for different mesh sizes, $Re = 200,000$, *Angle of Attack* = 0

> **Why did we do mesh refinement?** Any CFD study should be accompanied by a mesh independence study where you refine the mesh until the results converge. The ANSYS Student version has a maximum allowed number of elements and it can therefore sometimes be difficult to achieve mesh independence.

H. Theory

15. The Reynolds number for the flow over an airfoil is determined by

$$Re = \frac{Uc}{v} = 200,000 \qquad (4.1)$$

where U (m/s) is the magnitude of the free stream velocity, c (m) is the chord length of the airfoil, and v (m²/s) is the kinematic viscosity of air at room temperature. The angle of attack *AoA* is defined as the angle between the free stream and the chord line (the line between the leading edge and the trailing edge of the airfoil). For the flow case that we studied in this chapter the angle of attack was zero degrees.

The pressure distribution over the airfoil is expressed in non-dimensional form by the pressure coefficient

$$C_p = \frac{p_i - p}{\frac{1}{2}\rho U^2} \qquad (4.2)$$

where ρ (kg/m³) is the free stream density and where p_i (Pa) is the surface pressure at location i and p (Pa) is the pressure in the free stream. The pressure coefficient can alternatively be related to the velocity distribution u (m/s) around the airfoil through

$$C_p = 1 - \left(\frac{u}{U}\right)^2 \qquad (4.3)$$

The lift and drag coefficients can be determined from

$$C_l = \frac{F_l}{\frac{1}{2}\rho U^2 bc} \tag{4.4}$$

$$C_d = \frac{F_d}{\frac{1}{2}\rho U^2 bc} \tag{4.5}$$

, where F_l and F_d (N) are lift and drag forces and b (m) is the wingspan.

I. References

1. Jacobs E.N., Stack J. and Pinkerton R.M. Airfoil Pressure Distribution Investigation in the Variable Density Wind Tunnel, NACA Report No. 353, 1930.
2. Marchman III J.F and Werme T.D., Clark-Y Airfoil Performance at Low Reynolds Numbers, AIAA-84-0052, 1984.
3. Stern F., Muste M., Houser D., Wilson M. and Ghosh S., Measurement of Pressure Distribution and Forces acting on an Airfoil, Laboratory Experiment #3, 57:020 Mechanics of Fluids and Transfer Processes (http://css.engineering/uiowa.edu/fluidslab/pdfs/57-020/airfoil.doc)
4. Warner E.P., Airplane Design: Performance, McGraw-Hill, New York, 1936
5. Matsson J., A Student Project on Airfoil Performance, Annual Conference & Exposition, Honolulu, Hawaii, 2007.
6. Matsson J., Voth J., McCain C. and McGraw C., Aerodynamic Performance of the NACA 2412 Airfoil at Low Reynolds Number, ASEE 123rd Annual Conference & Exposition, New Orleans, LA, June 26-29, 2016.
7. Anderson, J., Fundamentals of Aerodynamics, 6st Edition, McGraw-Hill, 2017.
8. Abbott, I.H., Von Doenhoff, A.E., Theory of Wing Sections: Including a Summary of Airfoil Data, Dover Publications, 1959.
9. Saha, P. and Kaberwal, H.S., Numerical Analysis of Aerodynamic Performance Characteristics of NACA 2312 and NACA 2412, International Journal of Innovative Science and Research Technology, Volume 5, Issue 3, 2020.

J. Exercises

4.1 Determine contours of pressure and velocity for the flow field around an airfoil with a different NACA number than the one covered in this chapter. Include the corresponding mesh refinement level, number of iterations, refine threshold, number of cells refined, drag coefficient, lift coefficient, Reynolds number and the angle of attack as shown in Table 4.1 for the NACA 2412 airfoil. Include the residuals window after the solution has converged and include a printout of the mesh around the airfoil.

4.2 Continue with Exercise 4.1 using different angles of attack $AoA = 2, 4, 6, 8, 10, 12$ degrees for the same Reynolds number. For each AoA, refine the mesh until you get a converged solution. Plot the drag and lift coefficients versus AoA and compare with experimental results for the chosen airfoil using data from Abbott and Von Doenhoff[8].

4.3 Create an O-mesh mesh around the airfoil instead of the C-mesh that was modeled in this chapter. Run the simulations for the O-mesh and compare results with the mesh used in this chapter. Use the same radius for the O-mesh as for the C-mesh in this chapter.

K. Notes

CHAPTER 5. RAYLEIGH-BENARD CONVECTION

A. Objectives

- Using Ansys Workbench to Create the Mesh for 3D Rayleigh-Bénard Convection
- Inserting Wall Boundary Conditions
- Running Laminar Steady 3D Ansys Fluent Simulations
- Using Contour Plots for Visualizations
- Compare Results with Neutral Stability Theory

B. Problem Description

Bénard[1] was the first scientist to experimentally study convection in a fluid layer with a free surface heated from below. Rayleigh[2] used linear stability analysis to theoretically explain and study the stability of the fluid motion between horizontal parallel plates with the hotter plate at the bottom. Chandrasekhar[3] completed the linear stability analysis for Rayleigh-Bénard convection and Koschmieder[4] showed the development of the research in this area during the following couple of decades.

There are only a few experiments including Koschmieder and Pallas[5], Hoard et al.[6] and to some extent Matsson[6] that have produced the concentric ring cells in Rayleigh-Bénard convection that according to stability theory should appear with a rigid upper boundary condition. This instability with an upper rigid wall in experiments is very sensitive to the surface roughness and flatness of the upper wall. Torres et al.[7] completed 3D numerical simulations of Rayleigh Benard convection in a cylindrical container and verified the existence of roll cells in the form of concentric circles.

In this chapter, we will study Rayleigh-Bénard convection and we will analyze the problem using Ansys Fluent in 3D. The enclosure is filled with water and the temperature of the cold wall is 296.5 K while the hot wall is at 298 K. The diameter of the enclosure is 100 mm and the depth 4.8 mm.

C. Launching Ansys Workbench and Selecting Fluent

1. Start by launching Ansys Workbench. Double click on Fluid Flow (Fluent).

Figure 5.1 Selecting Fluid Flow

D. Launching Ansys DesignModeler

2. Right-click on Geometry in Project Schematic and select Properties. Select Geometry and 3D Analysis Type under Advanced Geometry Options in Properties of Schematic A2: Geometry. Right click on Geometry and open a New DesignModeler Geometry. Select **Millimeter as the Unit** from the menu in DesignModeler.

Fluid Flow (FLUENT)

Figure 5.2a) Geometry

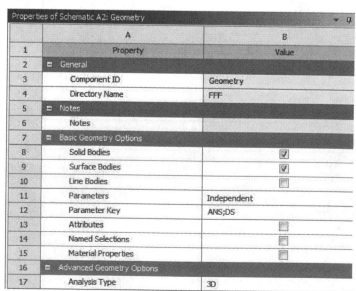

Figure 5.2b) Selecting 3D Analysis Type

3. Next, we will be creating the mesh region for the simulation. Select XYPlane from the Tree Outline. Select Look at Sketch . Select the Sketching tab and draw a circle from the origin. Select the Dimensions tab in Sketching Toolboxes. Click on the circle and enter a diameter of 100 mm. Select Create>>Extrude from the menu. Select Sketch1 under XYPlane in in the Tree Outline and Apply the sketch as Geometry in Details View. Enter 4.8 mm as FD1, Depth (>0). Click on Generate [⚡ Generate] and Close DesignModeler.

Figure 5.3a) Selection of XYPlane

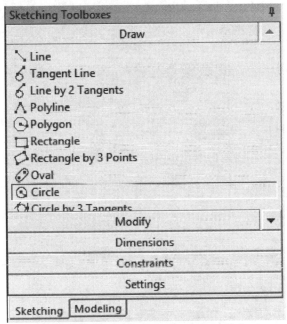

Figure 5.3b) Selection of circle sketch tool

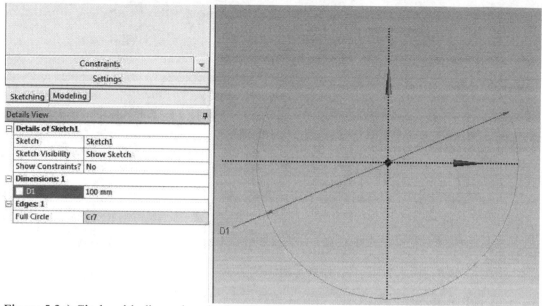

Figure 5.3c) Circle with dimensions in details view

Figure 5.3d) Selecting extrusion

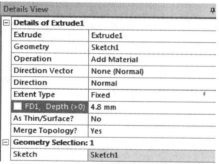

Figure 5.3e) Details of extrusion

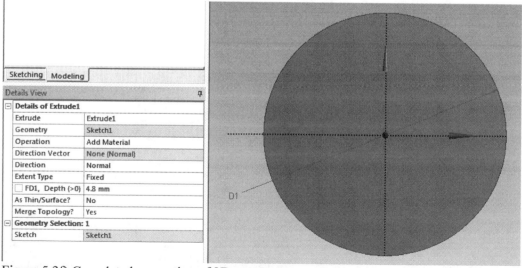

Figure 5.3f) Completed generation of 3D geometry

E. Launching Ansys Meshing

4. We are now going to double click on Mesh under Project Schematic in Ansys Workbench to open the meshing window. Select Mesh in the Outline. Select Unit Systems>>Metric (mm, kg, N …) from the bottom of the graphics window. Right click on Mesh in the Outline and select Update. A coarse mesh is created.

Figure 5.4a) Starting Mesh

Figure 5.4b) Selection of Mesh in Outline

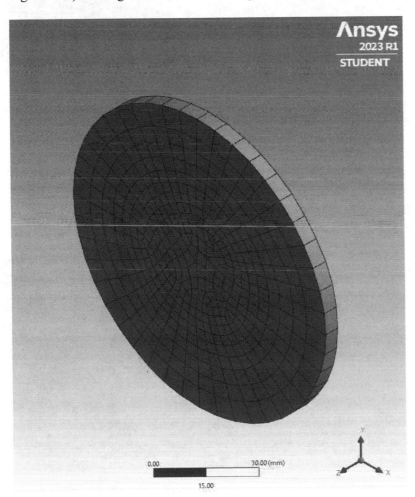

Figure 5.4c) Coarse mesh

5. Select Mesh>>Controls>>Face Meshing from the menu. Click on the cylindrical face of the mesh region in the graphics window. The face turns green. Click on the Apply button for Geometry in Details of "Face Meshing".

Select Mesh>>Controls>>Sizing from the menu and select Edge ⬚. Control click on the two circular edges of the mesh region. Click on Apply for the Geometry in "Details of Edge Sizing". Under Definition in "Details of Edge Sizing", select Number of Divisions for Type, 500 for Number of Divisions, set Capture Curvature to No and Hard as Behavior.

Repeat the selection of Mesh>>Controls>>Sizing from the menu and select Face ⬚. Control select the two plane faces and the cylindrical face of the mesh region. You will need to rotate the model to select both faces. Enter 0.5 mm for Element Size, set Capture Curvature to No and Hard Behavior.

Repeat the selection of Mesh>>Controls>>Sizing from the menu and select Body ⬚. Select and Apply the body of the mesh region. Enter 5 mm for Element Size, set Capture Curvature to No and Hard Behavior. Click on Update and select Mesh in the Outline. Right click in the graphics window and select Isometric View. The finished mesh is shown in the graphics window.

Figure 5.5a) Generation of Face Meshing

Figure 5.5b) Face for Face Meshing

Figure 5.5c) Mesh Control Sizing

Figure 5.5d) Round edges for sizing

Details of "Edge Sizing" - Sizing	▾ ⊣ □ ×
⊟ **Scope**	
Scoping Method	Geometry Selection
Geometry	2 Edges
⊟ **Definition**	
Suppressed	No
Type	Number of Divisions
☐ Number of Divisions	500
⊟ **Advanced**	
Behavior	Hard
Capture Curvature	No
Capture Proximity	No
Bias Type	No Bias

Figure 5.5e) Details of edge sizing

Figure 5.5f) Faces for face sizing

Details of "Face Sizing" - Sizing	▾ ⊣ □ ×
⊟ **Scope**	
Scoping Method	Geometry Selection
Geometry	3 Faces
⊟ **Definition**	
Suppressed	No
Type	Element Size
☐ Element Size	0.5 mm
⊟ **Advanced**	
☐ Defeature Size	Default (3.5376e-002 mm)
Influence Volume	No
Behavior	Hard
Capture Curvature	No
Capture Proximity	No

Figure 5.5g) Details of face sizing

Details of "Body Sizing" - Sizing	▾ ⊣ □ ×
⊟ **Scope**	
Scoping Method	Geometry Selection
Geometry	1 Body
⊟ **Definition**	
Suppressed	No
Type	Element Size
☐ Element Size	5.0 mm
⊟ **Advanced**	
☐ Defeature Size	Default (3.5376e-002 mm)
Behavior	Hard
Capture Curvature	No
Capture Proximity	No

Figure 5.5h) Details of body sizing

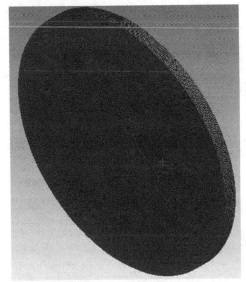

Figure 5.5i) Finished mesh

Details of "Mesh"	▾ ⊣ □ ×
⊟ **Display**	
Display Style	Use Geometry Setting
⊟ **Defaults**	
Physics Preference	CFD
Solver Preference	Fluent
Element Order	Linear
☐ Element Size	Default (7.0751 mm)
Export Format	Standard
Export Preview Surface Mesh	No
⊞ **Sizing**	
⊞ **Quality**	
⊞ **Inflation**	
⊞ **Advanced**	
⊟ **Statistics**	
☐ Nodes	342584
☐ Elements	309720
Show Detailed Statistics	No

Figure 5.5j) Details of mesh

6. Select Face ⬚ and select the cylindrical face of the mesh region, right click and select Create Named Selection. Enter *side wall* as the name and click on the OK button. Repeat this step for the upper and lower faces in the Z-direction of the mesh region and name these *cold wall* and *hot wall*, respectively.

Select File>>Export>>Mesh>>FLUENT Input File>>Export from the menu and save the mesh with the name *rayleigh-benard-3D.msh*. Select File>>Save Project from the menu and name the project *3D Rayleigh-Benard Convection*. Select File>>Close Meshing to close the meshing window. Right click on Mesh in Ansys Workbench and select Update.

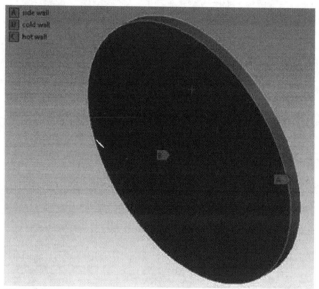

Figure 5.6a) Named selections for the mesh

Figure 5.6b) Updating the mesh

F. Launching Ansys Fluent

7. Double click on Setup under Project Schematic in Ansys Workbench. Check the box for Double Precision. Select the number of Solver Processes to the same number as the number of cores for your computer. To check the number of physical cores, press the Ctrl + Shift + Esc keys simultaneously to open the Task Manager. Go to the Performance tab and select CPU from the left column. You'll see the number of physical cores on the bottom-right side. Close the Task Manager window. Click Start to launch Fluent. Click Start to launch Fluent. Click OK when you get the Key Behavioral Changes window.

Figure 5.7 Ansys Fluent Launcher

8. Check the mesh in Ansys Fluent by selecting the Check button under Mesh in General Setup. Check the scale of the mesh by selecting the Scale button. Make sure that the Domain Extent is correct and Close the Scale Mesh window.

Figure 5.8a) Checking the mesh

```
Domain Extents:
   x-coordinate: min (m) = -5.000000e-02, max (m) = 5.000000e-02
   y-coordinate: min (m) = -5.000000e-02, max (m) = 5.000000e-02
   z-coordinate: min (m) = 0.000000e+00, max (m) = 4.800000e-03
Volume statistics:
   minimum volume (m3): 6.348532e-12
   maximum volume (m3): 3.542288e-10
      total volume (m3): 3.769812e-05
Face area statistics:
   minimum face area (m2): 1.058089e-08
   maximum face area (m2): 5.903813e-07
Checking mesh.......................................
Done.
```

Figure 5.8b) Console output generated by mesh check

Figure 5.8c) Scale Mesh window

9. Check the Gravity box and enter -9.81 as the Gravitational Acceleration in the Z direction. Double click on Models under Setup in Outline View. Double click on the Energy model and check the box for the Energy equation. Click OK to exit the Energy Model window.

Figure 5.9a) Gravity settings

Figure 5.9b) Energy equation

10. Double click on Materials under Setup in the Outline View. Click on the Create/Edit button for Fluid on the Task Page to open the Create/Edit Materials window. Select Fluent Database…. Scroll down in the Fluent Fluid Materials window and select *water-liquid (h2o<l>)*. Click on the Copy button and Close the Fluent Database Materials window. Select *boussinesq* using the drop-down menu next to Density under Properties. Enter the value 997 for the Density [kg/m3]. Enter the value 4180 for Cp (Specific Heat) [J/(kg K)] , 0.607 for Thermal Conductivity [W/(m K)], and 0.000891for Viscosity [kg/(m s)]. Scroll down and enter the value 0.000247 for the Thermal Expansion Coefficient [K^{-1}]. Click on the Change/Create button and Close the window.

> **Why did we choose the boussinesq setting?** The Boussinesq approximation is used for buoyancy driven flows. The reason for using boussinesq for density is that this will give faster convergence. Density is treated as a constant in all equations except for the buoyancy term in the momentum equation.

Double click on Boundary Conditions under Setup in the Outline View. Click on Operating Conditions on the Task Page. Set the value for the Operating Temperature [K] to 297.395 and click on OK to close the Operating Conditions box.

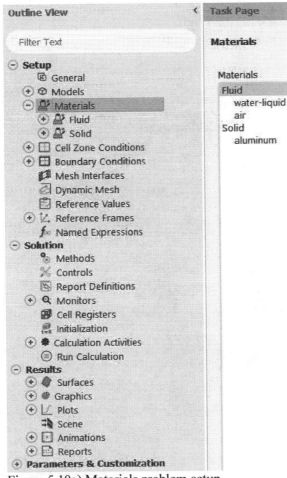

Figure 5.10a) Materials problem setup

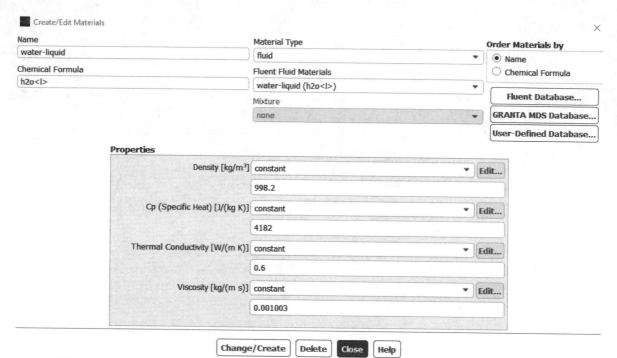

Figure 5.10b) Create/Edit Materials window

Figure 5.10c) Operating Conditions settings

11. Open Boundary Conditions under Setup in the Outline View. Double click on *cold_wall* under Boundary Conditions on the Task Page. Select the Thermal tab and select Temperature for Thermal Conditions. Enter the value 296.79 as Temperature [K]. Click on Apply and Close the window. Double click *hot_wall* under Boundary Conditions on the Task Page and set the Temperature [K] boundary condition to 298. Double click on *side_wall* under Boundary Conditions on the Task Page and set the Temperature [K] boundary condition to 296.79. Double click on Cell Zone Conditions under Setup in the Outline View. Double click on solid under Cell Zone Conditions on the Task Page. Select *water-liquid* from the Material Name drop-down menu in the Fluid window. Click on Apply and Close the window.

Figure 5.11a) Cold wall, hot wall and side wall temperature settings

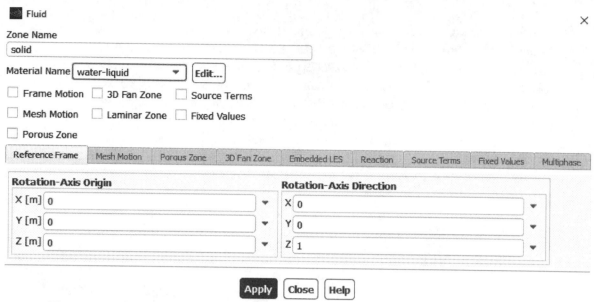

Figure 5.11b) Fluid selection

12. Double click on Models under Setup in Outline View. Double click on Viscous (SST k-omega) model and check the box for the Laminar model. Click OK to exit the Viscous Model window. Double click on Initialization under Solution in the Outline View. Select Standard Initialization under Initialization Methods. Enter 0.01 for Z Velocity [m/s], enter 297.395 as Temperature [K] and click on Initialize.

Figure 5.12a) Laminar Viscous Model

Figure 5.12b) Solution Initialization settings

13. Double click on Monitors under Solution in the Outline View and double click on Residual under Monitors. Set the Absolute Criteria to 1e-6 for all equations and click OK to close the window. Double click on Methods under Solution in the Outline View and select Body Force Weighted from the Pressure drop down menu under Spatial Discretization. Select SIMPLE as Pressure-Velocity Coupling Scheme.

Figure 5.13a) Residual Monitors Settings

Figure 5.13b) Solution Methods Settings

Why did we choose Body Force Weighted and SIMPLE Solution Methods? The reason for using SIMPLE as Pressure-Velocity Coupling algorithm is that we are studying steady laminar flow. SIMPLEC may alternatively be used as it can provide a faster solution. PISO is used for transient solutions. Coupled can also give an efficient steady state solution for single phase flows. The Body Force Weighted pressure scheme is recommended when we have large body forces such as gravity in Rayleigh-Benard convection.

14. Select File>>Save Project from the menu. Select File>>Export>>Case… from the menu. Save the Case File with the name *Rayleigh-Benard Convection.cas.h5*. Double click Run Calculation under Solution in the Outline View. Set the number of iterations to 6000. Click on the Calculate button. Click on the OK button in the Information window when the calculations are complete.

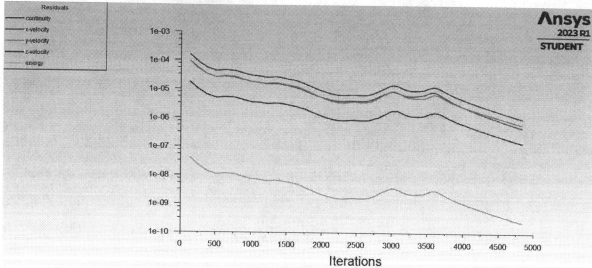

Figure 5.14a) Scaled Residuals for Rayleigh-Bénard convection after 4062 iterations

Click on Copy Screenshot of Active Window to Clipboard, see Figure 5.14b). The Scaled Residuals can be pasted into a Word document.

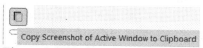

Figure 5.14b) Copying screenshot

G. Post-Processing

15. Double click on Graphics under Results in the Outline View. Double click on Contours in the Graphics section. Check the Filled and Contour Lines boxes under Options and select Contours of Velocity and Z Velocity.

    Select New Surface>>Plane… from the drop-down menu in the Contours window. Select Three Points as Method. For Point 1, set (X (m), Y (m), Z (m)) to (0, -0.05, 0.0024), for Point 2 set (X (m), Y (m), Z (m)) to (0, 0.05, 0.0024) and for Point 3 set (X (m), Y(m), Z (m)) to (0.05, 0, 0.0024). Enter the name *mid-plane-z=0.0024m* as the New Surface Name. Click on Create and Close the window.

    Select the new plane under Surfaces in the Contours window. Click on the Save/Display button and Close the window. Select the x-y plane view by clicking on the blue Z axis of the coordinate system in the graphics window. Repeat this step but select Contours of Temperature and Static Temperature. Click on Colormap Options. Select float as Type for Number Format and set Precision to 2. Click on Apply and Close the Colormap window. Click on the Save/Display button. Select the x-y plane view.

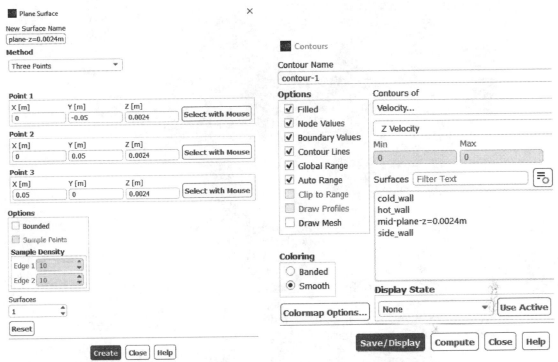

Figure 5.15a) Creation of a new x-y plane   Figure 5.15b) Contours settings

Figure 5.15c) Contours of Z Velocity in x-y plane, $Ra = 2491$

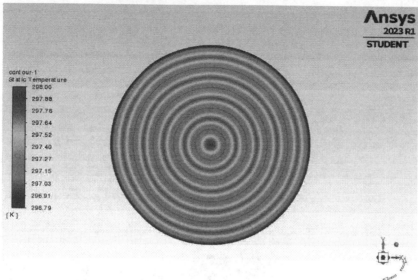

Figure 5.15d) Contours of Static Temperature in x-y plane, $Ra = 2491$

Click on Copy Screenshot of Active Window to Clipboard, see Figure 5.14b). The Contours can be pasted into a Word document.

16. Select New Surface>>Plane... from the drop-down menu in the Contours window. Select Three Points as Method. For Point 1, set (X (m), Y (m), Z (m)) to (0, 0, 0), for Point 2 set (X (m), Y (m), Z (m)) to (0, 0, 0.0048) and for Point 3 set (X (m), Y (m), Z (m)) to (0.05, 0, 0). Enter the name *x-z-plane-y=0* as the New Surface Name. Click on Create and Close the window. Select the new plane under Surfaces in the Contours window. Click on Colormap Options. Select float as Type for Number Format and set Precision to 2. Set Colormap Alignment to Top. Click on Apply and close the Colormap window. Click on the Save/Display button to display the Temperature field. Select the x-z plane view by clicking on the green Y axis in the graphics window. Finally, display the Z Velocity field in the *x-z* plane, see Figure 5.16c). You will need to change Colormap Options. Select float as Type for Number Format and set Precision to 5.

Figure 5.16a) Creation of *x-z* plane   Figure 5.16b) Colormap window

Figure 5.16c) Contours of static temperature and Z velocity in x-z plane, $Ra = 2491$

Click on Copy Screenshot of Active Window to Clipboard, see Figure 5.14b). The Contours can be pasted into a Word document.

Figure 5.16d) Photo of experimental Rayleigh-Benard convection at $Ra = 2490$ with a rigid upper wall, Matsson[7].

Why do the experiments have a different appearance than the concentric roll cells from ANSYS Fluent simulations? In the experiments with a rigid upper wall, it is difficult to get the perfectly concentric circular ring pattern as shown from simulations. The reason for this is that the experiments are sensitive to the non-flatness and imperfections of the bottom and top surfaces. The upper lid in experiments is an acrylic plate that is not perfectly flat.

17. Double click on Boundary Conditions under Setup in the Outline View. Double click on *cold_wall* under Boundary Conditions on the Task Page. Select the Momentum tab and select Marangoni Stress as Shear Condition. Enter the value -2e-5 as Surface Tension Gradient [N/ (m K)] for Marangoni Stress. Click Apply and Close to exit the window. Uncheck the Gravity box under General in the Setup.

Double click on Initialization under Solution in the Outline View. Select Standard Initialization under Initialization Methods. Enter 0.01 for Z Velocity [m/s], enter 297.395 as Temperature [K] and click on Initialize. Click OK when you get a Question box. Double click Run Calculation under Solution in the Outline View. Set the number of iterations to 5000. Click on the Calculate button. Click on the OK button in the Information window that appears when the calculations are complete. Repeat steps **15 – 16**.

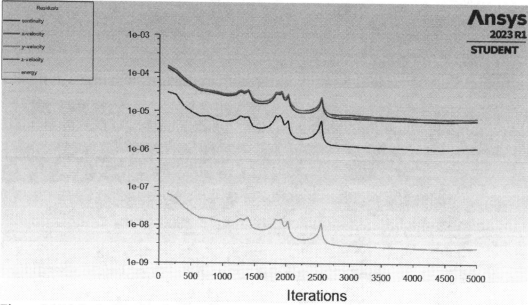

Figure 5.17a) Scaled Residuals for Marangoni convection

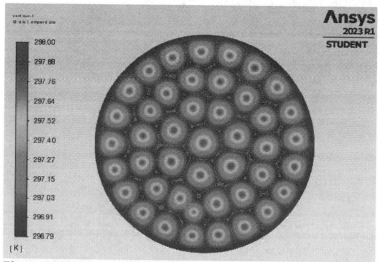

Figure 5.17b) Contours of Static Temperature in x-y plane, M = 895

Click on Copy Screenshot of Active Window to Clipboard, see Figure 5.14b). The Contours can be pasted into a Word document.

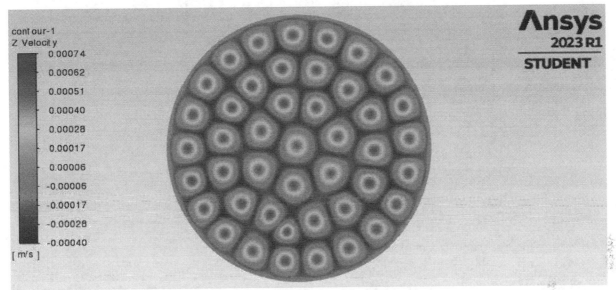

Figure 5.17c) Contours of Z Velocity in x-y plane, $M = 895$

Figure 5.17d) Contours of static temperature and Z velocity in x-z plane, $M = 895$

Click on Copy Screenshot of Active Window to Clipboard, see Figure 5.14b). The Contours can be pasted into a Word document.

H. Theory

<u>Comparison with Neutral Stability Theory for Rayleigh-Bénard Convection</u>

18. The instability of the fluid layer between two parallel plates heated from below is governed by the Rayleigh number Ra.

$$Ra = \frac{g\beta\rho^2 C_p (T_1 - T_2) L_c^3}{\mu \, k} \tag{5.1}$$

where g is acceleration due to gravity, β is the coefficient of volume expansion, ρ is density of the fluid, C_p is the specific heat, T_1 and T_2 are the temperatures of the hot and cold surfaces respectively, L_c is the distance between the surfaces (fluid layer thickness), k is thermal conductivity of the fluid, and μ is the dynamic viscosity of the fluid. Below the critical $Ra_{crit} = 1708$ for a rigid upper surface, the flow is stable but convective currents will develop above this Rayleigh number. For the case of a free upper surface, the theory predicts a lower critical Rayleigh number at 1101 and for two free boundaries the critical Rayleigh number is 657. The non-dimensional wave number α of this instability is determined by

$$\alpha = \frac{2\pi L_c}{\lambda} \tag{5.2}$$

where λ is the wave length of the instability. The determination of the average wave-length is determined from Figure 5.15d) and shown in Figure 5.18a). The diameter of the computational domain is 100 mm. The length for eight wave-lengths can be measured in Figure 5.18a) as a percentage of the 100 mm diameter using a ruler. In this case $8\lambda = 88.3\ mm$ which gives an average wave-length $\lambda = 11.0375\ mm$.

Figure 5.18a) Determination of average wave-length

The critical wave numbers are $\alpha_{crit} = 3.12, 2.68$ for rigid and free surface boundary conditions, respectively. The wave number can be determined to be

$$\alpha = \frac{2\pi \cdot 0.0048}{0.0110375} = 2.73 \tag{5.3}$$

The Rayleigh number in the flow simulation for rigid-rigid boundaries is

$$Ra = \frac{g\beta\rho^2 C_p (T_1 - T_2) L_c^3}{\mu\, k} = \frac{9.81*0.000247*997^2*4180*1.5*0.0048\char`^3}{0.000891*0.607} = 2491 \tag{5.4}$$

The equations governing the stability of Rayleigh-Benard convection are given by

$$(D^2 - \alpha^2)^2 W = \alpha^2 Ra\, \theta \tag{5.5}$$

$$(D^2 - \alpha^2)\theta = -W \tag{5.6}$$

, where $D = d/dz$ and W, θ are disturbances of the velocity z-component and temperature while a is the wave number. For rigid-rigid boundaries, the boundary conditions are

$$W = DW = \theta - 0,\ z = 0, 1 \tag{5.7}$$

The neutral stability curve for Rayleigh-Benard convection with rigid-rigid boundaries can be given to the first approximation, see Figure 5.18b).

$$Ra = \frac{(\pi^2 + \alpha^2)^3}{\alpha^2 \left\{ 1 - \frac{16\alpha\pi^2 \cosh^2\left(\frac{\alpha}{2}\right)}{\left[(\pi^2 + \alpha^2)^2 (\sinh\alpha + \alpha)\right]} \right\}} \tag{5.8}$$

For rigid-free boundaries, the boundary conditions are

$$W = DW = \theta = 0,\ z = 0 \tag{5.9a}$$
$$W = D^2 W = \theta = 0,\ z = 1 \tag{5.9a}$$

The corresponding expression the neutral stability curve to the first approximation for rigid-free boundary conditions is

$$Ra = \frac{(\pi^2 + \alpha^2)^3}{\alpha^2 \left\{ 1 - \frac{8\alpha\pi^2 \sinh^2(\alpha)}{\left[(\pi^2 + \alpha^2)^2 (\sinh 2\alpha - 2\alpha)\right]} \right\}} \tag{5.10}$$

For free-free boundaries, the boundary conditions are

$$W = D^2 W = \theta = 0,\ z = 0, 1 \tag{5.11}$$

The neutral stability curve for Rayleigh-Benard convection with free-free boundaries is

$$Ra = \frac{(\pi^2 + \alpha^2)^3}{\alpha^2} \tag{5.12}$$

Boundaries	$R_{c,\ exact}$	$R_{c,\ 1st\ approx.}$	α_c	α_{Fluent}	% diff.
Free-free	657.5	----------	2.22		
Rigid-free	1100.7	1112.7	2.68		
Rigid-rigid	1707.8	1715.1	3.12	2.73	12.5

Table 5.1 Critical parameters for Rayleigh-Benard convection

Figure 5.18b) Neutral stability curve for Rayleigh-Bénard convection. The filled circle represents result from ANSYS Fluent simulation.

Here is Mathematica code that plots the neutral stability curves and determines minimum values for Rayleigh-Benard convection with different types of boundary conditions, see Li and Steiner[9].

```
(* Solving (D^2-α^2)^2 W = α^2 Ra θ and (D^2-α^2)θ = -W with three types of boundary conditions *)
(* α: wave number, Ra: Rayleigh number, D = d/dz, W: perturbation in z-direction velocity *)
(* θ: temperature perturbation *)
(* Two free boundaries: W = D^2 W = θ = 0 at z = 0,1 *)
(* Two rigid boundaries: W = DW = θ = 0 at z = 0,1 *)
(* One free and one rigid boundary: W = DW = θ = 0 at z = 0 and W = D^2 W = θ = 0 at z = 1 *)

m[α_, R_] := {{0, 1, 0, 0, 0, 0}, {0, 0, 1, 0, 0, 0}, {0, 0, 0, 1, 0, 0},
  {-α^4, 0, 2*α^2, 0, R*α^2, 0}, {0, 0, 0, 0, 0, 1}, {-1, 0, 0, 0, α^2, 0}}
fa[α_, Ra_] := Det[MatrixExp[m[α, Ra]][[{1, 3, 5}, {2, 4, 6}]]]
fb[α_, Ra_] := Det[MatrixExp[m[α, Ra]][[{1, 2, 5}, {3, 4, 6}]]]
fc[α_, Ra_] := Det[MatrixExp[m[α, Ra]][[{1, 3, 5}, {3, 4, 6}]]]
ContourPlot[{fa[α, Ra] == 0, fb[α, Ra] == 0, fc[α, Ra] == 0}, {α, 0.1, 8}, {Ra, 500, 7000}, Contours → {0},
  ContourShading → False, PlotLegends → {"Two free boundaries", "Two rigid boundaries", "Mixed boundaries"},
  AxesOrigin → Automatic, Axes → {True, True}, FrameLabel → {"α", "Ra"},
  LabelStyle → {FontFamily → "Times New Roman", FontSize → 12}, RotateLabel → False]

SetSystemOptions["CheckMachineUnderflow" → False];
ga[α_, Ra_] := (1/2)*10^10*(fa[α + 10^(-10), Ra] - fa[α - 10^(-10), Ra])
FindRoot[{fa[α, Ra] == 0, ga[α, Ra] == 0}, {α, 2., 2.2}, {Ra, 500, 700}, WorkingPrecision → 25];
NumberForm[Chop[%, 10^(-10)], 6]
gb[α_, Ra_] := (1/2)*10^10*(fb[α + 10^(-10), Ra] - fb[α - 10^(-10), Ra])
FindRoot[{fb[α, Ra] == 0, gb[α, Ra] == 0}, {α, 3., 3.2}, {Ra, 1500, 1600}, WorkingPrecision → 25];
NumberForm[Chop[%, 10^(-10)], 6]
gc[α_, Ra_] := (1/2)*10^10*(fc[α + 10^(-10), Ra] - fc[α - 10^(-10), Ra])
FindRoot[{fc[α, Ra] == 0, gc[α, Ra] == 0}, {α, 2.5, 2.7}, {Ra, 1000, 1200}, WorkingPrecision → 25];
NumberForm[Chop[%, 10^(-10)], 6]
```

Figure 5.18c) Mathematica code for neutral stability curves for Rayleigh-Benard convection

Comparison with Neutral Stability Theory for Surface-Tension-Driven Benard Convection (Marangoni Convection)

The instability of the fluid layer heated from below and affected by surface tension gradients is governed by the Marangoni number M.

$$M = \frac{\sigma(T_1 - T_2)L_c}{\mu\,\kappa} \tag{5.13}$$

where $\sigma = -\frac{\partial S}{\partial T}$ is the surface tension S rate of change with temperature T, T_1 and T_2 are the hot and cold temperatures respectively, L_c is the fluid layer thickness, κ is thermal diffusivity, and μ is the dynamic viscosity of the fluid.

Below the critical $M_{crit} = 79.6$, the flow is stable. The non-dimensional wave number α of this instability is determined by

$$\alpha = \frac{2\pi L_c}{\lambda} \tag{5.14}$$

where λ is the wave length of the instability. The determination of the average wave-length is determined from Figure 5.17d) and shown in Figure 5.18c). The diameter of the computational domain is 100 mm. The length for five wave-lengths can be measured in Figure 5.18c) as a percentage of the 100 mm diameter using a ruler. In this case $5\lambda = 84.2\ mm$ which gives an average wave-length $\lambda = 16.84\ mm$.

100 mm

Figure 5.18d) Determination of average wave-length

The critical wave number is $\alpha_{crit} = 1.99$. The wave number can be determined to be

$$\alpha = \frac{2\pi \cdot 0.0048}{0.01684} = 1.79 \tag{5.15}$$

The Marangoni number in the flow simulation is

$$M = \frac{\sigma(T_1 - T_2)L_c}{\mu\,\kappa} = \frac{0.00002*1.5*0.0048}{0.000891*1.457E-7} = 895 \tag{5.16}$$

The equations governing the stability of Marangoni convection are given by

$$(D^2 - \alpha^2)^2 W = 0 \tag{5.17}$$

$$(D^2 - \alpha^2)\theta = -W \tag{5.18}$$

The boundary conditions are

$$W = DW = \theta = 0\,,\ z = 0 \tag{5.19a}$$
$$W = D^2 W - \alpha^2 M\theta = D\theta + B\theta = 0\,,\ z = 1 \tag{5.19b}$$

The neutral stability curve for Marangoni convection can be given by

$$M = \frac{8a(a \cosh \alpha + B \sinh \alpha)(\alpha - \sinh \alpha \cosh \alpha)}{a^3 \cosh \alpha - \sinh^3 \alpha}$$

(5.17)

where $B = qL_c/k$ is the Biot number and $q = \frac{\partial Q}{\partial T}$ is the heat flux Q rate of change with temperature T.

B	M_c	α_c	α_{Fluent}	% diff.
0	79.6	1.99	1.79	10
5	250.6	2.60		
10	413.4	2.74		
15	575.0	2.81		
20	736.0	2.85		
25	896.8	2.88		

Table 5.2 Critical parameters for Marangoni convection

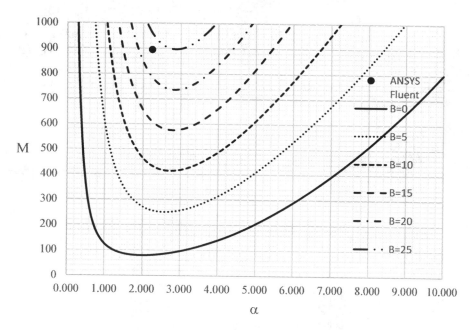

Figure 5.18e) Neutral stability curve for Marangoni convection.

I. References

1. Bénard, H. "Les tourbillons cellulaires dans une nappe liquide", *Rev. Gen. Sciences Pure Appl.* **11**, 1261-1271, 1309-1328, 1900
2. Rayleigh, L. "On convection currents in a horizontal layer of fluid when the higher temperature is on the underside.", Phil. Mag. **32**, 529-546, 1916.
3. Chandrasekhar, S. "Hydrodynamic and Hydromagnetic Stability", Dover, 1981.
4. Koschmieder, E.L. "Bénard cells and Taylor vortices", Cambridge University Press, 1993.
5. Koschmieder, E.L. and Pallas, S.G. "Heat transfer through a shallow, horizontal convecting fluid layer.", Int. J. Heat Mass Transfer **17**, 991-1002, 1974.

6. Hoard, C.Q., Robertson, C.R., and Acrivos, A. "Experiments on the cellular structure in Bénard convection", Int. J. Heat Mass Transfer **13**, 849-856, 1970.

7. Matsson J. "A Student Project on Rayleigh-Benard Convection", Annual ASEE Conference & Exposition, June 22-25, Pittsburg, Pa, 2008.

8. Sánchez Torres N.Y., López Sánches E.J., Hernández Zapata S., and Ruiz Chavarría G., 3D Numerical Simulation of Rayleigh-Bénard Convection in a Cylindrical Container. *Selected Topics of Computational and Experimental Fluid Mechanics*, Environmental Science and Engineering, J. Klapp et al.(eds.), Spring International Publishing, Switzerland, 2015.

9. Li, Nian and Steiner, J.M,"Recent Progress in the Theory of Thermal Instability Using: Mathematica", Buletinul ştiinţific al Universitatii Baia Mare, Seria B, Fascicola matematică-informatică, Vol. 12, No 1, 1996

J. Exercises

5.1 Run ANSYS Fluent simulations with a rigid upper boundary condition for Rayleigh number $Ra - 3088$ and compare the results with Figure 5.15d) by plotting contours of Z velocity and contours of temperature. Determine the wave number α in the simulations and compare with the neutral stability curve as shown in Figure 5.18b).

5.2 Run ANSYS Fluent simulations with a free upper boundary condition for Rayleigh number $Ra - 2120$ and compare the results with Figure 5.18e) by plotting contours of Z velocity and contours of temperature.

Figure 5.18e) Photo of experimental Rayleigh-Benard convection at $Ra = 2120$ with a free surface upper boundary condition (copied from Matsson[7]).

K. Notes

CHAPTER 6. CHANNEL FLOW

A. Objectives

- Using Ansys Workbench to Model Geometry and Mesh
- Creating a User Defined Temperature Profile at Inlet
- Inserting Boundary Conditions
- Running Laminar Steady 2D Planar Ansys Fluent Simulations
- Using Contour Plots for Visualizations of Pressure, Temperature and Velocity Fields
- Using XY Plots for Comparison of Inlet and Outlet Velocity and Temperature Profiles
- Comparing with Theoretical Solution Using Mathematica Code

B. Problem Description

We will study the development of the laminar flow of aqueous glycerin solution through a channel. The channel has a height of 90 mm and a length of 3,000 mm. The inlet temperature profile is linear where the bottom wall has a temperature of 0 °C and the upper wall is at 100 °C. The inlet velocity profile is laminar with a parabolic velocity profile in between the walls.

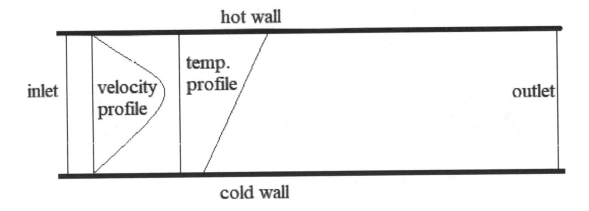

C. Launching Ansys Workbench and Selecting Fluent

1. Start by launching the Ansys Workbench. Double click on Fluid Flow (Fluent).

Figure 6.1 Selecting fluid flow Fluent

D. Launching Ansys DesignModeler

2. Right click Geometry and select Properties. In Properties of Schematic A2: Geometry, select Analysis Type 2D under Advanced Geometry Options. Right-click on Geometry in the Project Schematic window and select New DesignModeler Geometry. Select **Units>>Meter** meter as the desired length unit from the menu in DesignModeler.

Figure 6.2a) Selecting properties

16	⊟ Advanced Geometry Options	
17	Analysis Type	2D
18	Use Associativity	☑

Figure 6.2b) Advanced geometry options

	A	
1	🗘 Fluid Flow (Fluent)	
2	🗋 Geometry	? ◢
3	🍲 Mesh	? ◢
4	🐾 Setup	? ◢
5	🗐 Solution	? ◢
6	🗐 Results	? ◢

Fluid Flow (Fluent)

Figure 6.2c) Selecting the geometry

3. Select XY plane in the Tree Outline and select the Sketching tab. Select Look at Sketch [icon]. Draw a rectangle from the origin and set the horizontal length of the rectangle in Details View to 3 m and the vertical height of the rectangle to 0.09 m, see Figure 6.3a). Click on the Modeling tab in the Tree Outline and highlight Sketch1 under XYPlane. Select Concept>>Surfaces from Sketches from the menu. Apply Sketch1 as the Base Object in Details View and click on Generate. Close DesignModeler.

Figure 6.3a) Surface from a sketch

Concept Tools Units View
 Lines From Points
 Lines From Sketches
 Lines From Edges
 3D Curve
 Split Edges
 Surfaces From Edges
 Surfaces From Sketches

Figure 6.3b) Surface from a sketch

Details View	
⊟ **Details of SurfaceSk1**	
Surface From Sketches	SurfaceSk1
Base Objects	1 Sketch
Operation	Add Material
Orient With Plane Normal?	Yes
Thickness (>=0)	0 m

Figure 6.3c) Details view

129

E. Launching Ansys Meshing

4. Next, we will be creating the mesh for the flow field. Double click on Mesh under Project Schematic in Ansys Workbench. Right click on Mesh under Outline in the Meshing window and select Generate Mesh. Select Unit Systems>>Metric (mm, kg, N …) from the bottom of the graphics window.

Figure 6.4a) Coarse mesh

Select Mesh>>Controls>>Face Meshing from the menu. Select the mesh region in the graphics window and apply it as Geometry under Scope in Details of "Face Meshing".

Figure 6.4b) Face meshing

Select Mesh>>Controls>>Sizing from the menu. Select Edge ⬚ , control select the four edges of the rectangle and apply them as the Geometry under Scope in Details of "Edge Sizing". Set the element size to 5.0 mm, set Capture Curvature to No and the Behavior to Hard. Click on Generate Mesh and select Mesh in the Outline.

Figure 6.4c) Refined mesh

Select the left vertical edge of the mesh region, right click and select Create Named Selection. Name this edge *inlet*. Select the right vertical edge, right click and select Create Named Selection. Name this edge *outlet*. Select the lower horizontal edge, right click and select Create Named Selection. Name this edge *cold wall*. Finally, select the upper horizontal edge, right click and select Create Named Selection. Name this edge *hot wall*.

Figure 6.4d) Named selections

Select File>>Save Project from the menu and name the project *Channel Flow*. Select File>>Export...>>Mesh>>FLUENT Input File>>Export from the menu in the Meshing window and name the mesh *channel-flow-mesh.msh*. Close the meshing window. Right click on Mesh under the Project Schematic in Ansys Workbench and select Update.

Figure 6.4e) Updating the mesh

F. Creating a UDF

5. We are now going to create a Used Defined Function for the inlet temperature profile. Start Notepad located in the Windows Accessories folder and include the following text as listed below in Figure 6.5. Save the file as a text document with the name *udf-inlet-temp-velocity-profiles.txt*.

```
#include "udf.h"
DEFINE_PROFILE(inlet_t_temperature, thread, position)
{
real x[ND_ND]; /* this will hold the position vector */
real y;
face_t f;
begin_f_loop(f, thread)
{
F_CENTROID(x,f,thread);
y = x[1];
F_PROFILE(f, thread, position) =273.15+(100*y/0.09);
}
end_f_loop(f, thread)
}
DEFINE_PROFILE(inlet_u_velocity, thread, position)
{
real x[ND_ND]; /* this will hold the position vector */
real y;
face_t f;
begin_f_loop(f, thread)
{
F_CENTROID(x,f,thread);
y = x[1];
F_PROFILE(f, thread, position) =-6*0.03*(y*y-y*0.09)/(0.09*0.09*0.09);
}
end_f_loop(f, thread)
}
```

Figure 6.5 Code for UDF

> **What is a UDF?** A UDF is a user defined function that can be used to define your own boundary conditions, initial conditions or material properties. The C programming language is used for user defined functions and the UDF is saved with the extension .c. All user defined functions has the first line #include "udf.h". UDFs are defined using DEFINE macros and can be executed as compiled or interpreted functions. In this chapter we are using the UDF to define the temperature profile at the inlet.

G. Launching Ansys Fluent

6. Double click Setup under Project Schematic in Ansys Workbench. Check the Solver Options box for Double Precision. Set the number of Solver Processes to the number of cores for your computer processor. To check the number of physical cores, press the Ctrl + Shift + Esc keys simultaneously to open the Task Manager. Go to the Performance tab and select CPU from the left column. You'll see the number of physical cores on the bottom-right side. Close the Task Manager window. Notice the location of the Working Directory and copy the file *udf-inlet-temp-velocity-profiles.txt* to this location. Click on Start to close the Fluent Launcher window.

Figure 6.6a) Fluent launcher window

Select the Pressure-Based Solver under General on the Task Page and click on Units... under General, see Figure 6.6b). Scroll down the Quantities and select temperature. Set the Units to *C* for temperature and do the same for temperature-difference. Close the Set Units window. Open Models under Setup in the Outline View. Double click Energy-Off under Models and check the box for Energy Equation. Click OK to close the Energy window.

Figure 6.6b) General solution setup

Figure 6.6c) Setting the units

Figure 6.6d) Energy equation

7. Double click Materials under Setup in the Outline View. Click on Create/Edit... for Fluid on the Task Page and select Fluent Database in the new window. Scroll down in the Fluent Fluid Materials and select glycerin (c3h8o3). Click on Copy and Close the window.

Figure 6.7a) Selection of glycerin as fluid

Select piecewise-linear for Density [kg/m³] under Properties in the Create/Edit Materials window. Increase the number of points to 11 in the Piecewise-Linear Profile window. Set the Temperature [C] for Point 1 to 0 and Value (kg/m3) to 1268.4. Set the values for the remaining points as shown in Table 6.1. Click on the OK button to close the window.

Select piecewise-linear for Viscosity (kg/m-s) under Properties. Increase the number of points to 11. Set the Temperature [C] for Point 1 to 0 and Value (kg/m-s) to 7.2162. Set the values for the remaining points as shown in Table 6.1. Click on the OK button to close the piecewise-linear profile window. Click on Change/Create and Close the Create/Edit Materials window.

Figure 6.7b) Selection of piecewise-linear density and viscosity for glycerin

Figure 6.7c) Temperature and density for data points

Point	Temperature (c)	Density (kg/m3)	Viscosity (kg/m-s)
1	0	1268.4	7.2162
2	10	1262.2	2.3857
3	20	1256.0	0.92181
4	30	1249.8	0.40505
5	40	1243.7	0.19801
6	50	1237.6	0.10591
7	60	1231.5	0.06113
8	70	1225.5	0.03767
9	80	1219.5	0.024552
10	90	1213.5	0.016809
11	100	1207.6	0.012009

Table 6.1 Properties of glycerin (98 percent weight) at different temperatures

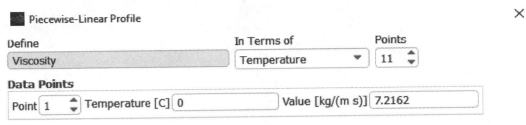

Figure 6.7d) Temperature and viscosity for data points

8. Open Cell Zone Conditions and Fluid under Setup in the Outline View. Double click *surface_body*. Select glycerin as the Material Name. Click on Apply and Close the window.

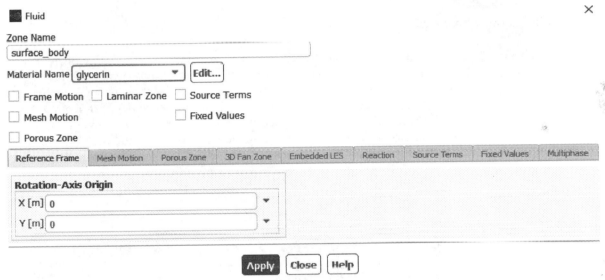

Figure 6.8a) Selection of glycerin as fluid

Select User-Defined>>Functions>>Interpreted… from the menu. Browse for the source file in your Working Directory and Select Files of type: All Files (*). Find the file named *udf-inlet-temp-velocity-profiles.txt* in the working directory and click OK. Click on Interpret and Close the Interpreted UDFs window.

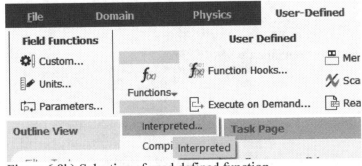

Figure 6.8b) Selection of used defined function

Figure 6.8c) Interpreted UDFs

9. Open Models under Setup in the Outline View. Double click on the Viscous (SST k-omega) model. Select the Laminar Model and click OK to close the Viscous Model window. Open Boundary Conditions under Setup in the Outline View. Double click on *inlet* under Boundary Conditions. Set the Velocity Specification Method to Components. Select *udf inlet_u_velocity* as X-Velocity [m/s]. Click on the Thermal tab and select *udf inlet_t_temperature* from the Temperature drop down menu. Click Apply and Close the Velocity Inlet window.

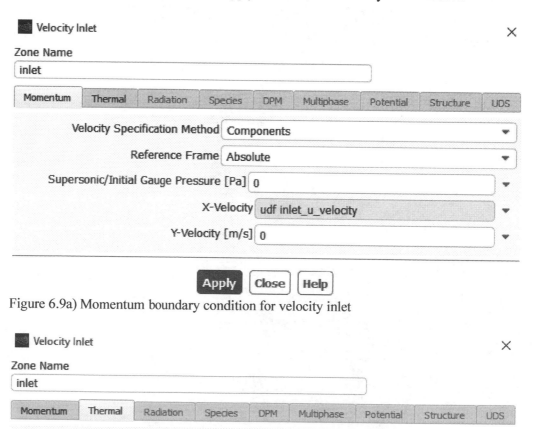

Figure 6.9a) Momentum boundary condition for velocity inlet

Figure 6.9b) Selection of *udf* for velocity inlet as thermal boundary condition

Double click on *cold_wall* under Boundary Conditions and Wall in the Outline View. Select the Thermal tab in the Wall window and select Temperature as Thermal Condition. Set the Temperature [C] to 0. Click Apply and Close the Wall window. Repeat this step for the *hot_wall* and set the Temperature [C] to 100.

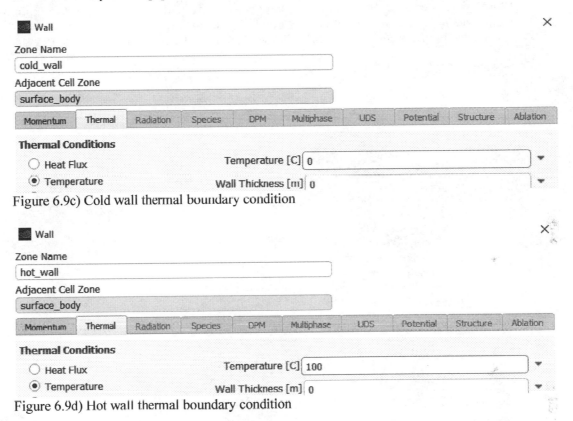

Figure 6.9c) Cold wall thermal boundary condition

Figure 6.9d) Hot wall thermal boundary condition

10. Open Monitors under Solution in the Outline View and double click on Residual under Monitors. Set Absolute Criteria to 1e-06 for all four Residuals. Click OK to close the Residual Monitors window.

Figure 6.10a) Residual monitors window

Double click on Reference Values under Setup in the Outline View. Select *Compute from inlet* and select *surface_body* as Reference Zone under Reference Values on the Task Page. Double click on Initialization under Solution in the Outline View. Select Standard Initialization on the Task Page as Initialization Method and *Compute from inlet*. Click on *Initialize*. Select File>>Save Project from the menu. Select File>>Export>>Case… from the menu. Save the Case File with the name *Channel Flow.cas.h5*. Double click on Run Calculation under Solution in the Outline View. Enter 10000 as *Number of Iterations*. Click on *Calculate*. Click OK when calculation is complete.

Figure 6.10b) Reference values

Figure 6.10c) Solution initialization

Figure 6.10d) Running the calculations

Figure 6.10e) Residuals window

Click on Copy Screenshot of Active Window to Clipboard, see Figure 6.10f). The Scaled Residuals can be pasted into a Word document.

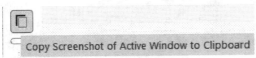

Figure 6.10f) Copying screenshot

H. Post-Processing

11. Open Graphics under Results in the Outline View. Double click on Contours under Graphics. Select all Surfaces and click on Colormap Options. Select float as Type and 0 for Precision under Number Format. Select Fixed Font Behavior and set Font Size to 12 under Font. Select Top as Colormap Alignment, click on Apply and Close the Colormap window. Click on Save/Display. Display another plot of temperature-contours by selecting Contours of Temperature and Static Temperature and click on Save/Display. Finally, select Contours of Velocity and X Velocity and set the Precision to 2 in the Colormap window. Click on Save/Display.

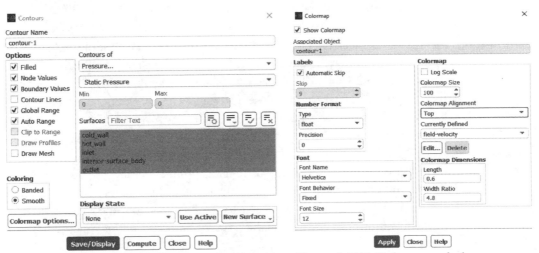

Figure 6.11a) Contours window   Figure 6.11b) Colormap window

Figure 6.11c) Contours of static pressure, static temperature and X velocity for glycerin flow

Click on Copy Screenshot of Active Window to Clipboard, see Figure 6.10f). The contours can be pasted into a Word document.

12. Double click on XY Plot under Results and Plots in the Outline View. Uncheck Position on X Axis under Options. Select Mesh and Y-Coordinate as Y Axis Function. Select Velocity and X Velocity as X Axis Function. Click on Load File and load the file with the name *theory channel flow.dat* from the working directory. This file can be downloaded from *sdcpublications.com*. Select *inlet* and *outlet* as the Surfaces and click on Curves. Select the first Pattern for Curve # 0 and no Symbol. Click on Apply. Select the next available Pattern for Curve # 1 and no Symbol. Click on Apply. Select the next available Pattern for Curve # 2 with no Symbol and blue color. Click on Apply and Close the window.

Click on Axes... and uncheck the box for Auto Range under Options for the X Axis. Set Minimum to 0 and Maximum to 1 under Range. Set Type as float and set Precision to 1 under Number Format. Click on Apply. Check the Y Axis and set the Precision to 2 and Type to float under Number Format. Click on Apply. Close the window and click on Save/Plot.

Repeat this step and create another XY Plot. Select Temperature and Static Temperature as X Axis Function. Select inlet and outlet as the Surfaces and unselect Theory under File Data. Click on Axes... and uncheck the box for Auto Range under Options for the X Axis. Set Minimum to 0 and Maximum to 100 under Range. Set Type as float and set Precision to 0 under Number Format. Click on Apply. Check the Y Axis and set the Precision to 2 and Type to float under Number Format. Click on Apply and close the Axes – Solution XY Plot window. Click on Save/Plot.

Figure 6.12a) Solution XY Plot window

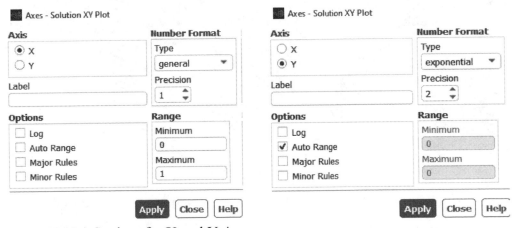

Figure 6.12b) Solution XY Plot window

Figure 6.12c) Settings for X and Y Axes

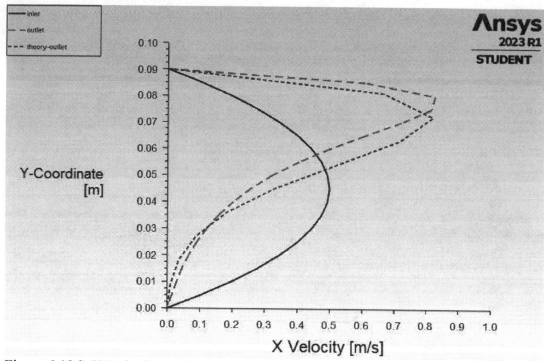

Figure 6.12d) X Velocity profiles for glycerin flow at the inlet and outlet

Figure 6.12e) Temperature profiles for glycerin flow at inlet and outlet

Click on Copy Screenshot of Active Window to Clipboard, see Figure 6.10f). The XY Plots can be pasted into a Word document.

I. Theory

13. The equation of motion for the fluid for fully developed flow is given by the following equation

$$\mu \frac{d^2 u}{dy^2} = \frac{dp}{dx} \tag{6.1}$$

where u (m/s) is streamwise velocity, x (m) is streamwise coordinate, y (m) is the wall-normal coordinate, μ (kg/ms) is dynamic viscosity, and p (Pa) is the pressure. The boundary conditions are $u(0) = u(h) = 0$ where h is the height of the channel.

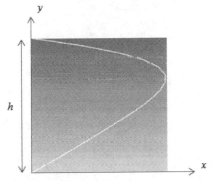

Figure 6.13a) Geometry for laminar flow of glycerin

Element i	Density (kg/m3)	Viscosity μ_i (kg/ms)	Pressure gradient $\left(\frac{dp}{dx}\right)_i$ (Pa/m)
1	1265.3	4.80095	-78.9249
2	1259.1	1.653755	-78.9249
3	1252.9	0.66343	-78.9249
4	1246.75	0.30153	-78.9249
5	1240.65	0.15196	-78.9249
6	1234.55	0.08352	-78.9249
7	1228.5	0.0494	-78.9249
8	1222.5	0.031111	-78.9249
9	1216.5	0.020681	-78.9249
10	1210.55	0.014409	-78.9249

Table 6.2 Average properties of glycerin (98 percent weight) and pressure gradient for each element

We can now compute the resistance or stiffness matrix for each element, see Moaveni[1]

$$K_i = \frac{10\mu_i}{h} \begin{pmatrix} 1 & -1 \\ -1 & 1 \end{pmatrix} \tag{6.2}$$

, and the forcing vector will be the following for each element

$$F_i = -\frac{h}{20} \left(\frac{dp}{dx}\right)_i \begin{pmatrix} 1 \\ 1 \end{pmatrix} \tag{6.3}$$

Next, we assemble the stiffness matrices into the following matrix

$$K = \frac{10}{h} \begin{pmatrix} \mu_1 & -\mu_1 & 0 & 0 & 0 & 0 & 0 & 0 & 0 & 0 & 0 \\ -\mu_1 & \mu_1+\mu_2 & -\mu_2 & 0 & 0 & 0 & 0 & 0 & 0 & 0 & 0 \\ 0 & -\mu_2 & \mu_2+\mu_3 & -\mu_3 & 0 & 0 & 0 & 0 & 0 & 0 & 0 \\ 0 & 0 & -\mu_3 & \mu_3+\mu_4 & -\mu_4 & 0 & 0 & 0 & 0 & 0 & 0 \\ 0 & 0 & 0 & -\mu_4 & \mu_4+\mu_5 & -\mu_5 & 0 & 0 & 0 & 0 & 0 \\ 0 & 0 & 0 & 0 & -\mu_5 & \mu_5+\mu_6 & -\mu_6 & 0 & 0 & 0 & 0 \\ 0 & 0 & 0 & 0 & 0 & -\mu_6 & \mu_6+\mu_7 & -\mu_7 & 0 & 0 & 0 \\ 0 & 0 & 0 & 0 & 0 & 0 & -\mu_7 & \mu_7+\mu_8 & -\mu_8 & 0 & 0 \\ 0 & 0 & 0 & 0 & 0 & 0 & 0 & -\mu_8 & \mu_8+\mu_9 & -\mu_9 & 0 \\ 0 & 0 & 0 & 0 & 0 & 0 & 0 & 0 & -\mu_9 & \mu_9+\mu_{10} & -\mu_{10} \\ 0 & 0 & 0 & 0 & 0 & 0 & 0 & 0 & 0 & -\mu_{10} & -\mu_{10} \end{pmatrix} \tag{6.4}$$

, and we assemble the forcing vector into the following

$$F = 6h \begin{pmatrix} 1 \\ 2 \\ 2 \\ 2 \\ 2 \\ 2 \\ 2 \\ 2 \\ 2 \\ 2 \\ 1 \end{pmatrix} \tag{6.5}$$

After inclusion of boundary conditions, we get system of equations (6.6) with the solution (6.7).

$$\begin{pmatrix} 1 & 0 & 0 & 0 & 0 & 0 & 0 & 0 & 0 & 0 & 0 \\ -\mu_1 & \mu_1+\mu_2 & -\mu_2 & 0 & 0 & 0 & 0 & 0 & 0 & 0 & 0 \\ 0 & -\mu_2 & \mu_2+\mu_3 & -\mu_3 & 0 & 0 & 0 & 0 & 0 & 0 & 0 \\ 0 & 0 & -\mu_3 & \mu_3+\mu_4 & -\mu_4 & 0 & 0 & 0 & 0 & 0 & 0 \\ 0 & 0 & 0 & -\mu_4 & \mu_4+\mu_5 & -\mu_5 & 0 & 0 & 0 & 0 & 0 \\ 0 & 0 & 0 & 0 & -\mu_5 & \mu_5+\mu_6 & -\mu_6 & 0 & 0 & 0 & 0 \\ 0 & 0 & 0 & 0 & 0 & -\mu_6 & \mu_6+\mu_7 & -\mu_7 & 0 & 0 & 0 \\ 0 & 0 & 0 & 0 & 0 & 0 & -\mu_7 & \mu_7+\mu_8 & -\mu_8 & 0 & 0 \\ 0 & 0 & 0 & 0 & 0 & 0 & 0 & -\mu_8 & \mu_8+\mu_9 & -\mu_9 & 0 \\ 0 & 0 & 0 & 0 & 0 & 0 & 0 & 0 & -\mu_9 & \mu_9+\mu_{10} & -\mu_{10} \\ 0 & 0 & 0 & 0 & 0 & 0 & 0 & 0 & 0 & 0 & 1 \end{pmatrix} \begin{pmatrix} u_1 \\ u_2 \\ u_3 \\ u_4 \\ u_5 \\ u_6 \\ u_7 \\ u_8 \\ u_9 \\ u_{10} \\ u_{11} \end{pmatrix} = \frac{6h^2}{10} \begin{pmatrix} 0 \\ 2 \\ 2 \\ 2 \\ 2 \\ 2 \\ 2 \\ 2 \\ 2 \\ 2 \\ 0 \end{pmatrix} \tag{6.6}$$

$$u = \begin{pmatrix} 0 \\ 0.0051 \\ 0.0178 \\ 0.0447 \\ 0.0930 \\ 0.1677 \\ 0.2647 \\ 0.3631 \\ 0.4153 \\ 0.3371 \\ 0 \end{pmatrix} \tag{6.7}$$

```
h = 0.09; n = 10; w = 1; dpdx = -78.9249; a = -dpdx * h / n;
μ = {4.80095, 1.653755, 0.66343, 0.30153, 0.15196, 0.08352, 0.0494, 0.031111, 0.020681, 0.014409};
ρ = {1265.3, 1259.1, 1252.9, 1246.75, 1240.65, 1234.55, 1228.5, 1222.5, 1216.5, 1210.55};
```

$$\{a1, b1, c1\} = \left\{\text{Array}\left[-\mu[[\#]] \&, n-1\right], \text{Array}\left[\mu[[\#-2]]+\mu[[\#-1]] \&, n, \{2, 11\}\right], \text{Array}\left[-\mu[[\#-1]] \&, n, \{2, 11\}\right]\right\};$$

```
K = (n / h) * SparseArray[{Band[{1, 2}] → c1, Band[{1, 1}] → b1, Band[{2, 1}] → a1}, 11];
K[[1, 1]] = h / n;
K[[1, 2]] = 0;
K[[n + 1, n + 1]] = h / n;
MatrixForm[Normal[K]];

s = LinearSolve[K, {0, a, a, a, a, a, a, a, a, a, a, 0}];
```

$$\dot{m} = \sum_{i=1}^{n} \rho[[i]] * w * (h/n) * (s[[i]] + s[[i+1]]) / 2;$$

$$\text{ListLinePlot}\left[\text{Array}\left[\{s[[\#]], (\#-1)*h/n\} \&, n+1, \{1, n+1\}\right], \text{AxesLabel} \to \{\text{"u (m/s)"}, \text{"y (m)"}\}, \text{PlotLabels} \to \{\text{"Theory"}\}\right]$$

```
mylist = Array[{s[[#]], (# - 1) * h / n} &, n + 1, {1, n + 1}];
TableOfValues1 = Prepend[mylist, {"((xy/key/label \"theory-outlet\")"}];
TableOfValues1 = Prepend[TableOfValues1, {""}];
TableOfValues1 = Prepend[TableOfValues1, {"(labels \"X Velocity\" \"Y-Coordinate\")"}];
TableOfValues1 = Prepend[TableOfValues1, {"(title \"Theory\")"}];
TableOfValues1 = Append[TableOfValues1, {")"}];
Grid[TableOfValues1]
Export["theory.dat", TableOfValues1]
```

Figure 6.13b) Mathematica code for theoretical solution

J. References

1.   S. Moaveni, Finite Element Analysis: Theory and Applications with ANSYS, 3rd Ed., Prentice Hall, 2007.

K. Exercises

6.1 Replace the fluid with engine oil from the Fluent Database and enter the piecewise-linear values for the four points as listed in the table below. Increase the Gage Total Pressure at the inlet to 500 Pa. Include residuals window, contour plots of Static Pressure, Static Temperature, X Velocity, and the X Velocity profile.

Point	Temperature (c)	Density (kg/m3)	Viscosity (kg/m-s)
1	20	881.5	0.23939
2	30	875.4	0.12842
3	40	869.3	0.07455
4	50	863	0.04643

Table 6.3 Properties for engine oil (SAE 30)

6.2 Change the size of the computational domain to 6 m in length and 0.2 m in height and change the mesh size to 0.015 m in horizontal direction and 0.01 m in the vertical direction. Modify the temperature of the cold lower wall to 0 (c) and 100 (c) for the upper hot wall. Rewrite and reinterpret the *udf* file for the new temperatures of the cold and hot wall and modify the *udf* file for the change in height of the computational domain. Replace the fluid with engine oil from the Fluent Database and enter the piecewise-linear values for the 11 points as listed in the table below. Increase the Gage Total Pressure at the inlet to 1000 pascal. Include *udf* code, zoomed in mesh that shows details, residuals window, contour plots of Static Pressure, Static Temperature, X Velocity, and the X Velocity profile.

Point	Temperature (c)	Density (kg/m3)	Viscosity (kg/m-s)
1	0	867.4	0.75352
2	10	860.9	0.37865
3	20	854.5	0.20689
4	30	848.3	0.12190
5	40	842.1	0.076551
6	50	835.8	0.050861
7	60	829.5	0.035409
8	70	823.2	0.025631
9	80	817	0.019181
10	90	810.6	0.014742
11	100	804.5	0.011619

Table 6.4 Properties for engine oil (SAE 5W-40)

CHAPTER 7. ROTATING FLOW IN A CAVITY

A. Objectives

- Using Ansys Fluent to Study Rotating Flow in a Cavity
- Inserting Boundary Conditions and Rotation
- Running Laminar 2D Axi-Symmetric Ansys Fluent Simulations with Swirl
- Using XY Plots for Profiles of Velocity Components
- Comparing with Results Using Mathematica Code
- Using Contour Plots for Visualizations of Streamlines and Velocity Components
- Using Pathlines to Visualize the Flow Patterns and Compare with Experiments

B. Problem Description

We will study the rotating flow in a cavity and we will analyze the problem using Ansys Fluent. The rotating cavity filled with water has a cylindrical geometry with a diameter of 95 mm and a height with the same dimension. The top lid is rotating with a rotational speed of 0.445 rad/s while the remaining walls of the cavity are stationary.

C. Launching Ansys Workbench and Selecting Fluent

1. Start by launching Ansys Workbench. Double click on Fluid Flow (Fluent) under Analysis Systems in Toolbox. Right click on Geometry in the Project Schematic and select Properties. Select 2D Analysis Type in Advanced Geometry Options under Properties of Schematic A2: Geometry.

Figure 7.1a) Selecting Fluid Flow

	A	B
1	Property	Value
2	⊟ General	
3	Component ID	Geometry
4	Directory Name	FFF
5	⊟ Notes	
6	Notes	
7	⊟ Basic Geometry Options	
8	Solid Bodies	☑
9	Surface Bodies	☑
10	Line Bodies	☐
11	Parameters	Independent
12	Parameter Key	ANS;DS
13	Attributes	☐
14	Named Selections	☐
15	Material Properties	☐
16	⊟ Advanced Geometry Options	
17	Analysis Type	2D

Properties of Schematic A2: Geometry

Figure 7.1b) 2D Analysis Type

D. Launching Ansys DesignModeler

2. Right click on Geometry under Project Schematic in Ansys Workbench and select New DesignModeler Geometry. Select **Units Millimeter** from the menu in DesignModeler.

    Select the XYPlane in Tree Outline. Look At Face/Plane/Sketch. Select the Sketching tab and Rectangle under the Draw tab. Draw a rectangle from the origin in the first quadrant. Select the Dimensions tab. Click on the left vertical edge of the rectangle and the lower horizontal edge. Enter 47.5 mm as the vertical dimension and 95 mm as the horizontal dimension. Select Concept>>Surfaces from Sketches from the menu. Select Sketch 1 under XY Plane in the Tree Outline. Apply the Sketch as Base Object in the Details View. Click on Generate and close DesignModeler.

Figure 7.2 Surface sketch of the rectangle

E. Launching Ansys Meshing

3. Double click on Mesh under Project Schematic in Ansys Workbench. In the Meshing window, right-click on Mesh under Project and select Update. Select Mesh>>Controls>>Face Meshing from the menu. Select the rectangle in the graphics window and Apply it as Geometry in Details of Face Meshing.

Figure 7.3a) Coarse mesh   Figure 7.3b) Face meshing

Select Mesh>>Controls>>Sizing from the menu. Select the Edge tool, control click on the two vertical edges of the rectangle and Apply them as Geometry in Details of Sizing. Select Number of Divisions as Type and enter 20. Set Capture Curvature to No and select Hard as Behavior. Select the third Bias type from the drop-down menu and enter 5.0 as Bias Factor. Repeat this step but select the two horizontal edges and 40 as the Number of Divisions. Set Capture Curvature to No and select Hard as Behavior. Use the same Bias type and the same Bias Factor 5.0. Right-click on Mesh and select Update.

Figure 7.3c) Details of edge sizing for vertical edges

Figure 7.3d) Details of edge sizing for horizontal edges

4. Control-select the upper horizontal edge and left vertical edge, right-click and select Create Named Selection. Enter the name *wall* and click OK to close the window. Select the right vertical edge, right click and select Create Named Selection. Enter the name *lid*. Finally, name the lower horizontal edge as *symmetry*. Select File>>Save Project from the menu and enter the name *Rotating Lid Flow in Cylinder*. Select File>>Export…>>Mesh>>FLUENT Input File>>Export from the menu in Ansys meshing window and save the mesh with the name *rotating-lid-cylinder-flow-mesh.msh*. Close the Meshing window. Right-click on Mesh in the Project Schematic of Ansys Workbench and select Update.

Figure 7.4 Named selections

F. Launching Ansys Fluent

5. Double-click on Setup under Project Schematic in Ansys Workbench. Select Double Precision. Set the number of Processes equal to the number of processor cores on your computer. To check the number of physical cores, press the Ctrl + Shift + Esc keys simultaneously to open the Task Manager. Go to the Performance tab and select CPU from the left column. You'll see the number of physical cores on the bottom-right side. Close the Task Manager window. Click on the Start button in the Fluent Launcher window.

Figure 7.5a) Fluent launcher window

Click on Check and Scale… under Mesh in General on the Task Page to verify the Domain Extents. Close the Scale Mesh window. Select Axisymmetric Swirl as 2D Space Solver. Select Relative Velocity Formulation. Check the box for Gravity. Enter -9.81 as Gravitational Acceleration in the X direction.

> **Why did we select Axisymmetric Swirl and Relative Velocity Formulation?** This problem is axisymmetric and has a rotating lid boundary condition that will cause a swirling flow in the container. We choose Relative Velocity Formulation since we are using a small container where the lid is rotating and this will cause most of the flow in the container to be rotating.

Scale Mesh

Domain Extents

Xmin [m] 0	Xmax [m] 0.095
Ymin [m] 0	Ymax [m] 0.0475

View Length Unit In
m

Figure 7.5b) Scale mesh window

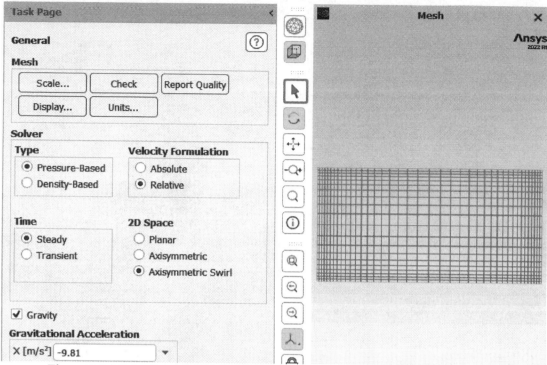

Figure 7.5c) Selecting axisymmetric swirl and gravitational acceleration

6. Next, we double click on Materials under Setup in the Outline View. Select Fluid under Materials on the Task Page and click on Create/Edit…. Click on Fluent Database… in the Create/Edit Materials window. Scroll down in the Fluent Fluid Material section and select water-liquid (h2o<l>). Click on Copy at the bottom of the Fluent Database Materials window. Close the two windows.

Figure 7.6a) Selecting the Fluent database

Figure 7.6b) Selection of water-liquid as fluid material

7. Open Cell Zone Conditions and Fluid under Setup in the Outline View. Double click on *surface_body* under Fluid. Select *water-liquid* as Material Name. Check the box for Frame Motion and enter Speed (rad/s) -0.4449 as Rotational Velocity. Click Apply and Close the Fluid window.

Figure 7.7 Selection of water-liquid and frame motion for the surface body

8. Double click on Boundary Conditions under Setup in the Outline View. Select *symmetry* under Zone in Boundary Conditions on the Task Page. Choose axis from the Type drop-down menu, rename the Zone Name to *axis*, click Apply and Close the Axis window.

Figure 7.8a) Selection of symmetric boundary condition

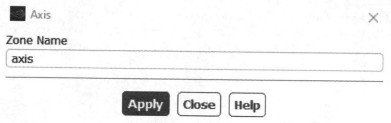

Figure 7.8b) Selection of axis boundary condition

Select *lid* under Zone in Boundary Conditions on the Task Page. Click on the Edit… button for *wall* Type. Select *Moving Wall* under Wall Motion. Check Absolute and Rotational under Motion and enter 0.4449 as Speed [rad/s]. Click Apply and Close the window. Repeat this step for the *wall* Zone, select *Moving Wall* under Wall Motion, select Absolute and Rotational Motion and enter 0 as Speed [rad/s], see Figure 7.8d). Click Apply and Close the window.

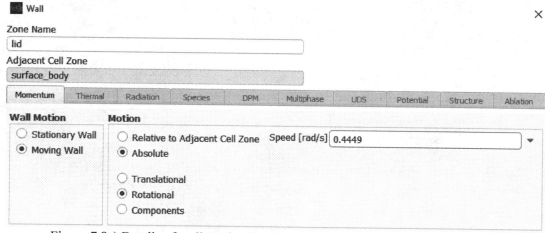

Figure 7.8c) Details of wall motion for lid

Figure 7.8d) Details of wall motion for wall

9. Double click on Models and Viscous (SST k-omega) under Setup in the Outline View. Choose Laminar as the Viscous Model and click OK to close the window. Double click on Methods under Solution in the Outline View and choose PRESTO! for Pressure Spatial Discretization on the Task Page. Select SIMPLEC as the Pressure-Velocity Coupling Scheme. Select Green-Gauss Node Based Gradient under Spatial Discretization. Select QUICK for both Momentum and Swirl Velocity under Spatial Discretization. Double click on Controls under Solution in the Tree. Set all Under-Relaxation Factors to 1 except Momentum that is set to 0.7.

> **Why did we choose Green-Gauss Node Based, QUICK, PRESTO! and SIMPLEC Solution Methods?** The reason for using SIMPLEC as Pressure-Velocity Coupling algorithm is that we are studying steady laminar flow and SIMPLEC is used instead of SIMPLE as it provides a faster solution and a higher under-relaxation. PISO is used for transient solutions. Coupled can alternatively give an efficient steady state solution for single phase flows. The PRESTO! pressure scheme is recommended when we have rotating, swirling flows with curvature. QUICK is used with quadrilateral meshes to provide better accuracy for rotating and swirling flows. The Green-Gauss Node Based gradient is more accurate than the cell based gradient.

Figure 7.9a) Solution method

Figure 7.9b) Under-relaxation factors

10. Double click on Monitors and Residual under Solution in the Outline View. Make sure that the box for Plot under Options is checked. Set the Absolute Criteria to 1e-12 for all four residuals and click OK to close the window.

Figure 7.10 Residual monitors window

11. Double click on Initialization under Solution in the Outline View and select Standard Initialization as Initialization Method on the Task Page. Set Swirl Velocity [m/s] to 0.02. This value is approximately the speed of rotation 0.4449 rad/s times the radius 0.0475m. Click on Initialize. Select File>>Export>>Case… from the menu. Save the Case file with the name *Rotating Flow in a Cavity.cas.h5*. Double click on Run Calculation under Solution in the Outline View and set the Number of Iterations to 2000. Click on the *Calculate* button on the Task Page. Select OK in the Information window when the calculation is complete.

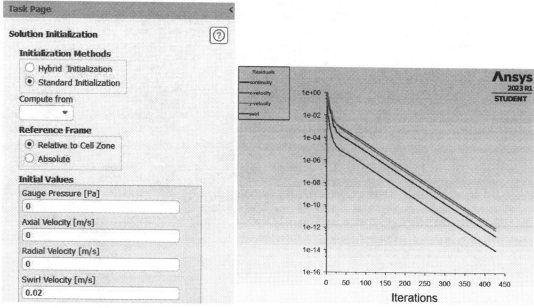

Figure 7.11a) Initialization   Figure 7.11b) Scaled residuals

Click on Copy Screenshot of Active Window to Clipboard, see Figure 7.11c). The Scaled Residuals can be pasted into a Word document.

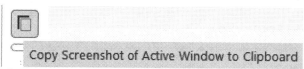

Figure 7.11c) Copying screenshot

G. Post-Processing

12. Open Graphics under Results in Outline View and double click on Contours under Graphics. Select Contours of Velocity and Stream Function. Deselect all surfaces including *interior-surface_body*. Select Colormap Options… in the Contours window. Set Colormap Alignment to Top, set Font Behavior to Fixed and set Font Size to 12 under Font. Set Type to exponential and Precision to 1 under Number Format. Set Length to 0.8 under Colormap Dimensions. Select Apply and Close the Colormap window. Click on Save/Display and Close the Contours window.

Figure 7.12a) Contours window

Figure 7.12b) Streamlines at Re = 1,000 and H/R = 2

Click on Copy Screenshot of Active Window to Clipboard, see Figure 7.11c). The Contours can be pasted into a Word document.

13. Open Plots and double click on XY Plot under Results in the Outline View. Uncheck Position on X Axis and Position on Y Axis under Options. Select Velocity... and Swirl Velocity as Y Axis Function. Select Mesh... and X-Coordinate as X Axis Function. Select New Surface>>Line/Rake.... Enter x0 (m) 0 and x1 (m) 0.095. Enter y0 (m) 0.0285 and y1 (m) 0.0285. Enter *y=0.0285m* as the New Surface Name and click on Create. Close the Line/Rake Surface window.

Select *y=0.0285m* under Surfaces in the Solution XY Plot window and click on Axes.... Set Type as *float* and set Precision to 2 under Number Format for X Axis. Click on Apply. Set Type as *float* and set Precision to 3 under Number Format for Y Axis. Click on Apply. Close the Axes window. Click on Curves.... Select the first available Line Style Pattern for Curve # 0 and use no symbol. Click on Apply. Select the next available Line Style Pattern for Curve # 1 and use no symbol. Click on Apply and Close the Curves – Solution XY Plot window.

Copy the file *swirl-velocity-y = 0.0285m.dat* to the Working Directory. This file and other files can be downloaded from *sdcpublications.com*. Click on Load File... and read the file *swirl-velocity-y = 0.0285m.dat* from the working directory. You will need to set the Files of Type: to All Files (*) to find the file in the Working Directory. Click on Save/Plot. Repeat this step twice but instead load *radial-velocity-y=0.0285m.dat* and *axial-velocity-y=0.0285m.dat*. Plot the Radial Velocity and the Axial Velocity components, respectively.

Create another Line/Rake... with x0 (m) 0.0475 and x1 (m) 0.0475. Enter y0 (m) 0 and y1 (m) 0.0475. Enter *x=0.0475m* as the New Surface Name. Plot the swirl velocity component versus Y-Coordinate and compare with loaded file *swirl-velocity-x = 0.0475m.dat*. Repeat this for the remaining two velocity components. Close the Solution XY Plot window.

Figure 7.13a) Line/rake surface window

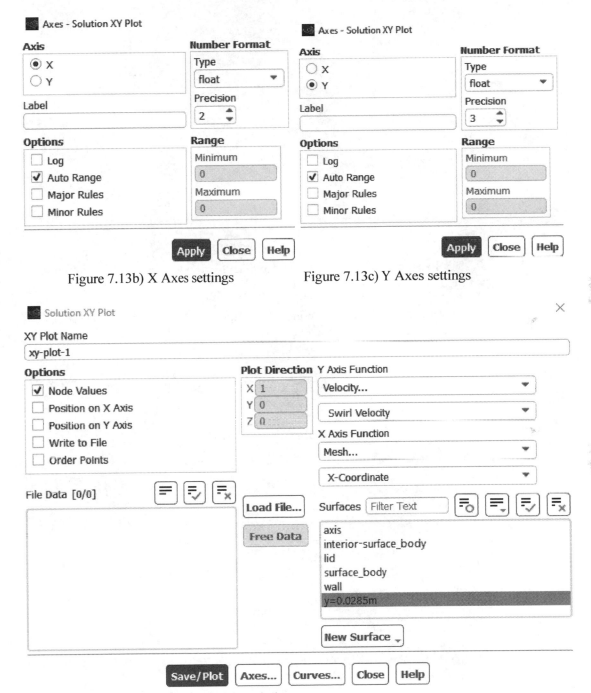

Figure 7.13b) X Axes settings   Figure 7.13c) Y Axes settings

Figure 7.13d) Solution XY Plot window

Figure 7.13e) Swirl velocity versus *x* (Re = 1,000 and H/R = 2) at *y* = 0.0285 m, full line – Ansys Fluent, dashed line - Mathematica

Click on Copy Screenshot of Active Window to Clipboard, see Figure 7.11c). The XY plot can be pasted into a Word document.

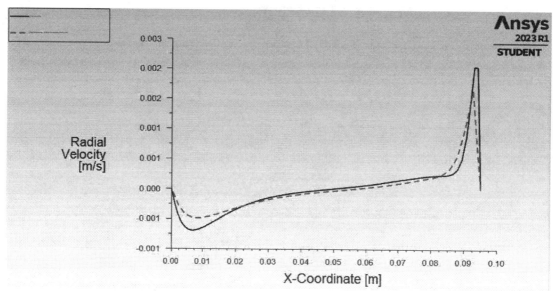

Figure 7.13f) Radial velocity versus *x* (Re = 1,000 and H/R = 2) at *y* = 0.0285 m, full line – Ansys Fluent, dashed line - Mathematica

Click on Copy Screenshot of Active Window to Clipboard, see Figure 7.11c). The XY plot can be pasted into a Word document.

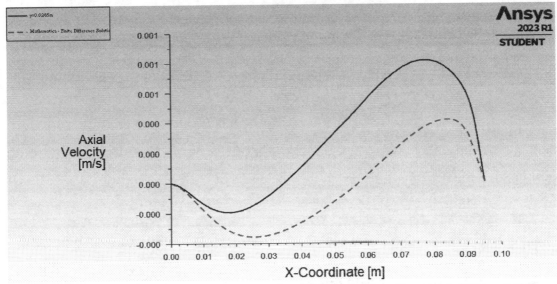

Figure 7.13g) Axial velocity versus x (Re = 1,000 and H/R = 2) at y = 0.0285m, full line – Ansys Fluent, dashed line - Mathematica

Click on Copy Screenshot of Active Window to Clipboard, see Figure 7.11c). The XY plot can be pasted into a Word document.

Figure 7.13h) Swirl velocity versus y (Re = 1,000 and H/R = 2) at x = 0.0475m, full line – Ansys Fluent, dashed line - Mathematica

Click on Copy Screenshot of Active Window to Clipboard, see Figure 7.11c). The XY plot can be pasted into a Word document.

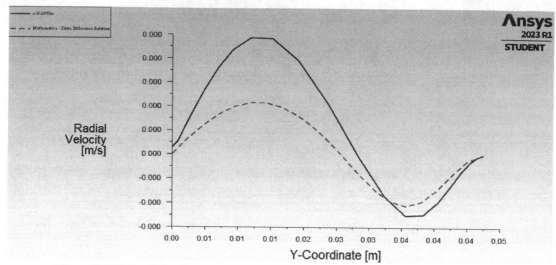

Figure 7.13i) Radial velocity versus *y* (Re = 1,000 and H/R = 2) at *x* = 0.0475m, full line – Ansys Fluent, dashed line - Mathematica

Click on Copy Screenshot of Active Window to Clipboard, see Figure 7.11c). The XY plot can be pasted into a Word document.

Figure 7.13j) Axial velocity versus *y* (Re = 1,000 and H/R = 2) at *x* = 0.0475m, full line – Ansys Fluent, dashed line - Mathematica

Click on Copy Screenshot of Active Window to Clipboard, see Figure 7.11c). The XY plot can be pasted into a Word document.

14. Double click on Graphics and Contours under Results in the Outline View. Select Contours of Velocity and Stream Function. Deselect all Surfaces in the Contours window. Click on Save/Display.

    Select the View tab in the menu and click on Views…. Select *axis* as Mirror Plane and click on Apply. Select the Camera… button in the Views window. Use your left mouse button to rotate the dial 90-degrees counter-clockwise until the cavity appears upright. Close the Camera Parameters window. Click on the Save button under Actions in the Views window and close the windows. Click on Copy Screenshot of Active Window to Clipboard, see Figure 7.11c). The Contour Plot can be pasted into a Word document.

Figure 7.14a) Views window

Figure 7.14b) Camera parameters

Double click on Graphics and Contours under Results in the Outline View. Select Contours of Velocity… and Swirl Velocity in the Contours window. Deselect all surfaces. Select Colormap Options… and set Colormap Alignment to Left. Click on Save/Display. Select the View tab in the menu and click on Views…. Select *view-0* under Views, select axis under Mirror Planes and click on Apply. Close the Views window. Repeat this for Radial Velocity and Axial Velocity. Click on Copy Screenshot of Active Window to Clipboard, see Figure 7.11c). The Contour Plot can be pasted into a Word document.

Figure 7.14c) Streamlines and swirl, radial and axial velocities at Re=1000

15. Right click on Fluid Flow (Fluent) in the Workbench Project Schematic and select Duplicate. Right click on Setup in Project Schematic for Copy of Fluid Flow Fluent and select Edit…. In the Fluent Launcher window, select Double Precision. Set the number of Processes equal to the number of processor cores on your computer. Click on the Start button in the Fluent Launcher window.

Copy the file *rotating-flow-mesh-lambda=2.5.msh* to the Working Directory. This file is available for download at *sdcpublications.com*. Select File>>Import>>Mesh from the Fluent menu. Answer *Yes* to the first question that you get on the screen. Select Replace Mesh under Options in the Read Mesh Options window and select Continue…. Select the file *rotating-flow-mesh-lambda=2.5.msh*. Open Cell Zone Conditions and Fluid under

Setup in the Outline View. Double click on *surface_body* under Fluid. Enter Speed (rad/s) -1 as Rotational Velocity. Click Apply and Close the Fluid window. Double click on Boundary Conditions under Setup in the Outline View. Select *lid* under Zone in Boundary Conditions on the Task Page. Click on the Edit… button for *wall* Type. Enter 1 as Speed [rad/s] corresponding to $Re = 2,245$. Click Apply and Close the window.

Double click on Initialization under Solution in the Outline View and select Standard Initialization as Initialization Method on the Task Page. Set Swirl Velocity [m/s] to 0.0475. Click on Initialize. Select File>>Export>>Case… from the menu. Save the Case file with the name *Rotating Flow in a Cavity Lambda = 2.5.cas.h5*. Double click on Run Calculation under Solution in the Outline View. Click on the *Calculate* button on the Task Page. Select OK in the Information window when the calculation is complete. Click on Copy Screenshot of Active Window to Clipboard, see Figure 7.11c). The Contour Plot can be pasted into a Word document.

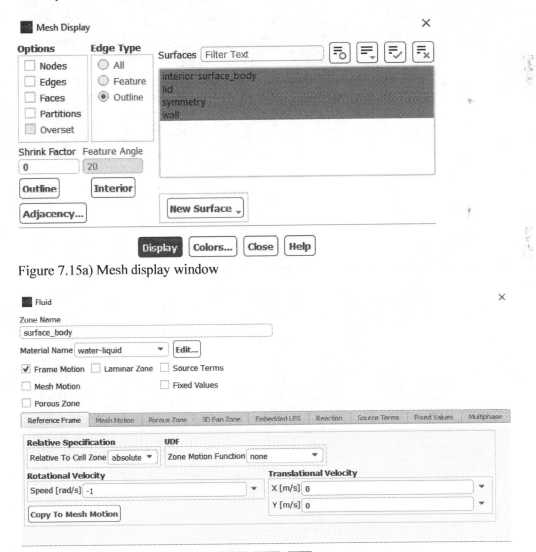

Figure 7.15a) Mesh display window

Figure 7.15b) Settings for rotational velocity of the surface body

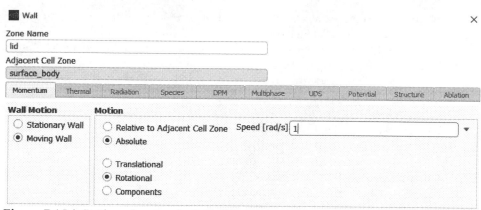

Figure 7.15c) Settings for speed of the lid corresponding to $Re = 2,245$

Figure 7.15d) Settings for solution initialization

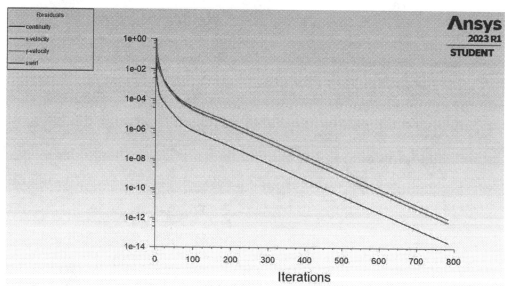

Figure 7.15e) Scaled residuals for rotating flow in a cavity at $Re = 2245$

16. Double click on Graphics and Contours under Results in the Outline View. Select Contours of Velocity and Stream Function. Deselect all Surfaces and select Colormap Options… in the Contours window. Set Font Behavior to Fixed and Font Size to 12 under Font in the Colormap window. Select Apply and Close the window. Click on Save/Display. Select the View tab in the menu and click on Views…. Select *axis* as Mirror Plane and click on Apply. Click on the Camera… button in the Views window. Use your left mouse button to rotate the dial counter-clockwise until the cavity appears upright. Close the Camera Parameters window. Select the Save button under Actions in the Views window and close the windows. Click on Copy Screenshot of Active Window to Clipboard, see Figure 7.11c).

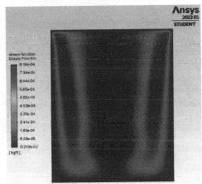

Figure 7.16a) Mirrored and rotated streamlines for Re = 2245 and H/R – 2.5

Double click on Graphics and Pathlines under Results in the Outline View. Select Color by Velocity and Velocity Magnitude. Select *interior-surface_body* under Release from Surfaces. Uncheck Relative Pathlines under Options, set the Tolerance to 1e-06, set the Steps to 20000, set Path Skip to 100 and Path Coarsen to 2. Click on Save/Display. Click on Copy Screenshot of Active Window to Clipboard, see Fig. 7.11c).

Figure 7.16b) Settings for pathlines

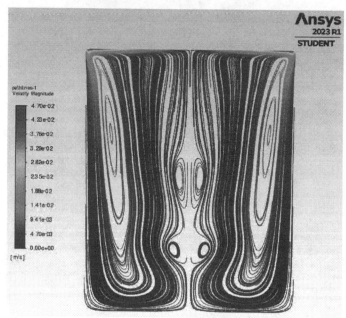

Figure 7.16c) Pathlines at Re = 2245 and H/R = 2.5

Figure 7.16d) Comparison between ANSYS Fluent simulations and experimental
visualizations at Re = 2245, H/R = 2.5

H. Theory

17. We define the Reynolds number as

$$Re = \Omega R^2 / \nu \qquad (7.1)$$

, where Ω (rad/s) is the angular velocity of the lid, R (m) is the radius of the cylindrical container and ν (m²/s) is kinematic viscosity. The aspect ratio of the cylindrical container is defined as

$$\lambda = H/R \qquad (7.2)$$

, where H (m) is the height of the cylindrical container.

Figure 7.17a) Geometry for cylinder with rotating lid

The incompressible Navier-Stokes equations in cylindrical coordinates (r, θ, z) with velocity components $\big(u_r(r, \theta, z, t), u_\theta(r, \theta, z, t), u_z(r, \theta, z, t)\big)$ are given by the following equations:

Continuity equation:
$$\frac{1}{r}\frac{\partial(r u_r)}{\partial r} + \frac{1}{r}\frac{\partial u_\theta}{\partial \theta} + \frac{\partial u_z}{\partial z} = 0 \qquad (7.3)$$

r-component:
$$\rho\left(\frac{\partial u_r}{\partial t} + u_r\frac{\partial u_r}{\partial r} + \frac{u_\theta}{r}\frac{\partial u_r}{\partial \theta} - \frac{u_\theta^2}{r} + u_z\frac{\partial u_r}{\partial z}\right) =$$
$$-\frac{\partial p}{\partial r} + \mu\left[\frac{1}{r}\frac{\partial}{\partial r}\left(r\frac{\partial u_r}{\partial r}\right) - \frac{u_r}{r^2} + \frac{1}{r^2}\frac{\partial^2 u_r}{\partial \theta^2} - \frac{2}{r^2}\frac{\partial u_\theta}{\partial \theta} + \frac{\partial^2 u_r}{\partial z^2}\right] \qquad (7.4)$$

θ-component:
$$\rho\left(\frac{\partial u_\theta}{\partial t} + u_r\frac{\partial u_\theta}{\partial r} + \frac{u_\theta}{r}\frac{\partial u_\theta}{\partial \theta} + \frac{u_r u_\theta}{r} + u_z\frac{\partial u_\theta}{\partial z}\right) =$$
$$-\frac{1}{r}\frac{\partial p}{\partial \theta} + \mu\left[\frac{1}{r}\frac{\partial}{\partial r}\left(r\frac{\partial u_\theta}{\partial r}\right) - \frac{u_\theta}{r^2} + \frac{1}{r^2}\frac{\partial^2 u_\theta}{\partial \theta^2} + \frac{2}{r^2}\frac{\partial u_r}{\partial \theta} + \frac{\partial^2 u_\theta}{\partial z^2}\right] \qquad (7.5)$$

z-component:
$$\rho\left(\frac{\partial u_z}{\partial t} + u_r\frac{\partial u_z}{\partial r} + \frac{u_\theta}{r}\frac{\partial u_z}{\partial \theta} + u_z\frac{\partial u_z}{\partial z}\right) =$$
$$-\frac{\partial p}{\partial z} + \mu\left[\frac{1}{r}\frac{\partial}{\partial r}\left(r\frac{\partial u_z}{\partial r}\right) + \frac{1}{r^2}\frac{\partial^2 u_z}{\partial \theta^2} + \frac{\partial^2 u_z}{\partial z^2}\right] \qquad (7.6)$$

where ρ (kg/m³) is density, g (m/s²) is acceleration due to gravity and μ (kg/m-s) is dynamic viscosity. For axisymmetric flow there is no θ dependence for the velocity components so they can be expressed as $\left(u_r(r,z), u_\theta(r,z), u_z(r,z)\right)$. This together with a steady flow will reduce the equations above to the following equations:

Continuity equation: $\quad \dfrac{1}{r}\dfrac{\partial(ru_r)}{\partial r} + \dfrac{\partial u_z}{\partial z} = 0$ $\hfill (7.7)$

r-component: $\quad \rho\left(u_r\dfrac{\partial u_r}{\partial r} - \dfrac{u_\theta^2}{r} + u_z\dfrac{\partial u_r}{\partial z}\right) = -\dfrac{\partial p}{\partial r} + \mu\left[\dfrac{1}{r}\dfrac{\partial}{\partial r}\left(r\dfrac{\partial u_r}{\partial r}\right) - \dfrac{u_r}{r^2} + \dfrac{\partial^2 u_r}{\partial z^2}\right]$ $\hfill (7.8)$

θ-component: $\quad \rho\left(u_r\dfrac{\partial u_\theta}{\partial r} + \dfrac{u_r u_\theta}{r} + u_z\dfrac{\partial u_\theta}{\partial z}\right) = \mu\left[\dfrac{1}{r}\dfrac{\partial}{\partial r}\left(r\dfrac{\partial u_\theta}{\partial r}\right) - \dfrac{u_\theta}{r^2} + \dfrac{\partial^2 u_\theta}{\partial z^2}\right]$ $\hfill (7.9)$

z-component: $\quad \rho\left(u_r\dfrac{\partial u_z}{\partial r} + u_z\dfrac{\partial u_z}{\partial z}\right) = -\dfrac{\partial p}{\partial z} + \mu\left[\dfrac{1}{r}\dfrac{\partial}{\partial r}\left(r\dfrac{\partial u_z}{\partial r}\right) + \dfrac{\partial^2 u_z}{\partial z^2}\right]$ $\hfill (7.10)$

We make the velocity components and pressure non-dimensional using the following relations:

$$u_r^* = u_r/(\Omega R), u_\theta^* = u_\theta/(\Omega R), u_z^* = u_z/\Omega R, p^* = p/(\rho\Omega^2 R^2) \qquad (7.11)$$

and we make the coordinates non-dimensional using the following relations:

$$r^* = r/R, \theta^* = \theta, z^* = z/R \qquad (7.12)$$

The continuity equation and the different component equations in non-dimensional form after skipping the $*$ superscript symbol:

$$\frac{\partial u_r}{\partial r} + \frac{u_r}{r} + \frac{\partial u_z}{\partial z} = 0 \qquad (7.13)$$

$$u_r\frac{\partial u_r}{\partial r} - \frac{u_\theta^2}{r} + u_z\frac{\partial u_r}{\partial z} = -\frac{\partial p}{\partial r} + \frac{1}{Re}\left[\frac{\partial^2 u_r}{\partial r^2} + \frac{1}{r}\frac{\partial u_r}{\partial r} - \frac{u_r}{r^2} + \frac{\partial^2 u_r}{\partial z^2}\right] \qquad (7.14)$$

$$u_r\frac{\partial u_\theta}{\partial r} + \frac{u_r u_\theta}{r} + u_z\frac{\partial u_\theta}{\partial z} = \frac{1}{Re}\left[\frac{\partial^2 u_\theta}{\partial r^2} + \frac{1}{r}\frac{\partial u_\theta}{\partial r} - \frac{u_\theta}{r^2} + \frac{\partial^2 u_\theta}{\partial z^2}\right] \qquad (7.15)$$

$$u_r\frac{\partial u_z}{\partial r} + u_z\frac{\partial u_z}{\partial z} = -\frac{\partial p}{\partial z} + \frac{1}{Re}\left[\frac{\partial^2 u_z}{\partial r^2} + \frac{1}{r}\frac{\partial u_z}{\partial r} + \frac{\partial^2 u_z}{\partial z^2}\right] \qquad (7.16)$$

The boundary conditions are the following:

$$u_r(0,z) = u_r(1,z) = u_r(r,0) = u_r(r,2) = 0 \qquad (7.17)$$

$$u_\theta(0,z) = u_\theta(1,z) = u_\theta(r,0) = 0,\ u_\theta(r,2) = \Omega r \qquad (7.18)$$

$$\frac{\partial u_z(0,z)}{\partial r} = u_z(1,z) = u_z(r,0) = u_z(r,2) = 0 \qquad (7.19)$$

```
Remove[XandYGrid,MakeVariables,FDMatrices];
XandYGrid[domain_List,pts_List]:=MapThread[N@Range[Sequence@@#1,Abs[Subtract@@#1]/#2]&,{domain,pts-1}];
BoundaryIndex[rgridlen_,zgridlen_]:=Module[{tmp,left,right,bot,top},tmp=Table[(n-1)zgridlen+Range[1,zgridlen],{n,1,rgridlen}];
  {left,right}=tmp[[{1,-1}]];{bot,top}=Transpose[{First[#],Last[#]}&/@tmp];{top,right[[2;;-2]],bot,left[[2;;-2]]}];
Attributes[MakeVariables]={Listable};MakeVariables[var_,n_]:=Table[Unique[var],{n}];
FDMatrices[deriv_,rzgrid_,difforder_]:=Map[NDSolve`FiniteDifferenceDerivative[#,rzgrid,"DifferenceOrder"→difforder]["Differen
tiationMatrix"]&,deriv];

Options[DrivenCylinderCavitySolver]={"InitialGuess"→1};
DrivenCylinderCavitySolver[Rey_,domain_List,pts_List,difforder_,OptionsPattern[]]:=Module[{rzgrid,rrgrid,rgrid,r2grid,nr,nz,top,r
ight,bot,left,ur,uz,uθ,p,urvar,uzvar,uθvar,pvar,dr,dz,dr2,dz2,eqnur,eqnuz,eqnuθ,eqncont,bcindx,sol,boundaries,deqns,dvars,grid}
,(*Get the grid and differentiation matrices*)rzgrid=XandYGrid[domain,pts];rrgrid=rzgrid[[1]];
  grid=Flatten[Outer[List,Sequence@@rzgrid],1];rgrid=grid[[All,1]];r2grid=rgrid*rgrid;
  {nr,nz}=Map[Length,rzgrid];{top,right,bot,left}=BoundaryIndex[nr,nz];
  {dr,dz,dr2,dz2}=FDMatrices[{{1,0},{0,1},{2,0},{0,2}},rzgrid,difforder];
  (*Get the discretized axisymmetric Navier Stokes equations*){urvar,uzvar,uθvar,pvar}=MakeVariables[{ur,uz,uθ,p},nr*nz];
  eqnur=r2grid urvar (dr.urvar)-rgrid uθvar uθvar+r2grid uzvar (dz.urvar)+r2grid(dr.pvar)-r2grid (1/Rey) (dr2+dz2).urvar-
rgrid(1/Rey) (dr.urvar)+(1/Rey) urvar;
  eqnuz=rgrid urvar(dr.uzvar)+rgrid uzvar (dz.uzvar)+rgrid(dz.pvar)-rgrid (1/Rey) (dr2+dz2).uzvar-(1/Rey) (dr.uzvar);
  eqnuθ=r2grid urvar (dr.uθvar)+rgrid urvar uθvar+r2grid uzvar (dz.uθvar)-r2grid (1/Rey) (dr2+dz2).uθvar-rgrid(1/Rey)
(dr.uθvar)+(1/Rey) uθvar;
  eqncont=rgrid (dr.urvar)+urvar+rgrid(dz.uzvar);
  (*Apply the boundary conditions and solve the
system*)boundaries=Join[top,right,bot,left];eqnur[[boundaries]]=urvar[[boundaries]];
  eqnuz[[boundaries]]=uzvar[[boundaries]];eqnuz[[left]]=(dr[[left]].uzvar);
  eqnuθ[[boundaries]]=uθvar[[boundaries]];eqnuθ[[top]]=uθvar[[top]]-(rrgrid-1);
  eqncont[[top[[1]]]]=pvar[[top[[1]]]];
  {deqns,dvars}={Join[eqnur,eqnuz,eqnuθ,eqncont],Join[urvar,uzvar,uθvar,pvar]};
  sol=dvars/.Quiet@FindRoot[deqns,Thread[{dvars,OptionValue["InitialGuess"]}]];
  (*Get Interpolating
functions*)grid=Flatten[Outer[List,Sequence@@rzgrid],1];Map[Interpolation@Join[grid,Transpose@List@#,2]&,Partition[sol,Len
gth[grid]]]];

(* Calculate Solution *)
Timing[res={ur,uz,uθ,p}=DrivenCylinderCavitySolver[1000,{{1,2},{0,2}},{30,60},4]]
StreamDensityPlot[{ur[r,z],uz[r,z]},{r,1,2},{z,0,2},AspectRatio→2,PlotLabel→Style[Text["Streamfunction at Re = 1000"]]]

(* Plot Results and Save Velocity Profiles for uθ *)
SetDirectory["C:\\Users\\johne"];r1=1;omega=0.4449;radius=0.0475;
Plot[uθ[0.6+r1,z]*omega*radius,{z,0,2},PlotRange→{{0,2},{0,0.014}},AxesLabel→{z,Uθ[m/s]},PlotLabel→Style[Text["r=1.6,
Re=1000"]]]
mylist=Table[{uθ[0.6+r1,z]*omega*radius,z*radius},{z,0,2,0.05}];
Plotex[mylist];Export["swirl-velocity-y=0.0285m.dat",TableOfValues1];

Plot[uθ[r,1]*omega*radius,{r,0+r1,1+r1},AxesLabel→{r+1,Uθ[m/s]},PlotLabel→Style[Text["z=1, Re=1000"]]]
mylist=Table[{uθ[r,1]*omega*radius,(r-r1)*radius},{r,0+r1,1+r1,0.05}];Plotex[mylist];Export["swirl-velocity-
x=0.0475m.dat",TableOfValues1];

Plotex[n_]:=(mylist[[All,{1,2}]]:=n[[All,{2,1}]];TableOfValues1=Prepend[mylist,{""}];
  TableOfValues1=Prepend[TableOfValues1,{"((xy/key/label \"Mathematica - Finite Difference Solution\")"}];
  TableOfValues1=Prepend[TableOfValues1,{""}];
  TableOfValues1=Prepend[TableOfValues1,{"(labels \"Swirl Velocity (m/s)\" \"Position (m)\")"}];
  TableOfValues1=Prepend[TableOfValues1,{"(title
\"Mathematica\")"}];TableOfValues1=Append[TableOfValues1,{")"}];Grid[TableOfValues1];)

(* Plot Results and Save Velocity Profiles for ur *)
SetDirectory["C:\\Users\\johne"];r1=1;omega=0.4449;radius=0.0475;
Plot[ur[0.6+r1,z]*omega*radius,{z,0,2},PlotRange→{{0,2},{-0.001,0.0025}},AxesLabel→{z,Ur[m/s]},PlotLabel→Style[Text["r=1.6,
Re=1000"]]]
mylist=Table[{ur[0.6+r1,z]*omega*radius,z*radius},{z,0,2,0.05}];
Plotex[mylist];Export["radial-velocity-y=0.0285m.dat",TableOfValues1];

Plot[ur[r,1]*omega*radius,{r,0+r1,1+r1},AxesLabel→{r+1,Ur[m/s]},PlotLabel→Style[Text["z=1, Re=1000"]]]
mylist=Table[{ur[r,1]*omega*radius,(r-r1)*radius},{r,0+r1,1+r1,0.05}];Plotex[mylist];Export["radial-velocity-
x=0.0475m.dat",TableOfValues1];
```

```
Plotex[n_]:=(mylist[[All,{1,2}]]=n[[All,{2,1}]];TableOfValues1=Prepend[mylist,{""}];
  TableOfValues1=Prepend[TableOfValues1,{"((xy/key/label \"Mathematica - Finite Difference Solution\")"}];
  TableOfValues1=Prepend[TableOfValues1,{""}];
  TableOfValues1=Prepend[TableOfValues1,{"(labels \"Radial Velocity (m/s)\" \"Position (m)\")"}];
  TableOfValues1=Prepend[TableOfValues1,{"(title
\"Mathematica\")"}];TableOfValues1=Append[TableOfValues1,{")"}];Grid[TableOfValues1];)

(* Plot Results and Save Velocity Profiles for uz *)
SetDirectory["C:\\Users\\johne"];r1=1;omega=0.4449;radius=0.0475;
Plot[uz[0.6+r1,z]*omega*radius,{z,0,2},PlotRange→{{0,2},{-0.0004,0.0010}},AxesLabel→{z,Uz[m/s]},PlotLabel→Style[Text["r=1.6,
Re=1000"]]]
mylist=Table[{uz[0.6+r1,z]*omega*radius,z*radius},{z,0,2,0.05}];
Plotex[mylist];Export["axial-velocity-y=0.0285m.dat",TableOfValues1];

Plot[uz[r,1]*omega*radius,{r,0+r1,1+r1},AxesLabel→{r+1,Uz[m/s]},PlotLabel→Style[Text["z=1, Re=1000"]]]
mylist=Table[{uz[r,1]*omega*radius,(r-r1)*radius},{r,0+r1,1+r1,0.05}];Plotex[mylist];Export["axial-velocity-
x=0.0475m.dat",TableOfValues1];

Plotex[n_]:=(mylist[[All,{1,2}]]=n[[All,{2,1}]];TableOfValues1=Prepend[mylist,{""}];
  TableOfValues1=Prepend[TableOfValues1,{"((xy/key/label \"Mathematica - Finite Difference Solution\")"}];
  TableOfValues1=Prepend[TableOfValues1,{""}];
  TableOfValues1=Prepend[TableOfValues1,{"(labels \"Axial Velocity (m/s)\" \"Position (m)\")"}];
  TableOfValues1=Prepend[TableOfValues1,{"(title
\"Mathematica\")"}];TableOfValues1=Append[TableOfValues1,{")"}];Grid[TableOfValues1];)
```

Figure 7.17b) Mathematica 8.0 code for cylinder with rotating lid

I. References

1. ANSYS Fluid Dynamics Verification Manual, Release 15.0, November 2013.
2. Granger, R.A., Experiments in Fluid Mechanics, Dryden Press, 1988.
3. Michelsen, J.A., Modeling of Laminar Incompressible Rotating Fluid Flow, AFM 86-05, Ph.D. thesis, Department of Fluid Mechanics, Technical University of Denmark, 1986.
4. Mokhasi, Paritosh, Using Mathematica to Simulate and Visualize Fluid Flow in a Box, 2013.
5. Sorensen, J.N. and Loc,T.P., Higher-Order Axisymmetric Navier-Stokes Code: Description and Evolution of Boundary Conditions, *International Journal For Numerical Methods in Fluids*, 9:1517-1537, 1989.
6. Sorensen, J.N. and Christensen, E.A., Direct Numerical Simulation of Rotating Fluid Flow in a Closed Cylinder, *Physics of Fluids* 7:764, 1995.

J. Exercises

7.1 Use ANSYS Fluent to study the rotating cavity filled with water that has a cylindrical geometry with a radius of 100 mm and a height 300 mm. The top lid is rotating with a Reynolds number $Re = \frac{\Omega R^2}{\nu} = 2800$ while the remaining walls of the cavity are stationary. Visualize pathlines, streamlines, swirl velocity, radial velocity and axial velocity for this flow case.

7.2 Use ANSYS Fluent to study the rotating cavity filled with water that has a cylindrical geometry with a radius of 100 mm and a height 150 mm. The top lid is rotating with a Reynolds number $Re = \frac{\Omega R^2}{\nu} = 1400$ while the remaining walls of the cavity are stationary. Visualize pathlines, streamlines, swirl velocity, radial velocity and axial velocity for this flow case.

7.3 Use ANSYS Fluent to study a rotating cavity filled with water that has a frustum geometry with a bottom radius of 50 mm, top radius of 100 mm and a height 150 mm. The top lid is rotating with a Reynolds number $Re = \frac{\Omega Rt^2}{\nu} = 2000$ while the remaining walls of the cavity are stationary. Visualize pathlines, streamlines, swirl velocity, radial velocity and axial velocity for this flow case.

7.4 Use ANSYS Fluent to study a rotating cavity filled with water that has a frustum geometry with a bottom radius of 50 mm, top radius of 100 mm and a height 150 mm. The top lid is rotating with a Reynolds number $Ret = \frac{\Omega t Rt^2}{\nu} = 2000$ and the bottom is rotating with the same Reynolds number $Reb = \frac{\Omega b Rb^2}{\nu} = 2000$ while the remaining walls of the cavity are stationary. Visualize pathlines, streamlines, swirl velocity, radial velocity and axial velocity for this flow case.

K. Notes

CHAPTER 8. SPINNING CYLINDER

A. Objectives

- Using Ansys Fluent to Study the Fluid in an Open Cylinder with a Free Surface and Solid Body Rotation
- Inserting Boundary Conditions and System Rotation
- Using Volume of Fluid Model for Multiphase Flow with Surface Tension
- Running Laminar 2D Axisymmetric Ansys Fluent Simulations with Swirl
- Using Contour Plots for Visualizations of Volume Fraction and Swirl Velocity
- Using Excel for Free Surface Plots

B. Problem Description

We will study the startup flow in a partially filled spinning open cylindrical container where the fluid has a free surface. We will analyze the problem using Ansys Fluent. The cylindrical container has a diameter of 304.8 mm and the same height. The rotational speed of the cylinder is 12 rad/s.

C. Launching Ansys Workbench and Selecting Fluent

1. Start by launching Ansys Workbench. Launch Fluid Flow (Fluent) that is available under Analysis Systems in Ansys Workbench. Select Geometry under Project Schematic, right click and select Properties. Select 2D Analysis Type under Advanced Geometry Options. Right click on Geometry in Project Schematic and select New DesignModeler Geometry to start DesignModeler.

Figure 8.1 Launching ANSYS Fluent and selecting 2D analysis type

D. Launching Ansys DesignModeler

2. Select **Units>>Millimeter** from the menu in DesignModeler. Select the XYPlane in the Tree Outline. Select Look At Face/Plane/Sketch ![icon]. Select the Sketching tab and Rectangle. Draw a rectangle from the origin in the first quadrant of the graphics window. Make sure that the letter P is showing when you move the cursor over the origin of the coordinate system and before you start drawing the rectangle in the first quadrant.

    Select Dimensions and click on the left vertical edge of the rectangle and the lower horizontal edge. Enter **304.8 mm** as the horizontal dimension and **152.4 mm** as the vertical dimension. Right click in the graphics window and select Zoom to Fit.

    Select Concept>>Surfaces from Sketches from the menu. Select Sketch 1 under XY Plane in the Tree Outline. Apply the sketch as a Base Object in Details View. Click on Generate and close DesignModeler.

Figure 8.2a) Rectangle with dimensions

Figure 8.2b) Surface sketch for the rectangle

E. Launching Ansys Meshing

3. Double click on Mesh under Project Schematic in Ansys Workbench. Right-click on Mesh under Project and Model (A3) in the Meshing window and select Update. Select Mesh>>Controls>Face Meshing from the menu. Click on the rectangle in the graphics window and Apply it as Geometry in Details of Face Meshing.

Figure 8.3a) Launching meshing window

Figure 8.3b) Coarse mesh

Figure 8.3c) Selecting face meshing

Figure 8.3d) Details of face meshing

Select Mesh>>Controls>>Sizing from the menu. Select the Edge tool , control click on the two vertical edges and Apply them as Geometry in Details of Sizing. Select Number of Divisions as Type and enter 60. Set Capture Curvature to No and select Hard as Behavior. Repeat this step but select the two horizontal edges and 120 as the Number of Divisions, set Capture Curvature to No and Hard as Behavior. Right click on Mesh and select Update. Open Statistics under Details of "Mesh". The number of Nodes are 7381 and the number of Elements are 7200.

Figure 8.3e) Details for sizing of horizontal edges

178

Figure 8.3g) Final mesh used for spinning cylinder

Select the lower horizontal edge, right click and select Create Named Selection. Enter the name *axis-2* and click OK to close the window. Control-select the right vertical edge and the upper horizontal edge, right click and select Create Named Selection. Enter the name *wall-1*. Name the left vertical edge as *pressure-inlet-4*. Finally, select the face tool , select the interior of the mesh, right click and name it *mesh-3*.

Figure 8.3h) Named selections

Select File>>Export...>>Mesh>>FLUENT Input File>>Export from the menu. Save the mesh with the name *spinning-cylinder.msh*. Select File>>Save Project from the menu and save the project with the name *Spinning Cylinder.wbpj*. Close the meshing window. Right click on Mesh in the Project Schematic and select Update.

F. Launching Ansys Fluent

4. Double click on Setup under Project Schematic in Ansys Workbench. Check the box for Double Precision and uncheck the box for Display Mesh After Reading under General Options tab. Set the number of Solver Processes equal to the number of processor cores for your computer. To check the number of physical cores, press the Ctrl + Shift + Esc keys simultaneously to open the Task Manager. Go to the Performance tab and select CPU from the left column. You'll see the number of physical cores on the bottom-right side. Close the Task Manager window. Click on the Start button in the Fluent Launcher window.

    Click on Display under General in Mesh on the Task Page in Ansys Fluent. Check the Edges box under Options in the Mesh Display window. Select all Surfaces and click on Display in the Mesh Display window and Close the same window.

Figure 8.4a) Fluent launcher

Figure 8.4b) Mesh display window

Figure 8.4c) Cylinder with mesh in Ansys Fluent

5. Select Axisymmetric Swirl for 2D Space in the General Solver settings on the Task Page in Ansys Fluent. Select Transient for Time in the General Solver settings.

Open Models under Setup in the Outline View and double click on Viscous (SST k-omega). Select Laminar Model and click on OK to close the Viscous Model window. Double click on Multiphase. Select the Volume of Fluid Model and check the box for Implicit Body Force under Body Force Formulation. Click on Apply and Close the Multiphase Model window.

> **Why did we select Axisymmetric Swirl and Transient for Time?** This problem is axisymmetric and has rotation. We choose Transient since we are interested in studying the shape of the free surface and how it deforms over time.

> **Why did we select Volume of Fluid as Multiphase model and Implicit Body Force?** The Volume of Fluid model is used when we have two or more immiscible fluids where the position of the interface between the fluids is of interest. The applications of the VOF Model includes free-surface flows, stratified flows and the tracking of liquid-gas interfaces in general. The VOF model is limited to the pressure based solver and doesn't allow for void regions without a fluid. Implicit Body Force is used when we have large body forces such as is the case when we have gravity, surface tension and rotation in multiphase flows. Checking the box for Implicit Body Force will make the solution more robust with better convergence.

Figure 8.5a) General settings

Figure 8.5b) Multiphase model

Figure 8.5c) Selecting volume of fluid model

6. Next, we double click on Materials under Setup in the Outline View. Select Fluid under Materials on the Task Page and click on Create/Edit…. Click on Fluent Database… in the Create/Edit Materials window. Scroll down in the Fluent Fluid Material section and select *water-liquid (h2o<l>)*. Click Copy at the bottom of the Fluent Database Materials window. Close the two windows.

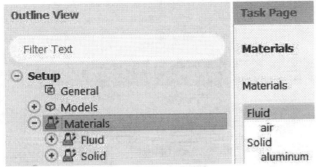

Figure 8.6a) Selecting fluid materials

Figure 8.6b) Selecting Fluent database

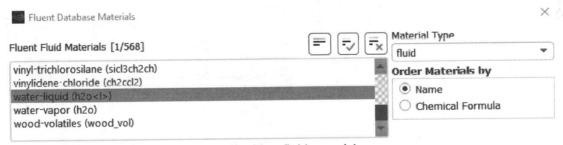

Figure 8.6c) Selection of water-liquid as fluid material

7. Select the Physics tab in the menu and select Models>>Multiphase Model. Select the Phases tab at the top of the Multiphase Model window. Select phase-1-Primary Phase and enter *air* as name for the Primary Phase. Click on Apply and select phase-2-Secondary Phase, select *water-liquid* as the Phase Material and enter *water* as the Name for the Secondary Phase. Click on Apply in the Multiphase Model window. Click on the Phase Interaction tab at the top of the Multiphase Model window. Select constant for Surface Tension Coefficients [N/m] and enter 0.07286 as the value. Click on Apply and Close the Multiphase Model window.

Figure 8.7a) Setting up the physics

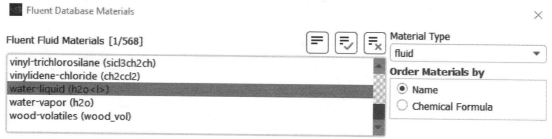

Figure 8.7b) Editing phase-1-Primary Phase

Figure 8.7c) Name for phase-1   Figure 8.7d) Name for phase-2

Figure 8.7e) Entering value for surface tension between water and air

8. Select the Physics tab in the menu and select Operating Conditions... under Solver. Check the box for Gravity and enter 9.81 [m/s²] for Gravitational Acceleration in the positive X direction. Select user-input as Operating Density Method. Enter 1.225 as Operating Density [kg/m³]. Click on the OK button to close the window.

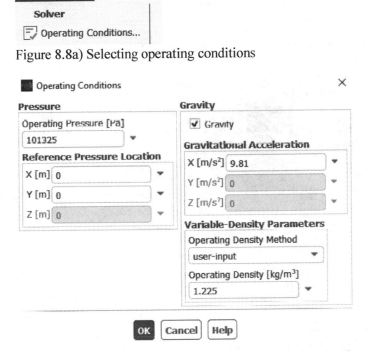

Figure 8.8a) Selecting operating conditions

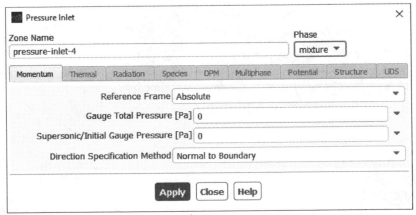

Figure 8.8b) Including gravity and specified operating density

9. Double click on Boundary Conditions under Setup in the Outline View. Select *pressure-inlet-4* under Zone in Boundary Conditions on the Task Page. Select *mixture* from the Phase drop-down menu under Boundary Conditions on the Task Page. Click on the Edit... button. Leave the Gauge Total Pressure at 0 and click Apply and Close the window.

Figure 8.9a) Pressure-inlet boundary

10. For *pressure-inlet-4*, select *water* from the Phase drop-down menu under Boundary Conditions on the Task Page. Click on the Edit… button, make sure that the value for the Volume Fraction is 0, select Apply and Close the window.

Figure 8.10 Details for pressure-inlet boundary condition

11. For *wall-1*, select *mixture* from the Phase drop-down menu under Boundary Conditions on the Task Page. Click on the Edit… button and select Moving Wall for Wall Motion. Choose Rotational Motion and set the Speed [rad/s] to 12. Click Apply and Close the Wall window.

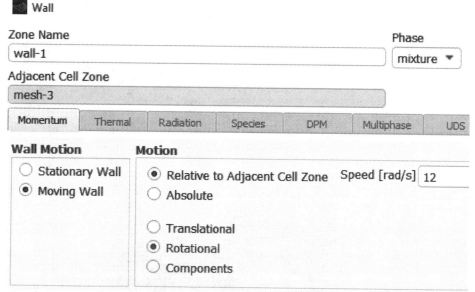

Figure 8.11 Selection of moving wall and rotational speed

12. Double click on Methods under Solution in the Outline View and choose Body Force Weighted for Pressure and First Order Upwind for Momentum under Spatial Discretization on the Task Page. Select SIMPLE as the Pressure-Velocity Coupling Scheme.

Figure 8.12 Solution methods

13. We are going to define a point close to the outer edge of the cylinder to enable the plotting of the swirl velocity over time. Select the Results tab from the menu and Surface>>Create Point. Set the x [m] and y [m] coordinates to 0.2339 and 0.1372, respectively. Click Create and Close the window.

Figure 8.13a) Creating a point surface

Figure 8.13b) Point surface window

14. Open Monitors and double click on Report Plots under Solution in the Outline View. Click on New… in the Report Plot Definitions window. Select New>>Surface Report>>Vertex Average… in the New Report Plot window. Check the Report File, Report Plot and Print to Console boxes under Create in the Surface Report Definition window. Select Vertex Average as Report Type. Select Velocity and Swirl Velocity as Field Variable. Select *point-4* in the Surfaces list. Enter *swirl-velocity.out* as Name. Select OK to exit the Surface Report Definition window. Click OK to close the New Report Plot window. Close the Report Plot Definitions window.

   Double click on Initialization under Solution in the Tree and select Compute from *pressure-inlet-4* on the Task Page. Click on Initialize.

Figure 8.14a) Report plot definitions window

Figure 8.14b) New report plot window

Figure 8.14c) Surface report definition   Figure 8.14d) Initialization

15. Select the *Domain* tab in the menu and select Adapt>>Manual…. Select Cell Registers>>New>>Region… from the Manual Mesh Adaption window. Set X Min [m] to 0.2058, X Max [m] to 0.3048, Y Min [m] to 0, and Y Max (m) to 0.1524. Click on Display Options…. Select red as Color and no Symbol. Check the box for Filled and uncheck the boxes for Wireframe and Marker under Options. Click on OK. Click on Save/Display in the Region Register window. Close the windows.

Figure 8.15a) Domain adaption

Figure 8.15b) Adaption controls

Figure 8.15c) Settings for region register

Figure 8.15d) Cell register display   Figure 8.15e) Filled bottom 33% of cylinder

Click on Copy Screenshot of Active Window to Clipboard, see Figure 8.15f). The contour plot can be pasted into a Word document.

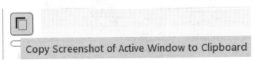

Figure 8.15f) Copying screenshot

16. Click on the Patch… button on the Solution Initialization Task Page. Select *water* as Phase in the Patch window and select Volume Fraction as Variable. Select region_0 as Registers to Patch and set the Value to 1. Click on the Patch button. You have now defined the lower third of the cylinder as filled with water. Select Mixture as Phase and choose Swirl Velocity as Variable. Set the Value [m/s] to 0 and click on the Patch button. Close the Patch window.

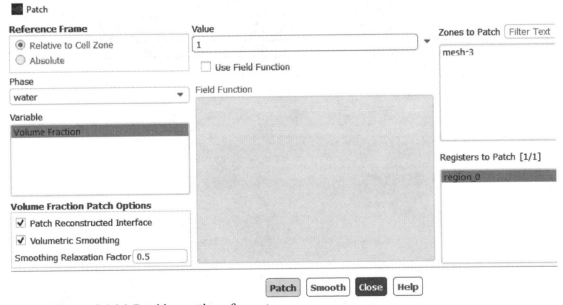

Figure 8.16a) Patching settings for water

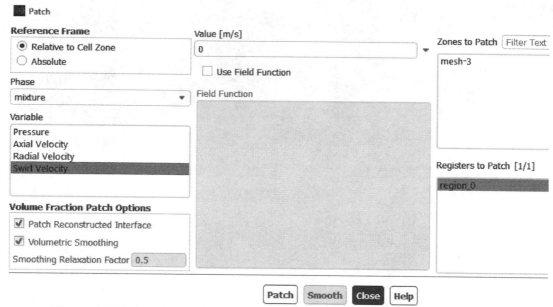

Figure 8.16b) Patching settings for mixture

17. Double click on Graphics and Contours under Results in the Outline View. Select Contours of Velocity and Swirl Velocity. Select Colormap Options... and set the Colormap Size to 30. Click on Apply and Close the Colormap window. Deselect all Surfaces and click on Save/Display. Close the Contours window.

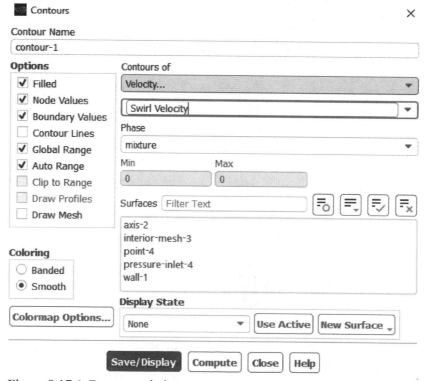

Figure 8.17a) Contours window

Select the View tab in the menu and click on Views…. Select *axis-2* as Mirror Plane and click on Apply. Zoom out and translate the view if needed so that the entire cylinder is visible in the graphics window. Click on the Camera… button in the Views window. Use your left mouse button to rotate the dial clockwise until the cylinder rotates 90 degrees clockwise and appears upright. Close the Camera Parameters window. Click on the Save button under Actions in the Views window and Close the Views window. Click on Copy Screenshot of Active Window to Clipboard, see Figure 8.15f). The swirl velocity contour plot can be pasted into a Word document.

Figure 8.17b) Displaying views

Figure 8.17c) Camera parameters window

Figure 8.17d) Views window

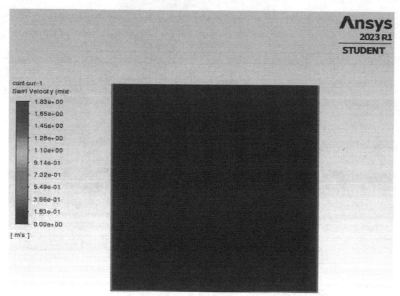

Figure 8.17e) Swirl velocity for cylinder

18. Double click on Contours under Results and Graphics in the Outline View. Select Contours of Phases... and Volume Fraction. Select *water* as the Phase. Deselect all Surfaces. Click on Save/Display and Close the Contours window. Click on Copy Screenshot of Active Window to Clipboard, see Figure 8.15f). The water volume fraction contour plot can be pasted into a Word document. Select File>>Export>>Case... from the menu. Save the Case File with the name *Spinning Cylinder.cas.h5*. Double click on Run Calculation under Solution in the Outline View and set the Time Step Size to 0.01 s. Set the Number of Time Steps to 10000. Click on the Calculate button on the Task Page. Click OK in the Information window when the calculation is complete. Click on Copy Screenshot of Active Window to Clipboard, see Figure 8.15f).

Figure 8.18a) Contours window   Figure 8.18b) Water volume fraction

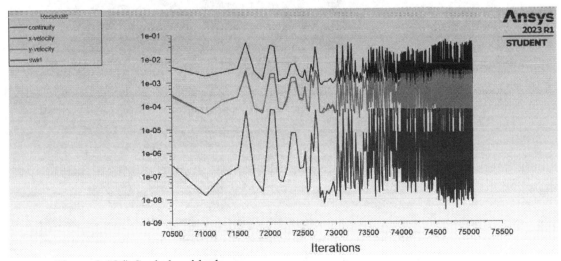

Figure 8.18c) Calculation settings

Figure 8.18d) Scaled residuals

G. Post-Processing

19. Select File>>Export>>Data... from the menu. Save the data file in the *Working Directory* folder with the name *t=100s.dat*. Select Files of type: Legacy Compressed Data Files (*.dat.gz) before you select OK in the Select File window.

Select the Results tab from the menu and select Surface>>Create>>Iso-Surface.... Select Surface of Constant Phases... and Volume fraction. Select *water* as the Phase and set Iso-Values to 0.5 by moving the slider with the left mouse button. Select the left and right arrows on each side of the slider as needed to get exactly the value 0.5. Select *mesh-3* under From Zones and *mesh-3* under From Surface. Enter *free-surface-t-100s* as New Surface Name and click on Create. Close the window.

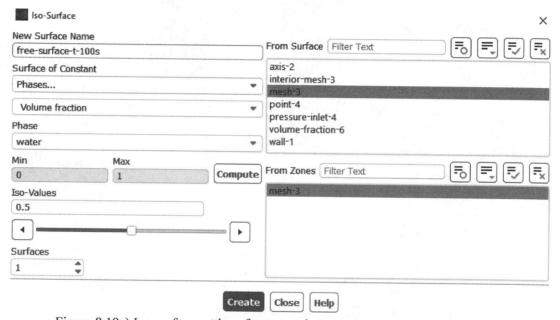

Figure 8.19a) Iso-surface settings for water phase

Select File>>Export>>Solution Data... from the menu. Select ASCII as File Type, Node under Location and Space as Delimiter. Select *free-surface-t-100s* under Surfaces. Select Volume fraction (water) under Quantities. Click on Write and save the ASCII File in the Working Directory with the name *free-surface-coordinates-t-100s*. Close the Export window.

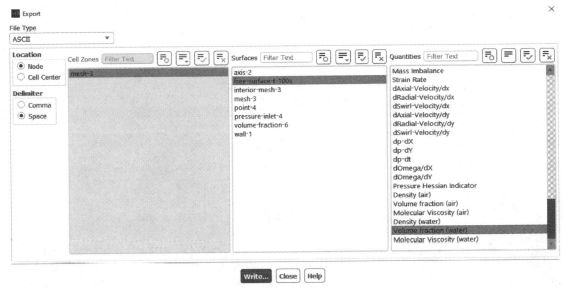

Figure 8.19b) Export settings

Open the saved file *free-surface-coordinates-t-100s* in Excel. You will need to select All Files (".") as file format in Excel to find and open the saved file. Click on Next in the Text Import Wizard – Step 1 of 3. Click on Next in Text Import Wizard – Step 2 of 3. Click on Finish in Text Import Wizard – Step 3 of 3. The opened file will contain 83 rows. If you get rows with other water-vof values than 0.5, then you will need to sort the rows by increasing or decreasing water-vof values and delete all rows except the rows with water-vof equal to 0.5. Plot the free surface in comparison with the initial undisturbed flat shape of the free surface and the final parabolic shape for the free surface, see Figures 8.19c), 8.19d). The process for this is described below and from the theory section we find the definition of the two coordinate systems (x, y) and (r, z):

$$z = Z(r,t) = H_w + H_a - x = 0.3048 - x, \; r = y \tag{8.1}$$

$$H(r,t) = Z(r,t) - H_w = z - H_w \tag{8.2}$$

Copy the y-coordinate column C in the Excel file and paste it to column E. Label this column *r (m)*. Create column F with the value 0.099 for all node numbers and label the column $t = 0$. This is the initial height of the water. Enter the equation =0.3048-B2 in cell G2, drag this cell down the G column and label the column *t = 100 s*. Enter the equation =0.099+12^2*(E2^2-0.1524^2/2)/(2*9.81) in cell H2 and drag this cell down the H column and label the column *parabola*. Select columns E – H in Excel and select the Insert tab from the menu.

Select Charts>>Scatter>>Scatter with Smooth lines in Excel, see Figure 8.19c). Click on the plus sign on the right hand side of the Excel graph and add Axis Titles, *r (m)* for the horizontal axis and *Z (r,t)* for the vertical axis. Double click on the horizontal axis scale on the Excel graph and set the Minimum and Maximum Bounds under Axis Options to 0 and 0.15, respectively. Set the Major Units under Axis Option to 0.01 and select the Number Category to Number with 2 Decimal places. Double click on the vertical axis scale on the Excel graph and set the Minimum and Maximum Bounds under Axis

Options to 0 and 0.2, respectively. Set the Major Units under Axis Option to 0.02 and select the Category under Number to Number with 2 Decimal places.

Figure 8.19c) Excel data and graph of dimensional free surface elevation

Enter the equation =E2/0.1524 in cell I2 and drag the cell down the I column and label this column *r/R*. Create column J with the value 0 for all node numbers and label the column *t = 0*. Enter the equation =9.81*(G2-0.099)/(12*0.1524)^2 in cell K2 and drag the cell down the K column and label this column *t = 100 s*. Enter the equation =9.81*(H2-0.099)/(12*0.1524)^2 in cell L2 and drag the cell down the L column and label this column *parabola*. Select the columns I – L in Excel and select the Insert tab from the menu.

Select Charts>>Scatter>>Scatter with Smooth lines in Excel, see Figure 8.19d). Click on the plus sign on the right hand side of the Excel graph and add Axis Titles, *r/R* for the horizontal axis and $gH(r,t)/(\Omega R)^2$ for the vertical axis. Double click on the horizontal axis scale on the Excel graph and set the Minimum and Maximum Bounds under Axis Options to 0 and 1.0, respectively. Set the Major Units under Axis Option to 0.1, set Label Position to Low and set Number Category to Number with 1 Decimal place. Double click on the vertical axis scale on the Excel graph and set the Minimum and Maximum Bounds under Axis Options to -0.3 and 0.3, respectively. Set the Major Units under Axis Option to 0.1 and select the Category under Number to General.

Select Excel Workbook (*.xlsx) as Save as type: and save the Excel file with the name *Free Surface Data and Plots for Spinning Cylinder.xlsx*.

Figure 8.19d) Excel data and graph of non-dimensional free surface elevation

198

Double click on contour-2 under Contours that is located under Results and Graphics in the Outline View. Select Colormap Options… and set Font Behavior to Fixed and Font Size to 12 under Font. Select Apply and Close the Colormap window. Click on Save/Display in the Contours window to display the volume fraction (water) at t = 100 s, see Figure 8.19e). Close the Contours window. Select the View tab in the menu and click on Views. Select *view-0*, click on Apply and Close the window.

Repeat this step but instead double click on contour-1 to display swirl velocity, see Figure 8.19e). Click on Copy Screenshot of Active Window to Clipboard, see Figure 8.15f). The water volume fraction contour plot and swirl velocity contour plot can be pasted into a Word document.

Continue to run the calculations to t = 200 s and t = 300 s, repeat step **19** and add to the free surface graphs, volume fraction of water and swirl velocity, see Figures 8.19f) to 8.19j).

Figure 8.19e) Contours of water volume fraction and swirl velocity at t = 100 s

Figure 8.19f) Dimensional free surface elevation at different times

Figure 8.19g) Non-dimensional free surface elevation at different times.

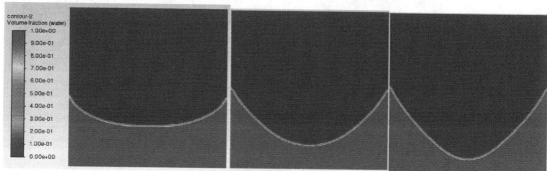

Figure 8.19h) Contours of water volume fraction at $t = 100$, 200, and 300 s

Figure 8.19i) Contours of swirl velocity at $t = 100$, 200, and 300 s

Figure 8.19j) Swirl velocity versus flow time at $(x, y) = (0.2339, 0.1372)$

H. Theory

20. We start by defining the Reynolds number for this flow case as

$$Re = \frac{\Omega R^2}{v} = \frac{12*0.1524^2}{1.0038*10^{-6}} = 277,650 \qquad (8.3)$$

where Ω (rad/s) is the angular velocity of the cylinder, R (m) is the inside radius of the cylindrical container and v (m²/s) is kinematic viscosity. Another non-dimensional number that is commonly used in the Ekman number that is the inverse of the Reynolds number. The initial height of water is $H_w = 0.099$ m and the initial height of air in the cylinder is $H_a = 0.2058$ m. The initial condition for the location of the free surface without rotation can be described as $Z(r,0) = H_w$. Next, we assume that the cylinder has been rotating long enough with constant angular velocity so that the fluid has attained solid-body rotation. We can then describe the shape of the surface as a parabola with the following function for the elevation of the free surface

$$Z(\text{r}, \infty) = H_w + \frac{\Omega^2}{2g}\left(r^2 - \frac{R^2}{2}\right) \qquad (8.4)$$

We can now define the deviation for the height of the free surface from the initial height H_w as

$$H(r, \infty) = Z(r, \infty) - H_w = \frac{\Omega^2}{2g}\left(r^2 - \frac{R^2}{2}\right) \qquad (8.5)$$

Then, we can modify equation (8.3) and define a non-dimensional variable related to the deviation of the free surface from the initial height, see Figure 8.19f).

$$\frac{gH(r,\infty)}{\Omega^2 R^2} = \frac{1}{2}\left[\left(\frac{r}{R}\right)^2 - \frac{1}{2}\right] \qquad (8.6)$$

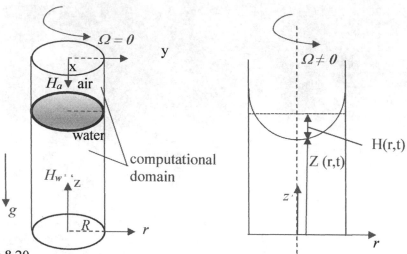

Figure 8.20
Geometry and shape of free surface for cylinder without and with solid body rotation

Generally, the incompressible Navier-Stokes equations in cylindrical coordinates (r, θ, z) with velocity components $\big(u_r(r, \theta, z, t), u_\theta(r, \theta, z, t), u_z(r, \theta, z, t)\big)$ are given by the following equations:

Continuity equation: $\quad \dfrac{1}{r}\dfrac{\partial(r u_r)}{\partial r} + \dfrac{1}{r}\dfrac{\partial u_\theta}{\partial \theta} + \dfrac{\partial u_z}{\partial z} = 0$ (8.7)

r-component: $\quad \rho\left(\dfrac{\partial u_r}{\partial t} + u_r\dfrac{\partial u_r}{\partial r} + \dfrac{u_\theta}{r}\dfrac{\partial u_r}{\partial \theta} - \dfrac{u_\theta^2}{r} + u_z\dfrac{\partial u_r}{\partial z}\right) =$
$-\dfrac{\partial p}{\partial r} + \mu\left[\dfrac{1}{r}\dfrac{\partial}{\partial r}\left(r\dfrac{\partial u_r}{\partial r}\right) - \dfrac{u_r}{r^2} + \dfrac{1}{r^2}\dfrac{\partial^2 u_r}{\partial \theta^2} - \dfrac{2}{r^2}\dfrac{\partial u_\theta}{\partial \theta} + \dfrac{\partial^2 u_r}{\partial z^2}\right]$ (8.8)

θ-component: $\quad \rho\left(\dfrac{\partial u_\theta}{\partial t} + u_r\dfrac{\partial u_\theta}{\partial r} + \dfrac{u_\theta}{r}\dfrac{\partial u_\theta}{\partial \theta} + \dfrac{u_r u_\theta}{r} + u_z\dfrac{\partial u_\theta}{\partial z}\right) =$
$-\dfrac{1}{r}\dfrac{\partial p}{\partial \theta} + \mu\left[\dfrac{1}{r}\dfrac{\partial}{\partial r}\left(r\dfrac{\partial u_\theta}{\partial r}\right) - \dfrac{u_\theta}{r^2} + \dfrac{1}{r^2}\dfrac{\partial^2 u_\theta}{\partial \theta^2} + \dfrac{2}{r^2}\dfrac{\partial u_r}{\partial \theta} + \dfrac{\partial^2 u_\theta}{\partial z^2}\right]$ (8.9)

z-component: $\quad \rho\left(\dfrac{\partial u_z}{\partial t} + u_r\dfrac{\partial u_z}{\partial r} + \dfrac{u_\theta}{r}\dfrac{\partial u_z}{\partial \theta} + u_z\dfrac{\partial u_z}{\partial z}\right) =$
$-\dfrac{\partial p}{\partial z} + \mu\left[\dfrac{1}{r}\dfrac{\partial}{\partial r}\left(r\dfrac{\partial u_z}{\partial r}\right) + \dfrac{1}{r^2}\dfrac{\partial^2 u_z}{\partial \theta^2} + \dfrac{\partial^2 u_z}{\partial z^2}\right]$ (8.10)

where ρ (kg/m^3) is density, g (m/s^2) is acceleration due to gravity and μ (kg/m-s) is dynamic viscosity. For axisymmetric flow there is no θ dependence for the velocity components so they can be expressed as $\big(u_r(r, z, t), u_\theta(r, z, t), u_z(r, z, t)\big)$. This will reduce the equations above to the following equations:

Continuity equation: $\quad \dfrac{1}{r}\dfrac{\partial(r u_r)}{\partial r} + \dfrac{\partial u_z}{\partial z} = 0$ (8.11)

r-component: $\quad \rho\left(\dfrac{\partial u_r}{\partial t} + u_r\dfrac{\partial u_r}{\partial r} - \dfrac{u_\theta^2}{r} + u_z\dfrac{\partial u_r}{\partial z}\right) =$
$-\dfrac{\partial p}{\partial r} + \mu\left[\dfrac{1}{r}\dfrac{\partial}{\partial r}\left(r\dfrac{\partial u_r}{\partial r}\right) - \dfrac{u_r}{r^2} + \dfrac{\partial^2 u_r}{\partial z^2}\right]$ (8.12)

202

θ-component: $\rho\left(\frac{\partial u_\theta}{\partial t} + u_r \frac{\partial u_\theta}{\partial r} + \frac{u_r u_\theta}{r} + u_z \frac{\partial u_\theta}{\partial z}\right) =$
$$\mu\left[\frac{1}{r}\frac{\partial}{\partial r}\left(r\frac{\partial u_\theta}{\partial r}\right) - \frac{u_\theta}{r^2} + \frac{\partial^2 u_\theta}{\partial z^2}\right] \tag{8.13}$$

z-component: $\rho\left(\frac{\partial u_z}{\partial t} + u_r \frac{\partial u_z}{\partial r} + u_z \frac{\partial u_z}{\partial z}\right) = -\frac{\partial p}{\partial z} + \mu\left[\frac{1}{r}\frac{\partial}{\partial r}\left(r\frac{\partial u_z}{\partial r}\right) + \frac{\partial^2 u_z}{\partial z^2}\right] \tag{8.14}$

We make the velocity components and pressure non-dimensional using the following relations:

$$u_r^* = u_r/\Omega R, u_\theta^* = u_\theta/\Omega R, u_z^* = u_z/\Omega R, p^* = p/\rho\Omega^2 R^2 \tag{8.15}$$

and we make the coordinates non-dimensional using the following relations:

$$r^* = \frac{r}{R}, \theta^* = \theta, \ z^* = \frac{z}{R}, \ t^* = t\,\Omega \tag{8.16}$$

The continuity equation and the different component equations in non-dimensional form after skipping the $*$ superscript symbol:

$$\frac{\partial u_r}{\partial r} + \frac{u_r}{r} + \frac{\partial u_z}{\partial z} = 0 \tag{8.17}$$

$$\frac{\partial u_r}{\partial t} + u_r \frac{\partial u_r}{\partial r} - \frac{u_\theta^2}{r} + u_z \frac{\partial u_r}{\partial z} = -\frac{\partial p}{\partial r} + \frac{1}{Re}\left[\frac{\partial^2 u_r}{\partial r^2} + \frac{1}{r}\frac{\partial u_r}{\partial r} - \frac{u_r}{r^2} + \frac{\partial^2 u_r}{\partial z^2}\right] \tag{8.18}$$

$$\frac{\partial u_\theta}{\partial t} + u_r \frac{\partial u_\theta}{\partial r} + \frac{u_r u_\theta}{r} + u_z \frac{\partial u_\theta}{\partial z} = \frac{1}{Re}\left[\frac{\partial^2 u_\theta}{\partial r^2} + \frac{1}{r}\frac{\partial u_\theta}{\partial r} - \frac{u_\theta}{r^2} + \frac{\partial^2 u_\theta}{\partial z^2}\right] \tag{8.19}$$

$$\frac{\partial u_z}{\partial t} + u_r \frac{\partial u_z}{\partial r} + u_z \frac{\partial u_z}{\partial z} = -\frac{\partial p}{\partial z} + \frac{1}{Re}\left[\frac{\partial^2 u_z}{\partial r^2} + \frac{1}{r}\frac{\partial u_z}{\partial r} + \frac{\partial^2 u_z}{\partial z^2}\right] \tag{8.20}$$

The boundary conditions excluding surface tension are the following:

$$u_r(0,z,t) = u_\theta(0,z,t) = \frac{\partial u_z(0,z,t)}{\partial r} = 0, \ , \ 0 \le z \le h(0,t) \tag{8.21}$$

$$u_r(1,z,t) = u_z(1,z,t) = 0, \ u_\theta(1,z,t) = 1, \ 0 \le z \le h(1,t) \tag{8.22}$$

$$p = \frac{\partial h}{\partial t} + u_r \frac{\partial h}{\partial r} - u_z = 0, \ z = h(r,t) \tag{8.23}$$

and the initial conditions are
$$u_r(r,z,0) = u_\theta(r,z,0) = u_z(r,z,0) = 0, \ 0 \le r \le 1, 0 \le z \le H_w/R \tag{8.24}$$

I. References

1. Greenspan, H.P. and Howard, L.N., "On a Time-Dependent Motion of a Rotating Fluid", *Journal of Fluid Mechanics,* **17**, 3, 385-404, (1963).
2. Kim, K.Y. and Hyun, J.M., "Solution for Spin-Up from Rest of Liquid with a Free Surface", *AIAA Journal,* **34**, 7, 1441-1446, (1996).
3. Van de Konijnenberg, J.A, van Heijst, G.J.F., "Nonlinear spin-up in a circular cylinder", *Phys. Fluids,* **7** (12), (1995).

J. Exercises

8.1 Use ANSYS Fluent to continue running the simulations in this chapter and include free surface elevation plots at $t = 150$ s and $t = 200$ s as shown in Figures 8.19f) and 8.19g). Include contours of water volume fraction and swirl velocity at $t = 150$ and 200 s.

8.2 Use ANSYS Fluent to rerun the calculations shown in this chapter but change the rotational speed to 8 rad/s. Include free surface elevation plots at $t = 100$, 150 and 200 s as shown in Figures 8.19f) and 8.19g). Include contours of water volume fraction and swirl velocity at $t = 100$, 150 and 200 s. Determine the Reynolds number.

8.3 Use ANSYS Fluent to rerun the calculations shown in this chapter but change the rotational speed to 16 rad/s. Include free surface elevation plots at $t = 100$, 150 and 200 s as shown in Figures 8.19f) and 8.19g). Include contours of water volume fraction and swirl velocity at $t = 100$, 150 and 200 s. Determine the Reynolds number.

8.4 Use ANSYS Fluent to rerun the calculations shown in this chapter but change the rotational speed to 10 rad/s. Also, change the radius of the cylinder to R = 0.2 m with everything else being the same. Include free surface elevation plots at $t = 100$, 150 and 200 s as shown in Figures 8.19f) and 8.19g). Include contours of water volume fraction and swirl velocity at $t = 100$, 150 and 200 s. Determine the Reynolds number.

8.5 Use ANSYS Fluent to rerun the calculations shown in this chapter but chnge the rotational speed to 10 rad/s. Also, change the geometry to a frustum shape with bottom radius of the cylinder $R_{bottom} = 0.1$ m and top radius $R_{top} = 0.2$ m with everything else being the same. Include free surface elevation plots at $t = 100$, 150 and 200 s as shown in Figures 8.19f) and 8.19g). Include contours of water volume fraction and swirl velocity at $t = 100$, 150 and 200 s. Determine the Reynolds number based on the average radius.

8.6 Use ANSYS Fluent to run the simulations in this chapter and determine the contour plots and graph at your own Flow Time as listed in the table below. Include contours of water volume fraction and swirl velocity at your Flow Time. You will need to change the number of time steps in Figure 8.18c) to your number of time steps as listed in the table below. Use the same time step 0.01 s as was used in this chapter. In this way the simulations will stop at $t =$ your flow time as listed in the table below.

Student Name	Number of Time Steps	Time step (s)	Flow Time (s)
Student A	5000	0.01	50
Student B	5500	0.01	55
Student C	6000	0.01	60
Student D	6500	0.01	65
Student E	7000	0.01	70
Student F	7500	0.01	75
Student G	8000	0.01	80
Student H	8500	0.01	85
Student I	9000	0.01	90
Student J	9500	0.01	95

Table 8.1 Data related to Exercise 8.6

CHAPTER 9. KELVIN-HELMHOLTZ INSTABILITY

A. Objectives

- Using Ansys Workbench to Create Geometry and Mesh for Kelvin-Helmholtz Instability
- Inserting Wall Boundary Conditions and an Acceleration Due to Gravity Vector Corresponding to a Tilt Angle of 4 degrees for the Tube.
- Using Volume of Fluid Model for Multiphase Flow with Surface Tension
- Running Laminar Transient 2D Planar Ansys Fluent Simulations
- Using Contour Plots of Volume Fraction and a Movie for Visualizations

B. Problem Description

The Kelvin-Helmholtz instability is a classical problem originally studied by Helmholtz[1] and Kelvin[2]. The mechanism causing the instability has been studied in detail by Lamb[3], Bachelor[4], Drazin and Reid[5], Chandrasekahr[6], Craik[7], and many others. The Kelvin-Helmholtz instability can appear at the interface of two fluid layers flowing with different velocities and with different densities. For a complete review of this instability, please see Thorpe[8]. Thorpe[9-12] found that the Kelvin-Helmholtz instability can be generated and visualized by tilting a tube that contains two fluids at different densities.

We will study the Kelvin-Helmholtz instability and analyze the problem using ANSYS Fluent. The length of the tube is 2,400 mm and the diameter 200 mm.

(fresh water) $\rho_1 = 998.2$ kg/m^3, $\mu_1 = 0.001002$ kg/ms, $\gamma_1 = 0.07274$ N/m
(salt water) $\rho_2 = 1074.9$ kg/m^3, $\mu_2 = 0.001259$ kg/ms, $\gamma_2 = 0.07422$ N/m

C. Launching Ansys Workbench and Selecting Fluent

1. Start by launching the Ansys Workbench. Double click on Fluid Flow (Fluent) under Analysis Systems in the Toolbox.

Figure 9.1 Fluid Flow (Fluent)

D. Launching Ansys DesignModeler

2. Right click on Geometry in the Project Schematic and select Properties. Select 2D Analysis Type under Advanced Geometry Options in Properties of Schematic A2: Geometry. Right click on Geometry and select New DesignModeler Geometry. Select **Units>>Millimeter** from the menu as the length unit in DesignModeler.

Fluid Flow (FLUENT)
Figure 9.2a) Selecting Geometry

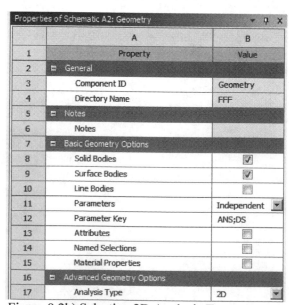

Figure 9.2b) Selecting 2D Analysis Type

3. Next, we will be creating the mesh for the simulation. Select XYPlane from the Tree Outline. Select Look at Sketch . Click on the Sketching tab on the left-hand side. Select Draw and sketch a Rectangle from the origin. Make sure that the letter P is visible as you move the cursor over the origin and before you start drawing the rectangle. Select the Dimensions tab in the Sketching Toolboxes. Select one of the vertical edges of the rectangle and enter a length of 200 mm. Select one of the horizontal edges of the rectangle and enter a length of 2400 mm.

Figure 9.3a) Selection of XYPlane

Figure 9.3b) Selection of Sketching Toolboxes and Rectangle

Figure 9.3c) Dimensions for rectangle in Details View

4. Click on the Modeling tab. Select Concept>>Surfaces from Sketches from the menu. Select Sketch1 under XYPlane in the Tree Outline on the left and select Apply as Base Objects in Details View. Click on Generate in the toolbar. The rectangle turns gray. Right click in the graphics window and select Zoom to Fit. Close the DesignModeler window.

Figure 9.4a) Selecting Surfaces from Sketches

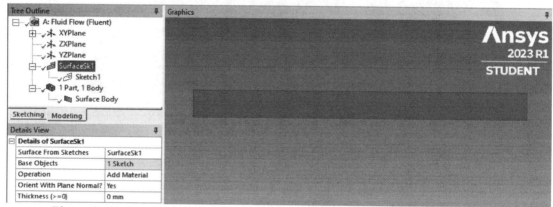

Figure 9.4b) Base Object and completed rectangle in DesignModeler

E. Launching Ansys Meshing

5. We are now going to double click on Mesh in Ansys Workbench to open the Meshing window. Select Mesh in the Outline. Click on Update under Mesh in the menu. A coarse mesh is created.

   Select Mesh>>Controls>>Face Meshing from the menu. Click on the rectangle in the graphics window. Select the yellow region next to Geometry that is labeled No Selection and click on the Apply button for Geometry in Details of "Face Meshing".

   Select Mesh>>Controls>>Sizing from the menu and select the Edge selection filter. Click on the upper horizontal edge of the rectangle and control select the lower horizontal edge. Click on Apply for the Geometry in "Details of Edge Sizing". Under Definition in "Details of Edge Sizing", select Number of Divisions as Type, 480 as Number of Divisions, set Capture Curvature to No and Hard as Behavior.

   Repeat the selection of Mesh>>Controls>>Sizing from the menu and control-select the two vertical edges of the rectangle. Enter 40 for Number of Divisions, set Capture Curvature to No and Hard Behavior. Choose the Bias Type —— — - — —— and enter 3.0 as Bias Factor. Select Mesh>>Update in the menu and select Mesh in the Outline. The finished mesh is shown in the graphics window.

Figure 9.5a) Starting Mesh

Figure 9.5b) Selection of Mesh in Outline

Figure 9.5c) Coarse mesh

Figure 9.5d) Face Meshing

Figure 9.5e) Applied Geometry in Details

Figure 9.5f) Mesh Control Sizing

Figure 9.5g) Details of edge sizing for horizontal edges

Figure 9.5h) Details of edge sizing for vertical edges

Figure 9.5i) Details of finished mesh

6. We are now going to rename the edges for the rectangle. Select the Edge filter ⬚ and control-select the four edges of the rectangle, right click and select Create Named Selection. Enter *wall* as the name for the selection group and click on the OK button.

Select File>>Export…>>Mesh>>FLUENT Input File>>Export from the menu and save the mesh with the name *khi-mesh.msh* in the working directory. Select File>>Save Project from the menu and save the project with the name *Kelvin-Helmholtz Instability.wbpj*. Close the Meshing window. Right click on Mesh in Ansys Workbench in the Project Schematic and select Update.

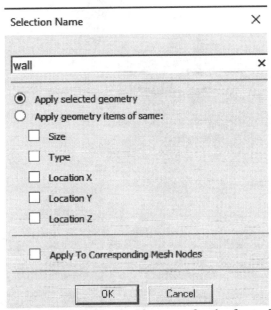

Figure 9.6a) Selection of a name for the four edges of the rectangle

Figure 9.6b) Named selections

F. Launching Ansys Fluent

7. Double-click on Setup under Project Schematic in Ansys Workbench. Check the box for Double Precision. Set the number of processes equal to the number of processor cores for your computer. To check the number of physical cores, press the Ctrl + Shift + Esc keys simultaneously to open the Task Manager. Go to the Performance tab and select CPU from the left column. You will see the number of physical cores on the bottom-right side. Close the Task Manager window. Click on the Start button in the Fluent Launcher window.

Figure 9.7a) Starting Setup

Figure 9.7b) Fluent launcher window

8. Check the mesh by selecting the Check button under Mesh in General on Task Page. Check the scale of the mesh by selecting the Scale button. Make sure that the Domain Extent is correct and close the Scale Mesh window.

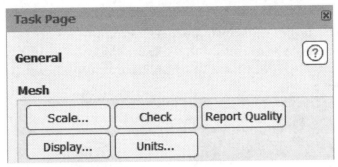

Task Page ⊠

General ⓘ

Mesh

Scale...	Check	Report Quality
Display...	Units...	

Figure 9.8a) Mesh Check

Scale Mesh ✕

Domain Extents

Xmin [m] 0   Xmax [m] 2.4

Ymin [m] 0   Ymax [m] 0.2

View Length Unit In

m

Scaling

◉ Convert Units

○ Specify Scaling Factors

Mesh Was Created In

<Select> ▼

Scaling Factors

X 1

Y 1

Scale   Unscale

Close   Help

Figure 9.8b) Scale Mesh window

9. Select Transient Time for the Solver in General on the Task Page. Check the Gravity box and enter -9.71453 m/s^2 as the Gravitational Acceleration in the Y direction. Enter -1.365288 m/s^2 as the Gravitational Acceleration in the X direction. These components correspond to a tilt angle of 8 degrees for the tube.

Open Models under Setup in the Outline View and double click on Viscous (SST k-omega). Select the Laminar Model and select OK to close the Viscous Model window. Double click on Multiphase (Off). Select the Volume of Fluid Model. Click on Apply and Close the Multiphase Model window.

Figure 9.9a) General Setup

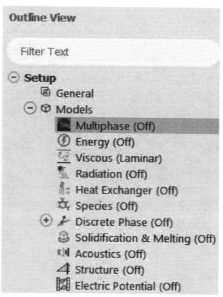

Figure 9.9b) Models Problem Setup

Figure 9.9c) Multiphase model

10. Double click on Materials under Setup in the Outline View. Click on the Create/Edit button for Fluid on the Task Page to open the Create/Edit Materials window. Click on Fluent Database. Scroll down in the Fluent Fluid Materials window and select *water-liquid (h2o<l>)*. Click on the Copy button and Close the Fluent Database Materials window. Click on the Change/Create button and the Close button in the Create/Edit Materials window.

Figure 9.10a) Create/Edit Materials window

Figure 9.10b) Selecting water-liquid as FLUENT Fluid Material

Repeat the steps above, select *water-liquid* and click on the Copy button in the Fluent Database Materials window. You will get a New Material Name window. Enter *salt-water-liquid* as the New Name and click OK. Close the Fluent Database Materials window. Click on the Change/Create button and the Close button in the Create/Edit Materials window.

Figure 9.10c) New Material Name

11. Open Models under Setup in the Outline View, double click on Multiphase (Volume of Fluid) and select the Phases tab in the Multiphase Model window. Select phase-1-Primary Phase under Phases. Select *water-liquid* from Phase Materials and enter *fresh-water* as the name of the Primary Phase. Click on the Apply button.

Figure 9.11a) Multiphase

Figure 9.11b) Primary Phase Setup

Repeat these steps for the Secondary Phase to select *salt-water-liquid* as Phase Material and enter the name *salt-water* under Phase Setup. Click on the Apply button.

Select the Phase Interactions tab at the top of the Multiphase Model window. Select the Surface Tension tab, check the Surface Tension Force Modeling box, select constant from the Surface Tension Coefficients [N/m] drop down menu, enter the value 0.00148 and click on the Apply button. Close the Multiphase Model window.

Select *salt-water-liquid* under Materials and Fluid on the Task Page and click on Create/Edit. Enter 1074.9 as Density [kg/m^3] and 0.001259 as Viscosity [kg/(m s)]. Select Change/Create and Close the window. For a reference of the dynamic viscosity value used for salt-water, see Sharqawy, Lienhard, and Zubair[13].

Figure 9.11c) Secondary Phase Setup

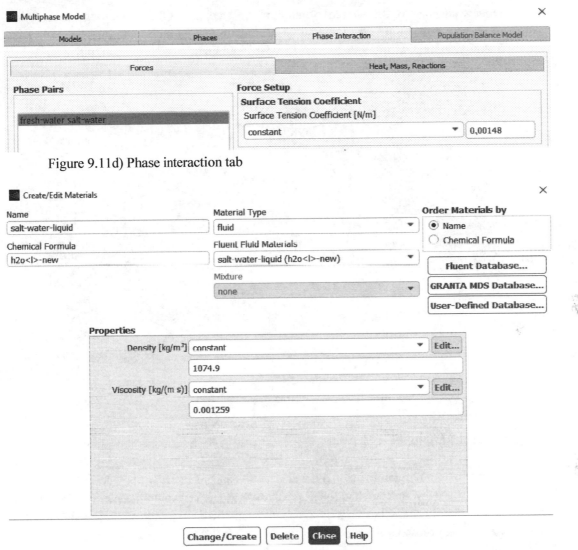

Figure 9.11d) Phase interaction tab

Figure 9.11f) Values of density and viscosity for salt-water

12. Double click on Methods under Solution in the Tree. Select Default settings on the Task Page except for First Order Upwind for Momentum. Double click on Initialization under Solution in the Outline View, select Standard Initialization and click on Initialize.

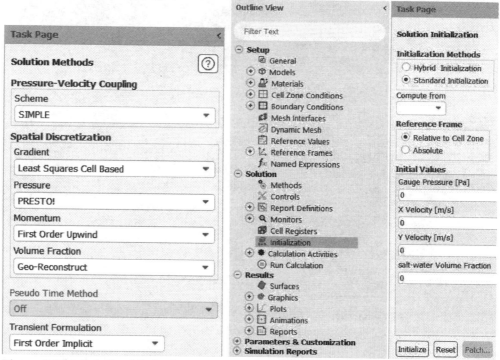

Figure 9.12a) Solution Methods   Figure 9.12b) Solution Initialization

13. Select the Domain tab in the menu and select Adapt>>Manual.... Select Cell Registers>>New>>Region in the Manual Mesh Adaption window. Enter 2.4 for X Max, and 0.1 for Y Max. Click on the Save button and Close the window. Close the Manual Mesh Adaption window.

Figure 9.13a) Manual Adaption   Figure 9.13b) Region Adaption Settings

Click on the Patch… button on the Task Page for Solution Initialization and select *salt-water* as the phase. Select Volume Fraction as the Variable and region_0 as Registers to Patch. Set the value to 1 and click on the Patch button.

Figure 9.13c) Patch Settings for Volume Fraction

Select mixture for the phase. Select X Velocity as the Variable and region_0 as Registers to Patch. Set the value to 0 and click on the Patch button. Close the Patch window.

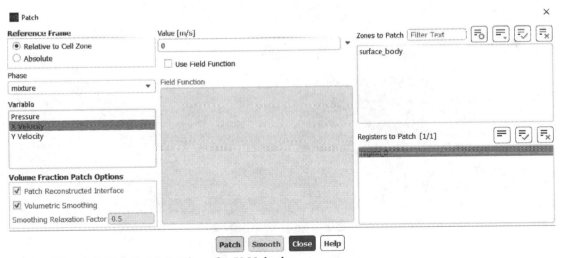

Figure 9.13d) Patch Settings for X Velocity

14. Open Graphics and double click on Contours under Results in the Outline View. Select Phases and Volume fraction under Contours of and select *fresh-water* as the Phase. Click on Colormap Options and select Top as Colormap Alignment. Set Type to float and Precision to 1 under Number Format. Set Font Behavior to Fixed and Font Size to 12 under Font. Click on Apply and Close the Colormap window. Click on the Save/Display button and Close the window. Click on Copy Screenshot of Active Window to Clipboard in the graphics window, see Figure 9.14c). The contour plot can be pasted into a Word document.

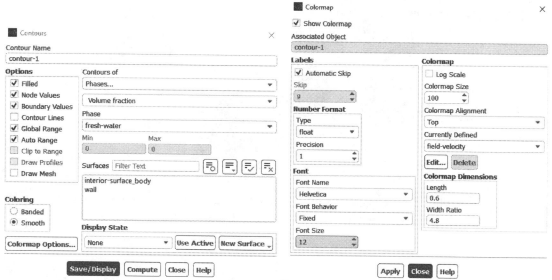

Figure 9.14a) Contours and colormap settings

Figure 9.14b) Contours of Volume Fraction (fresh-water)

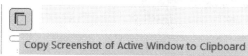

Figure 9.14c) Copying screenshot

15. Open Calculation Activities and double click on Solution Animations under Solution in the Outline View. Enter *khi-animation* as the name and select *contour-1* as Animation Object. Set Record after every to 10 time-step. Select the OK button to close the Animation Definition window.

Double click on Run Calculation under Solution, enter 0.01 as Time Step Size, and 200 for Number of Time Steps. Click on the Calculate button. Click OK when the calculation is complete. Open Graphics and double click on *contour-1* under Contours that is under Results in the Outline View. Click on the Save/Display button and Close the window.

Figure 9.15a) Animation definition window   Figure 9.15b) Running calculations

Figure 9.15c) Scaled residuals for the first two seconds.

Click on Copy Screenshot of Active Window to Clipboard in the graphics window, see Figure 9.14c). The contour plot can be pasted into a Word document.

Figure 9.15d) Contour of Volume Fraction for fresh-water after 2 s

221

G. Post-Processing

16. Enter 100 for Number of Time Steps and continue running the calculations twice followed by entering 50 for the Number of Time Steps. Continue running the calculations six more times for a total of 7 seconds. Ansys Fluent simulations can be compared with experiments of the instability, see Matsson and Boiselle[14].

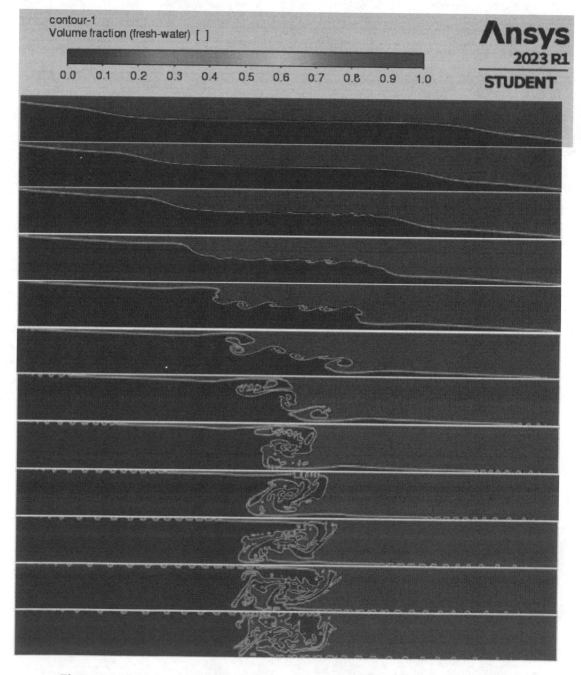

Figure 9.16a) Volume Fraction for fresh-water at 3, 4, 4.5, 5, 5.5, 6, 6.5, 7, 7.5, 8, 8.5, 9 s

Click on Copy Screenshot of Active Window to Clipboard in the graphics window, see Figure 9.14c). The contour plots can be pasted into a Word document.

Figure 9.16b) Experiments for Kelvin-Helmholtz instability at $t = 5.0$ s (8° tilt angle), from Matsson and Boisselle[1]

Select File>>Export>Case & Data... from the menu. Save the file with the name *khi-computation.cas.h5*.

Double click on Animations and Playback under Results in the Outline View. Set the Replay Speed to low and select Play Once as Playback Mode. Select Video File as Write/Record Format. Click on the Write button. This will create a movie with the name *khi-animation.mpeg*. Close the Playback window. The movie can be viewed using for example Windows Media Player.

H. Theory

17. We will look at the inviscid theory for the Kelvin-Helmholtz instability in terms of a sinusoidal disturbance between two fluids with velocities U_1, U_2 and densities ρ_1, ρ_2.

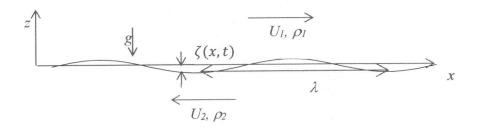

Figure 9.17 Development of Kelvin Helmholtz instability from a perturbation

The velocity vector in a Cartesian coordinate system can be described as

$$\bar{U} = u_x \bar{e}_x + u_y \bar{e}_y + u_z \bar{e}_z \tag{9.1}$$

where u_x, $u_y = 0$ and u_z are the velocity components in the streamwise, spanwise and normal directions, respectively. Euler equations of motion can be written as

$$\frac{\partial u_x}{\partial t} + u_x \frac{\partial u_x}{\partial x} + u_z \frac{\partial u_x}{\partial z} = -\frac{1}{\rho} \frac{\partial P}{\partial x} \tag{9.2}$$

$$\frac{\partial u_z}{\partial t} + u_x \frac{\partial u_z}{\partial x} + u_z \frac{\partial u_z}{\partial z} = -\frac{1}{\rho}\frac{\partial P}{\partial z} - g \tag{9.3}$$

$$\frac{\partial u_x}{\partial x} + \frac{\partial u_z}{\partial z} = 0 \tag{9.4}$$

The velocity components can be divided into mean flow components in the main streamwise direction plus a perturbation

$$\bar{U} = (U(z) + u)\bar{e}_x + w\bar{e}_z \tag{9.5}$$

and the pressure is $\qquad P = P_0 - \rho g z + p \tag{9.6}$

A perturbation is introduced in the form

$$u, w, p = [u_n(z), w_n(z), p_n(z)]e^{i(kx-\omega t)} \tag{9.7}$$

where k is the streamwise wave number, ω is the frequency and $u_n(z), w_n(z), p_n(z)$ are eigenfunctions. The linearized Euler equations can be written as

$$i(kU - \omega)u_n + \frac{ik}{\rho}p_n = 0 \tag{9.8}$$
$$i(kU - \omega)w_n + \frac{1}{\rho}p_n' = 0 \tag{9.9}$$
$$iku_n + w_n' = 0 \tag{9.10}$$

From equations (9.8) – (9.10) we can find the equation for w_n

$$w_n'' - k^2 w_n = 0 \tag{9.11}$$

with the general solution

$$w_n = c_1 e^{-kz} + c_2 e^{kz} \tag{9.12}$$

To get exponentially decaying solutions far from the interface, we set c_1 and c_2 equal to zero in the lower and upper layers, respectively. The interface location is defined by

$$z = \zeta(x, t) = ae^{i(kx-\omega t)} \tag{9.13}$$

The normal velocity w at the interface is given by the material derivative of $\zeta(x, t)$

$$w = w_n(z)e^{i(kx-\omega t)} = \frac{D\zeta}{Dt} = \frac{\partial \zeta}{\partial t} + U\frac{\partial \zeta}{\partial x} =$$
$$i(kU - \omega)\,\zeta = i(kU - \omega)ae^{i(kx-\omega t)} \tag{9.14}$$

, we find that $c_1 = i(kU_1 - \omega)a$, $c_2 = -i(kU_2 + \omega)a$

$$w_{n,1} = i(kU_1 - \omega)ae^{-kz}$$
$$w_{n,2} = -i(kU_2 + \omega)ae^{kz} \tag{9.15}$$

We can now find the pressure eigenfunctions in the upper and lower layers from equation (9.9).

$$p_{n,1} = -\rho_1 (kU_1 - \omega)^2 a e^{-kz}/k$$
$$p_{n,2} = \rho_2 (kU_2 + \omega)^2 a e^{kz}/k \tag{9.16}$$

Next, we use equation (9.6) and set the pressures identical at the interface to get the dispersion relation

$$P_0 - \rho_1 g\zeta + p_{n,1} e^{i(kx-\omega t)} = P_0 - \rho_2 g\zeta + p_{n,2} e^{i(kx-\omega t)} \tag{9.17}$$

$$(\rho_2 - \rho_1) gk = \rho_2 (kU_2 + \omega)^2 + \rho_1 (kU_1 - \omega)^2 \tag{9.18}$$

and we find the following equation for the frequency

$$\omega = \frac{k}{\rho_1 + \rho_2} \left[(\rho_1 U_1 - \rho_2 U_2) \pm \sqrt{\frac{2(\rho_2^2 - \rho_1^2)g}{k} - \rho_1 \rho_2 (U_1 + U_2)^2} \right] \tag{9.19}$$

The frequency is complex with the associated growth of Kelvin-Helmholtz instability waves when the expression under the square root is negative

$$\frac{2(\rho_2^2 - \rho_1^2)g}{k} < \rho_1 \rho_2 (U_1 + U_2)^2 \tag{9.20}$$

Using the definition of wave number as inversely proportional to the wave-length

$$k = \frac{2\pi}{\lambda} \tag{9.21}$$

We find that waves are amplified and develop when their wave lengths fulfill the following condition as shown in equation (9.22). A boundary below or above the interface can destabilize the flow as shown by Hazel[15].

$$\lambda < \frac{\pi \rho_1 \rho_2 (U_1 + U_2)^2}{g(\rho_2^2 - \rho_1^2)} \tag{9.22}$$

I. References

1. Helmholtz H. "On discontinuous movement of fluids", *Phil. Mag.* (4) **36**, 337, (1868).
2. Kelvin L. "Hydrokinetic solutions and observations", *Phil. Mag.* (4) **42**, 362, (1871).
3. Lamb H. "Hydrodynamics", *Cambridge University Press*, (1932).
4. Batchelor G.K. "An introduction to fluid dynamics", *Cambridge University Press*, (1967).
5. Drazin P.G. and Reid W.H. "Introduction to hydrodynamic instability", *Cambridge University Press*, (1981).
6. Chandrasekhar S. "Hydrodynamic and Hydromagnetic Stability", *Dover Publications, Inc.*, New York (1961).
7. Craik A.D.D. "Wave interactions and fluid flows", *Cambridge University Press*, (1985).
8. Thorpe S.A. "Transitional phenomena and the development of turbulence in stratified fluids: a review", *J. Geophys. Res.*, **92**, 5231 (1987).
10. Thorpe S.A. "A method of producing a shear flow in a stratified fluid", *J. Fluid Mech.*, **32**, 693 (1968).
11. Thorpe S.A. "Experiments on the instability of stratified shear flows: miscible

fluids", *J. Fluid Mech.*, **46**, 299 (1971).

11. Thorpe S.A. "Turbulence in stably stratified fluids: a review of laboratory experiments", *Boundary-Layer Meteorol.*, **5**, 95 (1973).

12. Thorpe S.A. "Experiments on the instability and turbulence in a stratified shear flow", *J. Fluid Mech.*, **61**, 731 (1973).

13. Sharqawy M.H, Lienhard J.H., and Zubair S.M. "Thermophysical Properties of Sea Water: A Review of existing Correlations and Data, Desalination, and Water Treatment", *Desalination and Water Treatment*, **16**, 354 (2010).

14. Matsson J. and Boisselle J. "A Senior Design Project on the Kelvin-Helmholtz Instability", 122nd ASEE Annual Conference and Exposition, June 14-17, 2015, Seattle, WA.

15. Hazel P. "Numerical Studies of the stability of inviscid stratified shear flows", *J. Fluid Mech.*, **51**, 39 (1972).

J. Exercises

9.1 Run the ANSYS Fluent simulations for a tube tilt angle of 4 degrees and show the development of the instability using contours of volume fraction for fresh water for the same fluids as used in this chapter.

9.2. Run the ANSYS Fluent simulations for a tube tilt angle of 12 degrees and show the development of the instability using contours of volume fraction for fresh water for the same fluids as used in this chapter.

9.2 Run the ANSYS Fluent simulations shown in this chapter but instead use a transient 3D geometry and mesh for the model. Use a tube tilt angle of 4 degrees and use contours of volume fraction for fresh water to visualize the flow for the same fluids as used in this chapter.

9.3 Run the ANSYS Fluent simulations shown in this chapter but instead use a transient 3D geometry and mesh for the model. Use a tube tilt angle of 8 degrees and use contours of volume fraction for fresh water to visualize the flow for the same fluids as used in this chapter.

9.4 Run the ANSYS Fluent simulations shown in this chapter but instead use a transient 3D geometry and mesh for the model. Use a tube tilt angle of 12 degrees and use contours of volume fraction for fresh water to visualize the flow for the same fluids as used in this chapter.

CHAPTER 10. RAYLEIGH-TAYLOR INSTABILITY

A. Objectives

- Using Ansys Workbench to Create Geometry and Mesh for Rayleigh-Taylor Instability
- Inserting Symmetry and Wall Boundary Conditions
- Using Volume of Fluid Model for Multiphase Flow
- Using a User Defined Function to Trigger the Rayleigh-Taylor Instability
- Running Laminar Transient 2D Planar Ansys Fluent Simulations
- Using Contour Plots of Volume Fractions and a Movie for Visualizations

B. Problem Description

We will study the Rayleigh-Taylor instability and will analyze the problem using Ansys. The Geometry for Rayleigh-Taylor instability (total height 1 m, width 0.25 m) is shown below. The heavier fluid is initially on top of the lighter fluid.

Salt-water
$\rho_1 = 1075$ kg/m^3
$\mu_1 = 0.00126$ kg/m-s

$H = 1$ m

Fresh-water
$\rho_2 = 998.2$ kg/m^3
$\mu_2 = 0.001$ kg/m-s

$D = 0.25$ m

C. Launching Ansys Workbench and Selecting Fluent

1. Start by launching Ansys Workbench. Double click on Fluid Flow (Fluent).

Figure 10.1 Selecting Fluent

D. Launching Ansys DesignModeler

2. Right click on Geometry under Project Schematic in Ansys Workbench and select Properties. Select 2D Analysis Type under Advanced Geometry Options in Properties of Schematic A2: Geometry. Right click on Geometry and select New DesignModeler Geometry to open DesignModeler. **Select Units>>Millimeter from the menu as the length unit in DesignModeler.**

Fluid Flow (FLUENT)

Figure 10.2a) Selecting Geometry

Figure 10.2b) Selecting 2D Analysis Type

3. Next, we will be creating the mesh for the simulation. Select XYPlane in the Tree Outline. Select Look at Sketch . Click on the Sketching tab and Draw a Rectangle from the origin. Make sure that you see the letter *P* as you move the cursor above the origin of the coordinate system and before you start drawing the rectangle. Select the Dimensions tab in the Sketching Toolboxes after completion of the rectangle. Click on one of the vertical edges of the rectangle and enter a vertical length of 1000 mm. Click on one of the horizontal edges and set the horizontal length of 250 mm.

Figure 10.3a) Selection of XYPlane in the Tree Outline

Figure 10.3b) Selection of Sketching Toolboxes and Rectangle

Figure 10.3c) Dimensions in Details View

4. Click on the Modeling tab and select Concept>>Surfaces from Sketches in the menu. Open XYPlane in the Tree Outline, select Sketch1 as Base Objects and select Apply in Details View. Click on ⚡Generate in the toolbar. The rectangle turns gray. Right click in the graphics window, select Zoom to Fit and close DesignModeler.

Figure 10.4 Completed rectangle in DesignModeler

E. Launching Ansys Meshing

5. We are now going to double click on Mesh in Ansys Workbench to open the Meshing window. Select Mesh in the Outline. Click on Update. A coarse mesh is created. Select Mesh>>Controls>>Face Meshing from the menu. Click on the rectangle in the graphics window. Apply the Geometry in Details of "Face Meshing".

Figure 10.5a) Starting Mesh

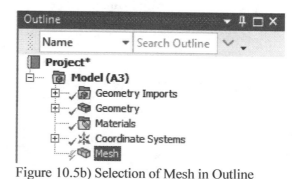

Figure 10.5b) Selection of Mesh in Outline

Figure 10.5c) Generation of Face Meshing

230

Details of "Face Meshing" - Mapped F: ▼ 🔲 ❌	
Scope	
Scoping Method	Geometry Selection
Geometry	1 Face
Definition	
Suppressed	No
Mapped Mesh	Yes
Method	Quadrilaterals
Constrain Boundary	No
Advanced	
Specified Sides	No Selection
Specified Corners	No Selection
Specified Ends	No Selection

Figure 10.5d) Applying Geometry in Details of "Face Meshing"

6. Select Mesh>>Controls>>Sizing from the menu and select Edge 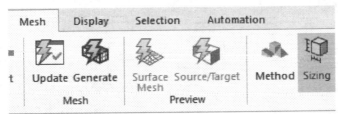. Control-select the four edges of the rectangle. Click on Apply for the Geometry in "Details of Edge Sizing". Under Definition in "Details of Edge Sizing", select Element Size as Type, 2.0 mm as Element Size, set Capture Curvature to No and Hard as Behavior. Click on Update and select Mesh in the Outline. The finished mesh is shown in Figure 10.6b).

Mesh	Display	Selection	Automation

Update Generate   Surface Source/Target   Method Sizing
                      Mesh
        Mesh   Preview

Figure 10.6a) Mesh Control Sizing

Details of "Edge Sizing" - Sizing ▼ 🔲 ❌	
Scope	
Scoping Method	Geometry Selection
Geometry	4 Edges
Definition	
Suppressed	No
Type	Element Size
☐ Element Size	2.0 mm
Advanced	
Behavior	Hard
Capture Curvature	No
Capture Proximity	No
Bias Type	No Bias

Figure 10.6b) Details of "Edge Sizing"

Figure 10.6c) Details of top portion of finished mesh

7. We are now going to rename the edges for the rectangle. Use the Edge tool 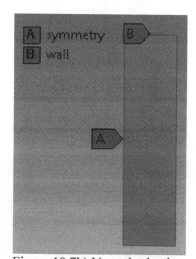, control-select the left and right edges of the rectangle, right click and select Create Named Selection. Enter *symmetry* as the name and click on the OK button.

Repeat this step for the upper and lower horizontal edges (control-select both edges) of the rectangle and name them *wall*. Select File>>Export…>>Mesh>>FLUENT Input File>>Export from the menu and save the mesh with the name *rt-mesh.msh*. Select File>>Save Project from the menu. Save the project with the name *Rayleigh-Taylor-Instability*. Select File>>Close Meshing. Right click on Mesh under Project Schematic in Ansys Workbench and select Update.

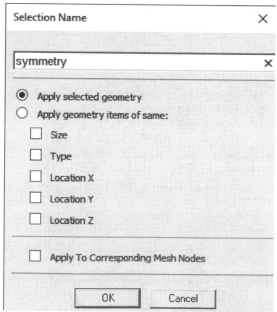

Figure 10.7a) Selecting a name for vertical edges

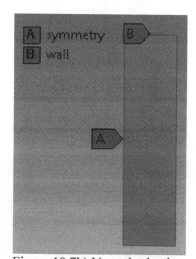

Figure 10.7b) Named selections

F. Launching Ansys Fluent

8. Double-click on Setup under Project Schematic in Ansys Workbench. Check the box for Double Precision. Set the number of processes equal to the number of processor cores for your computer. To check the number of physical cores, press the Ctrl + Shift + Esc keys simultaneously to open the Task Manager. Go to the Performance tab and select CPU from the left column. You will see the number of physical cores on the bottom-right side. Close the Task Manager window. Click on the Start button in the Fluent Launcher window.

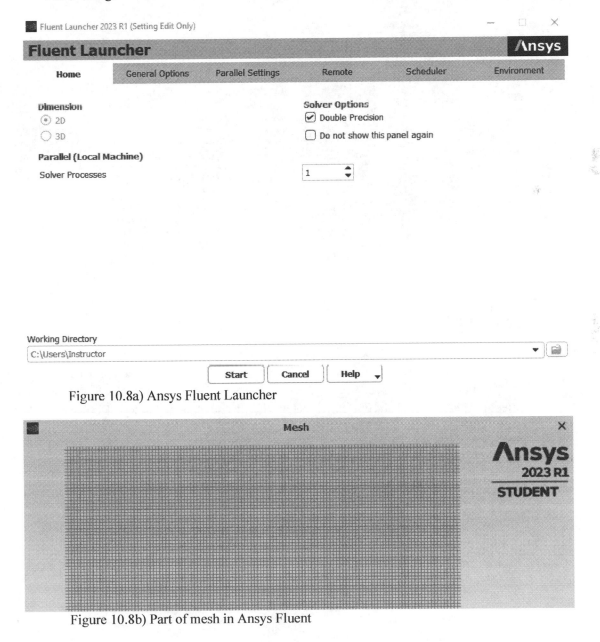

Figure 10.8a) Ansys Fluent Launcher

Figure 10.8b) Part of mesh in Ansys Fluent

9. Check the mesh in Ansys Fluent by selecting the Check button under Mesh in General Problem Setup. Check the scale of the mesh by selecting the Scale button. Make sure that the Domain Extent is correct and close the Scale Mesh window.

Figure 10.9a) Mesh check

Figure 10.9b) Scale Mesh window

10. Select Transient Time for the Solver in General Problem Setup. Check the Gravity box and enter -9.81 as the Gravitational Acceleration Y [m/s^2] in the Y direction. Open Models and double click on Viscous (SST k-omega) under Setup in the Outline View. Select Laminar Model and click OK to close the Viscous Model window. Double click on Multiphase (Off) under Setup in the Outline View. Select the Volume of Fluid Model. Check the box for Implicit Body Force. Click Apply and Close the Multiphase Model window.

Figure 10.10a) General Setup

234

Figure 10.10b) Volume of Fluid Model

11. Double click on Materials, Fluid and air under Setup in the Outline View. Change the Name from *air* to *light-fluid*. Change the Density [kg/m³] to 998.2 and the Viscosity [kg/m s]) to 0.001002. Click on the Change/Create button and answer *Yes* to the question that appears and Close the window. Click on Create/Edit for Fluid under Materials on the Task Page. Click on Fluent Database. Scroll down in the Fluent Fluid Materials window and select *water-liquid (h2o<l>)*. Click on the Copy button and Close the Fluent Database Materials window. Change the Name from *water-liquid* to *heavy fluid*. Change the Density [kg/m³] to 1074.9 and the Viscosity [kg/m s] to 0.001259. Erase the Chemical Formula. Click on the Change/Create button, answer *Yes* to the question that appears and Close the window.

Figure 10.11a) Settings for light-fluid

235

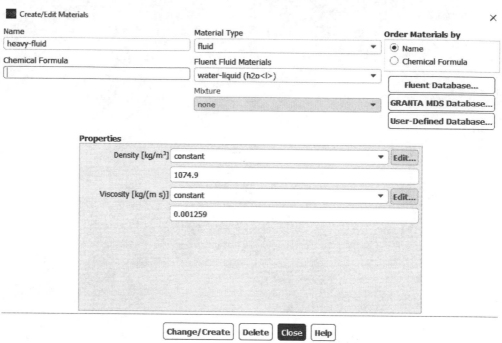

Figure 10.11b) Settings for heavy-fluid

12. Double click on Models and Multiphase (Volume of Fluid) under Setup in the Outline View. Select the Phases tab at the top of the Multiphase Model window. Select *light-fluid* as Phase Materials and enter *light-fluid* as the name of the Primary Phase. Click on Apply. Repeat this step to select and name *heavy-fluid* as the Secondary Phase. Click on Apply. Select the Phase Interaction tab at the top of the Multiphase Model window, check the Surface Tension Force Modeling box, select constant from the Surface Tension Coefficient [N/m] drop down menu and enter the value *0.00148*. Click on the Apply button and Close the Multiphase Model window.

Figure 10.12a) Primary Phase Setup

Figure 10.12b) Secondary Phase Setup

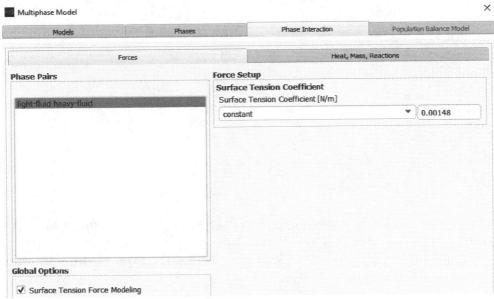

Figure 10.12c) Phase interaction toolbox

13. Copy the files *rayleigh-taylor.c* and *udfconfig.h* to your *Working Directory*. These files are available for download at *sdcpublications.com*. Select User-Defined>>Functions>>Interpreted… from the menu. Browse for the file *rayleigh-taylor.c,* select the file and click OK. Click on the Interpret button. Close the window.

```
#include "udf.h"
DEFINE_INIT(init_cos, domain)
{
cell_t c;
Thread*thread;
real xc[ND_ND];
thread_loop_c(thread,domain)
{
begin_c_loop_all(c,thread)
{
C_CENTROID(xc,c,thread);
if(xc[1] <= 0.5+0.05*cos(2*3.141592654/0.25*xc[0]))
{
C_V(c,thread)=0;
}
else
{
C_V(c,thread)=-0.1;
}
}
end_c_loop_all (c,thread)
}
}
```

Figure 10.13a) Code for the *rayleigh-taylor.c* file

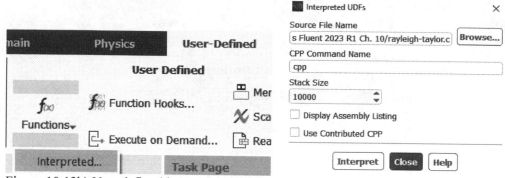

Figure 10.13b) User defined interpreted function Figure 10.13c) Interpreted UDFs

Select User-Defined>>Functions Hook from the menu. Click on the Edit button next to Initialization. Select *init_cos* from Available Initialization Functions and click on the Add button to move it over to the Selected Initialization Functions. Click on OK button to close the window.

Figure 10.13d) Selection of user defined function hooks

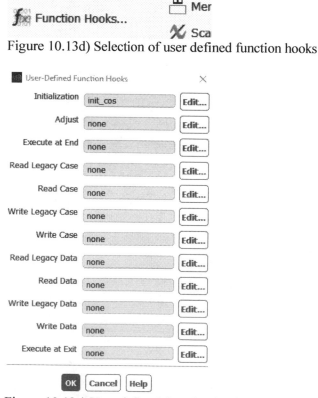

Figure 10.13e) User defined function hooks window

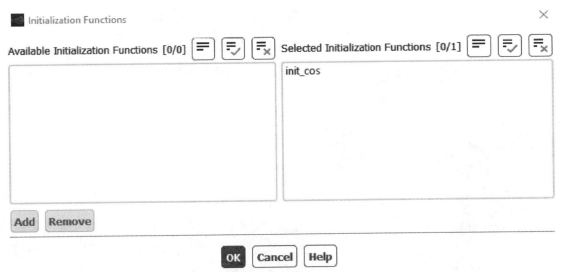

Figure 10.13f) Initializations functions window

14. Double click on Boundary Conditions under Setup in the Outline View. Click on the Operating Conditions on the Task Page in the Boundary Conditions section. Select user-input as Operating Density Method and enter the value 998.2 for Operating Density [kg/m^3]. This is the value for the less dense fluid. Click OK to close the Operating Conditions window.

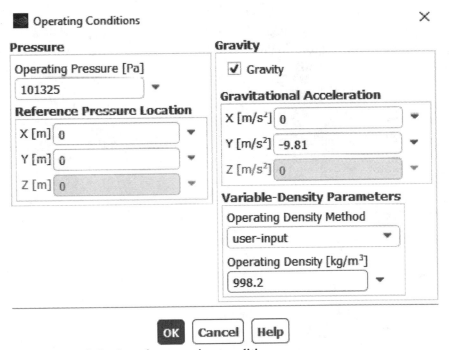

Figure 10.14a) Settings for operating conditions

Double click on Methods under Solution in the Outline View. Select Default settings except for First Order Upwind for Momentum and Body Force Weighted for Pressure.

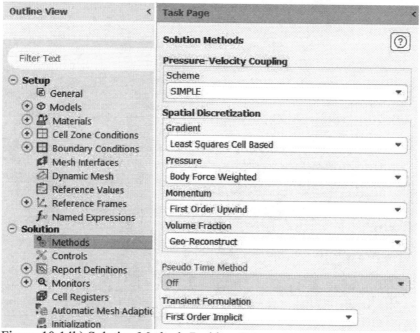

Figure 10.14b) Solution Methods Problem Setup

15. Double click on Initialization under Solution in the Outline View, select Standard Initialization under Solution Initialization on the Task Page, set the Y Velocity to -0.1 and select the Initialize button.

Figure 10.15 Solution initialization settings

16. Select the *Domain* tab in the menu and choose Adapt>>Manual….. Select Cell Registers>>New>>Region… in the Adaption Controls window. Enter 0.25 for X Max, 0.5 for Y Min, and 1 for Y Max. Click on the Save button and close the windows.

Figure 10.16a) Adaption   Figure 10.16b) Region Adaption Settings

17. Click on the Patch button on the Task Page and select *heavy-fluid* as the phase. Select Volume Fraction as the Variable and region_0 as Registers to Patch. Set the value to 1, click on the Patch button and close the window.

Click on the Patch button on the Task Page and select *mixture* for the phase. Select Y Velocity as the Variable and region_0 as Registers to Patch. Set the value to -0.1 and click on the Patch button and Close the window.

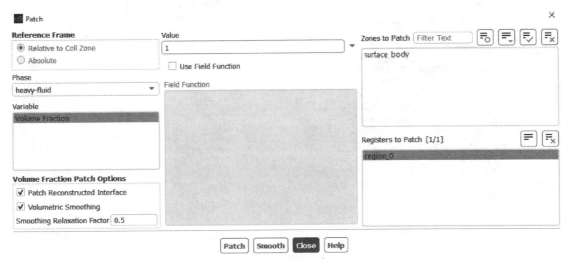

Figure 10.17a) Patch Settings for Volume Fraction

Figure 10.17b) Patch Settings for Y Velocity

18. Double click on Graphics and Contours under Results in the Outline View. Select Phases... and Volume fraction under Contours of and select *heavy-fluid* as the Phase. Uncheck the Boundary Values box under Options. Click on Colormap Options..., set Type to float and Precision to 1 under Number Format, set Font Behavior to Fixed and Font Size to 12, click on Apply and Close the Colormap window. Click on the Save/Display button and Close the Contours window. Click on Copy Screenshot of Active Window to Clipboard in the graphics window, see Figure 10.18d). The contour plot can be pasted into a Word document.

Figure 10.18a) Contours window

Figure 10.18b) Colormap window

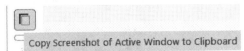

Figure 10.18c) Contours of Volume Fraction   Figure 10.18d) Copying screenshot

G. Calculations and Post-Processing

19. Double click on Calculation Activities and Solution Animations under Solution in the Outline View. Enter *rayleigh-taylor-instability-animation* as the Name, enter Record after every 10 and select time-step. Select contour-1 under Animation Object. Select OK to close the Animation Definition window. Double click on Run Calculation under Solution, enter 0.01 as Time Step Size, and 100 for Number of Time Steps. Click on the Calculate button. Click on Copy Screenshot of Active Window to Clipboard in the graphics window, see Figure 10.18d). The contour plot can be pasted into a Word document. Continue running the calculations with 20 as Number of Time Steps, 19 more times for a total of 4.8 seconds. Select File>>Export>Case & Data… from the menu. Save the file with the name *rt-computation.cas.h5*.

Figure 10.19a) Animation definition

Figure 10.19b) Calculations

Figure 10.19c) Scaled residuals for the first second

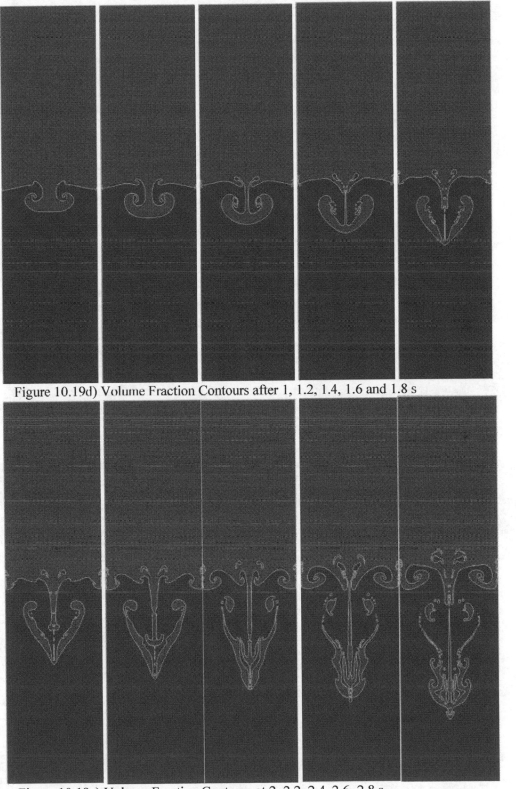

Figure 10.19d) Volume Fraction Contours after 1, 1.2, 1.4, 1.6 and 1.8 s

Figure 10.19e) Volume Fraction Contours at 2, 2.2, 2.4, 2.6, 2.8 s

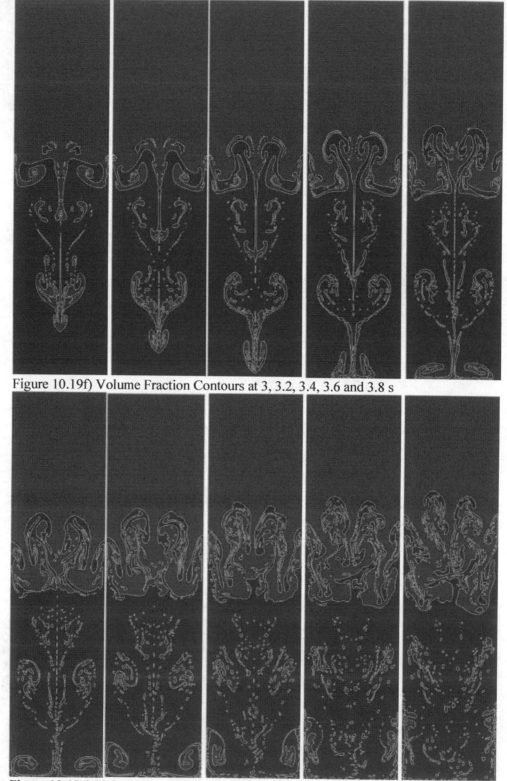

Figure 10.19f) Volume Fraction Contours at 3, 3.2, 3.4, 3.6 and 3.8 s

Figure 10.19g) Volume Fraction Contours at 4, 4.2, 4.4, 4.6 and 4.8 s

20. Double click on Animations and Playback under Results in the Outline View. Set the Replay Speed to low and select Play Once as Playback Mode. Select Video File as Write/Record Format. Click on the Write button. This will create the movie. Close the Playback window. The movie can be viewed using for example Windows Media Player.

Figure 10.20 Playback settings

H. Theory

21. We will look at the inviscid theory for the Rayleigh-Taylor instability in the same way as we did for the Kelvin-Helmholz instability in terms of a sinusoidal disturbance between two fluids with densities ρ_1, ρ_2.

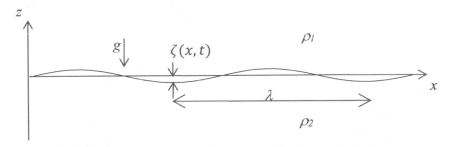

Figure 10.21 Development of Rayleigh-Taylor instability from a perturbation

The velocity vector in a Cartesian coordinate system can be described as

$$\bar{U} = u_x \bar{e}_x + u_y \bar{e}_y + u_z \bar{e}_z \qquad (10.1)$$

where u_x, $u_y = 0$, u_z are the velocity components in the streamwise, spanwise and normal directions, respectively. Euler equations of motion can be written as

247

$$\frac{\partial u_x}{\partial t} + u_x \frac{\partial u_x}{\partial x} + u_z \frac{\partial u_x}{\partial z} = -\frac{1}{\rho}\frac{\partial P}{\partial x} \tag{10.2}$$

$$\frac{\partial u_z}{\partial t} + u_x \frac{\partial u_z}{\partial x} + u_z \frac{\partial u_z}{\partial z} = -\frac{1}{\rho}\frac{\partial P}{\partial z} - g \tag{10.3}$$

$$\frac{\partial u_x}{\partial x} + \frac{\partial u_z}{\partial z} = 0 \tag{10.4}$$

The velocity components can be divided into mean flow components in the main streamwise direction plus a perturbation (no mean flow components in this case)

$$\bar{U} = u\bar{e}_x + w\bar{e}_z \tag{10.5}$$

and the pressure is $\quad P = P_0 - \rho g z + p \tag{10.6}$

A perturbation is introduced in the form

$$u, w, p = [u_n(z), w_n(z), p_n(z)]e^{i(kx-\omega t)} \tag{10.7}$$

where $k = 2\pi/\lambda$ is the wave number, λ is the wave length, ω is the frequency and $u_n(z), w_n(z), p_n(z)$ are eigenfunctions. The linearized Euler equations can be written as

$$-i\omega u_n + \frac{ik}{\rho}p_n = 0 \tag{10.8}$$
$$-i\omega w_n + \frac{1}{\rho}p_n' = 0 \tag{10.9}$$
$$iku_n + w_n' = 0 \tag{10.10}$$
From equations (10.8) – (10.10) we can find the equation for w_n

$$w_n'' - k^2 w_n = 0 \tag{10.11}$$

with the general solution

$$w_n = c_1 e^{-kz} + c_2 e^{kz} \tag{10.12}$$

To get exponentially decaying solutions far from the interface, we set c_1 and c_2 equal to zero in the lower and upper layers, respectively. The interface location is defined by

$$z = \zeta(x,t) = \mathcal{R}\{ae^{i(kx-\omega t)}\} \tag{10.13}$$

where $\mathcal{R}$ stands for the real part and a is the amplitude of the initial disturbance. The normal velocity w at the interface is given by the material derivative of $\zeta(x,t)$

$$w = w_n(z)e^{i(kx-\omega t)} = \frac{D\zeta}{Dt} = \frac{\partial\zeta}{\partial t} = -i\omega\,\zeta = -i\omega a e^{i(kx-\omega t)} \tag{10.14}$$

We find the following results: $c_1 = -i\omega a$, $c_2 = -i\omega a$

$$w_{n,1} = -i\omega a e^{-kz}, \; w_{n,2} = -i\omega a e^{kz} \tag{10.15}$$

We can now find the pressure eigenfunctions in the upper and lower layers from equation (10.9)

$$p_{n,1} = -\rho_1 \omega^2 a e^{-kz}/k \tag{10.16}$$
$$p_{n,2} = \rho_2 \omega^2 a e^{kz}/k \tag{10.17}$$

Next, we use equation (10.6) and set the pressures identical at the interface to get the dispersion relation

$$P_0 - \rho_1 g\zeta + p_{n,1} e^{i(kx-\omega t)} = P_0 - \rho_2 g\zeta + p_{n,2} e^{i(kx-\omega t)} - \sigma k^2 \zeta \tag{10.18}$$

$$(\rho_2 - \rho_1)gk = (\rho_2 + \rho_1)\omega^2 - \sigma k^3 \tag{10.19}$$

where σ is surface tension. We find the following equation for the frequency without surface tension

$$\omega = \sqrt{\frac{(\rho_2-\rho_1)gk}{\rho_1+\rho_2}} = i\sqrt{\frac{(\rho_1-\rho_2)gk}{\rho_1+\rho_2}} = i\sqrt{gkAt} \tag{10.20}$$

and including surface tension

$$\omega = \sqrt{\frac{(\rho_2-\rho_1)gk+\sigma k^3}{\rho_1+\rho_2}} = i\sqrt{\frac{(\rho_1-\rho_2)gk-\sigma k^3}{\rho_1+\rho_2}} = i\sqrt{gkAt(1-\frac{4\pi^2}{Bo})} \tag{10.21}$$

where $At = (\rho_1 - \rho_2)/\rho_1+\rho_2$ is non-dimensional Atwood number and $Bo = (\rho_1 - \rho_2)gD^2/\sigma$ is non-dimensional Bond number. The frequency is complex with the associated growth of waves when the expression under the square root without surface tension is negative for

$$\rho_2 - \rho_1 < 0 \qquad \text{or} \qquad \rho_2 < \rho_1 \tag{10.22}$$

and including surface tension we get

$$(\rho_2 - \rho_1)gk + \sigma k^3 < 0 \qquad \text{or} \qquad \lambda > 2\pi\sqrt{\frac{\sigma}{(\rho_1-\rho_2)g}} \tag{10.23}$$

The Bond number $Bo > 4\pi^2$ will give growth of waves when $\lambda = D$. The interface location can now be written as

$$z = \zeta(x,t) = \mathcal{R}\left\{ae^{\sqrt{gkAt(1-\frac{4\pi^2}{Bo})}\,t}e^{ikx}\right\} = ae^{\sqrt{gkAt(1-\frac{4\pi^2}{Bo})}\,t}\cos(kx) \tag{10.24}$$

and we see that surface tension has a very small but stabilizing effect. In ANSYS Fluent simulations we used $a = 0.05$ m, $k = 8\pi$ m^{-1}, $g = 9.81$ m/s^2, $\lambda = D = 0.25$ m, $\sigma = 0.00148$ N/m, $\rho_1 = 1074.9 \frac{kg}{m^3}$, $\rho_2 = 998.2 \frac{kg}{m^3}$. The Atwood and Bond numbers used in ANSYS Fluent are

$$At = \frac{1074.9-998.2}{1074.9+998.2} = 0.037 \tag{10.25}$$

$$Bo = \frac{(1074.9-998.2)*9.81*0.25^2}{0.00148} = 31,775 \tag{10.26}$$

There are a few more non-dimensional parameters in this problem that we can define such as the aspect ratio for the rectangular computational domain $/D = 4$, the density ratio $\rho_1/\rho_2 = 1.077$, the dynamic viscosity ratio $\mu_1/\mu_2 = 1.256$ and non-dimensional time $\tau = t\sqrt{gkAt(1 - \frac{4\pi^2}{Bo})}$. The interface location is

$$z = \zeta(x,t) \approx \frac{e^{3t}}{20}\cos(8\pi x) \tag{10.27}$$

Using the inviscid theory, we find that the wave-length with surface tension included according to equation (10.23) must be $\lambda > 0.0088\ m$ to get growth of waves.

I. References

1. Andrews, M.J. and Dalziel, S.B. "Small Atwood number Rayleigh-Taylor Experiments", *Phil. Trans. R. Soc. A*, **368**, 1663-1679, (2010).
2. Chandrasekhar S. "Hydrodynamic and Hydromagnetic Stability", *Dover Publications, Inc.*, New York (1961).
3. Drazin P.G. and Reid W.H. "Introduction to hydrodynamic instability", *Cambridge University Press*, (1981).
4. Sharqawy M.H, Lienhard J.H., and Zubair S.M. "Thermophysical Properties of Sea Water: A Review of existing Correlations and Data, Desalination, and Water Treatment", *Desalination and Water Treatment*, **16**, 354 (2010).
5. Strubelj, L., and Tiselj, I., "Numerical Simulations of Basic Interfacial Instabilities with Improved Two-Fluid Model", *International Conference Nuclear Energy for New Europe*, September 14–17, Bled, Slovenia, (2009).
6. Strubelj, L., and Tiselj, I., "Numerical Simulations of Rayleigh-Taylor Instability with Two-Fluid Model and Interface Sharpening", *Proceedings of ASME Fluids Engineering Division Summer Conference*, August 10-14, Jacksonville, FL, (2008).
7. Tryggvason, G., "Numerical Simulations of the Rayleigh-Taylor Instability", *Journal of Computational Physics*, **75**, 253-282, (1988).

J. Exercises

10.1 Use ANSYS Fluent to study the same problem as we completed in this chapter but instead complete transient 2D axisymmetric simulations. Visualize the instability using contours of volume fraction for the heavy fluid.

10.2 Use ANSYS Fluent to study the same problem as we completed in this chapter but instead use the following dimensions $H = 2$ m, $D = 0.25$ m. Visualize the instability using contours of volume fraction for the heavy fluid.

10.3 Use ANSYS Fluent to study the same problem as we completed in this chapter but instead use a frustum shaped geometry with the following dimensions $H = 1$ m, $D_{bottom} = 0.25$ m and $D_{top} = 0.5$ m. Visualize the instability using contours of volume fraction for the heavy fluid.

CHAPTER 11. FLOW UNDER A DAM

A. Objectives

- Using Ansys Workbench to Model Seepage Flow of Water under a Concrete Dam
- Inserting Boundary Conditions and a Porous Medium Zone with Viscous and Inertial Resistance
- Running Laminar Steady 2D Planar Ansys Fluent Simulations
- Using Contour Plots for Static Pressure and Velocity Magnitude
- Using XY Plots for Visualizations of Velocity and Pressure
- Comparing Ansys Fluent Results with Results Using Mathematica

B. Problem Description

We will study the seepage flow of water under a concrete dam using Ansys Fluent. The dimensions and water levels are shown below.

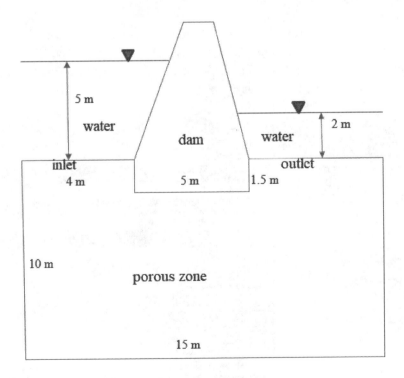

C. Launching Ansys Workbench and Selecting Fluent

1. Start by launching Ansys Workbench. Double click on Fluid Flow (Fluent) under Analysis Systems in the Toolbox.

Figure 11.1 Selecting Fluid Flow (Fluent)

D. Launching Ansys DesignModeler

2. Right click Geometry under Project Schematic in Ansys Workbench and select Properties. In Properties of Schematic A2: Geometry, select Analysis Type 2D under Advanced Geometry Options. Right click on Geometry in the Project Schematic window and select New DesignModeler Geometry.

Figure 11.2 Selecting properties, 2D analysis type and new DesignModeler geometry

3. Select the Sketching tab in the Tree Outline and select Line under Draw. Click on Look at Face/Plane/Sketch . Draw a horizontal line to the right from the origin of the coordinate system in the graphics window. Make sure you have the letters *P* at the origin, *H* along the line and *C* when you end the line. Click on the Dimensions tab under Sketching Toolboxes and click on the new line. Enter 15 m for the length of the line. Continue adding lines with lengths as shown in Fig. 11.3a).

Select Concepts>>Surfaces from Sketches from the menu. Click on Sketch 1 under XYPlane in the Tree Outline and select Apply for Base Objects in Details View. Click on Generate. Close DesignModeler.

Figure 11.3a) Geometry and dimensions for flow through a porous medium

Figure 11.3b) Creating surfaces from Sketches

Figure 11.3c) Finished model for seepage flow under a dam

253

E. Launching Ansys Meshing

4. Double click on Mesh under the Project Schematic in Ansys Workbench. Select Mesh under Model (A3) and Project in the Outline. Right click on Mesh and select Update.

Fluid Flow (Fluent)

Figure 11.4a) Mesh in Ansys

Figure 11.4b) Coarse mesh

Select Mesh>>Controls>>Face Meshing from the menu. Select the face of the mesh region. Apply the Geometry under Details of "Face Meshing".

Figure 11.4c) Face meshing

Select Mesh>>Controls>>Sizing from the menu and select the face of the mesh region. Apply the Geometry under Details of "Face Sizing". Set the Element Size to 50.0 mm, set Capture Curvature to No and set the Behavior to Hard. Right click on Mesh in the tree outline and select Generate Mesh.

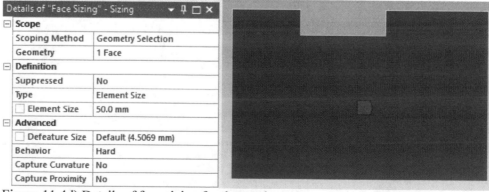

Figure 11.4d) Details of face sizing for the mesh

Figure 11.4e) A portion of the finished mesh for flow under a dam

Select Geometry under Project and Model (A3) in the Outline. Select the Edge Selection

Filter . Select the upper left horizontal edge, right click and select Create Named Selection. Name the edge *pressure-inlet* and click OK. Name the upper horizontal edge to the right *pressure-outlet* and control-select all the remaining six edges and enter the name *wall*.

Select File>>Export...>>Mesh>>FLUENT Input File>>Export from the menu and save the mesh with the name *seepage-flow.msh*. Select File>>Save Project from the menu and save the project with the name *Dam-Seepage-Flow.wbpj*. Close the meshing window. Right click on Mesh and select Update under Project Schematic in Ansys Workbench.

Figure 11.4f) Selection of Geometry

Figure 11.4g) Named selections

F. Launching Ansys Fluent

5. Double-click on Setup under Project Schematic in Ansys Workbench. Check the box for Double Precision. Set the number of processes equal to the number of processor cores for your computer. To check the number of physical cores, press the Ctrl + Shift + Esc keys simultaneously to open the Task Manager. Go to the Performance tab and select CPU from the left column. You will see the number of physical cores on the bottom-right side. Close the Task Manager window. Click on the Start button in the Fluent Launcher window.

Figure 11.5 Launching ANSYS Fluent

6. Double click on Models and Viscous (SST k-omega) under Setup in the Outline View. Select Laminar Model and select OK to close the Viscous Model window. Double click on Materials under Setup in the Outline View. Click on Create/Edit for Fluid under Materials on the Task Page. Click on Fluent Database in the Create/Edit Materials window. Select *water-liquid (h2o<l>)* as the Fluent Fluid Material. Click on Copy and Close the window. Set the Density to 992.2 [kg/m³] and the Viscosity to 0.00065 [kg/m-s]. Click Change/Create and Close the window.

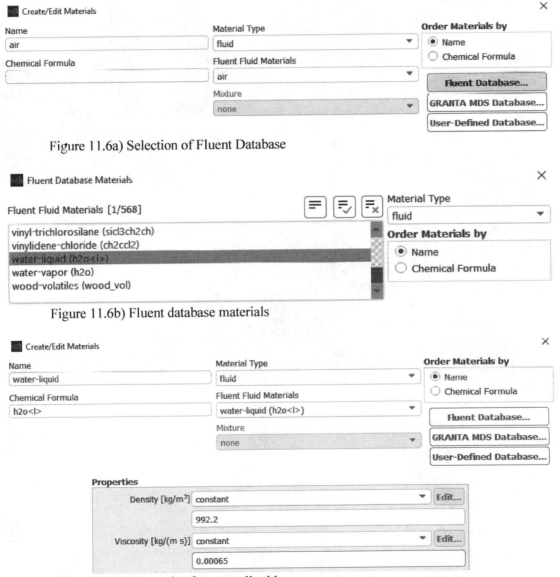

Figure 11.6a) Selection of Fluent Database

Figure 11.6b) Fluent database materials

Figure 11.6c) Properties for water-liquid

7. Double click on Cell Zone Conditions, Fluid and *surface_body* under Setup in the Outline View. Select *water-liquid* from the Material Name drop-down menu. Check the box for Porous Zone. Click on the Porous Zone tab and enter 1988070 [m^{-2}] for Viscous Resistance in Direction-1 and Direction-2. Enter 1719.4 [m^{-1}] for Inertial Resistance in Direction-1 and Direction-2. Set the Fluid Porosity to 0.5. Click on Apply button to Close the Fluid window.

Figure 11.7 Fluid window

8. Double click on Boundary Conditions under Setup in the Outline View. Select the *pressure-inlet* Zone on the Task Page. Click on Edit…. Enter 49050 as the Gauge Total Pressure [Pa]. Click on Apply and Close the window. Select the *pressure-outlet* Zone on the Task Page. Click on Edit…. Enter 19620 as the Gauge Pressure [Pa]. Select Apply and Close the window.

Figure 11.8a) Selection of pressure-inlet boundary condition

Figure 11.8b) Pressure inlet window

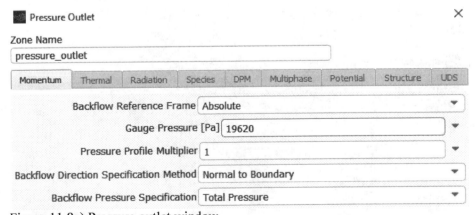

Figure 11.8c) Pressure outlet window

9. Double click on Initialization under Solution in the Outlet View. Check Standard Initialization as Initialization Method on the Task Page. Select Compute from *pressure-inlet* from the drop-down menu and click on Initialize.

Figure 11.9 Solution initialization

10. Double click on Monitors and Residual under Solution in the Outline View. Set the Absolute Criteria to 1e-12 for all Residuals and click on OK to exit the window.

Select File>>Export>Case… from the menu. Save the file with the name *flow-under-dam.cas.h5*. Double click on Run Calculation under Solution in the Outline View, enter 1000 for Number of Iterations on the Task Page and click on Calculate. Click OK when calculations are complete. Click on Copy Screenshot of Active Window to Clipboard in the graphics window, see Figure 11.10d). The scaled residuals can be pasted into a Word document.

Figure 11.10a) Residual monitors window

Figure 11.10b) Running calculations

Figure 11.10c) Residual iterations

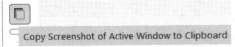

Figure 11.10d) Copying screenshot

G. Post-Processing

11. Double click on Graphics and Contours under Results in the Outline View. Select Contours of Velocity and Velocity Magnitude from the drop-down menu. Select Colormap Options…. Select Type as float and set Precision to 3 under Number Format. Set Font Behavior to Fixed and Font Size to 12 under Font. Set Colormap Alignment to Top. Click on Apply and Close the Colormap window. Deselect all Surfaces and click on Save/Display. Click on Copy Screenshot of Active Window to Clipboard in the graphics window, see Figure 11.10d). The contour plot can be pasted into a Word document. Select Colormap Options… and set Precision to 0 under Number Format. Click on Apply and Close the Colormap window. Select Contours of Pressure and Static Pressure and click on Save/Display. Close the window.

Figure 11.11a) Contours window

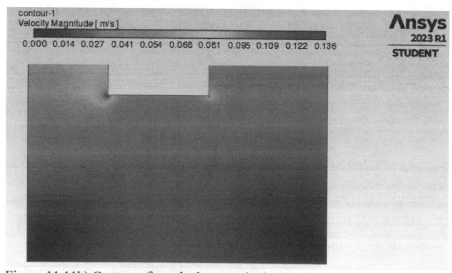

Figure 11.11b) Contours for velocity magnitude

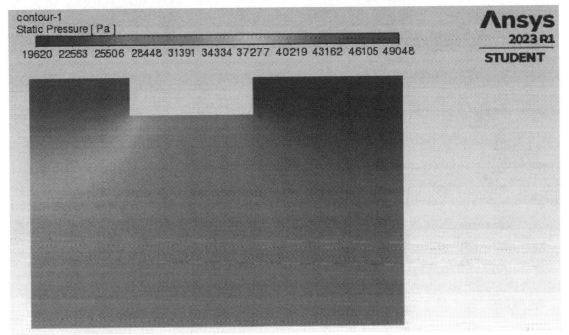

Figure 11.11c) Contours for static pressure

12. Double click on Plots and XY Plot under Results in the Outline View. Select Velocity and Velocity Magnitude for Y Axis Function and Direction Vector for X Axis Function. Select New Surface >> Line/Rake from the drop–down menu. Enter the following end points: x0 [m] = 9, y0 [m] = 10 and x1 [m] = 15, y1 [m] = 10. Click on Create and Close the window.

Click on Load File… and load the file with the name *theory-flow-under-dam.dat*. This and other files are available for download at *sdcpublications.com*. Select *line-5* in the Surfaces section, select *Theory* under File Data. Click on Curves…, select Curve #0 and select the first available pattern and no symbol. Click on Apply. Select Curve # 1, select the second available pattern and no symbol as shown in Figure 11.12c). Click on Apply and Close the Curves-Solution XY Plot window. Click on Axes… in the Solution XY Plot window. Select X Axis, set Type to float under Number Format, set Precision under Number Format to 0 and click on Apply. Select Y Axis, set Type under Number Format to General, set Precision under Number Format to 2 and click on Apply. Close the Axes – Solution XY Plot window. Click on Save/Plot in the Solution XY Plot window.

Click on Copy Screenshot of Active Window to Clipboard in the graphics window, see Figure 11.10d). The XY plot can be pasted into a Word document.

Select New Surface >> Line/Rake from the drop–down menu. Enter the following end points: x0 [m] = 9, y0 [m)] = 0 and x1 [m] = 9, y1 [m] = 8.5. Click on Create and Close the window. Select *line-6* in the Surfaces section, deselect *line-5,* uncheck Position on X Axis under Options and select Mesh and Y-Coordinate as Y Axis Function. Select Pressure and Static Pressure as X Axis Function. Click on Axes… and uncheck Auto

Range under Options for the X Axis. Set Minimum under Range to 25000 and Maximum to 30000. Set Precision under Number Format to 0. Select Y Axis and set the Precision to 0. Click on Apply and Close the Axes window. Click on Save/Plot. Click on Copy Screenshot of Active Window to Clipboard in the graphics window, see Figure 11.10d).

Figure 11.12a) Line/Rake Surface window

Figure 11.12b) Solution XY Plot

Figure 11.12c) Curves window   Figure 11.12d) Axes window

Figure 11.12e) Comparison of the variation for seepage velocity at the outlet from Ansys Fluent simulations (full line) and finite element calculations (dashed line).

Figure 11.12f) Static pressure distribution along a vertical line under the dam

265

H. Theory

13. Groundwater seepage can frequently occur in soils and has many engineering applications. We can relate the mean fluid velocity v (m/s) through the porous medium to the pressure gradient dP/dx (Pa) using Darcy's law

$$v = -\frac{\kappa}{\mu}\frac{dP}{dx} \tag{11.1}$$

where κ (m^2) is permeability of the porous medium and μ (kg/m-s) is the dynamic viscosity of the fluid. Furthermore, we define the hydraulic conductivity or coefficient of permeability k (m/s) as

$$k = \kappa\frac{\rho g}{\mu} \tag{11.2}$$

Where g (m/s^2) is acceleration due to gravity and ρ (kg/m^3) is the density of the fluid. We will also use the following equation from the ANSYS Fluent user manual

$$\frac{dP}{dx} = -\left(\frac{\mu v}{\alpha} + \frac{C_2 \rho v^2}{2}\right) \tag{11.3}$$

where $1/\alpha$ (1/m^2) is viscous resistance or inverse absolute permeability for the porous medium and C_2 (1/m) is inertial resistance of the porous medium. When we combine the equations we get the following expression for the permeability coefficient

$$k = \frac{g}{\frac{vC_2}{2} + \frac{\mu}{\alpha\rho}} \tag{11.4}$$

The following values were used in or determined from the ANSYS Fluent simulations.

C_2 (1/m)	1719.4
$1/\alpha$ (1/m^2)	$1.98807\cdot10^6$
ρ (kg/m^3)	992.2
μ (kg/m-s)	0.00065
v (m/s)	0.05
g (m/s^2)	9.81
k (m/s)	0.221508
p_1 (Pa)	29046.9
p_2 (Pa)	25124.4

Table 11.1 Parameters & coefficients used or determined by ANSYS Fluent simulations

The 2D flow through a porous medium is governed by

$$k_x \frac{\partial^2 \phi}{\partial x^2} + k_y \frac{\partial^2 \phi}{\partial y^2} = 0 \tag{11.5}$$

where ϕ is the hydraulic head and $k_x = k_y = k$ are coefficients of permeability in the x and y-directions. The seepage velocity components are related to the hydraulic head through

$$v_x = -k_x \frac{\partial \phi}{\partial x} \quad \text{and} \quad v_y = -k_y \frac{\partial \phi}{\partial y} \tag{11.6}$$

We follow the approach by Bhatti[3] using second order ($n = 2$) p-formulation. The equations for element **1**:

$$\begin{pmatrix} +0.2424 & -0.2021 & -0.1212 & +0.0809 & -0.0164 & +0.1320 & +0.0164 & -0.1320 \\ -0.2021 & +0.2424 & +0.0809 & -0.1212 & -0.0164 & -0.1320 & +0.0164 & +0.1320 \\ -0.1212 & +0.0809 & +0.2424 & -0.2021 & +0.0164 & -0.1320 & -0.0164 & +0.1320 \\ +0.0809 & -0.1212 & -0.2021 & +0.2424 & +0.0164 & +0.1320 & -0.0164 & -0.1320 \\ -0.0164 & -0.0164 & +0.0164 & +0.0164 & +0.4445 & 0 & +0.2021 & 0 \\ +0.1320 & -0.1320 & -0.1320 & +0.1320 & 0 & +0.1615 & 0 & -0.0809 \\ +0.0164 & +0.0164 & -0.0164 & -0.0164 & +0.2021 & 0 & +0.4445 & 0 \\ -0.1320 & +0.1320 & +0.1320 & -0.1320 & 0 & -0.0809 & 0 & +0.1615 \end{pmatrix} \begin{pmatrix} \phi_1 \\ \phi_4 \\ \phi_5 \\ \phi_2 \\ \delta_1^{(1,4)} \\ \delta_1^{(4,5)} \\ \delta_1^{(2,5)} \\ \delta_1^{(1,2)} \end{pmatrix} = \begin{pmatrix} 0 \\ 0 \\ 0 \\ 0 \\ 0 \\ 0 \\ 0 \\ 0 \end{pmatrix}$$

Equations for element **2**:

$$\begin{pmatrix} +0.1902 & +0.0380 & -0.0951 & -0.1331 & -0.0932 & +0.0233 & +0.0932 & -0.0233 \\ +0.0380 & +0.1902 & -0.1331 & -0.0951 & -0.0932 & -0.0233 & +0.0932 & +0.0233 \\ -0.0951 & -0.1331 & +0.1902 & +0.0380 & +0.0932 & -0.0233 & -0.0932 & +0.0233 \\ -0.1331 & -0.0951 & +0.0380 & +0.1902 & +0.0932 & +0.0233 & -0.0932 & -0.0233 \\ -0.0932 & -0.0932 & +0.0932 & +0.0932 & +0.1521 & 0 & -0.0380 & 0 \\ +0.0233 & -0.0233 & -0.0233 & +0.0233 & 0 & +0.3233 & 0 & |0.1331 \\ +0.0932 & +0.0932 & -0.0932 & -0.0932 & -0.0380 & 0 & +0.1521 & 0 \\ -0.0233 & +0.0233 & +0.0233 & -0.0233 & 0 & +0.1331 & 0 & +0.3233 \end{pmatrix} \begin{pmatrix} \phi_2 \\ \phi_5 \\ \phi_6 \\ \phi_3 \\ \delta_1^{(2,5)} \\ \delta_1^{(5,6)} \\ \delta_1^{(3,6)} \\ \delta_1^{(2,3)} \end{pmatrix} = \begin{pmatrix} 0 \\ 0 \\ 0 \\ 0 \\ 0 \\ 0 \\ 0 \\ 0 \end{pmatrix}$$

Equations for element **3**:

$$\begin{pmatrix} +0.2424 & -0.2021 & -0.1212 & +0.0809 & -0.0164 & +0.1320 & +0.0164 & -0.1320 \\ -0.2021 & +0.2424 & +0.0809 & -0.1212 & -0.0164 & -0.1320 & +0.0164 & +0.1320 \\ -0.1212 & +0.0809 & +0.2424 & -0.2021 & +0.0164 & -0.1320 & -0.0164 & +0.1320 \\ +0.0809 & -0.1212 & -0.2021 & +0.2424 & +0.0164 & +0.1320 & -0.0164 & -0.1320 \\ -0.0164 & -0.0164 & +0.0164 & +0.0164 & +0.4445 & 0 & +0.2021 & 0 \\ +0.1320 & -0.1320 & -0.1320 & +0.1320 & 0 & +0.1615 & 0 & -0.0809 \\ +0.0164 & +0.0164 & -0.0164 & -0.0164 & +0.2021 & 0 & +0.4445 & 0 \\ -0.1320 & +0.1320 & +0.1320 & -0.1320 & 0 & -0.0809 & 0 & +0.1615 \end{pmatrix} \begin{pmatrix} \phi_4 \\ \phi_7 \\ \phi_8 \\ \phi_5 \\ \delta_1^{(4,7)} \\ \delta_1^{(7,8)} \\ \delta_1^{(5,8)} \\ \delta_1^{(4,5)} \end{pmatrix} = \begin{pmatrix} 0 \\ 0 \\ 0 \\ 0 \\ 0 \\ 0 \\ 0 \\ 0 \end{pmatrix}$$

Equations for element **4**:

$$\begin{pmatrix} +0.1902 & +0.0380 & -0.0951 & -0.1331 & -0.0932 & +0.0233 & +0.0932 & -0.0233 \\ +0.0380 & +0.1902 & -0.1331 & -0.0951 & -0.0932 & -0.0233 & +0.0932 & +0.0233 \\ -0.0951 & -0.1331 & +0.1902 & +0.0380 & +0.0932 & -0.0233 & -0.0932 & +0.0233 \\ -0.1331 & -0.0951 & +0.0380 & +0.1902 & +0.0932 & +0.0233 & -0.0932 & -0.0233 \\ -0.0932 & -0.0932 & +0.0932 & +0.0932 & +0.1521 & 0 & -0.0380 & 0 \\ +0.0233 & -0.0233 & -0.0233 & +0.0233 & 0 & +0.3233 & 0 & +0.1331 \\ +0.0932 & +0.0932 & -0.0932 & -0.0932 & -0.0380 & 0 & +0.1521 & 0 \\ -0.0233 & +0.0233 & +0.0233 & -0.0233 & 0 & +0.1331 & 0 & +0.3233 \end{pmatrix} \begin{pmatrix} \phi_5 \\ \phi_8 \\ \phi_9 \\ \phi_6 \\ \delta_1^{(5,8)} \\ \delta_1^{(8,9)} \\ \delta_1^{(6,9)} \\ \delta_1^{(5,6)} \end{pmatrix} = \begin{pmatrix} 0 \\ 0 \\ 0 \\ 0 \\ 0 \\ 0 \\ 0 \\ 0 \end{pmatrix}$$

where $\phi_1 \ldots, \phi_9$ are hydraulic heads at the node points for the four elements and δ are the p-mode parameters, subscripts indicate mode number for the side modes and superscripts indicate node numbers on the corresponding line segments for an element. The boundary conditions are the following:

$$(\phi_1, \phi_2, \phi_3, \phi_6, \phi_9, \delta_1^{(1,2)}, \delta_1^{(3,6)}, \delta_1^{(6,9)}) = (2.98422, 2.58123, 2, 2, 2, 0, 0, 0) \tag{11.7}$$

The hydraulic head boundary conditions ϕ_1, ϕ_2 were determined from the static pressures

$$p_1, p_2 = \rho g \phi_1, \rho g \phi_2 = 29046.9, 25124.4 \; Pa \tag{11.8}$$

at nodes and 1 and 2, respectively. We can now assemble the global stiffness matrix.

Figure 11.13a) Four element model for seepage flow

$$
\begin{pmatrix}
+0.485 & +0.162 & -0.202 & -0.121 & -0.016 & 0 & +0.016 & -0.264 & -0.016 & 0 & +0.016 & +0.132 & 0 \\
+0.162 & +0.865 & -0.121 & -0.164 & +0.016 & +0.023 & -0.110 & -0.264 & +0.016 & -0.047 & -0.110 & +0.132 & +0.023 \\
-0.202 & -0.121 & +0.242 & +0.081 & 0 & 0 & 0 & +0.132 & -0.016 & 0 & +0.016 & -0.132 & 0 \\
-0.121 & -0.164 & +0.081 & +0.433 & 0 & 0 & 0 & +0.132 & +0.016 & +0.023 & -0.110 & -0.132 & -0.023 \\
-0.016 & +0.016 & 0 & 0 & +0.445 & 0 & +0.202 & 0 & 0 & 0 & 0 & 0 & 0 \\
0 & +0.023 & 0 & 0 & 0 & +0.323 & 0 & 0 & 0 & +0.133 & 0 & 0 & 0 \\
+0.016 & -0.110 & 0 & 0 & +0.202 & 0 & +0.597 & 0 & 0 & 0 & 0 & 0 & 0 \\
-0.264 & -0.264 & +0.132 & +0.132 & 0 & 0 & 0 & +0.323 & 0 & 0 & 0 & -0.081 & 0 \\
-0.016 & +0.016 & -0.016 & +0.016 & 0 & 0 & 0 & 0 & +0.445 & 0 & +0.202 & 0 & 0 \\
0 & -0.047 & 0 & +0.023 & 0 & +0.133 & 0 & 0 & 0 & +0.647 & 0 & 0 & +0.133 \\
+0.016 & -0.110 & +0.016 & -0.110 & 0 & 0 & 0 & 0 & +0.202 & 0 & +0.597 & 0 & 0 \\
+0.132 & +0.132 & -0.132 & -0.132 & 0 & 0 & 0 & -0.081 & 0 & 0 & 0 & +0.161 & 0 \\
0 & +0.023 & 0 & -0.023 & 0 & 0 & 0 & 0 & 0 & +0.133 & 0 & 0 & +0.323
\end{pmatrix}
\begin{pmatrix}
\phi_4 \\ \phi_5 \\ \phi_7 \\ \phi_8 \\ \delta_1^{(1,4)} \\ \delta_1^{(2,3)} \\ \delta_1^{(2,5)} \\ \delta_1^{(4,5)} \\ \delta_1^{(4,7)} \\ \delta_1^{(5,6)} \\ \delta_1^{(5,8)} \\ \delta_1^{(7,8)} \\ \delta_1^{(8,9)}
\end{pmatrix} =
$$

$$
\begin{pmatrix}
+0.916 \\ +1.698 \\ 0 \\ +0.456 \\ +0.007 \\ +0.060 \\ -0.139 \\ -0.735 \\ 0 \\ -0.060 \\ -0.373 \\ 0 \\ 0
\end{pmatrix}
,\text{ with the solution }
\begin{pmatrix}
\phi_4 \\ \phi_5 \\ \phi_7 \\ \phi_8 \\ \delta_1^{(1,4)} \\ \delta_1^{(2,3)} \\ \delta_1^{(2,5)} \\ \delta_1^{(4,5)} \\ \delta_1^{(4,7)} \\ \delta_1^{(5,6)} \\ \delta_1^{(5,8)} \\ \delta_1^{(7,8)} \\ \delta_1^{(8,9)}
\end{pmatrix} =
\begin{pmatrix}
+2.699 \\ +2.192 \\ +2.703 \\ +2.193 \\ -0.012 \\ +0.038 \\ +0.100 \\ -0.316 \\ +0.027 \\ -0.024 \\ +0.023 \\ -0.155 \\ +0.010
\end{pmatrix}
\qquad (11.9)
$$

We can summarize the solution at the outflow for elements 2 and 4 with the following table where v_y velocity component is compared with ANSYS Fluent in Figure 11.12e).

	x	y	ϕ	$\partial\phi/\partial x$	$\partial\phi/\partial y$	v_y
2	0	10	2.	0.	-0.32565825428980166	0.07432144875149277
2	0.75	10	2.	0.	-0.2555161785079309	0.05831368410289106
2	1.5	10	2.	0.	-0.2057152107246451	0.04694814975476898
2	2.25	10	2.	0.	-0.1762553509399443	0.04022484570712651
4	3	10	2.	0.	-0.16713659915382845	0.03814377195996365
4	3.75	10	2.	0.	-0.14647549864168652	0.03342851324124527
4	4.5	10	2.	0.	-0.13050473334316898	0.029783678820436723
4	5.25	10	2.	0.	-0.1192243032582763	0.0272092686975 38133
4	6	10	2.	0.	-0.11263420838700836	0.025705282872549466

Table 11.2 Solution summary at the outflow region

```
(*Seepage through soil, p=q=0,
2nd order p-formulation (n=2), wi = wj = 1, kx = ky = 0.2282191462137928` *)
Clear["Global`*"]; wi = 1; wj = 1; kx = 0.2282191462137928`; ky = 0.2282191462137928`;
NT = {(1 - s) * (1 - t) / 4, (1 + s) * (1 - t) / 4, (1 + s) * (1 + t) / 4, (1 - s) * (1 + t) / 4,
    (3 * s^2 / 2 - 3 / 2) * (1 - t) / (2 * Sqrt[6]), (3 * t^2 / 2 - 3 / 2) * (1 + s) / (2 * Sqrt[6]),
    (3 * s^2 / 2 - 3 / 2) * (1 + t) / (2 * Sqrt[6]), (3 * t^2 / 2 - 3 / 2) * (1 - s) / (2 * Sqrt[6])};
gcd[x0_, y0_] := Module[{x = x0, y = y0}, {J = {{D[x, s], D[x, t]}, {D[y, s], D[y, t]}};
    BT = Inverse[J].{D[NT, s], D[NT, t]}; k = wi * wj * Transpose[BT].{{kx, 0}, {0, ky}}.BT * Det[J];
    k1 = k //. {s -> -0.57735, t -> -0.57735}; k2 = k //. {s -> -0.57735, t -> 0.57735};
    k3 = k //. {s -> 0.57735, t -> -0.57735}; k4 = k //. {s -> 0.57735, t -> 0.57735}; kk = k1 + k2 + k3 + k4;}]

(*Element 1: 0<x<3, 0<y<8.5, -1<s<1, -1<t<1, Element 2: 0<x<3, 8.5<y<10, -1<s<1, -1<t<1,
Element 3: 3<x<6, 0<y<8.5, -1<s<1, -1<t<1, Element 4: 3<x<6, 8.5<y<10, -1<s<1, -1<t<1 *)
gcd[3 * s / 2 + 3 / 2, 8.5 * t / 2 + 8.5 / 2]; kk1 = kk; kk1 // MatrixForm;
gcd[3 * s / 2 + 3 / 2, 3 * t / 4 + 37 / 4]; kk2 = kk; kk2 // MatrixForm;
gcd[3 * s / 2 + 9 / 2, 8.5 * t / 2 + 8.5 / 2]; kk3 = kk; kk3 // MatrixForm;
gcd[3 * s / 2 + 9 / 2, y = 3 * t / 4 + 37 / 4]; kk4 = kk; kk4 // MatrixForm;

(*Global stiffness matrix and Boundary conditions *)
kG = SparseArray[{}, {21, 21}]; r1 = {1, 4, 5, 2, 11, 15, 13, 10}; kG[[r1, r1]] = kG[[r1, r1]] + kk1;
r2 = {2, 5, 6, 3, 13, 17, 14, 12};
kG[[r2, r2]] = kG[[r2, r2]] + kk2; r3 = {4, 7, 8, 5, 16, 20, 18, 15};
kG[[r3, r3]] = kG[[r3, r3]] + kk3; r4 = {5, 8, 9, 6, 18, 21, 19, 17}; kG[[r4, r4]] = kG[[r4, r4]] + kk4;
kG[[1]] = UnitVector[21, 1]; kG[[2]] = UnitVector[21, 2]; kG[[3]] = UnitVector[21, 3];
kG[[6]] = UnitVector[21, 6]; kG[[9]] = UnitVector[21, 9]; kG[[10]] = UnitVector[21, 10];
kG[[14]] = UnitVector[21, 14]; kG[[19]] = UnitVector[21, 19];

(*Global solution and element solutions) *)
RHS = {2.984224966974819`, 2.581234546896989`, 2, 0, 0, 2, 0, 0, 2, 0, 0, 0, 0, 0, 0, 0, 0, 0, 0, 0, 0};
ϕ = LinearSolve[kG, RHS]
gce[x0_, y0_, s0_, t0_] :=
  Module[{x = x0, y = y0}, {J = {{D[x, s], D[x, t]}, {D[y, s], D[y, t]}}; NT1 = NT //. {s -> s0, t -> t0};
    dNTds = D[NT, s] //. {s -> s0, t -> t0}; dNTdt = D[NT, t] //. {s -> s0, t -> t0};
    dNTdsodNTdt = {dNTds, dNTdt} //. {s -> s0, t -> t0}; BT = Inverse[J].dNTdsodNTdt;
    B = Transpose[BT]; BTx = BT[[1]]; BTy = BT[[2]]; ϕxy = NT1.dT; dϕdx = BTx.dT; dϕdy = BTy.dT;}]

(*Solution for element 2 at location (x,y) = (0,10), (0.75,10), (1.5,10), (2.25,10) *)
dT = {ϕ[[2]], ϕ[[5]], ϕ[[6]], ϕ[[3]], ϕ[[13]], ϕ[[17]], ϕ[[14]], ϕ[[12]]};
gce[3 * s / 2 + 3 / 2, 3 * t / 4 + 37 / 4, -1, 1]; mxy1 = {0, 10, ϕxy, dϕdx, dϕdy, -ky * dϕdy};
gce[3 * s / 2 + 3 / 2, 3 * t / 4 + 37 / 4, -0.5, 1]; mxy2 = {0.75, 10, ϕxy, dϕdx, dϕdy, -ky * dϕdy};
gce[3 * s / 2 + 3 / 2, 3 * t / 4 + 37 / 4, 0, 1]; mxy3 = {1.5, 10, ϕxy, dϕdx, dϕdy, -ky * dϕdy};
gce[3 * s / 2 + 3 / 2, 3 * t / 4 + 37 / 4, 0.5, 1]; mxy4 = {2.25, 10, ϕxy, dϕdx, dϕdy, -ky * dϕdy};

(*Solution for element 4 at location (x,y) = (3,10), (3.75,10), (4.5,10), (5.25,10), (6,10) *)
dT = {ϕ[[5]], ϕ[[8]], ϕ[[9]], ϕ[[6]], ϕ[[18]], ϕ[[21]], ϕ[[19]], ϕ[[17]]};
gce[3 * s / 2 + 9 / 2, 3 * t / 4 + 37 / 4, -1, 1]; mxy5 = {3, 10, ϕxy, dϕdx, dϕdy, -ky * dϕdy};
gce[3 * s / 2 + 9 / 2, 3 * t / 4 + 37 / 4, -0.5, 1]; mxy6 = {3.75, 10, ϕxy, dϕdx, dϕdy, -ky * dϕdy};
gce[3 * s / 2 + 9 / 2, 3 * t / 4 + 37 / 4, 0, 1]; mxy7 = {4.5, 10, ϕxy, dϕdx, dϕdy, -ky * dϕdy};
gce[3 * s / 2 + 9 / 2, 3 * t / 4 + 37 / 4, 0.5, 1]; mxy8 = {5.25, 10, ϕxy, dϕdx, dϕdy, -ky * dϕdy};
gce[3 * s / 2 + 9 / 2, 3 * t / 4 + 37 / 4, 1, 1]; mxy9 = {6, 10, ϕxy, dϕdx, dϕdy, -ky * dϕdy};
```

```
(* Nodal solution summary and write to  file*)
mm = {mxy1, mxy2, mxy3, mxy4, mxy5, mxy6, mxy7, mxy8, mxy9}; MatrixForm[mm,
 TableHeadings → {{"2", "2", "2", "2", "4", "4", "4", "4", "4"}, {"x", "y", "ϕ", "∂ϕ/∂x", "∂ϕ/∂y", "vy"}}]
SetDirectory["C:\\Users\\jmatsson"]; mylist = Table[
   {{9, mxy1[[6]]}, {9.75, mxy2[[6]]}, {10.5, mxy3[[6]]}, {11.25, mxy4[[6]]}, {12, mxy5[[6]]},
    {12.75, mxy6[[6]]}, {13.5, mxy7[[6]]}, {14.25, mxy8[[6]]}, {15, mxy9[[6]]}}];
TableOfValues1 = Prepend[mylist, {"((xy/key/label \"theory-outlet\")"}];
TableOfValues1 = Prepend[TableOfValues1, {""}];
TableOfValues1 = Prepend[TableOfValues1, {"(labels \"X-Coordinate\" \"Velocity\")"}];
TableOfValues1 = Prepend[TableOfValues1, {"(title \"Theory\")"}];
TableOfValues1 = Append[TableOfValues1, {")"}]; Grid[TableOfValues1]
Export["theory.dat", TableOfValues1]
```

Figure 11.13b) Mathematica code for seepage flow under a dam

I. References

1. Ochs et al., "Soil-Water Pit Heat Store with Direct Charging System", Ecostock 2005 Conference Proceedings.
2. Moaveni S., "Finite Element Analysis: Theory and Applications with ANSYS", 3rd Ed., Prentice Hall, 2007.
3. Bhatti, M.A., "Fundamental Finite Element Analysis and Applications", John Wiley & Sons, Inc., 2005.

J. Exercise

11.1 Use ANSYS Fluent to study the seepage flow of water through the porous medium under the dam as shown in Figure 11.14. Include contour plots for velocity magnitude and static pressure together with an XY Plot of the velocity magnitude at the outlet. Use the same values for viscous and inertial resistance and mesh size as was used in this chapter.

Figure 11.14 Geometry for Exercise 11.1

CHAPTER 12. WATER FILTER FLOW

A. Objectives

- Using Ansys Workbench to Model Flow Through Water Filter
- Inserting Boundary Conditions and Porous Medium Zone for Filter with Viscous and Inertial Resistance
- Using Volume of Fluid Model for Multiphase Flow with Surface Tension
- Running Laminar Transient 2D Axisymmetric Ansys Fluent Simulations
- Using Volume Fraction Contour Plots for Visualizations and Movie
- Comparing Ansys Fluent Results with Theory

B. Problem Description

We will study the flow through a ceramic gravity driven clay pot water filter using Ansys Fluent. The dimensions and initial water levels are shown below. The thickness of the filter is $d = 20$ mm and the radius $R = 100$ mm while the initial height of water $h_0 = 200$ mm. The height $h(t)$ of the water in the filter will vary over time as the fluid seeps through the filter.

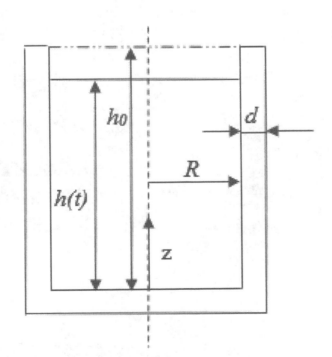

C. Launching Ansys Workbench and Selecting Fluent

1. Start by launching Ansys Workbench. Double click on Fluid Flow (Fluent) under Analysis Systems in the Toolbox.

Figure 12.1 Selecting Fluent

D. Launching Ansys DesignModeler

2. Right click Geometry under Project Schematic in Ansys Workbench and select Properties. In Properties of Schematic A2: Geometry, select Analysis Type 2D under Advanced Geometry Options. Right click on Geometry in the Project Schematic window and select New DesignModeler Geometry.

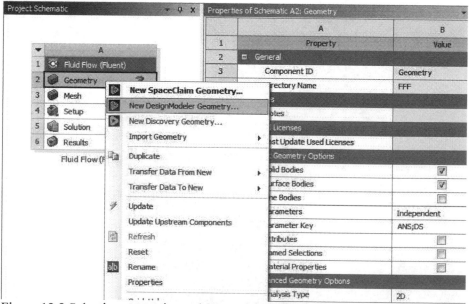

Figure 12.2 Selecting properties and 2D analysis type

3. Select Units>>Millimeter from the menu in DesignModeler. Select XY Plane in the Tree Outline and click on Look at Face/Plane/Sketch 🔧 . Select Sketching tab and Draw>>Polyline from Sketching Toolboxes. Start at origin with a *P* and draw the Sketch with Dimensions as shown in Figure 12.3c). Make sure that you have a *V* when drawing vertical lines and an *H* when drawing horizontal lines. Right click and select Closed End when you have completed the sketch. Enter the dimensions as shown in Figure 12.3c).

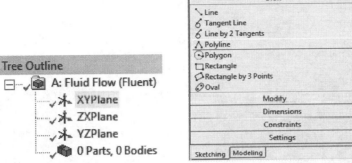

Figure 12.3a) Tree outline   Figure 12.3b) Sketching toolboxes

Figure 12.3c) Sketch of cylindrical water filter with dimensions

4. Select the Modeling tab and select Concepts>>Surfaces from Sketches from the menu. Select Sketch1 under XYPlane in the Tree Outline and Apply the sketch in Details View. Click on Generate.

   Select the XYPlane once again in the Tree Outline and click on New Sketch 📝 . Select the new sketch (Sketch2) under XYPlane and select the Sketching tab. Select Rectangle under Draw in Sketching Toolboxes and draw a rectangle inside the filter from the origin, see Figure 12.4. Select Concepts>>Surfaces from Sketches from the menu, select Sketch 2 and Apply the new sketch in Details View. Select Add Frozen as Operation in Details View. Click on Generate and Close DesignModeler.

Figure 12.4 Finished model for cylindrical water filter

273

E. Launching Ansys Meshing

5. Double click on Mesh under Project Schematic in Ansys Workbench. Select Mesh in the Outline. Set the Element Size to 2.0 mm under Defaults in Details of "Mesh". Select Yes for Capture Curvature under Sizing and set Smoothing to High under Quality. Right click on Mesh in the Outline and select Generate Mesh. The mesh has 6922 Nodes and 6600 Elements.

Figure 12.5a) Ansys Meshing

Figure 12.5b) Finished mesh for cylindrical water filter

6. Select Edge selection filter and control select the two edges on the right side of the computational domain, see Figure 12.6a). Right click on the edges and select Create Named Selection. Name the edges *outlet*.

Figure 12.6a) Named selections for outlet

Control-select the upper horizontal edge and the left outer vertical edge, see Figure 12.6b). Right click and select Create Named Selection. Name the edges *outlet-filter*.

Figure 12.6b) Named selections for outlet-filter

Select Body ▣ and select the outer L-shaped filter region, right click and select Hide Body. Select Edge ▣ and control-select the upper horizontal edge and the left vertical edge, see Figure 12.6c). Right click and select Create Named Selection. Name the edges *interface1*. Right click in the graphics window and select Show All Bodies.

Figure 12.6c) Named selections for interface1

Select Body , left click on the inner fluid region, right click and select Hide Body.

Select Edge and control-select the inner horizontal edge and the inner vertical edge, see Figure 12.6d). Right click and select Create Named Selection. Name the edges *interface2*. Right click in the graphics window and select Show All Bodies.

Figure 12.6d) Named selections for interface2

Control select the two lower horizontal edges and name these *axis*, see Figure 12.6e).

Figure 12.6e) Named selections for axis

Figure 12.6f) Named selections for cylindrical water filter

Open Geometry in the Outline, select the first Surface Body, right click and select Rename. Name this body *Porous zone*. Rename the next Surface Body and call it *Inlet zone*.

Figure 12.6g) Outline

Select File>>Export...>>Mesh>>FLUENT Input File>>Export from the menu and save the mesh with the name *cylindrical-water-filter.msh*. Select File>>Save Project from the menu and save the project with the name *Cylindrical Water Filter*. Close the Ansys Meshing window. Right click on Mesh under Project Schematic in Ansys Workbench and select Update.

F. Launching Ansys Fluent

7. Double-click on Setup under Project Schematic in Ansys Workbench. Check the box for Double Precision. Set the number of processes equal to the number of processor cores for your computer. To check the number of physical cores, press the Ctrl + Shift + Esc keys simultaneously to open the Task Manager. Go to the Performance tab and select CPU from the left column. You will see the number of physical cores on the bottom-right side. Close the Task Manager window. Click on the Start button in the Fluent Launcher window.

Figure 12.7a) Fluent launcher settings

Select Transient Time and Axisymmetric 2D Space for the Solver under General on the Task Page. Check the box for Gravity and enter -9.81 as Gravitational Acceleration in the X direction. Double click on Models and Viscous (SST k-omega) and select Laminar Model. Click on OK to close the Viscous Model window. Double click on Models and Multiphase (Off) under Setup in the Outline View. Select Volume of Fluid model and check the box for Implicit Body Force. Click on Apply and Close the window.

Figure 12.7b) Volume of Fluid Multiphase Model

8. Double click on Materials under Setup in the Outline View. Select Create/Edit for Fluid under Materials on the Task Page. Select Fluent Database and scroll down to *water-liquid (h2o<l>)* and select this as Fluent Fluid Materials. Click on Copy and Close the window. Close the Create/Edit Materials window.

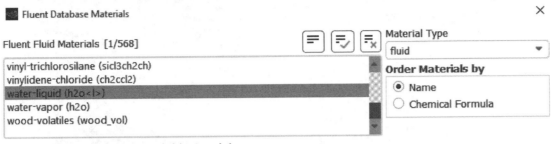

Figure 12.8 Fluent Fluid Materials

9. Double click on Models and Multiphase (Volume of Fluid) under Setup in the Outline View. Select the Phases tab at the top of the Multiphase Model window. Select phase-1 – Primary Phase under Phases. Select *water-liquid* as the Primary Phase Material and enter *water* as the name under Phase Setup. Click on Apply.

   Select phase-2 – Secondary Phase under Phases and select *air* as the Secondary Phase Material. Enter *air* as the name under Phase Setup. Click on Apply. Select the Phase Interaction tab at the top of the Multiphase Model window. Check the box for Surface Tension Force Modeling. Select constant for the Surface Tension Coefficients [N/m] and enter 0.07286 as the value. Click on Apply and Close the Multiphase Model window.

Figure 12.9a) Primary phase

Figure 12.9b) Secondary phase

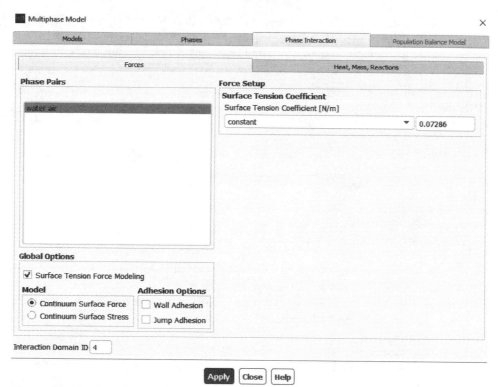

Figure 12.9c) Phase interaction

10. Double click on Cell Zone Conditions under Setup in the Outline View. Select the *porous_zone* under Cell Zone Conditions on the Task Page. Select *fluid* as Type and *mixture* as Phase and click on the Edit button on the Task Page. Check the Porous Zone box. Select the Porous Zone tab and make sure that Porosity is 0.5. Click Apply and Close the Fluid window.

Select *water* as the Phase for the *porous_zone* and click on Edit. Enter 9.76305e+08 as the Viscous Resistance in both directions. Click Apply and Close the Fluid window. Click on Operating Conditions on the Task Page under Cell Zone Conditions. Select user-input as Operating Density Method and enter the value 1.225 for Operating Density [kg/m³]. Enter X [m] 0.2 as Reference Pressure Location. Click OK to close the Operating Conditions window.

Figure 12.10a) Porous zone mixture phase

Fluid ✕

Zone Name
porous_zone

Phase
water ▾

☐ Source Terms

☐ Mesh Motion ☐ Fixed Values

☑ Porous Zone

Reference Frame	Mesh Motion	Porous Zone	3D Fan Zone	Embedded LES	Reaction	Source Terms	Fixed Values	Multiphase

Update From Line Tool

Direction-1 Vector

X 1 ▾

Y 0 ▾

☑ Relative Velocity Resistance Formulation

Viscous Resistance (Inverse Absolute Permeability)

Direction-1 [m⁻¹] 976305000 ▾

Direction-2 [m⁻²] 976305000 ▾

Figure 12.10b) Porous zone water phase

Operating Conditions ✕

Pressure

Operating Pressure [Pa]

101325 ▾

Reference Pressure Location

X [m] 0.2 ▾

Y [m] 0 ▾

Z [m] 0 ▾

Gravity

☑ Gravity

Gravitational Acceleration

X [m/s²] -9.81 ▾

Y [m/s²] 0 ▾

Z [m/s²] 0 ▾

Variable-Density Parameters

Operating Density Method

user-input ▾

Operating Density [kg/m³]

1.225 ▾

OK Cancel Help

Figure 12.10c) Operating conditions

11. Double click on Boundary Conditions under Setup in the Outline View. Select the *outlet-surface-body* Boundary Condition on the Task Page and make sure that Type is *pressure-outlet*. Select *mixture* as Phase and click on Edit…. Make sure that Gauge Pressure [Pa] is 0, Backflow Pressure Specification as Total Pressure and Backflow Direction Specification Method is set to Normal to Boundary. Select Apply and Close the window.

Select *air* as Phase and click on the Edit… button on the Task Page under Boundary Conditions. Set the Backflow Volume Fraction to 1, click Apply and Close the window.

Repeat this step 11 for the *outlet-porous_zone* Boundary Condition.

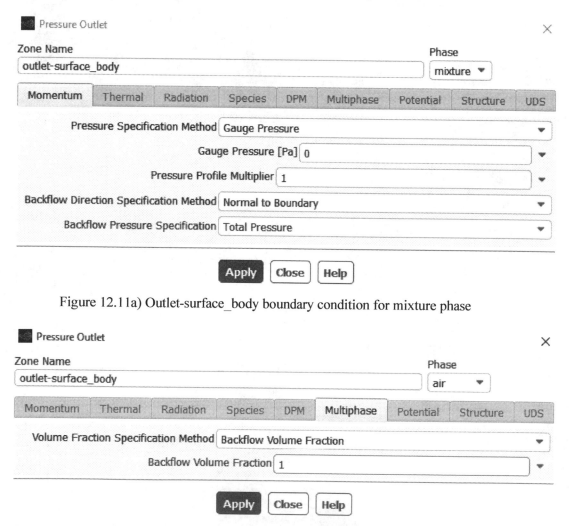

Figure 12.11a) Outlet-surface_body boundary condition for mixture phase

Figure 12.11b) Outlet-inlet_zone boundary condition for air phase

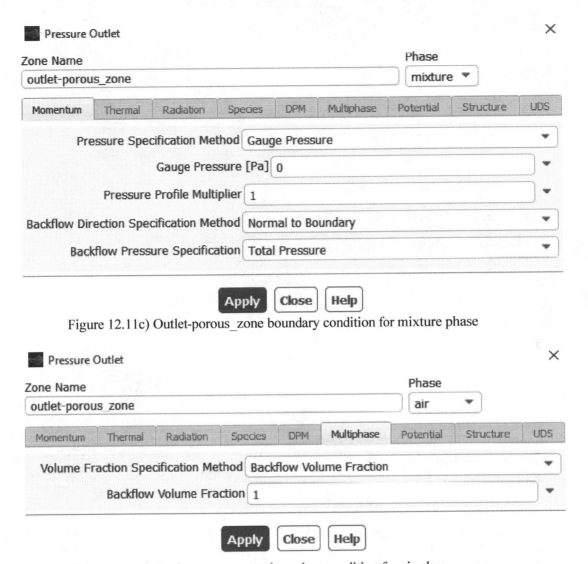

Figure 12.11c) Outlet-porous_zone boundary condition for mixture phase

Figure 12.11d) Outlet-porous_zone boundary condition for air phase

Continue with the *outlet-filter* Boundary Condition. Make sure that Type is *pressure-outlet* on the Task Page and select *mixture* as Phase and click on Edit.... Gauge Pressure [Pa] is 0, Backflow Pressure Specification as Total Pressure and Backflow Direction Specification Method is set to Normal to Boundary. Click Apply and Close the window. Select *air* as Phase and click on the Edit... button. Select *From Neighboring Cell* as Volume Fraction Specification Method, click Apply and Close the window.

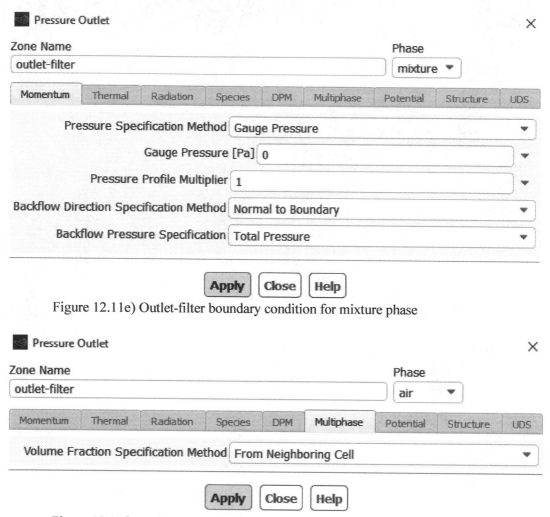

Figure 12.11e) Outlet-filter boundary condition for mixture phase

Figure 12.11f) Outlet-filter boundary condition for air phase

12. Double click on Initialization under Solution in the Outline View. Select Hybrid Initialization Method on the Task Page and click on Initialize. Click on Patch… on the Task Page under Solution Initialization. Select *air* as the Phase, select Volume Fraction as Variable, select *surface_body* and *porous_zone* as Zones to Patch, set the Value to 0 and click on Patch. Close the window.

Figure 12.12a) Initialization of solution

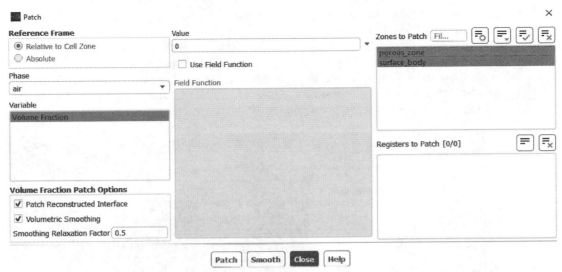

Figure 12.12b) Patch settings

13. Double click on Graphics and Contours under Results in the Outline View. Select Contours of Phases… and Volume fraction for *water* as the Phase. Scroll down and deselect all Surfaces and uncheck Auto Range Options. Set Min to 0 and Max to 1. Select Colormap Options…, set Font Behavior to Fixed and Font Size to 11 under Font. Set Type to float and Precision to 1 under Number Format. Click on Apply and Close the Colormap window. Click on Save/Display and Close the Contours window.

Figure 12.13a) Contours settings

Select the View tab in the menu and click on Views…. Select *axis-porous_zone* and *axis-surface_body* as Mirror Planes and click on Apply. Select the Camera… button in the Views window. Use your left mouse button to rotate the dial counter-clockwise 90 degrees until the bowl is upright. Close the Camera Parameters window. Click on the Save button under Actions in the Views window and Close the same window. Click on Copy Screenshot of Active Window to Clipboard in the graphics window, see Figure 12.13c). The contour plot can be pasted into a Word document.

Figure 12.13b) Water volume fraction after rotation and mirroring

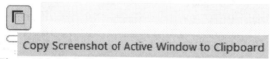

Figure 12.13c) Copying screenshot

Double click on Calculation Activities and Solution Animations under Solution in the Outline View. Enter *water-filter* as the Name, enter Record after every 10 and select time-step. Select *contour-1* as Animation Object. Select OK to close the Animation Definition window.

Figure 12.13d) Animation definition

14. Double click on Report Definitions under Solution in the Outline View. Select New>>Surface Report>>Volume Flow Rate…. Enter *volume-flow-rate* as the name. Select *water* as the Phase and select *outlet-filter* under Surfaces. Check the boxes for Report File, Report Plot and Print to Console under Create. Click OK to close the window.

Figure 12.14a) Surface report definition for volume flow rate

Select New>>Expression in the Report Definitions window. Select Functions>>Mathematical>>abs and select *volume-flow-rate* as Report Definition, see Figure 12.14b). Enter *absolute-volume-flow-rate* as Name and check the boxes for Report File, Report Plot and Print to Console under Create. Click on OK to close the window. Close the Report Definitions window. Select File>>Export>Case & Data… from the menu. Save the file with the name *water-filter-flow.cas.h5*.

Expression Report Definition ✕

Name | absolute-volume-flow-rate

abs(volume-flow-rate)

Functions ▼
Variables ▼
Constants ▼
Report Definitions ▼
Locations ▼

Current Value: Refresh value

Average Over 1

Report Files [0/1]   Report Plots [0/1]   **Create**
volume-flow-rate-rfile   volume-flow-rate-rplot   ✔ Report File
                                                    ✔ Report Plot
                                                    Frequency 1
                                                    ✔ Print to Console

☐ Create Output Parameter

OK | Compute | Cancel | Help

Figure 12.14b) Expression report definition for absolute flow rate

15. Double click on Run Calculation under Solution in the Outline View and set the Time Step Size [s] to 0.005. Set the Number of Time Steps to 50 and Max Iterations/Time Step to 30. Click on the Calculate button on the Task Page. Click OK in the Information window when the calculations have finished. Click on Copy Screenshot of Active Window to Clipboard in the graphics window, see Figure 12.13c). The scaled residuals can be pasted into a Word document.

Figure 12.15a) Calculation settings

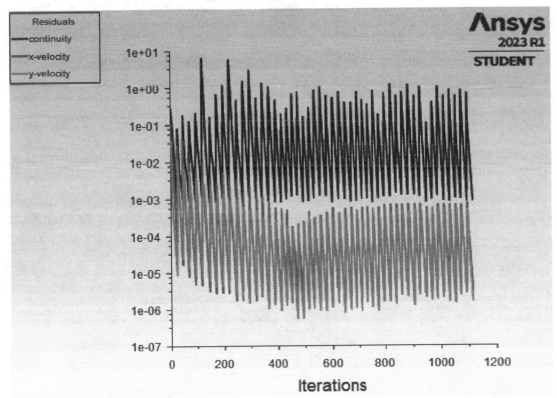

Figure 12.15b) Residuals during the first 0.25 s

G. Post-Processing and Continued Simulations

16. Select the tab Contours of Volume fraction (water) in the graphics window. Select the View tab in the menu and click on Views…. Select *view-0* in Views, click on Apply in the Views window and Close the window. Click on Copy Screenshot of Active Window to Clipboard in the graphics window, see Figure 12.13c). The scaled residuals can be pasted into a Word document.

Select the Results tab from the menu and select Surface>>Create>>Iso-Surface…. Select Surface of Constant Phases… and Volume Fraction. Select *water* as the Phase and set Iso-Values to 0.5. Use the horizontal slider to set the Iso-Value. Select *surface_body* under From Zones. Enter *free-surface-t=0.25s* as New Surface Name and click on Create. The new surface will now be available under From Surface, see Figure 12.16a). Close the Iso-Surface window.

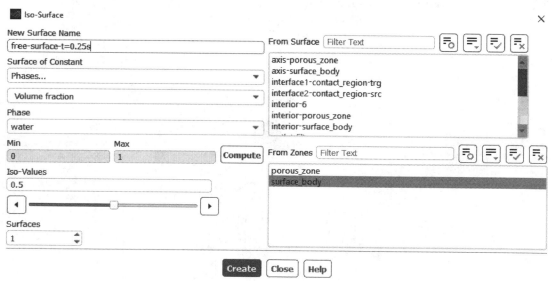

Figure 12.16a) Iso-surface settings

Select File>>Export>>Solution Data… from the menu. Select ASCII as File Type, Node under Location and Space as Delimiter. Select *Iso-surface>> free-surface-time-t=0.25s* under Surfaces. Select *Volume fraction (water)* under Quantities. Click on Write and save the ASCII File in the *Working Directory* with the name *free-surface-coordinates-t=0.25s*. Close the Export window.

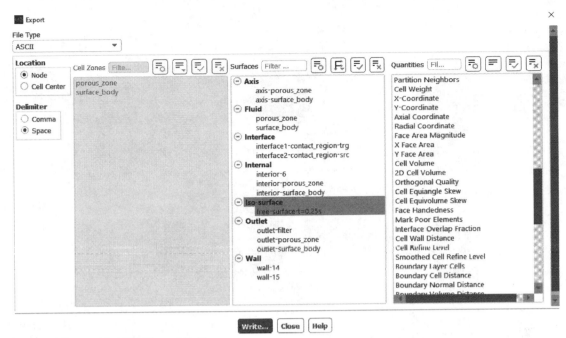

Figure 12.16b) Export settings

Open the saved file *free-surface-coordinates-t=0.25s* in Excel. Click on Next twice in the Text Import Wizard and click on Finish. Copy the 50 first values from the *C* column and paste the data in the *E* column. Label the new column *r (m)*. Copy the 50 first values from the *B* column and paste the data in the *F* column. Label the new column *h (m)*. Plot the free surface and take the average free surface height for the first 50 points. Repeat steps **15 – 16** seven more times for a total of 2 seconds.

Figure 12.16c) Free surface height h (m) versus radius r (m) at $t = 0.25$ s

t (s)	Average h (m), ANSYS Fluent	h (m), Theory
0	0.2	0.2
0.25	0.13927	0.1429
0.5	0.09868	0.1080
0.75	0.07290	0.08457
1	0.05487	0.06788
1.25	0.04156	0.05548
1.5	0.03185	0.04597
1.75	0.02417	0.03849
2	0.01769	0.03249

Table 12.1 Data for average free surface height over time

Figure 12.16d) Average free surface height h (m) versus time

Figure 12.16e) Water volume fraction at t = 0 − 2 s with an increment of 0.25 s

Double click on Animations and Playback under Results in the Outline View. Uncheck the box for Use Stored View, set the Replay Speed to low and select Play Once as Playback Mode. Select Video File as Write/Record Format. Click on the Write button for your chosen Animation Sequence. This will create the movie. Close the Playback window. The movie can be viewed using for example Windows Media Player.

Figure 12.16f) Playback settings

Double click on Data Sources under Results and Plots in the Outline View. Select Load File… under Data Sources and load the file *absolute-volume-flow-rate-rfile.out*. Select *flow-time* as X Axis Variable and *absolute-volume-flow-rate* as Y Axis Variable.

Next, load the file *volume-flow-rate.dat*. This and other files are available for download from *sdcpublications.com*. Select *Time(s)* as X Axis Variable and *Volume Flow Rate (m3/s)* as Y Axis Variable. Click on Axes… at the bottom of the Plot Data Sources window. Uncheck Auto Range under Options for X Axis. Set Minimum to 0 and Maximum to 2 under Range. Click on Apply and select Y Axis. Select float as Type under Number Format and set Precision to 3. Click on Apply and Close the Axes – Plot Data Sources window. Change the X Axis Label under Plot to *t* and the Y Axis Label to *dV/dt* (m3/s). Delete the Title and Legend Label under Plot.

Click on Curves… and select the first available Pattern under Line Style for Curve # 0. Select no (blank) Symbol under Marker Style and click on Apply. Select Curve # 1, select the second available Pattern under Line Style. Select no (blank) Symbol under Marker Style and click on Apply. Close the Curves – Plot Data Sources window.

Click on Plot at the bottom of the Plot Data Sources window. Click on Copy Screenshot of Active Window to Clipboard in the graphics window, see Figure 12.13c). The curves can be pasted into a Word document.

Open the saved file *absolute-volume-flow-rate-rfile.out* in Excel. Select Delimited instead of Fixed width in Text Import Wizard – Step 1 of 3. Click on Next. Uncheck Tab as Delimiters and check Space as Delimiters in Text Import Wizard – Step 2 of 3. Click on Next. Click on Finish in Text Import Wizard – Step 3 of 3. Use the trapezoidal rule to find the area under the volume flow rate curve (2nd column *B* in Excel file) for each time interval (3rd column *C*) and add these areas in order to show the volume produced by the filter.

Figure 12.16g) Plot data sources settings

Figure 12.16h) Comparing Ansys Fluent and theory-dashed curve for volume flow rate

294

This is accomplished by first deleting the title rows in the file and relabeling the three first columns $B - C$ with dV/dt and t as shown in Figure 12.16i).

Next, add the D column by inserting the equation =(B2+B3)/2 in cell $D3$ and drag this cell down the whole column D. Add $(f(a)+f(b))/2$ as the title to column D.

Add the E column by inserting the equation =C3-C2 in cell $E3$ and drag this cell down the whole column E. Add $b-a$ as the title to column E.

Add the F column by inserting the equation =D3*E3 in cell $F3$ and drag this cell down the whole column F. Add $(b-a)(f(a)+f(b))/2$ as the title to column F.

Copy column C to column G. Add t as the title to column G.

Enter 0 in $H2$. Enter the equation =H2+F3 in cell $H3$ and drag this cell down the whole column H. Add the title $V(t)$, Fluent as the title to column H.

Enter equation (12.8) in column I and add equation (12.10) in column J as shown in the theory section. Equation (12.8) is =0.2*0.1/((0.1+0.2)*Exp(0.01*G2/0.02)-0.2) that is entered in cell $I2$. Drag this equation down the whole column I. Add $h(t)$ as the title to column I.

Equation (12.10) is =PI()*0.1^2*(0.2-I2) that is entered in cell $J2$. Drag this cell down the whole column J. Add $V(t)$, Theory as the title to column J.

	C	D	E	F	G	H	I	J
1	t	(f(a)+f(b))/2	b-a	(b-a)(f(a)+f(b))/2	t	V(t), Fluent	h(t)	V(t), Theory
2	0				0	0	0.2	8.71967E-19
3	0.005	0.000763814	0.005	3.81907E-06	0.005	3.81907E-06	0.198509	4.68312E-05
4	0.01	0.002161201	0.005	1.0806E-05	0.01	1.46251E-05	0.197037	9.3084E-05
5	0.015	0.003351484	0.005	1.67574E-05	0.015	3.13825E-05	0.195583	0.000138769
6	0.02	0.004395387	0.005	2.19769E-05	0.02	5.33594E-05	0.194146	0.000183897
7	0.025	0.005299026	0.005	2.64951E-05	0.025	7.98546E-05	0.192727	0.000228477
8	0.03	0.006077387	0.005	3.03869E-05	0.03	0.000110241	0.191325	0.000272519
9	0.035	0.006743404	0.005	3.3717E-05	0.035	0.000143959	0.18994	0.000316033
10	0.04	0.007308995	0.005	3.6545E-05	0.04	0.000180503	0.188572	0.000359028
11	0.045	0.007786363	0.005	3.89318E-05	0.045	0.000219435	0.187219	0.000401513
12	0.05	0.008185225	0.005	4.09261E-05	0.05	0.000260361	0.185883	0.000443497
13	0.055	0.008514868	0.005	4.25743E-05	0.055	0.000302936	0.184562	0.000484989
14	0.06	0.008783506	0.005	4.39175E-05	0.06	0.000346853	0.183257	0.000525997
15	0.065	0.008998345	0.005	4.49917E-05	0.065	0.000391845	0.181967	0.00056653
16	0.07	0.009165558	0.005	4.58278E-05	0.07	0.000437673	0.180691	0.000606595
17	0.075	0.009291487	0.005	4.64574E-05	0.075	0.00048413	0.179431	0.000646201
18	0.08	0.009381159	0.005	4.69058E-05	0.08	0.000531036	0.178184	0.000685355
19	0.085	0.009439362	0.005	4.71968E-05	0.085	0.000578233	0.176952	0.000724065
20	0.09	0.009468448	0.005	4.73422E-05	0.09	0.000625575	0.175734	0.000762338

Figure 12.16i) Data for absolute volume flow rate

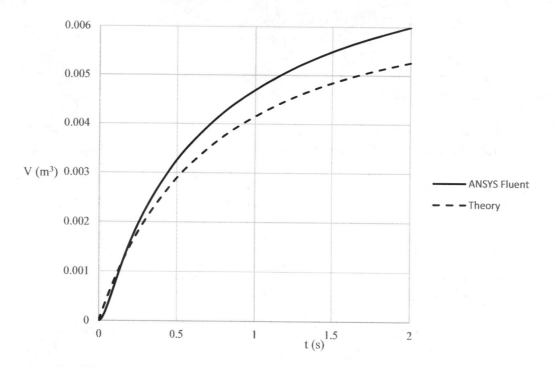

Figure 12.16j) Comparison Ansys Fluent and theory for volume produced

H. Theory

17. We can relate the mean fluid velocity v (m/s) through the porous medium to the pressure gradient dP/dx (Pa) using Darcy's law in the same way as we did in the former chapter

$$v = -\frac{\kappa}{\mu}\frac{dP}{dx} \qquad (12.1)$$

where κ (m^2) is permeability of the porous medium and μ (kg/m-s) is the dynamic viscosity of the fluid. Furthermore, we define the hydraulic conductivity or coefficient of permeability k (m/s) as

$$k = \kappa\frac{\rho g}{\mu} \qquad (12.2)$$

Where g (m/s^2) is acceleration due to gravity and ρ (kg/m^3) is the density of the fluid. We will also use the following equation from the ANSYS Fluent user manual

$$\frac{dP}{dx} = -(\frac{\mu v}{\alpha} + \frac{C_2\rho v^2}{2}) \qquad (12.3)$$

where $1/\alpha$ (1/m^2) is viscous resistance or inverse absolute permeability for the porous medium and C_2(1/m) is inertial resistance of the porous medium. When we combine these equations we get this expression for the permeability coefficient

$$k = \frac{g}{\frac{v C_2}{2} + \frac{\mu}{\alpha\rho}} \qquad (12.4)$$

The following values were used in the ANSYS Fluent simulations in this chapter.

Parameter	Value
C_2 (1/m)	0
$1/\alpha$ (1/m^2)	$9.76305 \cdot 10^8$
ρ (kg/m^3)	998.2
μ (kg/m-s)	0.001003
g (m/s^2)	9.81
k (m/s)	0.01
R (m)	0.1
h_0 (m)	0.2
d (m)	0.02
γ	2

Table 12.2 Parameters used in ANSYS Fluent simulations

For a cylindrical filter with a constant thickness d and constant radius R we have the schematic diagram as shown in Figure 12.17a).

Figure 12.17a) Geometry for cylindrical ceramic water filter

The height of the free surface $h(t)$ will decrease over time as water is filtered through the filter. The volumetric flow rate through the filter can be described by

$$\dot{V}(t) = \frac{k}{d}\left[\pi R^2 h + R \int_0^{2\pi} \int_0^h (h-z)dzd\theta\right] = \frac{k\pi R h}{d}(R + h) \tag{12.5}$$

We can alternatively describe the volumetric flow rate as

$$\dot{V}(t) = -\pi R^2 \frac{dh}{dt} \tag{12.6}$$

Combining these two equations we get the following first order ordinary differential equation

$$\frac{dh}{dt} + \frac{kh}{d}\left(1 + \frac{h}{R}\right) = 0 \tag{12.7}$$

with the following solution where we have used the initial condition $h(0) = h_0$

$$h(t) = \frac{h_0 R}{e^{kt/d}(R+h_0) - h_0}$$ (12.8)

The volume flow rate using equation (12.6) will be

$$\dot{V}(t) = \frac{(R+h_0)\pi R^3 k h_0 e^{kt/d}}{d(e^{kt/d}(R+h_0) - h_0)^2}$$ (12.9)

The total water volume produced can be expressed as

$$V(t) = \pi R^2 (h_0 - h(t))$$ (12.10)

and with the use of equation (12.8) we get

$$V(t) = \pi R^2 h_0 \left(1 - \frac{R}{e^{kt/d}(R+h_0) - h_0}\right)$$ (12.11)

We can alternatively make equation (12.7) non-dimensional using non-dimensional variables $h^* = \frac{h}{h_0}$ and $t^* = \frac{kt}{d}$ where h_0 is the initial height of fluid in the cylinder. This will transform equation (12.7) to the following equation after skipping*

$$\frac{dh}{dt} + h(1 + \gamma h) = 0$$ (12.12)

where $\gamma = \frac{h_0}{R} = 2$ is the aspect-ratio.

Using initial condition $h(0) = 1$ we get the following solution

$$h(t) = \frac{1}{e^t(1+\gamma) - \gamma} = \frac{1}{3e^t - 2}$$ (12.13)

Now we can express the volumetric flow rate in non-dimensional form as (after introducing $V^* = \frac{V}{\pi R^2 h_0}$ and skipping*)

$$\dot{V}(t) = -\frac{dh}{dt} = \frac{e^t(1+\gamma)}{[\gamma - e^t(1+\gamma)]^2} = \frac{3e^t}{(2-3e^t)^2}$$ (12.14)

The total water volume produced can be expressed as

$$V(t) = \pi R^2 (h_0 - h(t))$$ (12.15)

and in non-dimensional form this expression will be the following after skipping*

$$V(t) = 1 - h(t) = 1 - \frac{1}{e^t(1+\gamma) - \gamma} = \frac{3e^t - 3}{3e^t - 2}$$ (12.16)

For a frustum filter with a constant thickness d and varying radius r we have the schematic diagram as shown in Figure 12.17b).

Figure 12.17b) Geometry for frustum water filter

The volumetric flow rate in this case can be described as

$$\dot{V}(t) = \frac{k}{d}\left[\pi R^2 h + \int_0^{2\pi}\int_0^h (h-z)rdzd\theta\right] \tag{12.17}$$

Using $r = R + z\,tan\alpha$, we can express equation (12.13) as

$$\dot{V}(t) = \frac{k\pi h}{d}(R^2 + hR + \frac{h^2 tan\alpha}{3}) \tag{12.18}$$

and equation (12.6) for a frustum shaped filter will be modified to

$$\dot{V}(t) = -\pi R(t)^2 \frac{dh}{dt} \tag{12.19}$$

where $R(t) = R + h(t)\,tan\alpha$. The differential equation for $h(t)$ will then be

$$\frac{dh}{dt} + \frac{kh}{d}\frac{(R^2 + hR + \frac{h^2 tan\alpha}{3})}{(R + htan\alpha)^2} = 0 \tag{12.20}$$

and in non-dimensional form after skipping* we have

$$\frac{dh}{dt} + \frac{h}{(1+\delta h)^2}(1 + \gamma h + \frac{1}{3}\delta\gamma h^2) = 0 \tag{12.21}$$

where $\delta = \gamma\,tan\alpha$. There is no analytical solution available except in the case $\delta = 0$ corresponding to the cylindrical geometry. Equation (11.17) can to be solved numerically in the general case $\delta \neq 0$ using the initial condition $h(0) = 1$.

We express the volumetric flow rate in non-dimensional form after skipping*

$$\dot{V}(t) = -(1 + \delta h)^2 \frac{dh}{dt} \tag{12.22}$$

The total water volume produced can be expressed as

$$V(t) = \pi\left[R^2 + R(h_0 + h)tan\alpha + (h_0^2 + h_0 h + h^2)\frac{tan^2\alpha}{3}\right](h_0 - h) \tag{12.23}$$

and in non-dimensional form this expression will be the following after skipping[*]

$$V(t) = (1 - h) + \delta(1 - h^2) + \frac{\delta^2}{3}(1 - h^3)$$

(12.24)

I. References

1. Kelly, A.C., "Finite Element Modeling of Flow Through Ceramic Pot Filters.", *Master's Thesis in Civil and Environmental Engineering*, MIT, (2013).
2. Schweitzer, R.W., Cunningham, J.A., Mihelcic, J.R., "Hydraulic Modeling of Clay Ceramic Water Filters for Point-of-Use Water Treatment.", *Environ. Sci. Technol.*, **47**, 429-435, (2013).
3. Wald, I., "Modeling Flow Rate to Estimate Hydraulic Conductivity in a Parabolic Ceramic Water Filter.", *Undergraduate Journal of Mathematical Modeling*, **4**, 2, (2012).

J. Exercise

12.1 Complete ANSYS Fluent simulation of the flow through a frustum-shaped filter using the values listed in Table 12.3. Include graphs of your results showing a comparison between ANSYS Fluent and Theory for free-surface height, volume flow rate and volume produced by the filter. Use a time step size of 0.005 s and 200 time-steps with 50 time steps for each run. Include contours of volume fraction for water at t = 0, 0.25, 0.5, 0.75 and 1 second.

Parameter	Value
C_2 (1/m)	0
$1/\alpha$ (1/m^2)	$9.76305 \cdot 10^8$
ρ (kg/m^3)	998.2
μ (kg/m-s)	0.001003
g (m/s^2)	9.81
k *(m/s)*	0.01
R *(m)*	0.1
h_0 *(m)*	0.3
d *(m)*	0.01
γ	3
α *(degrees)*	10

Table 12.3 Parameter values for frustum shaped filter

CHAPTER 13. MODEL ROCKET FLOW

A. Objectives

- Model Geometry and Mesh
- Run Simulations for Turbulent Steady 3D Flow Past a Model Rocket in Ansys Fluent
- Determine Drag Coefficient for Model Rocket
- Visualize Flow Around Model Rocket

B. Problem Description

We will study the flow past the model rocket Estes Firestreak SST with a free stream velocity of 30 m/s.

C. Launching Ansys Workbench and Importing the Geometry

1. Start by launching Ansys Workbench. Select File>>Import from the menu. Select Geometry File as format and open the file *Estes Model Rocket.IGS*. You will need to select Geometry File as file format in order to open the file. Download this and other files from *sdcpublications.com*.

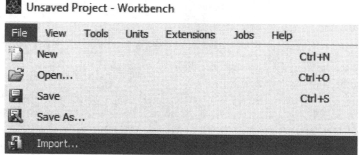

Figure 13.1 Import of geometry file

D. Launching Ansys DesignModeler

2. Right click on the imported Geometry in Ansys Workbench and select Edit Geometry in DesignModeler. Click on Generate in DesignModeler.

   Select Tools>>Enclosure from the menu. Set the Number of Planes to 2 in Details View. Select ZXPlane from the Tree Outline as Symmetry Plane 1 and Apply the plane in Details View. Select YZPlane from the Tree Outline as Symmetry Plane 2 and Apply it in Details View. Click on Generate. Right click in graphics window and select Zoom to Fit.

Figure 13.2a) Editing geometry

Figure 13.2b) Computational domain around the model rocket

3. Select Create>>Boolean from the menu. Select Subtract as Operation. Click on the plus sign next to 2 Parts, 2 Bodies in the Tree Outline. Select the second Solid and Apply it as Target Body in Details of Boolean1. Select the first Solid and Apply it as Tool Body in Details of Boolean1. Click on Generate. Right click on the Solid in the Tree Outline and Rename it to *Air*. Select File>>Save Project with the name *Model Rocket Flow Study* and close DesignModeler.

Figure 13.3 Details of body

E. Launching Ansys Meshing

4. Double click on Fluid Flow (Fluent) under Analysis Systems in Toolbox in Ansys
    Workbench. Drag the Geometry to the Geometry under Fluid Flow (Fluent). Double click
    on Mesh under Fluid Flow (Fluent). In the Meshing window, right click on Mesh in the
    Outline and select Update. Click on the plus sign next to Sizing under Details of Mesh.
    Change Capture Proximity and Capture Curvature both to Yes. Click on the plus sign
    next to Quality under Details of Mesh and set Smoothing to High. Update the mesh and
    rotate the Mesh to see the details of the mesh around the model rocket.

Figure 13.4a) Shared geometry with Fluent

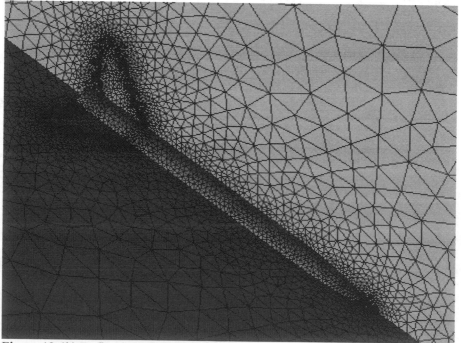

Figure 13.4b) Refined mesh around model rocket

5. Right click on Mesh in the Outline, select Insert>> Sizing. Select Face and control select the 5 faces of the model rocket as shown in figure 13.5a). Apply the faces as the Geometry under Scope in Details of "Face Sizing". Select the Home tab in the menu and Tools>>Units>>Metric (mm, kg, ...). Set the Element Size to 0.84347 mm. Generate the mesh once again.

Details of "Face Sizing" - Sizing	
Scope	
Scoping Method	Geometry Selection
Geometry	5 Faces
Definition	
Suppressed	No
Type	Element Size
Element Size	0.84347 mm
Advanced	
Defeature Size	Default (0.42174 mm)
Influence Volume	No
Behavior	Soft
Growth Rate	Default (1.2)
Capture Curvature	No
Capture Proximity	No

Figure 13.5a) Face sizing for the five faces of the model rocket

Figure 13.5b) Further refinement of mesh around model rocket

6. Click on the plus sign next to Inflation in Details of "Mesh". Set Use Automatic Inflation to Program Controlled and Generate the mesh once again. Select Mode Box Select from the menu of the graphics window, see Figure 13.6a). Select Face

     and make a Box Select around the model rocket, see Figure 13.6a). This will select all surfaces, see Figure 13.6b).

Figure 13.6a) Box Select   Figure 13.6b) Selection of surfaces

Right click in the graphics window and select Create Named Selection. Enter the name *model-rocket* and click OK. Select *model-rocket* under Named Selections and select *Include* from the drop-down menu for Program Controlled Inflation under Definition in Details of "model rocket". Generate the mesh once again.

Details of "model rocket"	
Scope	
Scoping Method	Geometry Selection
Geometry	18 Faces
Definition	
Send to Solver	Yes
Protected	Program Controlled
Visible	Yes
Program Controlled Inflation	Include
Statistics	
Type	Manual
☐ Total Selection	18 Faces
☐ Surface Area	5864.7 mm²
Suppressed	0
Used by Mesh Worksheet	No

Figure 13.6c) Details for model rocket named selection

Right click in the graphics window and select View>>Front. Right click in the graphics window and select Zoom to Fit. Select Mode Single Select from the menu of the graphics window, see Figure 13.6a), and select the square velocity-inlet face for the computational domain in the graphics window, see Figure 13.6d). Right click in the graphics window and select Create Named Selection. Enter the name *velocity-inlet* and click OK.

Right click in the graphics window and select View>>Back. Select Face 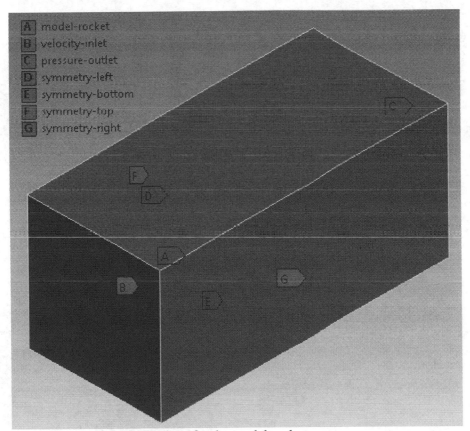 and select the square pressure-outlet face. Right click in the graphics window and select Create Named Selection. Enter the name *pressure-outlet* and click OK.

Right click in the graphics window and select View>>Left. Select Face and select the rectangular symmetry-left face. Right click in the graphics window and select Create Named Selection. Enter the name *symmetry-left* and click OK.

Right click in the graphics window and select View>>Bottom. Select Face and select the rectangular symmetry-bottom face. Right click in the graphics window and select Create Named Selection. Enter the name *symmetry-bottom* and click OK.

Right click in the graphics window and select View>>Top. Select Face and select the rectangular symmetry-top face. Right click in the graphics window and select Create Named Selection. Enter the name *symmetry-top* and click OK.

Right click in the graphics window and select View>>Right. Select Face and select the rectangular symmetry-right face. Right click in the graphics window and select Create Named Selection. Enter the name *symmetry-right* and click OK.

Figure 13.6d) Named selections for the model rocket

7. Go back to DesignModeler by double clicking on the Geometry under A in the Project Schematic in Ansys Workbench. Select Bottom view and select YZ Plane under Tree Outline in DesignModeler. Right click on YZ Plane and select Look At. Create a new sketch ![sketch icon]. Use your mouse wheel to zoom in on the model rocket.

Click on the Sketching tab in the Tree Outline and select Rectangle. Draw a rectangle around the model rocket. Make sure you start drawing the rectangle on the vertical axis and that you get a *C* when you move the cursor on the vertical axis. This means that one of the sides will be positioned on the same vertical axis. Select Dimensions and click on the horizontal and vertical sides of the new rectangle. Enter 0.6375 m as the vertical height of the rectangle and 0.1275 m as the horizontal width. Select the Vertical dimensioning tool, click on the horizontal coordinate axis and the lower side of the rectangle and enter 0.275 m as the vertical dimension.

Select the Modeling tab in the Tree Outline and select ![Extrude icon] Extrude. Select Sketch1 under YZ plane in the Tree Outline and Apply it as Geometry in Details of Extrude. Enter 0.1275 m as the FD1, Depth (>0) of the Extrusion. Click on Generate. Go to 2 Parts, 2 Bodies in the Tree Outline, right click on Solid and select Rename. Rename the Solid to *ModelRocketBox*.

Go back to Ansys Workbench, double click on Mesh and answer *Yes* to the question whether you would like to read upstream data.

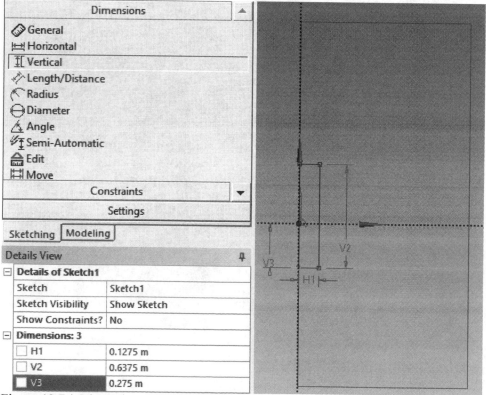

Figure 13.7a) Dimensions and location for rectangle around the model rocket

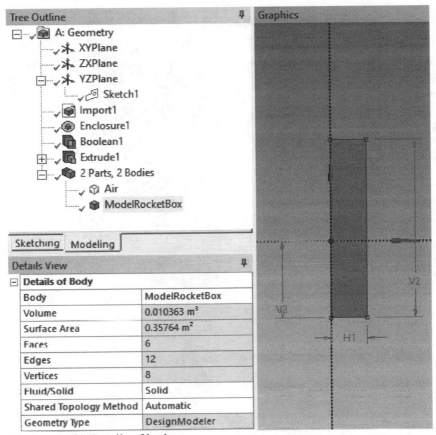

Figure 13.7b) Details of body

8. Right click Mesh under Project in Outline and select Update. Right click Mesh under Project in Outline, select Insert>>Sizing. Select Body 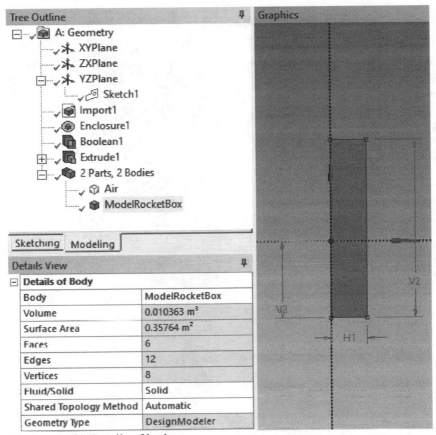 and select the larger computational domain in the graphics window. Apply the computational domain in Geometry under Scope for Details of Body Sizing. Select *Body of Influence* as Type under Definition.

Select Bottom view in the graphics window, zoom in on the model rocket, select the *ModelRocketBox* in the graphics window and Apply it as the Body of Influence. You will need to select the second blue small rectangle in the lower left corner of the graphics window as shown in Figure 13.8a). *ModelRocketBox* turns from green to red when you apply it as the *Body of Influence*. Update the Mesh.

Right click on *ModelRocketBox* under Geometry in Project Outline and select Hide Body. Select Mesh in the Outline and click on the plus sign next to Quality in Details of "Mesh". Select Skewness as Mesh Metric. The max skewness should not be above 0.9. In this study the max value is 0.89995, see Figure 13.8b).

Select File>>Save Project. Select File>>Export...>>Mesh>>FLUENT Input File>>Export and save the mesh with the name *model-rocket-flow-mesh.msh*. Close the Meshing window.

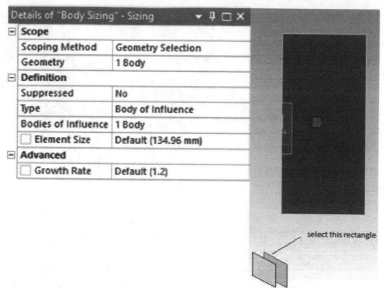

Figure 13.8a) Selection of ModelRocketBox as Body of Influence

Quality	
Check Mesh Quality	Yes, Errors
Target Skewness	Default (0.9)
Smoothing	High
Mesh Metric	Skewness
Min	6.8015e-004
Max	0.39995
Average	0.22903
Standard Deviation	0.13223

Figure 13.8b) Details on quality of the mesh

F. Launching Ansys Fluent

9. Double-click on Setup under Project Schematic in Ansys Workbench. Check the box for Double Precision. Set the number of Solver Processes equal to the number of processor cores for your computer. Ansys Student is limited to four Solver Processes. To check the number of physical cores, press the Ctrl + Shift + Esc keys simultaneously to open the Task Manager. Go to the Performance tab and select CPU from the left column. You will see the number of physical cores on the bottom-right side. Close the Task Manager window. Click on the Start button in the Fluent Launcher window.

   Click on Display… under Mesh and General on the Task Page. Select all Surfaces in the Mesh Display window. Check the boxes for Edges and Faces under Options. Select All as Edge Type and click on Display.

Figure 13.9a) Fluent launcher

Figure 13.9b) Displayed mesh with inflow and outflow

Double click on Models and Viscous (SST k-omega) under Setup in the Outline View.
Set the k-omega Model to Standard and click OK to close the Viscous Model window.

Viscous Model ✕

Model
- ○ Inviscid
- ○ Laminar
- ○ Spalart-Allmaras (1 eqn)
- ○ k-epsilon (2 eqn)
- ● k-omega (2 eqn)
- ○ Transition k-kl-omega (3 eqn)
- ○ Transition SST (4 eqn)
- ○ Reynolds Stress (7 eqn)
- ○ Scale-Adaptive Simulation (SAS)
- ○ Detached Eddy Simulation (DES)
- ○ Large Eddy Simulation (LES)

k-omega Model
- ● Standard
- ○ GEKO
- ○ BSL
- ○ SST

k-omega Options
- ☐ Low-Re Corrections
- ☑ Shear Flow Corrections

Options
- ☐ Curvature Correction
- ☐ Corner Flow Correction
- ☐ Production Kato-Launder
- ☑ Production Limiter

Model Constants

Alpha*_inf
`1`

Alpha_inf
`0.52`

Beta*_inf
`0.09`

Beta_i
`0.072`

TKE Prandtl Number
`2`

User-Defined Functions

Turbulent Viscosity
`none ▼`

Prandtl Numbers

TKE Prandtl Number
`none ▼`

SDR Prandtl Number
`none ▼`

[OK] [Cancel] [Help]

Figure 13.9c) Details for the viscous model

312

10. Double click on Boundary Conditions under Setup in the Tree and select *velocity-inlet* under Zone for Boundary Conditions on the Task Page. Click on the Edit button. Select Components as Velocity Specification Method. Enter -30 as Z-Velocity (m/s). Set the Turbulent Intensity to 1%. Select Apply and Close the Velocity Inlet window.

Figure 13.10 Velocity inlet boundary condition

11. Double click Projected Areas under Results and Reports in the Outline View. Select *model-rocket* as the Surface, enter 0.0001 as Min Feature Size [m], select Z as Projection Direction and click on Compute. The computed frontal Area [m²] is 0.0001021013. Close the Projected Surface Areas window.

Figure 13.11a) Projected surface areas

Double click on Reference Values under Setup in the Outline View and change the Area [m²] to 0.0001021013. Select Compute from *velocity-inlet* and set Reference Zone to *air*.

Double click on Methods under Solution in the Outline View and set the Pressure-Velocity Coupling scheme to Coupled. Set the Spatial Discretization Gradient to Least Squares Cell Based, and Pressure to Standard. Set the remaining three (Momentum, Turbulent Kinetic Energy and Specific Dissipation Rate) to First Order Upwind. Uncheck the Auto Select box for Flux Type and select Rhie-Chow: distance-based Flux Type.

Figure 13.11b) Reference values

Figure 13.11c) Solution Methods

12. Double click on Controls under Solution in the Outline View and set the Pressure and Momentum Pseudo Transient Explicit Relaxation Factors to 0.25. Set the Turbulent Viscosity Explicit Relaxation Factor to 0.8. Click on Limits… and set the Maximum Turb. Viscosity Ratio to 1e+07. Click on the OK button to close the Solution Limits window.

Figure 13.12a) Solution controls

Figure 13.12b) Solution limits

314

13. Double click on Monitors and Residual under Solution. Check the box for Show Advanced Options. Set the Convergence Criterion to none. Click on OK to close Residual Monitors.

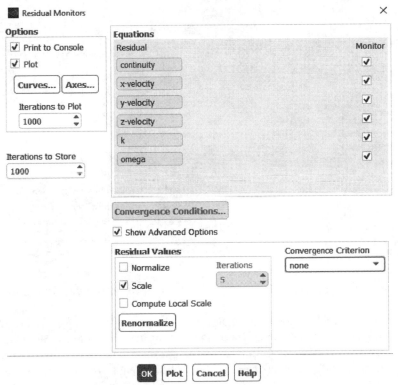

Figure 13.13 Settings for Residual Monitors

14. Double-click on Report Definitions under Solution in the Outline View. Select New>>Force Report>>Drag… from the drop-down menu. Select *model-rocket* as Zone. Check the boxes for Report File, Report Plot and Print to Console under Create. Set the Force Vector to X = 0, Y = 0, and Z = -1. Click on the OK button to close the Drag Report Definition window. Close the Report Definitions window.

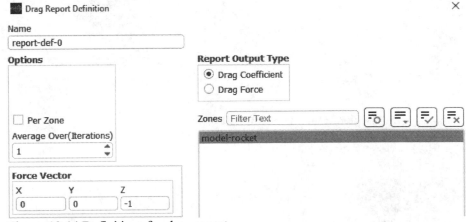

Figure 13.14 Definitions for drag report

15. Double click on Initialization under Solution in the Outline View and select *Hybrid Initializa*tion. Click on the Initialize button. Double click on Calculation Activities under Solution in the Outline View and click on Edit... for Autosave Every (Iterations). Set Save Data File Every (Iterations) to 100. Click OK to close the Autosave window.

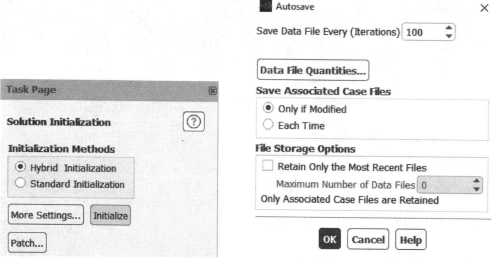

Figure 13.15a) Solution initialization   Figure 13.15b) Autosave window

Double click on Run Calculation under Solution in the Outline View. Set the Number of Iterations to 100. Click on Calculate. When calculations are complete, click on Copy Screenshot of Active Window to Clipboard in the graphics window, see Figure 13.15d). The scaled residuals and drag coefficient plots can be pasted into a Word document.

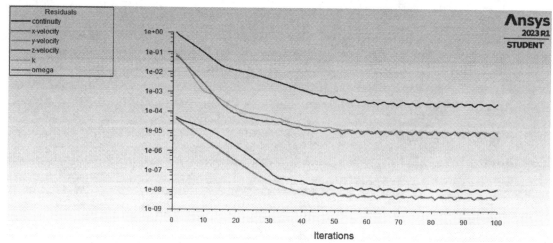

Figure 13.15c) Scaled residuals for model rocket

Figure 13.15d) Copying screenshot

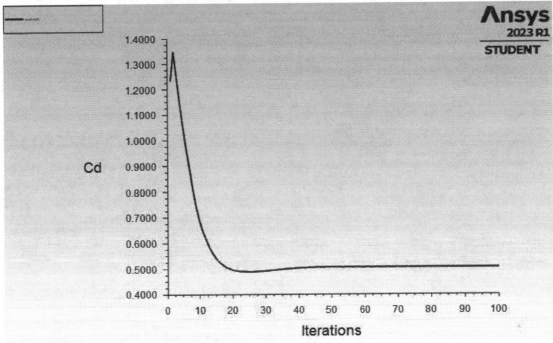

Figure 13.15e) Drag coefficient for model rocket

G. Post-Processing

16. Double click on Methods under Solution in the Outline View and replace First Order Upwind with Second Order Upwind for Momentum, Turbulent Kinetic Energy and Specific Dissipation Rate.

    Double click on Controls under Solution and set the Turbulent Viscosity value to 0.95. Double click on Run Calculation under Solution and set Number of Iterations to 300. Click on Calculate and Click OK in the window that appears.

    Double-click on Reports under Results in the Outline View and double-click on Forces under Reports. Set Direction Vector for X to 0, Y to 0, and Z to -1. Select the *model-rocket* Wall Zone and click on the Print button. The total drag-coefficient is 0.43666. Close the Force Reports window.

```
Forces - Direction Vector (0 0 -1)
                         Forces [N]                                    Coefficients
Zone                     Pressure        Viscous         Total         Pressure        Viscous         Total
model-rocket             0.003924703     0.020651738     0.024576441   0.069731164     0.36692452      0.43665568
-------------------      ---------       ----------      ----------    ----------      ----------      ----------
---------------
Net                      0.003924703     0.020651738     0.024576441   0.069731164     0.36692452      0.43665568
```

Figure 13.16a) Drag-coefficient from console

When calculations are complete, click on Copy Screenshot of Active Window to Clipboard in the graphics window, see Figure 13.15d). The scaled residuals and drag coefficient plots can be pasted into a Word document.

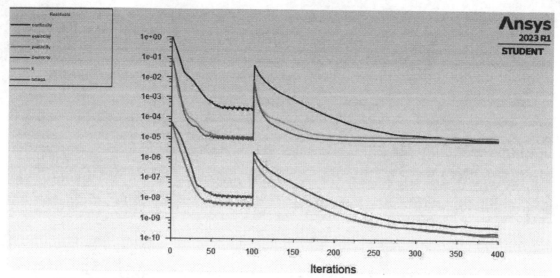

Figure 13.16b) Scaled residuals for model rocket

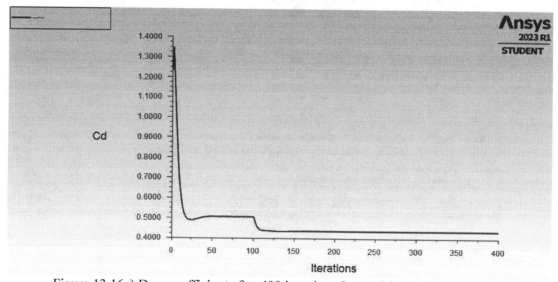

Figure 13.16c) Drag coefficient after 400 iterations for model rocket

17. Double click on Graphics and Contours under Results. Select the settings as shown in Figure 13.17a) and Apply the Colormap settings as shown in Figure 13.17b). Click on Save/Display. Select the Y-Z plane view in the graphics window. Zoom in on the rocket. Close the Contours window.

Figure 13.17a) Contours settings

Figure 13.17b) Colormap settings

Figure 13.17c) Velocity magnitude for the flow around the model rocket at 30 m/s

Double click on Pathlines under Results and Graphics in the Outline View. Select the settings and Apply the same Colormap Options and settings as shown in Figures 13.17b) and 13.17d). Click on Save/Display. Close the Pathlines window.

Figure 13.17d) Settings for pathlines

Figure 13.17e) Pathlines around model rocket at 30 m/s

Double click on Contours under Results and Graphics in the Outline View. Select the settings as shown in Figure 13.17f) for static pressure. Apply the same Colormap settings as shown in Figures 13.17b). Click on Save/Display. Close the Contours window.

Figure 13.17f) Settings for static pressure

Figure 13.17g) Static pressure for the flow past the model rocket at 30 m/s

H. Batch Jobs

18. You will need to create the case and scheme files to complete a batch job. Select File>>Export>>Case & Data… from the menu. Save the case and data files in the *Working Directory* folder with the name ***model_rocket.casII.h5***. Open Notebook and create and save the scheme file with the name ***model_rocketII.scm*** for the batch job. The scheme file is shown in Figure 13.18a) and is available for download from *sdcpublications.com*. Select File>>Read>>Scheme… from the menu in Ansys Fluent. Open the scheme file to start the batch job. Table 13.1 and Figure 13.18b) are showing the results from the batch job in comparison with theory.

```scheme
(Do ((x 70 (- x 5))) ((< x 5))
(begin
    (Ti-menu-load-string (format #f "file read-case model-rocket.cas.h5"))
    (Ti-menu-load-string (format #f "define boundary-conditions velocity-inlet velocity-inlet n y y n 0 y n 0 n 0 n -~a n n y 1 10" x))
    (Ti-menu-load-string (format #f "report reference-values velocity ~a" x))
    (Ti-menu-load-string (format #f "solve set discretization-scheme mom 0"))
    (Ti-menu-load-string (format #f "solve set discretization-scheme k 0"))
    (Ti-menu-load-string (format #f "solve set discretization-scheme omega 0"))
    (Ti-menu-load-string (format #f "solve set pseudo-time-method relaxation-factors turb-viscosity 0.8"))
    (Ti-menu-load-string (format #f "solve initialize hyb-initialization yes"))
    (Ti-menu-load-string (format #f "solve iterate 100"))
    (Ti-menu-load-string (format #f "solve set discretization-scheme mom 1"))
    (Ti-menu-load-string (format #f "solve set discretization-scheme k 1"))
    (Ti-menu-load-string (format #f "solve set discretization-scheme omega 1"))
    (Ti-menu-load-string (format #f "solve set pseudo-time-method relaxation-factors turb-viscosity 0.95"))
    (Ti-menu-load-string (format #f "solve iterate 1 ok 300"))
    (Ti-menu-load-string (format #f "report forces wall-forces y 0 0 -1 y model_rocket_drag_coefficient_data_U=~amps" x))
    (Ti-menu-load-string (format #f "display objects display contour-1"))
    (Ti-menu-load-string (format #f "display views restore-view right"))
    (Ti-menu-load-string (format #f "display save-picture model_rocket_static_pressure_contour_plot_U=~amps.jpg" x))
    (Ti-menu-load-string (format #f "display objects display contour-2"))
    (Ti-menu-load-string (format #f "display views restore-view right"))
    (Ti-menu-load-string (format #f "display save-picture model_rocket_velocity_magnitude_contour_plot_U=~amps.jpg" x))
    (Ti-menu-load-string (format #f "display objects display pathlines-1"))
    (Ti-menu-load-string (format #f "display views restore-view right"))
    (Ti-menu-load-string (format #f "display save-picture model_rocket_path_lines_U=~amps.jpg" x))
    (Ti-menu-load-string (format #f "plot residuals y y y y y y"))
    (Ti-menu-load-string (format #f "display save-picture model_rocket_residuals_plot_U=~amps.jpg" x))
    (Ti-menu-load-string (format #f "plot file report-def-0-rfile.out"))
    (Ti-menu-load-string (format #f "display save-picture model_rocket_drag_coefficient_plot_U=~amps.jpg" x))
))
```

Figure 13.18a) Scheme file for model rocket batch job

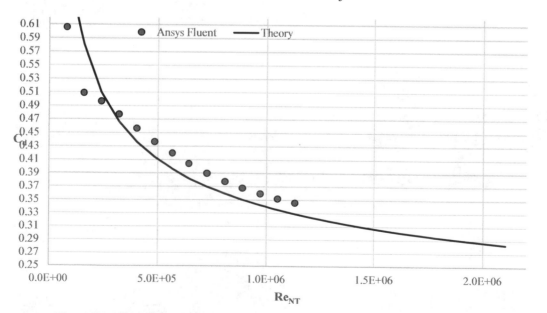

Figure 13.18b) Drag coefficient versus Reynolds number for Ahmed body

I. Theory

19. The total drag-coefficient for a model rocket[1,2] can be expressed using equation (13.1).

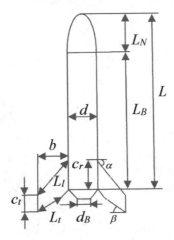

Figure 13.18 Geometry for model rocket

$$C_{D,T} = C_{D,NT} + C_{D,B} + C_{D,F} + C_{D,I} + C_{D,L} \tag{13.1}$$

, where

$$C_{D,NT} = 1.02 \frac{C_{f,turb} A_{w,NT}}{A_c} \left(1 + \frac{3}{2}\left(\frac{L}{d}\right)^{-3/2}\right) \tag{13.2}$$

$$C_{D,B} = \frac{0.029}{\sqrt{C_{D,NT}}} \left(\frac{d_b}{d}\right)^3 \tag{13.3}$$

$$C_{D,F} = \frac{2C_{f,lam} A_{w,F}}{A_c} \left(1 + 2\frac{t}{c_r}\right) \tag{13.4}$$

$$C_{D,I} = \frac{C_{f,lam} n d c_r}{A_c} \left(1 + 2\frac{t}{c_r}\right) \tag{13.5}$$

The different drag coefficients and parameters are explained in Table 13.1. We define the Reynolds number for the nose cone and body tube as

$$Re_{NT} = \frac{UL\rho}{\mu} = \frac{30*0.2365*1.225}{0.000017894} = 485,714 \tag{13.6}$$

, and the Reynolds number for the fin as

$$Re_F = \frac{U c_{ave} \rho}{\mu} = \frac{30*0.0271525*1.225}{0.000017894} = 55,765 \tag{13.7}$$

, where $c_{ave} = S/b$. The surface area for the fin can be determined from the geometry in Figure 13.19.

$$S = (c_t + L_l \, sin\alpha)L_l \, cos\alpha - \frac{1}{2}\{L_l^2 \, sin\alpha \, cos\alpha + L_t^2 \, sin\beta \, cos\beta + (L_l \, cos\alpha - L_t \, cos\beta)[(c_t - c_r + L_l \, sin\alpha - L_t \, sin\beta) + 2L_t \, sin\beta]\} \tag{13.8}$$

323

The skin friction coefficient can be approximated assuming fully turbulent flow on nose cone and body tube

$$C_{f,turb} = \frac{0.0315}{Re_{NT}^{1/7}} = \frac{0.0315}{485,714^{1/7}} = 0.00485255 \tag{13.9}$$

, and for a laminar boundary layer on the fin as

$$C_{f,lam} = \frac{1.328}{Re_F^{1/2}} = \frac{1.328}{55,765^{1/2}} = 0.00562365 \tag{13.10}$$

For an ogive nose cone we have the following relation

$$\frac{A_{w,NT}}{A_c} = \frac{8L_N + 12L_B}{3d} = 39.2012 \tag{13.11}$$

The different drag coefficient and the total value becomes

$$C_{D,NT} = 1.02 * 0.00485255 * 39.2012 \left(1 + \frac{3}{2}(10.8986)^{-\frac{3}{2}}\right) = 0.2021 \tag{13.12}$$

$$C_{D,B} = \frac{0.029}{\sqrt{0.202}} \left(\frac{0.0164}{0.0217}\right)^3 = 0.0278 \tag{13.13}$$

$$C_{D,F} = \frac{2 * 0.00562 * 0.00419}{0.00037} \left(1 + 2\frac{0.0007}{0.0038}\right) = 0.1321 \tag{13.14}$$

$$C_{D,I} = \frac{0.00562 * 0.038 * 4 * 0.0217}{0.00037} \left(1 + 2\frac{0.0007}{0.0038}\right) = 0.0520 \tag{13.15}$$

$$C_{D,T} = 0.2021 + 0.0278 + 0.1321 + 0.0520 = 0.414 \tag{13.16}$$

The value for the total drag coefficient 0.414 according to the theory can be compared with the value 0.4354 from Ansys Fluent simulations, a difference of 5.2 %.

A_c (m^2): cross sectional area for body tube 0.000369836 m²
$A_{w,F}$ (m^2): total wetted surface area for all fins 0.00418879 m²
$A_{w,NT}$ (m^2): wetted surface area for nose and body tube 0.014498 m²
α : fin leading edge sweep angle 50°
β : fin trailing edge angle 15°
b (m): span for the fin 0.0385673 m
c_{ave} (m): average fin chord length 0.0271525 m
c_r (m): root fin chord length 0.038 m
c_t (m): tip fin chord length 0.010 m
$C_{D,NT}$: nose and body tube drag coefficient 0.2021
$C_{D,B}$: base drag coefficient 0.0278
$C_{D,F}$: fin drag coefficient 0.1321
$C_{D,I}$: interference drag coefficient 0.0520
$C_{D,T}$: total drag coefficient 0.414
$C_{f,lam}$: skin friction coefficient for laminar boundary layer 0.00562365
$C_{f,turb}$: skin friction coefficient for turbulent boundary layer 0.00485255
d (m): body tube diameter 0.0217 m
d_b (m): base diameter 0.0164 m

L (m): length of nose and body 0.2365 m	
L_B (m): length of body tube 0.165 m	
L_l (m): length of leading edge of fin 0.060 m	
L_N (m): length of nose 0.0715 m	
L_t (m): length of trailing edge of fin 0.033 m	
n: number of fins 4	
Re_F: Reynolds number for the fin 55,765	
Re_{NT}: Reynolds number for nose cone and body tube 485,714	
S (m²): area for fin 0.0010472 m²	
t (m): thickness of fin 0.0007 m	
$U(m/s)$: free stream velocity 30 m/s	
$\rho(kg/m^3)$: density of air 1.225 kg/m³	
$\mu(kg/ms)$: dynamic viscosity of air 0.000017894kg/m³	

Table 13.1 Drag coefficients and parameters

J. References

1. DeMar, J.S.,"Model Rocket Drag Analysis Using a Computerized Wind Tunnel", NARAM-37, 1995.
2. Gregorek, G.M., "Aerodynamic Drag of Model Rockets.", Estes Industries, Penrose, CO, 1970.
3. Cannon, B.,"Model Rocket Simulation with Drag Analysis", BYU, 2004.
4. Milligan, T.V, "Determining the Drag Coefficient of a Model Rocket Using Accelerometer Based Payloads", NARAM-54, 2012.

K. Exercise

13.1 Rerun ANSYS Fluent simulations in this chapter at 10, 20, 40 and 50 m/s and compare the drag coefficient with values using the theory section. Include a plot of total drag coefficient versus Reynolds number Re$_{NT}$ corresponding to free stream velocities 10, 20, 30, 40 and 50 m/s and fill out Table 12.2 below.

U (m/s)	Re_{NT}	Re_F	$C_{D,NT}$	$C_{D,B}$	$C_{D,F}$	$C_{D,I}$	$C_{D,T}$	$C_{D,Fluent}$	% Diff.
10									
20									
30	485714	55765	0.2021	0.0278	0.1321	0.0520	0.414	0.437	5.6
40									
50									

Table 13.2 Drag coefficients at different Reynolds numbers

L. Notes

CHAPTER 14. AHMED BODY

A. Objectives

- Model the Flow Past an Ahmed Body
- Run Simulations for Turbulent Steady 3D Flow Past the Ahmed Body
- Determine the Drag Coefficient for the Ahmed Body
- Visualize the Flow Around Ahmed Body using Pathlines, Pressure and Velocity

B. Problem Description

Studying the flow around the Ahmed body is a standard flow test case for wind tunnel experiments and CFD codes[1-6]. The geometry in *mm* that is being used for the Ahmed body in this chapter is shown below.

C. Ahmed Body Design

1. Start by launching Ansys Workbench. Double click on Fluid Flow (Fluent) under Analysis Systems in the Toolbox. Right-click on Geometry in the Project Schematic window and select New DesignModeler Geometry. Select Units>>Millimeter from the menu as the desired length unit.

    Select Look At Face ![icon]. Select XY Plane in the Tree Outline and New Sketch ![icon]. Select the Sketching tab, click on Draw and select Rectangle. Draw a rectangle from the origin in the first quadrant. Select Dimensions and set the height of the rectangle to 288 mm and the length to 1044 mm. Right click in the graphics window and select Zoom to Fit.

Figure 14.1 First step in the creation of the 3D Ahmed body

2. Select Modify and Fillet under Sketching Toolboxes. Set the Radius of the Fillet to 100 mm. Click on the two left hand side corners of the rectangle.

    Click on Chamfer under Modify. Set the Length of the Chamfer to 201.2 mm. Click on the upper right corner of the rectangle. The rear slant angle is 25 degrees. Select Dimensions and set the Angle to 25 degrees. You may need to right-click and select alternate angle when you define the angle. Select General under Dimensions and select the chamfer to set the length to 201.2 mm, see Figure 14.2a).

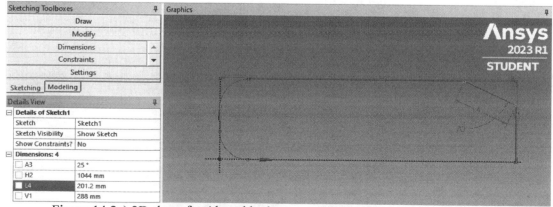

Figure 14.2a) 2D shape for Ahmed body

Click on Extrude and select Sketch 1 in the Tree Outline under XYPlane. Apply the sketch in Details View next to Geometry. Set the Direction for the Extrude to Both - Symmetric in Details View and set the depth of the extrusion to 194.5 mm. Click on Generate. Select Blend>>Fixed Radius from under the menu. Set the Radius to 100 mm. Control select the two vertical edges at the front of the Ahmed body and Apply the edges next to Geometry in Details View. Click on Generate.

Figure 14.2b) Extruded Ahmed body

Figure 14.2c) Extruded and blended Ahmed body

3. Select ZXPlane in the Tree Outline. Right click in the graphics window and select View>>Bottom View. Select the Sketching tab in Tree Outline and select Circle. Draw a circle and use General under Dimensions to define a 30 mm diameter circle that you position 202 mm from the front of the Ahmed body and 163.5 mm from the center plane.

Create three more circles with the same diameter and position them according to Figure 14.3a).

Figure 14.3a) Locations for the four cylindrical supports

Click on **Extrude** and select Sketch 2 in the Tree Outline under ZXPlane. Apply the sketch in Details View next to Geometry. Set the Direction under Details View to Reversed and set the depth of the extrusion to 50 mm. Click on Generate.

Figure 14.3b) Details view for extrusion and finished Ahmed model

Select Create>> New Plane from the menu. Select in Details View From Plane as Type, ZXPlane as the Base Plane and Offset Z as Transform 1 (RMB).
Set the Offset Z value to -50 mm. Click on Generate.

Select Tools>>Enclosure from the menu. Select Box as the Shape in Details View and set the Number of Planes to 2. Select XYPlane as Symmetry Plane 1, Plane4 as Symmetry Plane 2 and Apply them in Details View. Set the Model Type to Full Model and the Cushion to Non-Uniform. Set the different Cushion values FD1 – FD4 as listed in Figure 14.3c). Click on Generate.

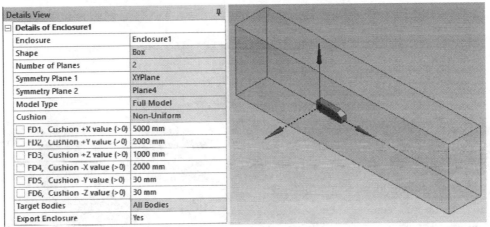

Figure 14.3c) Details for enclosure

4. Select Create>>Boolean from the menu. Select Subtract as Operation. Click on the plus sign next to 2 Parts, 2 Bodies in the Tree Outline. Select the second Solid and Apply it as Target Body in Details of Boolean1. Select the first Solid and Apply it as Tool Body in Details of Boolean1. Click on Generate. Right click on the Solid in the Tree Outline and Rename it to Air. Select File>>Save Project and name it 3D Ahmed Body Study and close DesignModeler.

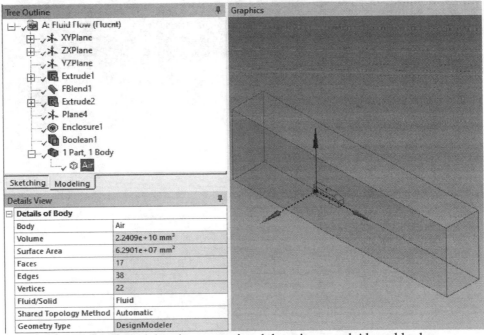

Figure 14.4 Finished model and computational domain around Ahmed body

D. Ahmed Body Mesh

5. Double click on Mesh under Project Schematic in Ansys Workbench. In the Meshing window, right click on Mesh in the Outline and select Update. Select the Home tab in the menu. Select Layout>>Reset Layout under the Home tab.

    Click on the plus sign next to Sizing under Details of "Mesh". Change Capture Proximity and Capture Curvature both to Yes. Click on the plus sign next to Quality under Details of "Mesh". Set Smoothing under Quality in Details of Mesh to High. Update the mesh once again. Right click in the graphics window and select View>>Back.

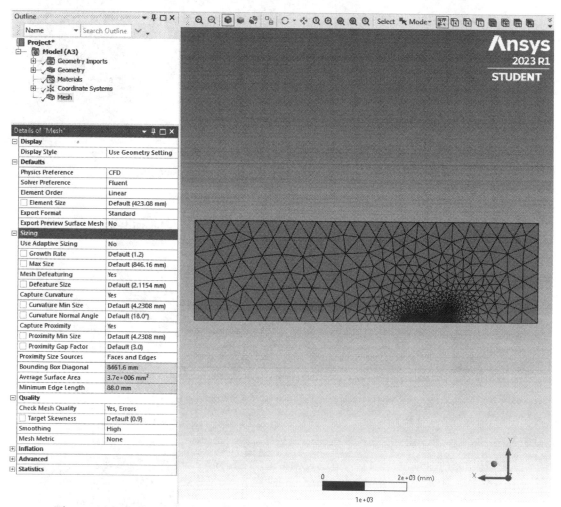

Figure 14.5 Refined mesh around Ahmed body

6. Right click on Mesh in the Outline, select Insert>> Sizing. Right click in the graphics window and select Cursor Mode>>Face. Select Mode>>Box Select from above the graphics window. Create a box around the Ahmed body to select all 11 faces of the Ahmed body as shown in figure 14.6a). Apply the faces as the Geometry under Scope in Details of "Face Sizing". Set the Element Size to 7.5 mm. Update the mesh once again.

Figure 14.6a) Face sizing for the eleven faces of the Ahmed body

Figure 14.6b) Refined mesh after face sizing

7. Click on the plus sign next to Inflation in Details of Mesh. Set Use Automatic Inflation to Program Controlled. Right click in the graphics window and select Cursor Mode>>Vertex. Create a box around the Ahmed body to select the 14 vertices as shown in Figure 14.7a).

Figure 14.7a) Program controlled inflation and selected vertices for Ahmed body

Right click in the graphics window and select Create Named Selection. Enter the name *ahmed body* and click OK. Select *ahmed body* under Named Selections and select Include from the drop-down menu for Program Controlled Inflation under Definition in Details of *ahmed body*, see Figure 14.7b). Select Mode>>Single Select from above the graphics window.

Figure 14.7b) Program controlled inflation set to include for Ahmed body

8. Right click in the graphics window and select View>>Left. Right click in the graphics window and select Cursor Mode>>Face. Select the face of the computational domain that is available from the Left View. Right click in the graphics window and select Create Named Selection. Enter the name *velocity-inlet* and select OK.

Right click in the graphics window and select View>>Right. Select the face of the computational domain that is available from the Right View. Right click in the graphics window and select Create Named Selection. Enter the name *pressure-outlet* and select OK.

Right click in the graphics window and select View>>Back. Select the face of the computational domain that is available from the Back View. Right click in the graphics window and select Create Named Selection. Enter the name *symmetry1* and select OK.

Right click in the graphics window and select View>>Bottom. Select the face of the computational domain that is available from the Bottom View. Right click in the graphics window and select Create Named Selection. Enter the name *street* and click OK.

Right click in the graphics window and select View>>Top. Select the face of the computational domain that is available from the Top View. Right click in the graphics window and select Create Named Selection. Enter the name *symmetry-top* and click OK.

Right click in the graphics window and select View>>Front. Select the face of the computational domain that is available from the Front View. Right click in the graphics window and select Create Named Selection. Enter the name *symmetry-side* and click OK. Update the mesh.

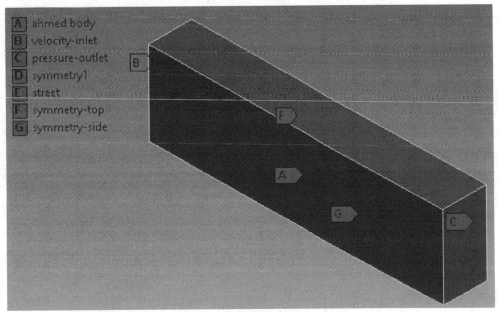

Figure 14.8 Named selections for the Ahmed body

9. Go back to DesignModeler by double clicking on Geometry under Project Schematic in Ansys Workbench. Right-click in the graphics window and select Back View. Select XYPlane in the Tree Outline and create a new Sketch.

Click on the Sketching tab in the Tree Outline and select Rectangle. Draw a rectangle around the Ahmed body from the bottom of the computational domain. Select Dimensions and click on the vertical and horizontal sides of the new rectangle. Enter 900 mm as the vertical height of the rectangle and 2700 mm as the width. Set the horizontal distance from the front of the Ahmed body to the right-hand side vertical edge of the rectangle to 552 mm. Set the vertical distance between the lower horizontal edge of the rectangle and the bottom of the Ahmed body to 50 mm.

Select the Modeling tab in the Tree Outline and select Extrude. Select the new Sketch3 under XYPlane in the Tree Outline and Apply it as Geometry in Details of Extrude3. Enter 450 mm as the FD1, Depth (>0) of the Extrusion. Click on Generate. Click on plus sign next to 2 Parts, 2 Bodies in the Tree Outline, right click on Solid and select Rename. Rename Solid to *AhmedBodyBox*. Close the DesignModeler window.

Go back to Ansys Workbench, double click on Mesh and answer *Yes* to the question whether you would like to read upstream data.

Figure 14.9a) Dimensions for rectangle surrounding the Ahmed body

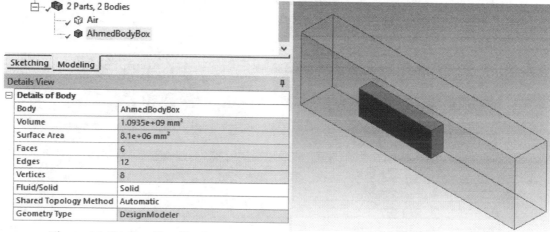

Figure 14.9b) Details of body

10. Right click Mesh under Project in Outline, select Insert>>Sizing. Right click in the graphics window and select Cursor Mode>>Body. Select the larger computational domain in the graphics window. Apply the computational domain in Geometry under Scope in Details of "Sizing". Select Body of Influence as Type under Definition.

Select Back view in the graphics window, zoom in on the Ahmed body, select the AhmedBodyBox and Apply it as the Body of Influence. You will need to select the second small rectangle in the lower left corner in Figure 14.10a) before you can select the AhmedRocketBox that turns from green to red when you apply it as the body of influence. Update the Mesh. If you click on the plus sign next to Statistics in Details of Mesh you will find that the number of Elements for the mesh is 496373 and the number of Nodes is 142843.

Right click on AhmedBodyBox under Geometry in Outline and select Hide Body. Select Mesh in the Outline and click on the plus sign next to Quality in Details of Mesh. Select Skewness as Mesh Metric. The max skewness should not be above 0.9 (max is 0.76867). Select File>>Save Project. Select File>>Export>>Mesh>>FLUENT Input File>>Export and save the mesh with the name *ahmed-body-flow-mesh.msh*. Close the Meshing window.

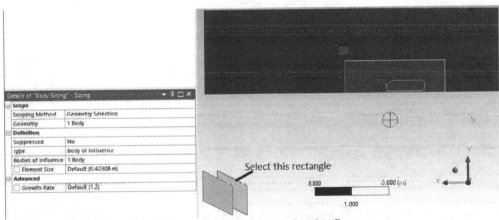

Figure 14.10a) Selection of AhmedBodyBox as Body of Influence

Figure 14.10b) Finished mesh for Ahmed body

337

E. Launching Ansys Fluent for 3D Ahmed Body

11. Double click on Setup in Ansys Workbench. Check Double Precision under Options and set the number of Processes equal to the number of cores for the computer. Click Start to launch Fluent.

    Double click on Models under Setup in the Outline View and double click on Viscous – SST k-omega under Models on the Task Page. Select the k-epsilon (2 eqn) turbulence model. Select Realizable as the k-epsilon Model and check Non-Equilibrium Wall Functions as Near-Wall Treatment. Click OK to exit the Viscous Model window.

    Double click on Boundary Conditions under Setup in the Outline View and select *velocity-inlet* under Zone for Boundary Conditions on the Task Page. Click on the Edit… button. Select Components as Velocity Specification Method. Enter 30 as X-Velocity (m/s). Click Apply and close the Velocity Inlet window.

Figure 14.11 Velocity inlet specifications for X-Velocity

12. Double click Projected Areas under Results and Reports in the Outline View. Select *wall-air* as the Surface, enter 0.030 as Min Feature Size (m), select X as Projection Direction and click on Compute. The computed frontal area is 0.06175669 m2. Close the window.

Figure 14.12a) Projected surface area

Double click on Reference Values under Setup in the Outline View and change the Area (m2) to 0.06175669. Select Compute from velocity-inlet and set Reference Zone to air. Double click on Methods under Solution in the Outline View and set the Pressure-Velocity Coupling scheme to Coupled. Uncheck the box for Auto-Select and select Flux Type as Rhie-Chow: distance based. Set the Spatial Discretization Gradient to Least Squares Cell Based, and Pressure to Standard. Set the remaining three for Momentum, Turbulent Kinetic Energy and Turbulent Dissipation Rate to First Order Upwind. Select Pseudo Time Method as Off.

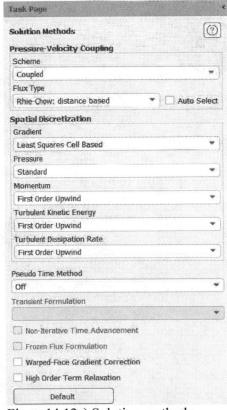

Figure 14.12b) Reference values   Figure 14.12c) Solution methods

13. Double click on Controls under Solution in the Outline View, set the Flow Courant Number to 50, set the Momentum and Pressure Explicit Relaxation Factors to 0.25 and set the Turbulent Viscosity Under-Relaxation Factor to 0.8. Click on Limits... and set the Maximum Turb. Viscosity Ratio to 10000000. Click on the OK button to close the window.

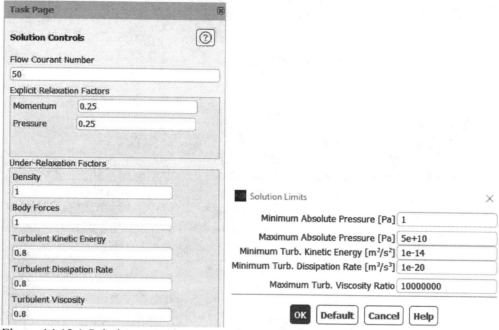

Figure 14.13a) Solution controls   Figure 14.13b) Solution limits

Double click on Monitors and Residual under Solution in the Outline View. Check the box for Show Advanced Options and set the Convergence Criterion to none. Click on OK to close Residual Monitors.

Figure 14.13c) Residual monitors

Double-click on Report Definitions on the left-hand side under Solution in the Tree. Select New>>Force Report>>Drag… from the drop-down menu. Select *wall-air* as Wall Zone. Check the boxes for Report File, Report Plot and Print to Console under Create. Set the Force Vector to X = 1, Y = 0, and Z = 0. Click on the OK button to close the window. Close the Report Definitions window.

Figure 14.13d) Drag report definition

14. Double click on Initialization under Solution in the Outline View and select *Hybrid Initializa*tion. Click on the Initialize button. Double click on Calculation Activities under Solution in the Outline View and click on Edit… for Autosave Every (Iterations). Set Save Data File Every (Iterations) to 100. Click OK to close the Autosave window. Double click on Run Calculation under Solution in the Outline View. Set the Number of Iterations to 200. Click on Calculate.

Double click on Methods under Solution in the Outline View and replace First Order Upwind with Second Order Upwind for Momentum, Turbulent Kinetic Energy and Turbulent Dissipation Rate. Double click on Controls under Solution and set the Turbulent Viscosity value to 0.95. Double click on Run Calculation under Solution and set Number of Iterations to 500. Click on Calculate.

Figure 14.14a) Scaled residuals during the first 700 iterations

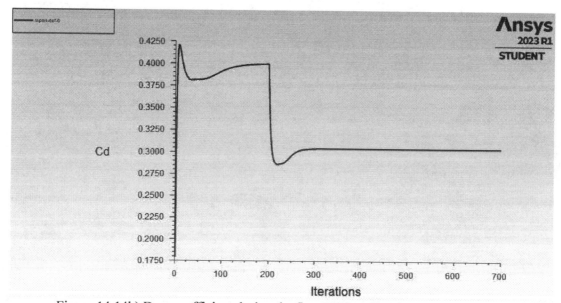

Figure 14.14b) Drag coefficient during the first 700 iterations

F. Post-Processing for 3D Ahmed Body

15. Double-click on Reports under Results in the Outline View and double-click on Forces under Reports. Set Direction Vector for X to 1, Y to 0, and Z to 0. Select the *wall-air* Wall Zone, deselect *street* Wall Zone and click on the Print button. The total drag coefficient is 0.30392925. Close the Force Reports window.

Double-click on Graphics and Contours under Results in the Outline View. Select Contours of Pressure… and Static Pressure. Select *symmetry1* as Surfaces. Select Colormap Options… and set Colormap Alignment to Top. Set Number Format Type to float and Precisions to 0. Set Font Behavior to Fixed and Font Size to 12. Select Apply and Close the Colormap window. Click on Save/Display.

Figure 14.15a) Static pressure for Ahmed body at 30 m/s.

Double-click on Graphics and Contours under Results in the Outline View. Select Contours of Velocity… and Velocity Magnitude. Select *symmetry1* as the Surface. Use the same Colormap Options… as listed above for the pressure plot and click on Save/Display.

Figure 14.15b) Velocity magnitude around Ahmed body at 30 m/s.

343

Double-click on Graphics and Pathlines under Results in the Outline View. Select the settings as shown in Figure 14.15c). Select *symmetry1* as the Surface. Use the same Colormap Options... as listed above for the pressure plot and click on Save/Display.

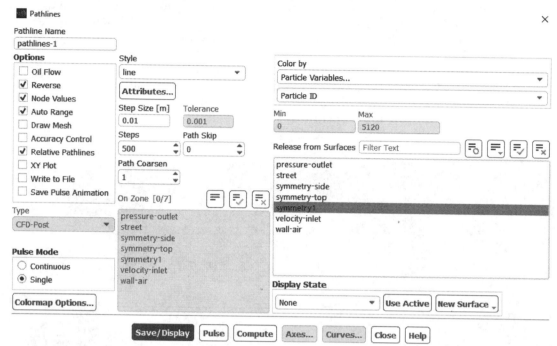

Figure 14.15c) Settings for pathlines

Figure 14.15d) Pathlines around Ahmed body at 30 m/s.

G. Batch Jobs

16. You will need to create the case and scheme files to complete a batch job. Select File>>Export>>Case & Data… from the menu. Save the case and data files in the *Working Directory* folder with the name ***ahmed_body.cas.h5***. Open Notebook and create and save the scheme file with the name ***ahmed_body.scm*** for the batch job. The scheme file is shown in Figure 14.16a) and is available for download from *sdcpublications.com*. Select File>>Read>>Scheme… from the menu in Ansys Fluent. Open the scheme file to start the batch job. Table 14.1 and Figure 14.16b) are showing the results from the batch job in comparison with Meile et al. (2011).

```
(Do ((x 70 (- x 5))) ((< x 5))
(begin
    (Ti-menu-load-string (format #f "file read-case ahmed_body.cas.h5"))
    (Ti-menu-load-string (format #f "define boundary-conditions velocity-inlet velocity-inlet n n y y n ~a n 0 n n y 5 10" x))
    (Ti-menu-load-string (format #f "report reference-values velocity ~a" x))
    (Ti-menu-load-string (format #f "solve set discretization-scheme mom 0"))
    (Ti-menu-load-string (format #f "solve set discretization-scheme k 0"))
    (Ti-menu-load-string (format #f "solve set discretization-scheme epsilon 0"))
    (Ti-menu-load-string (format #f "solve set under-relaxation turb-viscosity 0.8"))
    (Ti-menu-load-string (format #f "solve initialize hyb-initialization yes"))
    (Ti-menu-load-string (format #f "solve iterate 200"))
    (Ti-menu-load-string (format #f "solve set discretization-scheme mom 1"))
    (Ti-menu-load-string (format #f "solve set discretization-scheme k 1"))
    (Ti-menu-load-string (format #f "solve set discretization-scheme epsilon 1"))
    (Ti-menu-load-string (format #f "solve set under-relaxation turb-viscosity 0.95"))
    (Ti-menu-load-string (format #f "solve iterate 1 ok 500"))
    (Ti-menu-load-string (format #f "report forces wall-forces n 1 () 1 0 0 y ahmed_body_drag_coefficient_data_U=~amps" x))
    (Ti-menu-load-string (format #f "display objects display contour-1"))
    (Ti-menu-load-string (format #f "display views restore-view front"))
    (Ti-menu-load-string (format #f "display save-picture ahmed_body_static_pressure_contour_plot_U=~amps.jpg" x))
    (Ti-menu-load-string (format #f "display objects display contour-2"))
    (Ti-menu-load-string (format #f "display views restore-view front"))
    (Ti-menu-load-string (format #f "display save-picture ahmed_body_velocity_magnitude_contour_plot_U=~amps.jpg" x))
    (Ti-menu-load-string (format #f "display objects display pathlines-1"))
    (Ti-menu-load-string (format #f "display views restore-view front"))
    (Ti-menu-load-string (format #f "display save-picture ahmed_body_path_lines_U=~amps.jpg" x))
    (Ti-menu-load-string (format #f "plot residuals y y y y y"))
    (Ti-menu-load-string (format #f "display save-picture ahmed_body_residuals_plot_U=~amps.jpg" x))
    (Ti-menu-load-string (format #f "plot file report-def-0-rfile.out"))
    (Ti-menu-load-string (format #f "display save-picture ahmed_body_drag_coefficient_plot_U=~amps.jpg" x))
))
```

Figure 14.16a) Scheme file for batch job

U (m/s)	Re	Cd, Fluent	Cd, Meile *et al.*	Percent difference
5	357,354	0.34725594	0.35380	1.8
10	714,709	0.3180707	0.34027	6.5
15	1,072,063	0.31175134	0.32919	5.3
20	1,429,418	0.30911536	0.32010	3.4
25	1,786,772	0.30619663	0.31265	2.1
30	2,144,127	0.30392925	0.30655	0.9
35	2,501,481	0.30247244	0.30155	0.3
40	2,858,835	0.30108136	0.297444	1.2
45	3,216,190	0.30007601	0.294082	2.0
50	3,573,544	0.2989811	0.291326	2.6
55	3,930,899	0.29811084	0.289067	3.1
60	4,288,253	0.29739094	0.287216	3.5
65	4,645,607	0.29649293	0.285698	3.8
70	5,002,962	0.29597577	0.28445	4.1

Table 14.1 Comparison between Ansys Fluent and established data

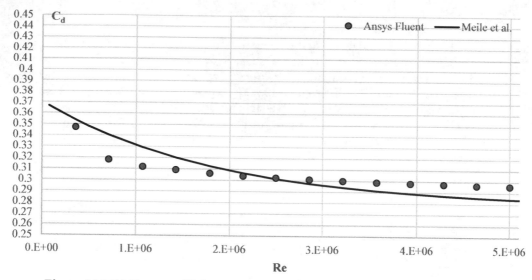

Figure 14.16b) Drag coefficient versus Reynolds number for Ahmed body

H. Theory

We start by defining the Reynolds number for the 3D case outlined in this chapter

$$Re = \frac{UL\rho}{\mu} \tag{14.1}$$

where U (m/s) is inlet velocity, L (m) is the length of the Ahmed body, ρ (kg/m³) is the density of the fluid and μ (m²/s) is dynamic viscosity of the fluid.
Next, we define the drag coefficient C_d as

$$C_d = \frac{F_d}{\frac{1}{2}\rho U^2 A} \tag{14.2}$$

where F_d(N) is the drag force and A (m²) is the frontal area of the Ahmed body. Meile *et al.*[6] determined the following relationship between the drag coefficient and the Reynolds number for the Ahmed body.

$$C_d = 0.2788 + 0.0915e^{-Re/1797100} \tag{14.3}$$

I. References

1. Ahmed, S.R., Ramm, G.,"Some Salient Features of the Time-Averaged Ground Vehicle Wake", *SAE-Paper 840300*, 1984.
2. Dogan, T., Conger, M., Kim, D-H., Mousaviraad, M., Xing, T., Stern, F.,"Simulation of Turbulent Flow over the Ahmed Body", *ME: 5160 Intermediate Mechanics of Fluids CFD Lab 4*, (2016).

3. Kalyan, D.K., Paul, A.R., "Computational Study of Flow around a Simplified 2D Ahmed Body", *International Journal of Engineering Science and Innovative Technology (IJESIT)*, **2**, 3, (2013).

4. Khan, R.S, Umale, S.,"CFD Aerodynamic Analysis of Ahmed Body", *International Journal of Engineering Trends and Technology (IJETT)*, **18**, 7, (2014).

5. Lienhart, H., Stoots, C. and Becker, S.,"Flow and Turbulence Structures in the Wake of a Simplified Car Model (Ahmed Model), *Notes on Numerical Fluid Mechanics*, **77**, 6, (2002).

6. Meile, W., Brenn, G., Reppenhagen, A., Lechner, B., Fuchs, A.,"Experiments and numerical simulations on the aerodynamics of the Ahmed body", *CFD Letters*, **3**, 1, (2011).

J. Exercise

14.1 Model the 3D flow past the Ahmed body with the dimensions as shown and units in mm from Figure 14.17. Follow the same steps as shown in this chapter. Find the drag curve corresponding to Figure 14.16 and compare the two rear slant angles 25 and 35 degrees in the same graph.

14.2 Model the 3D flow past the Ahmed body with the dimensions as shown and units in mm from Figure 14.18. Follow the same steps as shown in this chapter. Find the drag curve corresponding to Figure 14.16.

Figure 14.17 Dimensions of Ahmed body and computational domain

Figure 14.18 Dimensions of Ahmed body

CHAPTER 15. HOURGLASS

A. Objectives

- Using Ansys Workbench to Create Geometry and Mesh for the Hourglass
- Inserting Boundary Conditions
- Running the Calculations
- Using Contour Plots for Volume Fractions

B. Problem Description

We will study the flow in the hourglass and will analyze the problem using Ansys Fluent.

C. Launching Ansys Workbench and Selecting Fluent

1. Start by launching Ansys Workbench. Double click on Fluid Flow (Fluent) under Analysis Systems in the Toolbox.

Figure 15.1 Selecting Fluent

D. Launching Ansys DesignModeler

2. Right click Geometry under Project Schematic in Ansys Workbench and select
   Properties. In Properties of Schematic A2: Geometry, select Analysis Type 2D under
   Advanced Geometry Options. Right click on Geometry in the Project Schematic window
   and select New DesignModeler Geometry.

Figure 15.2 Selecting properties and 2D analysis type

3. Select Units>>Millimeter from the menu in DesignModeler. Select the XY Plane in the

   Tree Outline and click on Look at Face/Plane/Sketch [icon]. Select the Sketching tab
   and select Polyline under Draw from Sketching Toolboxes. Start at the origin (make sure
   that letter *P* is shown when the cursor is close to the origin and that letters *V*, *H* and *C* are
   showing when you are drawing vertical, horizontal and coinciding lines) and draw the
   sketch with the dimensions as shown in Figure 15.3a). Right click and select *Closed End*
   after drawing the last line. Select General under Dimensions and select the eight lines
   (H1, V2-V5 and L6-L9) as shown in Figure 15.3a). Select Angle under Dimensions and
   create angles A10 and A11 by selecting the lines next to the angle. You may need to right
   click and select *Alternate Angle* when creating the angle.

Figure 15.3a) Selection of units

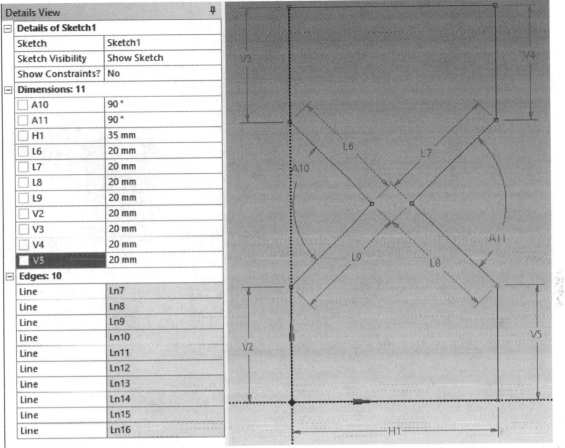

Figure 15.3b) Dimensions for hourglass

Select Concept>>Surfaces from Sketches from the menu. Select Sketch1 under XYPlane in Tree Outline and Apply the sketch as the Base Object in Details View. Click on Generate and close DesignModeler.

Figure 15.3c) Completed hourglass in DesignModeler

E. Launching Ansys Meshing

4. Double click on Mesh under Project Schematic in Ansys Workbench. Set the Element Size under Defaults in Details of Mesh to 0.5 mm. Right click on Mesh under Project in Outline and select Generate Mesh. Click on the plus sign next to Statistics in Details of "Mesh". The number of Nodes is 8310 and the number of Elements is 8090.

Figure 15.4a) Completed mesh for the hourglass

Right click in the graphics window and select Cursor Mode>> Edge and select the upper horizontal edge of the mesh. The edge turns green. Right click on the edge and select Create Named Selection. Enter *outlet* as the name and click OK to close the Selection Name window. Control select the remaining nine edges of the mesh, right click and select Create Named Selection. Name these edges *wall*.

Select File>>Export>>Mesh>>FLUENT Input File>>Export from the menu and save the mesh with the name *hourglass-mesh.msh*. Select File>>Save Project from the menu and save the project with the name *Hour-Glass.wbpj*. Close the Meshing window and right-click on Mesh under Project Schematic in ANSYS Workbench and select Update.

Figure 15.4b) Named selections for the hourglass

F. Launching Ansys Fluent

5. Double-click on Setup under Project Schematic in Ansys Workbench. Check the box for Double Precision under Solver Options in the Ansys Fluent Launcher. Set the number of processes equal to the number of processor cores for your computer. To check the number of physical cores, press the Ctrl + Shift + Esc keys simultaneously to open the Task Manager. Go to the Performance tab and select CPU from the left column. You will see the number of physical cores on the bottom-right side. Close the Task Manager window. Click on the Start button in the Fluent Launcher window.

Figure 15.5a) Fluent launcher

Click on Display… under Mesh and General on the Task Page. Select all Surfaces in the Mesh Display window. Check the boxes for Edges and Faces under Options. Select All as Edge Type, click on Display and Close the Mesh Display window. Select Transient under Time on the Task Page. Check the box for Gravity and enter -9.81 as the value for Y [m/s^2].

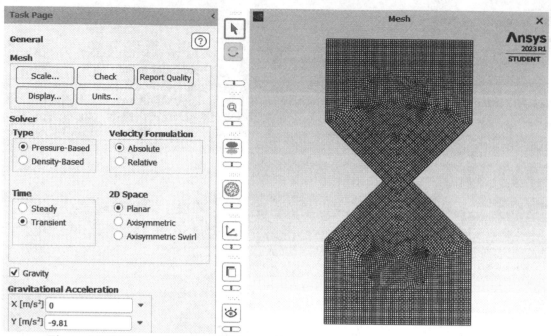

Figure 15.5b) General settings in Ansys Fluent

6. Double click on Models and Multiphase under Setup in the Outline View. Select Eulerian Inhomogeneous Model. Set the Number of Eulerian Phases to 2 and Implicit Formulation for Volume Fraction Parameters. Click Apply and Close the Multiphase Model window. Double click on Models and Viscous (SST, k-omega, Mixture) under Setup in the Outline View. Select Laminar Model and click OK to close the Viscous Model window.

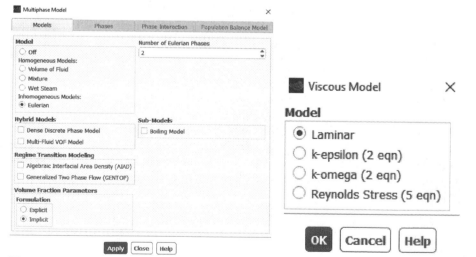

Figure 15.6a) Settings for the Eulerian model   Figure 15.6b) Selecting laminar model

7. Open Materials and Fluid under Setup in the Outline View and select *air*. Right click on *air* and select copy. Enter the name *wood* and set the Density [kg/m³] to 850. Click on Change/Create and answer Yes to the question that appears. Close the Create/Edit Materials window.

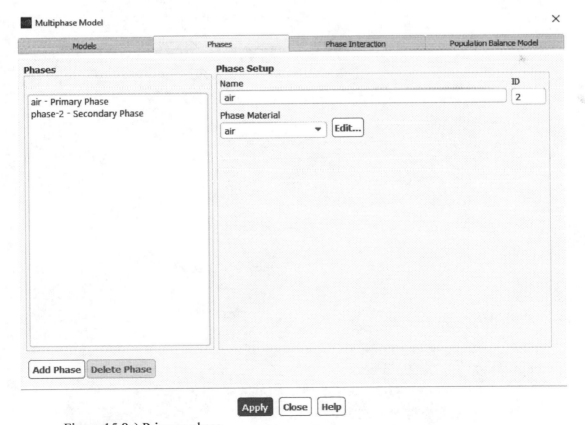

Figure 15.7 Settings for wood

8. Double click on Multiphase under Models in the Setup and select the Phases tab at the top of the Multiphase Model window. Select phase- 1 – Primary Phase under Phases. Select *air* as Phase Material and enter *air* as the Name for the Primary Phase. Click on Apply.

Figure 15.8a) Primary phase

Select phase- 2 – Secondary Phase under Phases. Select *wood* as Phase Material and enter *wood* as the Name for the Secondary Phase. Check the box for Granular and enter 0.0002 as Diameter [m]. Set the Granular Properties as shown in Table 15.1 and Figure 15.8b). Click on Apply and Close the window.

Figure 15.8b) Secondary phase settings

Diameter [m]	0.0002
Granular Viscosity [kg/m s]	gidaspow
Granular Bulk Viscosity [kg/m s]	lun-et-al
Frictional Viscosity [kg/m s]	schaeffer
Angle Of Internal Friction [deg]	45
Frictional Pressure [Pa]	based-ktgf
Frictional Modulus [Pa]	derived
Friction Packing Limit	0.45
Granular Temperature [m²/s²]	algebraic
Solids Pressure [Pa]	lun-et-al
Radial Distribution	lun-et-al
Elasticity Modulus [Pa]	derived
Packing Limit	0.63

Table 15.1 Settings for secondary phase

9. Select the Phase Interaction tab at the top of the Multiphase Model window. Select *air wood* under Phase Pairs and set Drag Coefficient, Lift Coefficient, Surface Tension Coefficient and Virtual Mass Coefficient to *none*. Select Apply and *select wood wood* under Phase Pairs and set the Restitution Coefficient to 0.001. Select Apply and Close the Multiphase Model window.

Figure 15.9a) Settings for air-wood phase pair

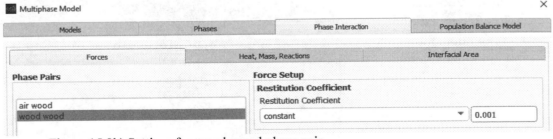

Figure 15.9b) Settings for wood-wood phase pair

10. Double-click on Boundary Conditions under Setup and click on Operating Conditions under Boundary Conditions on the Task Page. Select user-input as Operating Density Method. Set Reference Pressure Location X [m] to 0.0175 and Y [m] to 0.06828427. Click OK to close the Operating Conditions window.

Figure 15.10 Settings for operating conditions

11. Double-click on Methods under Solution and set the Spatial Discretization for Momentum to Second Order Upwind. Set the Spatial Discretization for Pressure to Second Order. Double-click on Controls under Solution and set the Under-Relaxation Factor for Momentum to 0.3.

Figure 15.11a) Solution methods

Figure 15.11b) Under-relaxation factors

Double click on Residual under Monitors and Solution in the Outline View. Set Absolute Criteria for continuity, u-air and v-air to 1e-05. Set the other Absolute Criteria to 0.001 and click OK to close the Residual Monitors window.

Figure 15.11c) Residual monitors

12. Double-click on Initialization under Solution in the Outline View and set Initializations Method to Hybrid Initialization on the Task Page. Click on Initialize, see Figure 15.12a).

Figure 15.12a) Hybrid initialization

Select Adapt>>Automatic... from the Domain tab in the menu, see Figure 15.12b). Select Cell Registers>>New>>Field Variable... in the Automatic Mesh Adaption window, see Figure 15.12c).

Figure 15.12b) Automatic adaption

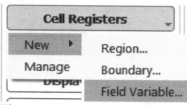

Figure 15.12c) Selecting new field variable

In the Field Variable Register, select Gradient as Derivative Option, select *wood* as Phase and select Gradient of Phases... and Volume Fraction. Select Cells Outside Standard Deviation as Type. Click on Save and Close the Field Variable Register window. Select gradient_0 as Refinement Criterion and set the Frequency (time-step) to 1 in the Automatic Mesh Adaption window, see Figure 15.12e).

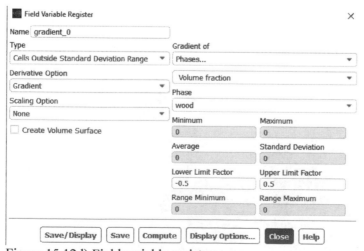

Figure 15.12d) Field variable register

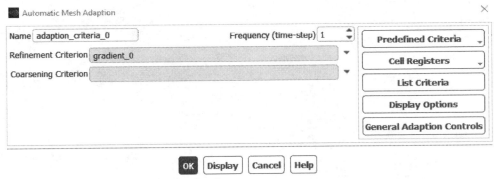

Figure 15.12e) Automatic mesh adaption

Select Cell Registers>>New>>Region… in the Adaption Mesh Adaption window. Set the X Max [m] value to 0.035 and Y Min [m] value to 0.034 and Y Max [m] value to 0.04. Click on Save and Close the Region Register window. Click OK to close the Automatic Mesh Adaption window. Click OK in the Information window.

Figure 15.12f) Region register

Click on Patch… under Solution Initialization on the Task Page. Select *wood* as the Phase, select Volume Fraction as Variable, region_0 as Registers to Patch and Value to 0.63. Select Patch and Close the Patch window.

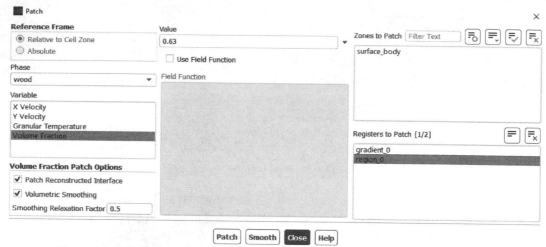

Figure 15.12g) Patch window

13. Double-click on Contours under Graphics and Results in the Outline View. Select wood as Phase and select Contours of Phases… and Volume fraction. Deselect all Surfaces. Select Colormap Options…. Set Font Behavior under Font to Fixed and Font Size to 12. Select float as Type under Number Format and set Precision to 2. Click on Apply and Close the Colormap window. Click on Save/Display and Close the Contours window. Click on Copy Screenshot of Active Window to Clipboard in the graphics window, see Figure 15.13c). The contour plot can be pasted into a Word document.

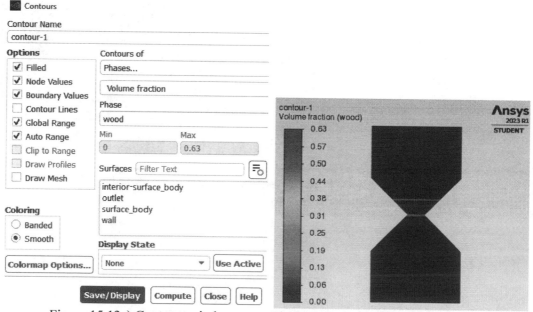

Figure 15.13a) Contours window   Figure 15.13b) Volume fraction (wood)

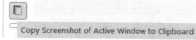

Figure 15.13c) Copying screenshot

14. Double-click on Solution Animations under Solution and Calculation Activities in the Outline View. Select Record after every 1 time-step. Select HSF File as Storage Type. Select contour-1 as Animation Object. Enter *hour-glass* as Name and click OK to close the Animation Definition window.

Figure 15.14 Contours window

15. Open Cell Registers under Solution in the Outline View and double click on gradient_0. Select Compute in the Field Variable Register window, see Figure 15.15a). Select Display Options in the Field Variable Register window. Use the settings as shown in Figure 15.15b). Select Save/Display in the Field Variable Register window to see the cells in the graphics window that have a gradient in this region. Close the Field Variable Register window.

Figure 15.15a) Computed values for field variable register

363

Figure 15.15B) Cell register display options   Figure 15.15c) Adaption cells

16. Double-click on Run Calculation under Solution in the Outline View. Enter 0.001 as Time Step Size [s] under Parameters in Time Advancement on the Task Page. Enter 40 as Number of Time Steps, set Max Iterations/Time Step to 30, see Figure 15.16a). Select File>>Export>>Case... from the menu and save the Case File with the name *hourglass.cas.h5*. Click on Calculate under Run Calculation on the Task Page. After the first set of calculations are complete, continue running the calculations four more times with the same settings, see Figure 15.16b). Click on Copy Screenshot of Active Window to Clipboard in the graphics window, see Figure 15.13c). The contour plot can be pasted into a Word document in between calculations.

Figure 15.16a) Calculations   Figure 15.16b) Scaled residuals for the first 40 ms.

G. Post-Processing

Double click on Animations and Playback under Results in the Outline View. Uncheck the box for Use Stored View, set the Replay Speed to low and select Play Once as Playback Mode. Select Video File as Write/Record Format. Click on the Write button for your chosen Animation Sequence. This will create the movie. Close the Playback window. The movie can be viewed using for example Windows Media Player.

$t = 0$ s $t = 0.04$ s $t - 0.08$ s $t = 0.12$s

$t = 0.16$ s $t = 0.2$ s $t = 0.24$ s $t = 0.28$ s

Figure 15.16b) Volume fraction (wood) at t = 0, 0.04, 0.08, 0.12, 0.16, 0.2 and 0.28 s.

H. Theory

17. We start with continuity, momentum and energy equations.

$$\frac{\partial}{\partial t}(\alpha_s \rho_s) + \nabla \cdot (\alpha_s \rho_s \bar{u}_s) = \dot{m}_{fs} \tag{15.1}$$

$$\frac{\partial}{\partial t}(\alpha_s \rho_s \bar{u}_s) + \nabla \cdot (\alpha_s \rho_s \bar{u}_s \bar{u}_s) = -\alpha_s \nabla p_f + \nabla \cdot \tau_s + \sum_{s=1}^{n}(\bar{R}_{fs} + \dot{m}_{fs}\bar{u}_{fs}) + \bar{F}_s \tag{15.2}$$

$$\frac{3}{2}\left[\frac{\partial}{\partial t}(\alpha_s \rho_s \theta_s) + \nabla \cdot (\alpha_s \rho_s \bar{u}_s \theta_s)\right] = \tau_s : \nabla \bar{u}_s + \nabla \cdot (\kappa_\theta \nabla \theta_s) - \gamma_s + \phi_{lm} + \phi_{fs} \tag{15.3}$$

where

$$\tau_s = -P_s \bar{I} + 2\alpha_s \mu_s \bar{S} + \alpha_s (\lambda_s - \frac{2}{3}\mu_s)\nabla \cdot \bar{u}_s \bar{I} \tag{15.4}$$

$$\bar{S} = \frac{1}{2}[\nabla \cdot \bar{u}_s + (\nabla \cdot \bar{u}_s)^T] \tag{15.5}$$

The total granular solids phase pressure is the sum of kinetic and frictional pressures

$$P_s = P_{s,friction} + P_{s,kinetic} \tag{15.6}$$

The Frictional Pressure (pascal) in ANSYS Fluent is the frictional part of equation (15.6). The following expression was given by Johnson and Jackson[4].

$$P_{s,friction} = \frac{\alpha_s}{10}\frac{(\alpha_s - \alpha_{s,min})^2}{(\alpha_{s,max} - \alpha_s)^5} \tag{15.7}$$

Alternatively, we have the expression by Syamlal et al[7].

$$P_{s,friction} = 2\alpha_s^2 \rho_s g_0 \theta_s (1 + e_s) \tag{15.8}$$

where the radial distribution function

$$g_0 = \frac{1}{1-(\frac{\alpha_s}{\alpha_{s,max}})^{1/3}} \tag{15.9}$$

The Solids Pressure (pascal) in ANSYS Fluent is the kinetic portion of granular solids phase pressure and can be described using the following relation by Lun et al[5].

$$P_{s,kinetic} = \alpha_s \rho_s \theta_s [1 + 2\alpha_s g_0 (1 + e_s)] \tag{15.10}$$

The granular phase shear viscosity can be described as the sum of the collisional, kinetic and frictional parts.

$$\mu_s = \mu_{s,collision} + \mu_{s,kinetic} + \mu_{s,friction} \tag{15.11}$$

The collisional part of the solids phase shear viscosity was modeled by Gidaspow et al.[1] and Syamlal et al[7].

$$\mu_{s,collision} = \frac{4}{5}\alpha_s^2\rho_s d_s g_0(1+e_s)\sqrt{\frac{\theta_s}{\pi}} \tag{15.12}$$

The Granular Viscosity (kg/m-s) in ANSYS Fluent is the kinetic part of equation (15.11). The following expression was given by Gidaspow et al[1].

$$\mu_{s,kinetic} = \frac{10\sqrt{\pi\theta_s}\rho_s d_s}{96(1+e_s)g_0}\left[1+\frac{4}{5}(1+e_s)g_0\alpha_s\right]^2 \tag{15.13}$$

Alternatively, we have the expression by Syamlal et al[7].

$$\mu_{s,kinetic} = \frac{\alpha_s\sqrt{\pi\theta_s}\rho_s d_s}{6(3-e_s)}\left[1+\frac{2}{5}(1+e_s)(3e_s-1)g_0\alpha_s\right] \tag{15.14}$$

The Frictional Viscosity (kg/m-s) in ANSYS Fluent is the frictional part of equation (15.11). The equation is given by Schaetter[6]

$$\mu_{s,friction} = \frac{P_{s,friction}\sin\varphi}{2\sqrt{I_2}} \tag{15.15}$$

The Granular Bulk Viscosity (kg/m-s) in ANSYS Fluent accounts for resistance of the granular material to compression and expansion. The equation is given by Lun et al[5].

$$\lambda_s = \frac{4}{3}\alpha_s^2\rho_s d_s g_0(1+e_s)\sqrt{\frac{\theta_s}{\pi}} \tag{15.16}$$

α_s	volume fraction for granular material
$\alpha_{s,min}$	friction packing limit
$\alpha_{s,max}$	packing limit
φ	internal angle of friction
γ_s	energy dissipation
κ_θ	granular temperature conductivity
λ_s	bulk viscosity
μ_s	granular shear viscosity
$\mu_{s,collision}$	collisional viscosity
$\mu_{s,friction}$	frictional viscosity
$\mu_{s,kinetic}$	kinetic viscosity
ϕ_{fs}	energy exchange between fluid and granular phase
ϕ_{lm}	energy exchange between granular phases
ρ_s	density
θ_s	temperature
τ_s	stress tensor
d_s	particle diameter
e_s	coefficient of restitution
g_0	radial distribution function
I_2	second invariant of the deviatoric stress tensor

$\dot{m}_{fs}$	unidirectional mass transfer
p_f	fluid pressure
P_s	granular pressure
$P_{s,friction}$	frictional pressure
$P_{s,kinetic}$	kinetic pressure
$\bar{S}$	strain rate
t	time
$\bar{u}_s$	granular velocity vector

Table 15.2 Parameters used in theory section

I. References

1. Gidaspow, D., Bezburuah, R., Ding, J. "Hydrodynamics of Circulating Fluidized Beds: Kinetic Theory Approach", *7th Fluidization Conference*, May 3, (1992).
2. Hirt, C.W.," A Flow-3D Continuum Model for Granular Media.", Flow Science Report 02-13, (2013).
3. Jansson, M., "CFD Simulations of Silos Content, An Investigation of Flow Patterns and Segregation Mechanisms.", *Master's Thesis within the Innovative and Sustainable Chemical Engineering Programme*, Chalmers, (2014).
4. Johnson, P.C., Jackson. R.," Frictional-Collisional Constitutive Relations for Granular Materials, with Application to Plane Shearing". *J. Fluid Mech.* **176**. (1987).
5. Lun, C.K.K., Savage, S.B., Jeffrey, D.J., Chepurniy, N., "Kinetic Theories for Granular Flow: Inelastic Particles in Couette Flow and Slightly Inelastic Particles in a General Flow Field". *J. Fluid Mech.* **140**, (1984).
6. Schaeffer, D.G., "Instability in the Evolution Equations Describing Incompressible Granular Flow.", *Journal of Differential Equations*, **66**, (1987).
7. Syamlal, M., Rogers, W., O'Brien, T.J. "MFIX Documentation Theory Guide", DOE/METC-94/1004 (1993).

J. Exercises

15.1 Use ANSYS Fluent to simulate an axisymmetric cylindrical heap of sand that deforms due to gravity into a conical pile. Use an initial 100 mm in diameter and 100 mm high heap of sand resting on a horizontal surface with a volume fraction packing limit of 0.63, see Hirt[2]. Use the density 1638 kg/m3 for the sand, a grain diameter of 0.45 mm and an internal angle of friction of 34 degrees. Determine the angle of repose for the conical pile of sand and show contours for the volume fraction of sand over time as it slumps.

15.2 Use ANSYS Fluent to simulate a 2-D hourglass. The shape of the hourglass is 200 mm wide at the top and the bottom, a total height of 500 mm and sides with a slope of 40.5 degrees. The opening at the waist of the hourglass is 10 mm. Start the simulation with sand (1638 kg/m3) and a grain diameter of 0.45 mm in the upper half of the hourglass with a packing limit of 0.63 for the volume fraction of sand. Use an internal angle of friction of 34 degrees. Determine the angle of repose for the final conical pile of sand in the bottom half of the hourglass and show contours for the volume fraction of sand over time as it moves due to gravity from the upper half to the lower half of the hourglass. See Hirt[2] for a similar simulation using Flow-3D.

15.3 Use ANSYS Fluent to simulate the flow of wood when filling a 2-D axisymmetric silo, see Jansson[3]. The shape of the silo is shown in Figure 15.16 with the following dimensions: $v1 = 9$ m, $h1 = 10$ m, $v2 = 17$ m, $h2 = 0.5$ m, $v3 = 0.5$ m, $v4 = 1.5$ m, $h3 = 0.5$ m. Use an inlet volume fraction for wood of 0.4 and other values as listed in Table 15.2.

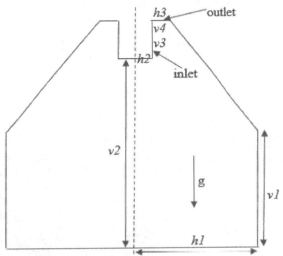

Figure 15.16 Geometry for silo

$\alpha_{s,min} = 0.45$	friction packing limit
$\alpha_{s,max} = 0.63$	packing limit
$\varphi = 45°$	internal angle of friction
$\rho_s = 850\ kg/m3$	wood density
$d_s = 0.02\ m$	particle diameter
$e_s = 0.001$	coefficient of restitution

Table 15.3 Parameters used in Exercise 15.3

K. Notes

CHAPTER 16. BOUNCING SPHERES

A. Objectives

- Use the Macroscopic Particle Model
- Create Pictures of the Resulting Motion of the Spheres

B. Problem Description

In this chapter we will be simulating a number of spheres falling through an hourglass shaped geometry. We will let the balls bounce with each other and they will bounce with the walls.

C. Launching Ansys Workbench and Selecting Fluent

1. Start by launching Ansys Workbench. Double click on Fluid Flow (Fluent) under Analysis Systems in the Toolbox.

Figure 16.1 Selecting Fluent

D. Launching Ansys DesignModeler

2. Right click on Geometry in the Project Schematic window in Ansys Workbench and select New DesignModeler Geometry. Select Units>>Millimeter from the menu in DesignModeler. Select the XY Plane in the Tree Outline and select Look at Face/Plane/Sketch . Select the Sketching tab and select Polyline under Draw from Sketching Toolboxes. Start at the origin and draw the sketch, right click and select *Closed End* and enter the dimensions as shown in Figure 16.2a). When you start drawing from the origin (make sure that letter *P* is shown when the cursor is close to the origin and that letters *V*, *H* and *C* are showing when you are creating vertical, horizontal and coinciding lines) and draw the sketch with the dimensions as shown in Figure 16.2a).

Select General under Dimensions and click on the eight lines (H1-H2, V3-V4 and L5-L6) as shown in Figure 16.2a). Select Angle under Dimensions and create angle A7 by selecting the lines next to the angle. Select the lower line first and then the upper line. Select Revolve and apply Sketch1 under XY Plane in the Tree Outline as Geometry in Details View. Select the left vertical side of the sketch and apply it as the Axis in Details View. Click on Generate and close DesignModeler.

Details View	
Details of Sketch1	
Sketch	Sketch1
Sketch Visibility	Show Sketch
Show Constraints?	No
Dimensions: 7	
A7	90 °
H1	17.5 mm
H2	17.5 mm
L5	20 mm
L6	20 mm
V3	20 mm
V4	20 mm
Edges: 7	
Line	Ln7
Line	Ln8
Line	Ln9
Line	Ln10
Line	Ln11
Line	Ln12
Line	Ln13

Figure 16.2a) Dimensions for sketch

Take the following steps if for some reason you need to reset the graphics window in DesignModeler: select Tools>>Options from the menu and select Addins under DesignModeler in the Options window. Set BladeModeler Licensing to No and select OK to close the window. Select File>>Close DesignModeler from the menu and double click on Geometry in Ansys Workbench to open DesignModeler once again. Select View>>Windows>>Reset Layoot from the menu in DesignModeler.

Figure 16.2b) Revolved sketch

E. Launching Ansys Meshing

3. Double click on Mesh under Project Schematic in Ansys Workbench. Select Mesh in the Outline and Set the Element Size under Defaults in Details of Mesh to 1.0 mm. Right click on Mesh under Project in Outline and select Generate Mesh. Under Statistics in Details of "Mesh" we can see that this mesh has 114,580 Nodes and 109,956 Elements. Right click in the graphics window and select Cursor Mode>>Face. Control select the 6 faces (push the mouse wheel and drag the mouse to turn around the model to select the bottom face) of the mesh, right click and select Create Named Selection. Enter the name *wall* and click on OK to close the Selection Name window.

    Select File>>Export...>>Mesh>>FLUENT Input File>>Export from the menu and save the mesh with the name *3d-enclosure-mesh.msh*. Select File>>Save Project from the menu and save the project with the name *3D-Enclosure.wbpj*. Close the Meshing window, right-click on Mesh under Project Schematic in Ansys Workbench and select Update.

Figure 16.3 Mesh for bouncing spheres

F. Launching Ansys Fluent

4. Double-click on Setup under Project Schematic in Ansys Workbench. Check the box for Double Precision under Options in the Ansys Fluent Launcher. Take a note of the location of the *Working Directory*. You will need this information later in step 9. Set the number of processes equal to the number of processor cores for your computer. To check the number of physical cores, press the Ctrl + Shift + Esc keys simultaneously to open the Task Manager. Go to the Performance tab and select CPU from the left column. You will see the number of physical cores on the bottom-right side. Close the Task Manager window. Click on the Start button in the Fluent Launcher window.

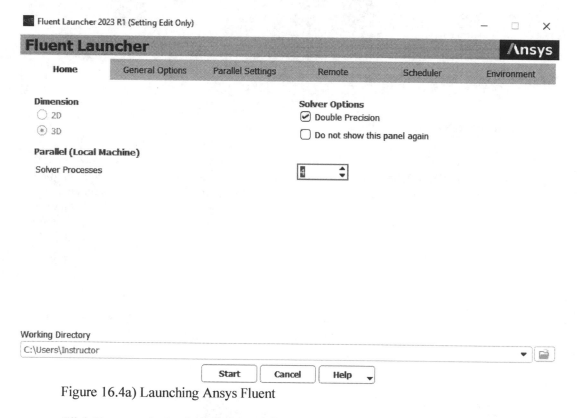

Figure 16.4a) Launching Ansys Fluent

Click Enter on the keyboard with the cursor in the Console section that is located under the graphics window. Enter the following commands in the Console to load the Macroscopic Particle Model (MPM), see Figure 16.4b).

>**define**
/define> **models**
/define/models/> **addon**
Enter Module Number: [0] **10**

Click on Display under Mesh in General on the Task Page. Check the boxes for Edges and Faces under Options in the Mesh Display window. Select wall under Surfaces, click on Display and Close the window, see Figure 16.4c).

374

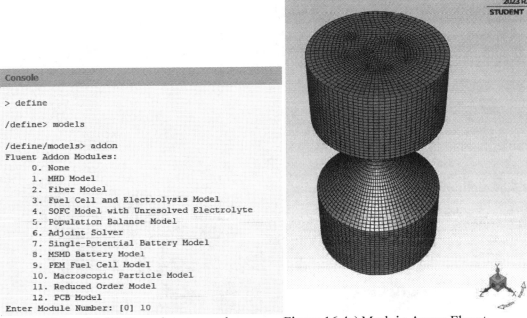

Figure 16.4c) Mesh in Ansys Fluent

```
Console

> define

/define> models

/define/models> addon
Fluent Addon Modules:
     0. None
     1. MHD Model
     2. Fiber Model
     3. Fuel Cell and Electrolysis Model
     4. SOFC Model with Unresolved Electrolyte
     5. Population Balance Model
     6. Adjoint Solver
     7. Single-Potential Battery Model
     8. MSMD Battery Model
     9. PEM Fuel Cell Model
     10. Macroscopic Particle Model
     11. Reduced Order Model
     12. PCB Model
Enter Module Number: [0] 10
```

Figure 16.4b) Console commands

5. Double click on General under Setup in the Outline View and select Transient under Time on the Task Page. Check the box for Gravity and enter -9.81 as the value for Y [m/s²]. Click on Scale… under Mesh in General on the Task Page. Close the window.

Figure 16.5a) General settings

Figure 16.5b) Scale mesh

375

6. Double click on Models and Macroscopic Particles under Setup in the Outline View and check the box to Enable Macroscopic Particle Model. Select the Drag tab and select Disable Drag Calculations under Drag Law Options. Select the Injections tab and choose Create…. Select plane as Injection Type. Set the Diameter [m] to 0.003. Set the Plane Shape to Circular and select Y as Axis Direction. Set Y [m] to 0.06828427 and Radius [m] to 0.0175 under Center Location/Radius of Circular Plane. Click on Create/Modify.

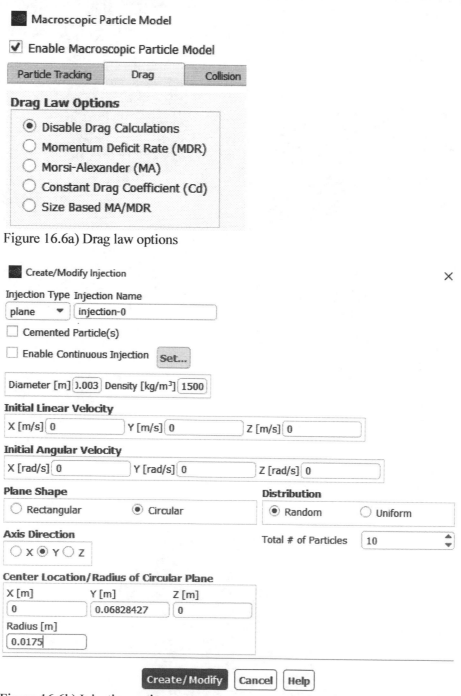

Figure 16.6a) Drag law options

Figure 16.6b) Injection options

Select the Initialize MPM tab in the MPM window. Check the box for Print Warning Messages and click on Initialize MPM Functions. Click on Initialize Particles, select Apply and Close the window.

Select Display Mesh ⊕ twice next to the graphics window to open the Mesh Display window. Select *wall* under Surfaces in the Mesh Display window. Uncheck the box for Faces under Options and select Outline as Edge Type, see Figure 16.6d). Click on Display and Close the Mesh Display window.

Double click on Macroscopic Particles under the Models on the Task Page. Click on Display Injections in the Macroscopic Particle Model MPM window. Select Particle ID under Display Particles Colored by, click on Display and Close the Display Particles window. Click on Apply in the MPM window and Close the same window.

Figure 16.6c) Macroscopic Particle Model

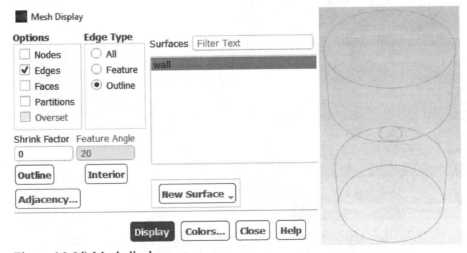

Figure 16.6d) Mesh display

Display Particles   ✕

Display Particles Coloured by

○ Particle Diameter
○ Particle Mass
⦿ Particle ID
○ Particle Velocity Magnitude

Velocity Magnitude Range

Min Velocity m/s   0

Max Velocity (m/s)   1

Display   Close   Help

Figure 16.6e) Display particles settings

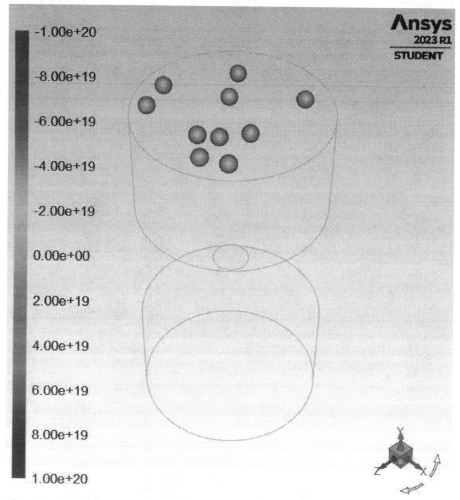

Figure 16.6f) Ten spheres randomly distributed on top of the mesh domain

7. Double click on Methods under Solution in the Outline View and select Second Order Implicit as Transient Formulation under Solution Methods on the Task Page. Select First Order Upwind for Turbulent Kinetic Energy and Specific Dissipation Rate under Spatial Discretization.

Figure 16.7a) Solution methods settings

Double click on Cell Zone Conditions and Fluid under Setup in the Outline View and uncheck the box for Source Terms in the Fluid window. Click on Apply and Close the Fluid window.

Double click on Monitors and Residual under Setup in the Outline View and uncheck the box for Plot under Options. Check the box for Show Advanced Options, select none as Convergence Criterion and click on OK to close the Residual Monitors window.

Figure 16.7b) Fluid window

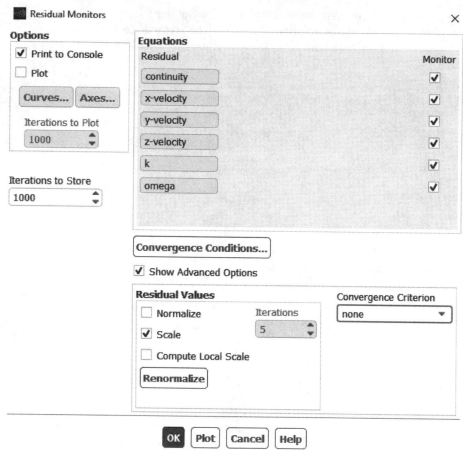

Figure 16.7c) Residual monitors window

8. Double click on Initialization under Solution in the Outline View and select Hybrid Initialization. Click on Initialize. Answer OK when you get a question.

Figure 16.8 Solution initialization

9. Create a new folder in your *Working Directory* called *jpg-files*. Double click on Calculation Activities and Execute Commands under Solution in the Outline View. Select New… in the Execute Commands Manager window and Enter the commands listed below and as shown with settings in Figure 16.9. You need to be careful to use the correct symbol in front of *particle-id* on the first command line. The key/symbol ` that is used is explained as acute, back quote, grave, grave accent, left quote, open quote, or a push. You find the symbol under the tilde ~ symbol in the upper left corner under Esc on a standard keyboard. You will need to change the second command so that it is in line with the name and location of your *Working Directory*. Select Time Step as When in the Execute Commands window. Click OK to close the Execute Commands window. Close the Execute Commands Manager window.

(display-mpm-injections 'particle-id 0 10)
dis hard C:/Users/Instructor/Desktop/An Introduction to Ansys Fluent 2023/Ansys Fluent 2023 R1 Ch. 16/jpg-files/mpm-%t.jpg

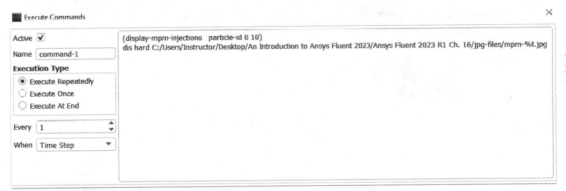

Figure 16.9 Executive commands

10. Select File>>Export>>Case… from the menu and save the Case File with the name *bouncing-spheres.cas.h5*. Double click on Run Calculation under Solution in the Outline View and set the Time Step Size [s] to 0.01 and Number of Time Steps to 5. Set Max Iterations/Time Step to 1 and click on Calculate.

Figure 16.10b) Running the calculations

G. Post-Processing

11. Select the *YZ* plane by clicking on the red *X* axis of the coordinate system in the lower right corner of the graphics window as shown in Figure 16.11a). Select Copy Screenshot of Active Window to Clipboard in the graphics window, see Figure 16.11b). The particle traces can be pasted into a Word document.

    Continue running the calculations two more times with the same settings. Next, change the Number of Time Steps to 10 and run the calculations once again.

    Next, change the Number of Time Steps to 25 and run the calculations once again. Change the Number of Time Steps to 50 and run the calculations once again. Answer OK to the question that you get. Next, change the Number of Time Steps to 100 and run the calculations. Change the Number of Time Steps to 300 and run the calculations.

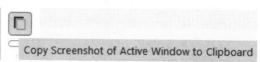

Figure 16.11a) Coordinate system   Figure 16.11b) Copying screenshot

$t = 0.05$ s $t = 0.1$ s $t = 0.15$ s $t = 0.25$ s

Figure 16.11c) Distribution of spheres during the first 0.25 s.

382

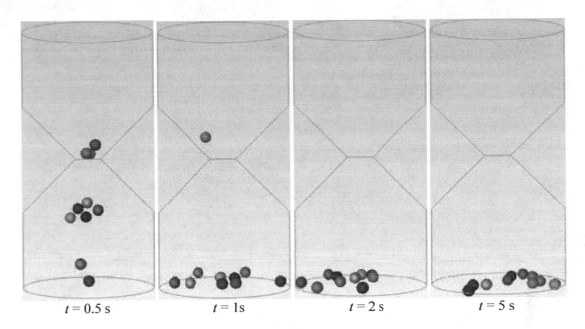

$t = 0.5$ s $t = 1$s $t = 2$ s $t = 5$ s

Figure 16.11d) Distribution of spheres over time up to 5 s.

H. Reference

1. Agrawal, M., Bakker, A., Prinkey, M.T. "Macroscopic Particle Model Tracking Big Particles in CFD", *AIChE Annual Meeting, Particle Technology Forum, Paper 268b, November 7-12, Austin, TX*, (2004).

I. Exercise

16.1   Use Ansys Fluent and the macroscopic particle model to perform a billiard table simulation of the first shot to break the balls including the collision of the cue ball with the 15 object balls initially racked together in a triangular pattern on a 50 in. x 100 in. billiard table. The diameter of the billiard balls is 57 mm.

J. Notes

CHAPTER 17. FALLING SPHERE

A. Objectives

- Using the Macroscopic Particle Model Module
- Determine the Terminal Velocity for a Falling Sphere

B. Problem Description

A 2 mm diameter sphere with a density of 1695 kg/m^3 is falling in a tube with a diameter of 20 mm and a height of 60 mm.

C. Launching Ansys Workbench and Selecting Fluent

1. You can move to step 4 if you would like to read the mesh directly in Ansys Fluent. Otherwise, start by launching Ansys Workbench. Double click on Fluid Flow (Fluent) under Analysis Systems in the Toolbox. Select **Units>>Metric (tonne, mm, s, …)** from the menu.

Figure 17.1 Selecting Fluent

D. Launching Ansys DesignModeler

2. Right click on Geometry in the Project Schematic window in Ansys Workbench and select New DesignModeler Geometry. Select **Units>>Millimeter** from the menu in DesignModeler. Select the XY Plane in the Tree Outline and click on Look at Face/Plane/Sketch ![icon]. Select the Sketching tab and select Rectangle under Draw from Sketching Toolboxes. Start at the origin (make sure you see letter P when the cursor is at

the origin and you are about to start drawing) and draw the rectangle in the first quadrant. Click on Dimensions and select the lower horizontal edge of the rectangle. Enter a length of 10 mm for the horizontal edge. Click on the right vertical edge of the rectangle. Enter a length of 60 mm for the vertical edge.

Select 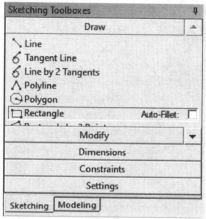 and apply Sketch1 under XYPlane in the Tree Outline as Geometry in Details View. Select the left vertical side of the rectangle sketch and apply it as the Axis in Details View. Click on Generate.

Figure 17.2a) Rectangle drawing tool

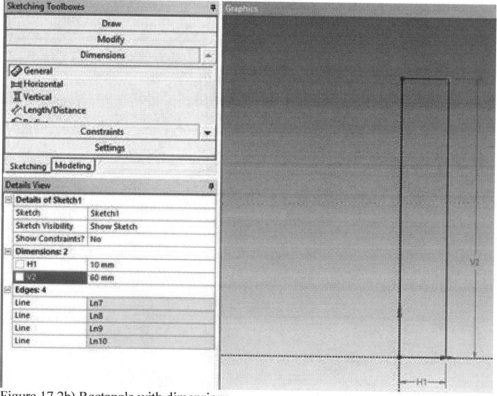

Figure 17.2b) Rectangle with dimensions

Figure 17.2c) Finished cylinder

Select ZXPlane in the Tree Outline and click on Look at Face/Plane/Sketch . Select the Draw tab and Rectangle under Sketching Toolboxes. Draw the rectangle inside the circle and use the Horizontal and Vertical Dimensions tools to create a square centered at the origin with a side length of 8 mm.

Figure 17.2d) Finished cylinder

Select Extrude, select Sketch2 and Apply the sketch as Geometry in Details of Extrude1. Set the Depth of the extrusion to 60 mm, select Slice Material as Operation and click on Generate.

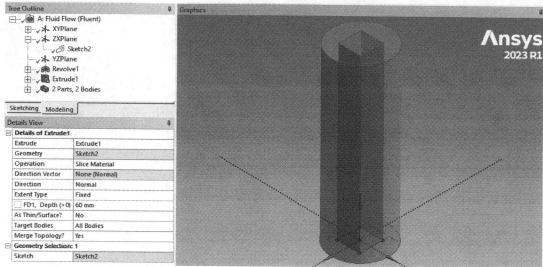

Figure 17.2e) Sliced cylinder with two parts

Select ZXPlane in the Tree Outline and click on Look at Face/Plane/Sketch ![icon]. Select

New Sketch ![icon]. Select the Draw tab and Rectangle under Sketching Toolboxes. Draw four radial outward lines from the four corners of the square. Set the angles between the lines and the vertical and horizontal axes to 45 degrees. Draw two horizontal lines between the endpoints of the lines as shown in Figure 17.2f). Use the Horizontal tool under Constraints in the Sketching Toolboxes to make the two lines horizontal.

Figure 17.2f) Radial lines from the four corners of the square

Select Extrude, select Sketch3 and Apply the sketch as Geometry in Details of Extrude2. Set the Depth of the extrusion to 60 mm, select Slice Material as Operation and click on Generate.

Figure 17.2g) Cylinder with five parts

Select the three sketches one by one in the Tree Outline, right click and select Hide Sketch. Click on plus sign next to 5 Parts, 5 Bodies in the Tree Outline. Control select the five part, right click and select Form New Part. Select Fluid in Details View. Close DesignModeler.

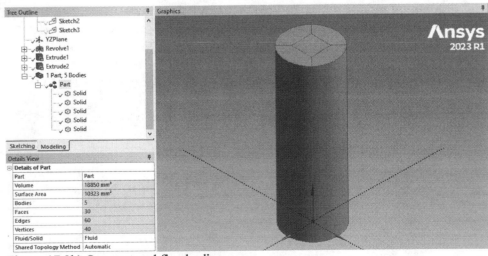

Figure 17.2h) One part and five bodies

E. Launching Ansys Meshing

3. Double click on Mesh under Project Schematic in Ansys Workbench, see Figure 17.3a). Select Mesh in the Outline under Project. Select Home>>Tools>>Units>>Metric (mm, kg, N, …) from the menu. Right click on Mesh under Project in Outline and select Generate Mesh. A coarse mesh is generated.

Right click on Mesh in the Outline and select Insert>>Sizing. Select the 8 edges as shown in Figure 17.3b) and Apply the edges as Geometry under Scope in Details of "Edge Sizing". Set Type to Number of Divisions under Definition and set the Number of Divisions to 40. Set Capture Curvature to No under Advanced and set Behavior to Hard.

Insert another Sizing and select the 8 edges as shown in Figure 17.3c). Set the Number of Divisions to 20. Set Capture Curvature to No under Advanced and set Behavior to Hard.

Insert a third Sizing and select the 8 edges as shown in Figure 17.3d). Set the Number of Divisions to 40. Set Capture Curvature to No under Advanced and set Behavior to Hard.

Insert a fourth Sizing and select the 8 edges as shown in Figure 17.3e). Set the Number of Divisions to 60. Set Capture Curvature to No under Advanced and set Behavior to Hard. Select the last Bias Type and set the Bias Factor to ten.

Right click on Mesh under Project in Outline and select Generate Mesh. Under Statistics in Details of Mesh we can see that this mesh has 297741 Nodes and 288000 Elements, see Figure 17.3f).

Select Home>>Insert>>Section Plane from the menu. Draw a line through the mesh as shown in Figure 17.3h). Select the Section Planes tab in the lower left corner and uncheck the box for the Section Plane.

Right click in the graphics window and select Cursor Mode>>Face. Control select the 14 faces of the mesh, right click and select Create Named Selection. Enter the name *wall* and click on OK to close the Selection Name window, see Figure 17.3i).

Select File>>Export…>>Mesh>>FLUENT Input File>>Export from the menu and save the mesh in the *Working Directory* with the name *cylindrical-mesh-lambda-0.1.msh*. Select File>>Save Project from the menu and save the project with the name *Cylinder Mesh Falling Sphere Lambda = 0.1.wbpj*. Close the Meshing window, right-click on Mesh under Project Schematic in Ansys Workbench and select Update.

Figure 17.3a) Starting Ansys Meshing

Figure 17.3b) First sizing for cylinder

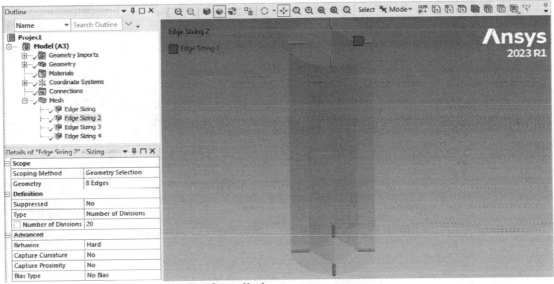

Figure 17.3c) Second sizing for cylinder

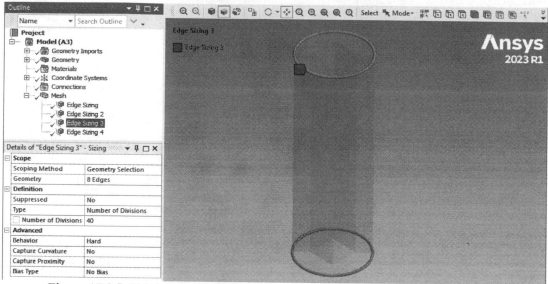

Figure 17.3d) Third sizing for cylinder

Figure 17.3e) Fourth sizing for cylinder

Figure 17.3f) Top view of finished mesh

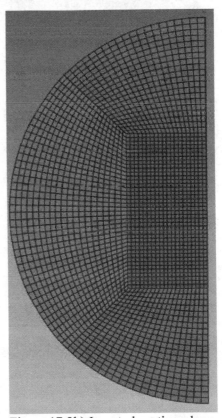

Figure 17.3g) Isometric view of finished mesh   Figure 17.3h) Inserted section plane

Figure 17.3i) Section plane view

Figure 17.3i) Named selection for cylinder

F. Launching Ansys Fluent

4.   If you are reading the mesh directly to Ansys Fluent instead of creating the mesh as

shown in steps 1-3, you can start Ansys Fluent in standalone mode that is available in the Ansys folder. Select the Solution tab under Home in Fluent Launcher, see Figure 17.4a). Read the Mesh file cylindrical-mesh-lambda-0.1.msh. Check the box for Double Precision under Solver Options in the Ansys Fluent Launcher and set the number of Solver Processes. Click on the Start button in the Fluent Launcher window. Enter the following commands in the Ansys Fluent Console that is located below the graphics window to load the Macroscopic Particle Model (MPM), see Figure 17.4c). Click in the Console window and select Enter on the keyboard to get to the prompt. You can now move to step 5 if you started Ansys Fluent in standalone mode as described above.

Otherwise, if you have created the mesh in steps 1-3, you can start Fluent from Workbench by double-clicking on Setup under Project Schematic in Ansys Workbench. Check the box for Double Precision under Solver Options in the Ansys Fluent Launcher, see Figure 17.4b). Set the number of processes equal to the number of processor cores for your computer. To check the number of physical cores, press the Ctrl + Shift + Esc keys simultaneously to open the Task Manager. Go to the Performance tab and select CPU from the left column. You will see the number of physical cores on the bottom-right side. Close the Task Manager window. Click on the Start button in the Fluent Launcher window.

Figure 17.4a) Ansys Fluent launcher from standalone mode

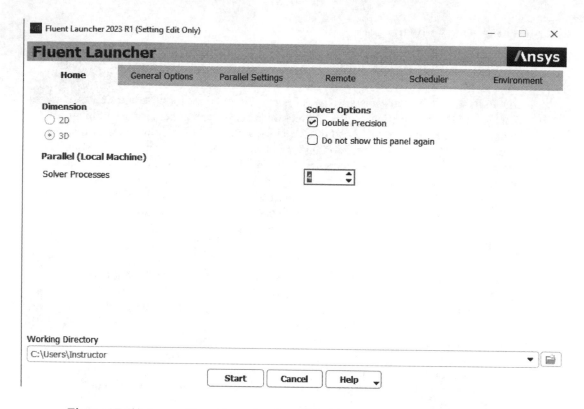

Figure 17.4b) Ansys Fluent launcher from Workbench

Enter the following commands in the Console that is located below the graphics window to load the Macroscopic Particle Model (MPM). Click in the Console window and select Enter on the keyboard to get to the prompt.

>**define**
/define> **models**
/define/models/> **addon**
Enter Module Number: [0] **10**

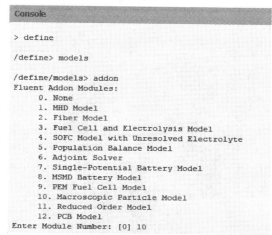

Figure 17.4c) Console commands to load MPM module

5. Double click on General under Setup in the Outline View and select Transient under Time on the Task Page. Check the box for Gravity and enter -9.81 as the value for Y [m/s²]. Click on Scale… under Mesh in General on the Task Page.

Figure 17.5a) General settings   Figure 17.5b) Scale mesh

6. Double click on Models and Viscous (SST k-omega) under Setup in the Outline View. Select Laminar Model and click OK to close the Viscous Model window. Double click on Materials under Setup in the Outline View. Click on Create/Edit for Fluid under Materials on the Task Page. Click on Fluent Database in the Create/Edit Materials window. Scroll down and select *water-liquid (h2o<l>)* as the Fluent Fluid Material. Click on Copy and Close the Fluent Database Materials window. Set the Density [kg/m³] to 1020 and Viscosity [kg/m s] to 0.49. Click on Change/Create and Close the Create/Edit Materials window.

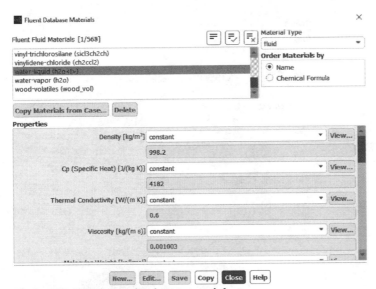

Figure 17.6a) Fluent database materials

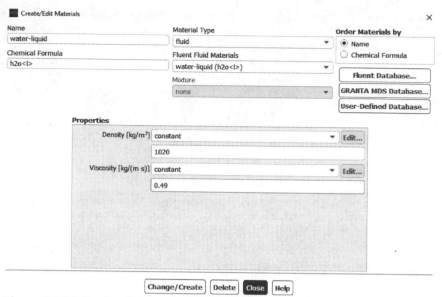

Figure 17.6b) Setting the density and viscosity for the fluid

Open Cell Zone Conditions and Fluid under Setup in the Outline View. Double click on *solid (fluid, id=2)* under Cell Zone Conditions and Fluid in the Outline View. Select *water-liquid* as Material Name and enter *water-liquid* as Zone Name. Check the box for Source Terms and click on the Source Terms tab. Click on Edit for X Momentum. Set Number of X Momentum Sources to 1, select *udf x_mom::mpm* and click OK. Repeat this for Y Momentum, Z Momentum and Mass. Select Apply and Close the Fluid window.

Figure 17.6c) Selection of source terms and water-liquid as fluid

Double click on Models and Macroscopic Particles under Setup in the Outline View and check the box to Enable Macroscopic Particle Model. Select the Particle Tracking tab and check the box for Enable Interaction with Continuous Phase. Set the Particle Sub-Timesteps within each Flow Timestep to 10 and Under-Relaxation Factor to Velocities in Touched Cells to 1. Click on Apply.

Figure 17.6d) Particle tracking

Select the Drag tab and select Momentum Deficit Rate (MDR) under Drag Law Options. Click on Apply.

Figure 17.6e) Drag law options

Select the Collision tab and uncheck Enable Particle-Particle Collision and uncheck Enable Particle-Wall Collision. Click on Apply.

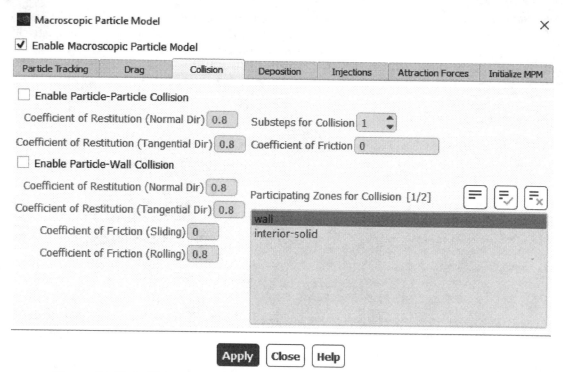

Figure 17.6f) Collision options

Select the Injections tab and click on Create.... Select *point* as Injection Type. Set the Diameter [m] to 0.002 and set the Density [kg/m^3] to 1695. Check the box for Cylindrical Coordinates and select Y as Axis Direction. Set Axial [m] to 0.03 under Initial Location. Click on Create/Modify in the Create/Modify Injection window. Select Apply in the MPM window.

Figure 17.6g) Injection options

Select the Initialize MPM tab and check the box for ASCII File under Write Particle Data. Click on Browse, enter the file name *lambda0p1.csv*, save the file in the *Working Directory* and click on OK. Check the box for Print Warning Messages. Select Initialize MPM Functions. Click on Initialize Particles. Select Apply in the MPM window.

Figure 17.6h) Initialize MPM

Select Display Mesh ⊚ twice next to the graphics window. Select *wall* under Surfaces. Check the box for Edges and uncheck the box for Faces under Options and select Outline as Edge Type, see Figure 17.6i). Click on Display and Close the Mesh Display window. Click on Apply in the MPM window.

Figure 17.6i) Mesh display

Click on Display Injections. Select Display Particles Coloured by Particle Diameter. Select Display and Close the Display Particles window. Click on Apply and Close the Macroscopic Particle Model window.

Figure 17.6j) Display particles

Figure 17.6k) Sphere in the middle of cylinder

7. Double click on Methods under Solution in the Outline View and select Second Order Implicit as Transient Formulation under Solution Methods on the Task Page. Check the box for Higher Order Term Relaxations and select Options…. Select Flow Variables Only and set the Relaxation Factor to 0.75. Click OK to close the window.

Figure 17.7a) Solution methods settings

Figure 17.7b) Relaxation options

Double click on Monitors and Residual under Setup in the Outline View. Check the boxes for Print to Console and Plot under Options. Click on OK to close the Residual Monitors window.

Figure 17.7c) Residual monitors window

8. Double click on Initialization under Solution in the Outline View and select Hybrid Initialization. Click on Initialize. Answer OK to the question that you get.

Figure 17.8 Solution initialization

9. Select File>>Save Project from the menu. Select File>>Export>>Case… from the menu. Save the Case File in the *Working Directory* with the name *falling-sphere-lambda-0.1.cas.h5*.

Double click on Run Calculation under Solution in the Outline View and set the Time Step Size (s) to 2e-5 and Number of Time Steps to 2000. Set Max Iterations/Time Step to 20 and click on Calculate.

Figure 17.9a) Running calculations for falling sphere

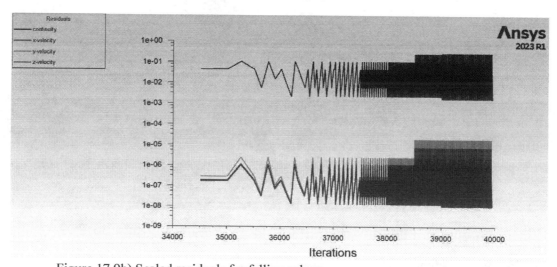

Figure 17.9b) Scaled residuals for falling sphere

G. Post-Processing

10. Open Excel and open the file *lambda0p1.csv*. You will need to select All Files format to open the file in Excel. Save the file as *lambda0p1.xlsx*. Delete the first row. Copy the first column *A* to column *N*. In position *O2*, enter =ABS(G2) and drag the *O2* cell down the whole *O* column to get the absolute *y*-velocity component. Select the data values in columns *N* and *O*. Label the *N* column as *t* (s) and the *O* column as *Ansys Fluent*. The theoretical solution for velocity is given by equation (17.7) in the theory section. Enter the following expression in cell *P2*:

= (0.791576*9.81*(1-(1020/1695)) / (18*0.49/ (1695*0.002^2))) *(1-Exp(-(18*0.49/ (1695*0.002^2)) *N2/0.791576)) 　　　(17.1)

The wall factor *f* = 0.791576 used in equation (17.1) corresponds to *d* = 0.002 m and *λ* − 0.1.

Drag the *P2* cell down the whole *P* column to get the theoretical velocity. Label the *P* column as *Theory*. Select columns *N − P* and select from the Excel menu Insert>>Charts>>Scatter with Smooth Lines, see Figures 17.10a).

Figure 17.10a) Sphere velocity versus time for *λ* = 0.1

Copy the column *N* to column *Q* and copy column *D* to column *R*. Label the *Q* column as *t* (s) and the *R* column as *Ansys Fluent*. The theoretical solution for position is given by equation (17.8) in the theory section. Enter the following expression in cell *S2*:

= 0.03+(0.791576*9.81*(1-(1020/1695)) / (18*0.49/ (1695*0.002^2)) ^2) * (0.791576*(1-Exp(-(18*0.49/ (1695*0.002^2)) *N2/0.791576))- (18*0.49/ (1695*0.002^2)) *N2)　　(17.2)

The wall factor *f* = 0.791576 used in equation (17.2) corresponds to *d* = 0.002 m and *λ* = 0.1.

Drag the *S2* cell down the whole *S* column to get the theoretical *y*-position. Label the *S* column as *Theory*. Select columns *Q − S* and select from the Excel menu Insert>>Charts>>Scatter with Smooth Lines, see Figures 17.10b).

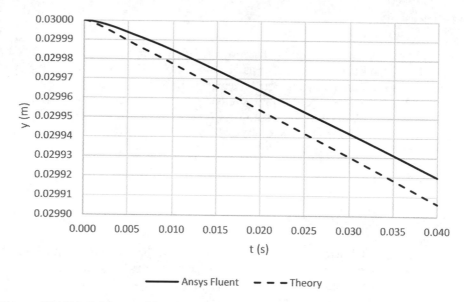

Figure 17.10b) Sphere position versus time for $\lambda = 0.1$

Copy the column Q to column T and copy column L to column U. Label the T column as t (s) and the U column as *Ansys Fluent*. The theoretical solution for drag force is given by equation (17.9) in the theory section. Enter the following expression in cell *V2*:

= 3*PI()*0.49***0.002***(9.81*(1-(1020/1695)) / (18*0.49/ (1695***0.002**^2))) * (1-Exp(-(18*0.49/ (1695***0.002**^2))
***N2/0.791576**))

$$(17.3)$$

The wall factor $f = 0.791576$ used in equation (17.3) corresponds to $d = 0.002$ m and $\lambda = 0.1$.

Drag the *V2* cell down the whole *V* column to get the theoretical drag force. Label the *V* column as *Theory*. Select columns $T - V$ and select from the Excel menu Insert>>Charts>>Scatter with Smooth Lines, see Figures 17.10c).

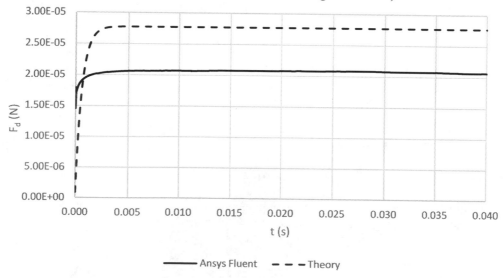

Figure 17.10c) Sphere drag force versus time for $\lambda = 0.1$

Double-click on Graphics and Contours under Results in the Outline View. Select Contours of Velocity… and Velocity Magnitude. Select New Surface>>Plane…. Select YZ Plane as Method and select Create. Close the Plane Surface window. Select *plane-1* as the Surface in the Contours window. Check and uncheck the boxes as shown in Figure 17.10d). Select Colormap Options…, set Type to float under Number Format and set Precision to 4. Set Font Behavior to Fixed under Font and set Font Size to 12. Select Apply and Close the Colormap window. Click on Save/Display. The velocity magnitude for the sphere at $t = 0.04$ s is shown in Figure 17.10e).

Figure 17.10d) Settings for contours

Figure 17.10e) Sphere velocity magnitude and Y Velocity at $t = 0.04$ s

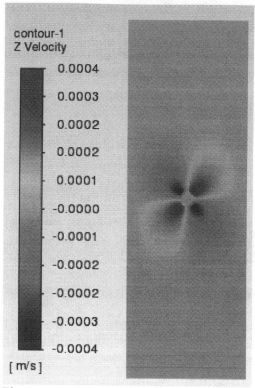

Figure 17.10f) Sphere velocity magnitude and Z Velocity at $t = 0.04$ s

diameter of sphere	d (m)	0.002
diameter of mesh	D (m)	0.02
diameter ratio	$\lambda = d / D$	0.1
height of mesh	H (m)	0.06
wall factor	f	0.791576
density of sphere	ρ_s (kg/m3)	1695
density of fluid	ρ_f (kg/m3)	1020
dynamic viscosity of fluid	μ (Pa-s)	0.49
terminal velocity	U_∞ (m/s)	0.003
modified terminal velocity	$U = f\,U_\infty$ (m/s)	0.002375
terminal drag force	F_d (N)	0.0000277
Reynolds number	Re	0.009887
upper limit for Reynolds number	Re_{max}	0.027
terminal drag coefficient	C_d	3067
relaxation time	τ_∞ (s)	0.001
upper limit for time step size	Δt (s)	0.00002
number of nodes for mesh	#Nodes	297741
number of elements for mesh	#Elements	288000
mesh count	N	288000

volume of sphere	V_s (m³)	4.1888E-09
total volume of mesh	V_{mesh} (m³)	1.8850E-05
number of cells per particle volume	N_{CPV} (m⁻³)	108
number of cells per particle diameter	N_{CPD} (m⁻¹)	5.9
modified terminal velocity from Ansys Fluent	$f\,U_\infty$(m/s), Fluent	0.002241
percent difference between Ansys Fluent and Theory	% Difference	5.6
wall factor from Ansys Fluent	f, Fluent	0.747
terminal drag force on sphere Ansys Fluent	F_d (N)	0.0000207
terminal Re for sphere Ansys Fluent	Re	0.008085
terminal drag coefficient for sphere Ansys Fluent	C_d	3974

Table 17.1 Parameter values for Ansys Fluent simulation.

H. Theory

11. For Stokes flow or creeping flow with Reynolds number $\ll 1$ we can define the terminal velocity U_∞ for a sphere falling in an infinite quiescent Newtonian fluid under the influence of gravity as

$$U_\infty = \frac{(\rho_s - \rho_f)d^2 g}{18\mu} \qquad (17.4)$$

, where $d\,(m)$ is the diameter of the sphere, ρ_s (kg/m³) is the density of the sphere, ρ_f (kg/m³) is the density of the fluid, μ (kg/m-s) is the dynamic viscosity of the fluid, and g (m/s²) is acceleration due to gravity. The relaxation time for an infinite fluid is defined by

$$\tau_\infty = \frac{\left(\rho_s + \frac{\rho_f}{2}\right)d^2}{18\mu} \qquad (17.5)$$

The terminal velocity will be reduced by a factor f by the presence of a surrounding cylindrical wall, see Figure 7.11a). Using Newton's second law we get

$$W - F_d - F_b = m\frac{d^2Y}{dt^2} = m\frac{dU}{dt} \qquad (17.6)$$

, where m is the mass of the sphere, W is the weight of the sphere, F_d is the drag force on the sphere and F_b is the buoyancy force. Using Stokes law in combination with the wall effect factor, we replace the drag force with $F_d = 3\pi\mu dU/f$. We also use the buoyancy force on the sphere $F_b = \rho_f g V_s$ and the weight of the sphere $W = \rho_s g V_s$ to get

$$\rho_s g V_s - 3\pi\mu dU/f - \rho_f g V_s = m\frac{dU}{dt} \qquad (17.7)$$

After expressing the volume of the sphere in terms of the diameter of the sphere we get the following

$$\rho_s g \frac{\pi d^3}{6} - 3\pi\mu dU/f - \rho_f g \frac{\pi d^3}{6} = \rho_s \frac{\pi d^3}{6}\frac{dU}{dt} \qquad (17.8)$$

This equation can be rewritten as an ordinary first order differential equation for the velocity of a falling sphere in a cylinder

$$\frac{dU}{dt} + bU/f - a = 0 \tag{17.9}$$

, where $a = g(1 - \rho_f/\rho_s)$ and $b = \frac{18\mu}{\rho_s d^2}$. The solution to equation 17.9 is of the form

$$U = \frac{fa}{b}\left(1 - e^{-bt/f}\right) \tag{17.10}$$

When $t \to \infty$, $U \to fU_\infty = fa/b$, equation 17.10 gives equation 17.4. The location of the sphere can be expressed as

$$y = y_0 + \frac{fa}{b^2}\left[f\left(1 - e^{-bt/f}\right) - bt\right] \tag{17.11}$$

The drag force can now be expressed as

$$F_d = \frac{3\pi\mu dU}{f} = 3\pi\mu d\frac{a}{b}\left(1 - e^{-bt/f}\right) \tag{17.12}$$

When $t \to \infty$, $F_d = \frac{3\pi\mu da}{b} = 3\pi\mu dU_\infty$. The terminal drag coefficient can be expressed as

$$C_d = \frac{F_d}{\frac{1}{2}\rho_f U^2 A} = \frac{3\pi\mu dU}{f\frac{1}{2}\rho_f U^2 \pi d^2/4} = \frac{24\mu}{f\rho_f Ud} = \frac{24}{fRe} \tag{17.13}$$

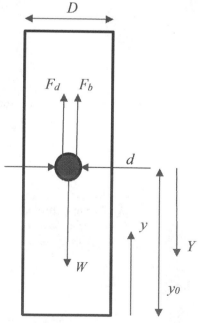

Figure 7.11a) Geometry for falling sphere

The wall factor for very low Reynolds numbers is according to Haberman and Sayre (1958)

$$f = \frac{1 - 2.105\lambda + 2.0865\lambda^3 - 1.7068\lambda^5 + 0.72603\lambda^6}{1 - 0.75857\lambda^5} \tag{17.14}$$

, where $\lambda = d/D$ and D is the mesh diameter or inner diameter of the clear glass graduated cylinder. The modified terminal velocity is defined as $U = fU_\infty$. The upper level of the Reynolds number for the Stokes regime and the wall factor is shown versus λ in Table 17.2 and Figure 17.11b), see Chhabra *et al.* (2003).

λ	f	Re_{max}
0.1	0.792	0.027
0.2	0.595	0.040
0.3	0.422	0.050
0.4	0.279	0.083
0.5	0.170	0.18
0.6	0.094	0.52
0.7	0.047	2.10
0.8	0.020	8.40
0.9	0.008	25.17

Table 17.2 Values of wall factor and max *Re* for different λ

Figure 17.11b) Wall factor f and maximum Reynolds number versus λ

The Reynolds number is defined as

$$Re = \frac{Ud\rho_f}{\mu} = \frac{fU_\infty d\rho_f}{\mu} \tag{17.15}$$

The time step size used in Ansys Fluent simulations is described by the following

$$\Delta t = \frac{2\tau_\infty}{100} \tag{17.16}$$

The volume for the spherical particle is given by

$$V_s = \frac{\pi d^3}{6} \tag{17.17}$$

The total volume of the mesh including the volume of the sphere used in Ansys Fluent simulations

$$V_{mesh} = \frac{\pi H D^2}{4}$$ (17.18)

, where H is the height of the mesh. This height needs to be a minimum of $H = 3D$ in the Ansys Fluent simulations. We can now determine the number of cells per particle volume

$$N_{CPV} = \frac{N V_s}{V_{mesh}}$$ (17.19)

, where N is the mesh count or number of elements for the mesh. The number of cells per particle diameter is defined as

$$N_{CPD} = \left(\frac{6 N_{CPV}}{\pi}\right)^{\frac{1}{3}} = \left(\frac{6 N V_s}{\pi V_{mesh}}\right)^{\frac{1}{3}}$$ (17.20)

, and was chosen to be at least 5 in the Ansys Fluent simulations. The modified terminal velocity will depend on this parameter in simulations.

I. References

1. Agrawal M., Bakker A. and Prinkey M.T. "Macroscopic particle model – tracking big particles in CFD", *Proceedings of AIChE 2004 Annual Meeting, Austin, TX,* (2004).
2. Agrawal M., Ookawara S. and Ogawa K. "Drag Force Formulation in Macroscopic Particle Model and its validation", *Proceedings of AIChE 2009 Annual Meeting,* (2009).
3. Chhabra R.P., Agarwal S. and Chaudhary K. "A note on wall effect on the terminal falling velocity of a sphere in quiescent Newtonian media in cylindrical tubes", *Powder Technology*, **129**, 53 - 58, (2003).
4. Haberman W.L and Sayre R.M. "David Taylor Model Basin Report No. 1143", *Department of Navy, Washington, DC* (1958).
5. Ookawara, S., M. Agrawal, D. Street and K. Ogawa. Modeling the motion of a sphere falling in a quiescent Newtonian liquid in a cylindrical tube by using the macroscopic particle model. In: The Seventh World Congress of Chemical Engineering. Glasgow, Scotland. C39-004, (2005).

J. Exercises

17.1 Run Ansys Fluent Simulations for $\lambda = 0.1 - 0.9$ as shown in Table 17.3. Fill out the missing results in the table based on your simulations. Use the element size as listed in the table when you create the mesh for each λ. Change the #Nodes, #Elements and Mesh Count in the table to the values that you get when creating the mesh for each λ. The value for Mesh Count is equal to the value for #Elements. #Cells/Particle Volume and #Cells/Particle Diameter will need to be recalculated based on your actual values. Plot the modified terminal velocity versus λ from Ansys Fluent simulations and compare with Haberman and Sayre (1958).

d (m)	0.002	0.004	0.006	0.008	0.01	0.012	0.014	0.016	0.018
D (m)	0.02	0.02	0.02	0.02	0.02	0.02	0.02	0.02	0.02
$\lambda = d / D$	0.1	0.2	0.3	0.4	0.5	0.6	0.7	0.8	0.9
H (m)	0.06	0.06	0.06	0.06	0.06	0.06	0.06	0.06	0.06
f	0.79157616	0.595336803	0.421995126	0.279200952	0.170357599	0.09440553	0.046675065	0.020398637	0.008245803
ρ_s (kg/m3)	1695	1695	1695	1695	1695	1695	1695	1695	1695
ρ_f (kg/m3)	1020	1020	1020	1020	1020	1020	1020	1020	1020
μ (Pa-s)	0.49	1.96	4.41	7.84	12.25	17.64	24.01	31.36	39.69
U_∞ (m/s)	0.003	0.003	0.003	0.003	0.003	0.003	0.003	0.003	0.003
fU_∞ (m/s)	0.00237473	0.00178601	0.001265985	0.000837603	0.000511073	0.000283217	0.000140025	6.11959E-05	2.47374E-05
Re	0.00988662	0.003717818	0.001756878	0.000871791	0.000425546	0.000196518	8.32803E-05	3.18469E-05	1.14432E-05
Re_{max}	0.027	0.04	0.05	0.083	0.18	0.52	2.1	8.4	25.17
τ_∞ (s)	0.001	0.001	0.001	0.001	0.001	0.001	0.001	0.001	0.001
Time Step Size (s)	0.00002	0.00002	0.00002	0.00002	0.00002	0.00002	0.00002	0.00002	0.00002
#Nodes	346,060	47,141	13,720	6,450	3,059	1,890	1,092	980	624
#Elements	335,787	44,776	12,753	5,829	2,684	1,600	840	741	465
Element Size(mm)	0.465	1.045	1.555	2.065	2.705	3.16	317	320	420
Mesh Count	335,787	44,776	12,753	5,829	2,684	1,600	840	741	465
Particle Volume (m^3)	4.1888E-09	3.3510E-08	1.1310E-07	2.6808E-07	5.2360E-07	9.0478E-07	1.4368E-06	2.1447E-06	3.0536E-06
Total Volume (m^3)	1.8850E-05	1.8850E-05	1.8850E-05	1.8850E-05	1.8850E-05	1.8850E-05	1.8850E-05	1.8850E-05	1.8850E-05
#Cells/Particle Volume	74.62	79.60	76.52	82.90	74.56	76.80	64.03	84.31	75.33
#Cells/Particle Diameter	5.2	5.3	5.3	5.4	5.2	5.3	5.0	5.4	5.2
fU_∞ (m/s), Fluent									
% Diff									
f, Fluent									

Table 17.3 Data for $\lambda = 0.1 - 0.9$

17.2 Run Ansys Fluent Simulations for $\lambda = 0.1$ as shown in Table 17.3 for different values of #Cells/Particle Diameter from 2 – 10. Include a plot of f versus #Cells/Particle Diameter.

413

K. Notes

CHAPTER 18. FLOW PAST A SPHERE

A. Objectives

- Using Ansys DesignModeler to Model the Sphere
- Inserting Boundary Conditions and Free Stream Velocity
- Running the Calculations
- Using Velocity and Pressure Plots for Visualizations
- Determining the Drag Coefficient

B. Problem Description

We will study the axisymmetric flow past a sphere and we will analyze the problem using Ansys Fluent. A 50 mm diameter sphere will be used and the Reynolds number will be set to $Re = 100$.

C. Launching Ansys Workbench and Selecting Fluent

1. Start by launching Ansys Workbench. Double click on Fluid Flow (Fluent) under Analysis Systems in the Toolbox. Select Units>>Metric (tonne, mm, s, …) from the menu.

Figure 18.1 Selecting Fluent

D. Launching Ansys DesignModeler

2. Right click Geometry and select Properties. In Properties of Schematic A2: Geometry, select Analysis Type 2D under Advanced Geometry Options. Right click on Geometry in the Project Schematic window and select New DesignModeler Geometry. Select millimeter as the desired length unit from the menu (Units>>Millimeter).

    Select the Sketching tab and click on Arc by Center. Click on Look at Face/Plane/Sketch
    . Draw in the upper two quadrants a half-circle centered at the origin of the coordinate system in the graphics window. Click on the Dimensions tab and select Radius. Click on the half-circle in the graphics window and enter 25 mm for the radius. Draw a line between the two end-points of the half-circle.

    Select Concepts>>Surfaces From Sketches from the menu. Control select the half-circle and the line in the graphics window, and select Apply for Base Objects in Details View. Click on Generate.

Figure 18.2a) Arc by center

Details View	⏚
Details of Sketch1	
Sketch	Sketch1
Sketch Visibility	Show Sketch
Show Constraints?	No
Dimensions: 1	
R1	25 mm

Figure 18.2b) Half-circle and line

Details View	⏚
Details of SurfaceSk1	
Surface From Sketches	SurfaceSk1
Base Objects	1 Sketch
Operation	Add Material
Orient With Plane Normal?	Yes
Thickness (>=0)	0 mm

Figure 18.2c) Finished half-circle

Next, we will be creating the 2D axisymmetric mesh region around the half-sphere. Click on the modeling tab and select the XYPlane in the Tree Outline. Click on ⎙ to generate a new sketch. Select Rectangle by 3 Points from the sketching tools. Click anywhere at the upper left corner of the graphics window, followed by clicking anywhere on the left-hand plane on the horizontal axis, and finally click on the right-hand plane on the horizontal axis. Select Dimensions, select the Horizontal dimensioning tool and click on the vertical edges of the rectangle. Enter 1575 mm as the horizontal length of the rectangle. Click on the vertical axis and the vertical edge of the rectangle on the left-hand side. Enter 575 mm as the length. Select the Vertical dimensioning tool and click on the horizontal edges of the rectangle. Enter 625 mm as the vertical height of the rectangle.

Details View	⏚
Details of Sketch2	
Sketch	Sketch2
Sketch Visibility	Show Sketch
Show Constraints?	No
Dimensions: 3	
H2	1575 mm
H3	575 mm
V4	625 mm

Figure 18.2d) Dimensions for rectangular mesh region around half-sphere

3. Select Concept>>Surface From Sketches. Click on Sketch2 in the Tree Outline and select Apply for Base Objects in Details View. Select Operation>>Add Frozen in Details View followed by Generate. Select Create>>Boolean from the menu. Select Subtract Operation from the Details View. Select the mesh region in the graphics window and click on Apply for Target Bodis. Select the Sphere as the Tool Body and click on Generate.

Figure 18.3 Mesh region around sphere

4. Select XYPlane in the tree outline and create a new sketch ⬚. Select the Sketching tab and select Line. Draw a vertical line from top to bottom and to the left of the sphere. The vertical line will intersect the entire mesh region. Position the line 75 mm from the vertical axis. Select Concepts>>Lines From Sketches from the menu. Select the newly created vertical line and Apply it as a Base Object in the Details View. Click on Generate.

Create another sketch in the XYPlane with a vertical line 75 mm to the right of the vertical axis and repeat all the different steps as listed above. Repeat these steps one more time with a horizontal line from left to right intersecting the mesh region and positioned 75 mm above the half-sphere centerline.

Figure 18.4a) Lines added to the mesh region

Create another sketch in the XYPlane with a vertical line towards the half-sphere as shown in Figure 18.4b). Repeat the steps completed earlier. Finally, create the last sketch with two diagonal lines at 45 degrees towards the half-sphere as shown in Figure 18.4b). Use the Angle tool under Dimensions to set the angle to 45 degrees for both lines.

Figure 18.4b) Lines added to the mesh region close to the half-sphere

5. Select Tools>>Projection from the menu. Select Edges on Face as type in Details View.

Use the Edge selection filter and control-select all 11 line-segments from the 6 lines that you have created and Apply as the Edges in the Details View. Select the mesh region and Apply it as the Target under Details View. Click Generate.

Figure 18.5 Selection of the projection tool

You should now have 9 different regions for meshing. Select File>>Save Project from the menu and save the project with the name *Axisymmetric Sphere Flow Study.wbpj*. Close DesignModeler.

E. Launching Ansys Meshing

6. Double click on Mesh under Project Schematic in Ansys Workbench. Select Mesh in the Outline under Project. Select Home>>Tools>>Units>>Metric (mm, kg, N, …) from the menu. Right click on Mesh under Project in Outline and select Generate Mesh. A coarse mesh is generated.

   Select Mesh under Model (A3) in Project under Outline on the left-hand side. Select the Mesh tab in the menu and Controls>>Face Meshing. Control-select all 9 faces of the mesh. Apply the Geometry under Details of "Face Meshing".

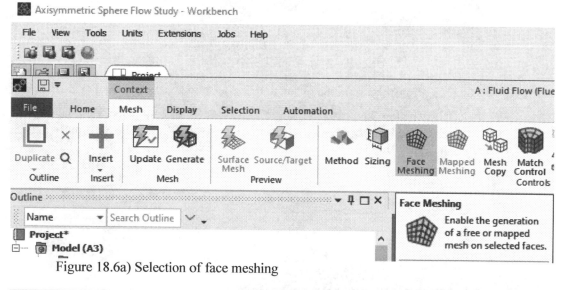

Figure 18.6a) Selection of face meshing

Figure 18.6b) Details of face meshing

Select Mesh>>Controls>>Sizing from the menu, right-click in the graphics window and select Cursor Mode>>Edge. Control-select the two edges 8 and 19, see Figure 18.6c).

Figure 18.6c) Numbered edges for mesh sizing

Apply the Geometry under Details of "Sizing". Select Number of Divisions as Type under Details of "Sizing". Set the number of Divisions to 100. Set Capture Curvature to No and set Behavior to Hard. Select the first Bias Type from the drop-down menu. Enter a Bias Factor of 10.

Figure 18.6d) Details for 1st mesh sizing

Select Mesh Control>>Sizing from the menu. Control-select edges 11 and 22. Apply the Geometry under Details of "Sizing". Select Number of Divisions as Type under Details of "Sizing". Set the number of Divisions to 150. Set Capture Curvature to No and set Behavior to Hard. Select the second Bias Type from the drop-down menu. Enter a Bias Factor of 10.

Figure 18.6e) Details for 2nd mesh sizing

Select Mesh Control>>Sizing from the menu. Select edge 1. Apply the Geometry under Details of "Sizing". Select Number of Divisions as Type under Details of "Sizing". Set the number of Divisions to 100. Set Capture Curvature to No and Behavior to Hard. Select the second Bias Type from the drop-down menu. Enter a Bias Factor of 10.

Figure 18.6f) Details for 3rd mesh sizing

Select Mesh Control>>Sizing from the menu. Select edge 3. Apply the Geometry under Details of "Sizing". Select Number of Divisions as Type under Details of "Sizing". Set the number of Divisions to 150. Set Capture Curvature to No and Behavior to Hard. Select the first Bias Type from the drop-down menu. Enter a Bias Factor of 10.

Figure 18.6g) Details for 4th mesh sizing

Select Mesh Control>>Sizing from the menu. Control select edge 2. Apply the Geometry under Details of "Sizing". Select Number of Divisions as Type under Details of "Sizing". Set the number of Divisions to 54. Set Capture Curvature to No and Behavior to Hard.

Figure 18.6h) Details for 5th mesh sizing

Select Mesh Control>>Sizing from the menu. Control-select edges 9, 10, 12, 13, 17, 18, and the four parts of the half-circle. Apply the Geometry under Details of "Sizing". Select Number of Divisions as Type under Details of "Sizing". Set the number of Divisions to 27. Set Capture Curvature to No and Behavior to Hard.

Figure 18.6i) Details for 6th mesh sizing

Select Mesh Control>>Sizing from the menu. Control-select the three edges 4, 5, and 6. Apply the Geometry under Details of "Sizing". Select Number of Divisions as Type under Details of "Sizing". Set the number of Divisions to 100. Set Capture Curvature to No and Behavior to Hard. Select the first Bias Type from the drop-down menu. Enter a Bias Factor of 20.

Figure 18.6j) Details for 7th mesh sizing

Select Mesh Control>>Sizing from the menu. Select edge 7. Apply the Geometry under Details of "Sizing". Select Number of Divisions as Type under Details of "Sizing". Set the number of Divisions to 100. Set Capture Curvature to No and Behavior to Hard. Select the second Bias Type from the drop-down menu. Enter a Bias Factor of 20.

Figure 18.6k) Details for 8th mesh sizing

Select Mesh Control>>Sizing from the menu. Select edges 14, 15, 16, 20, and 21. Apply the Geometry under Details of "Sizing". Select Number of Divisions as Type under Details of "Sizing". Set the number of Divisions to 18. Set Capture Curvature to No and Behavior to Hard.

Figure 18.6l) Details for 9th mesh sizing

Right click on Mesh in the tree outline and select Generate Mesh. The mesh has 39571 Nodes and 39094 Elements.

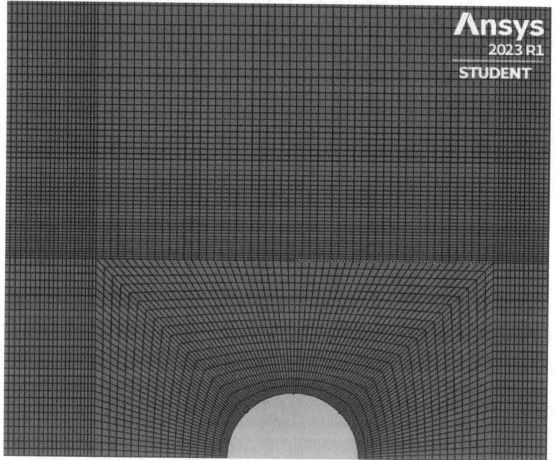

Figure 18.6m) Details of mesh around half-sphere

7. Select geometry in the Outline. Right-click in the graphics window and select Cursor Mode>>Edge. Control select the two vertical edges on the right-hand side, right click and select Create Named Selection. Name the edges *outlet* and click OK.

Name the two vertical edges to the left *inlet* and the four edges of the sphere with the name *sphere*.

Name the three horizontal edges at the top *symmetry* and name the four horizontal edges at the bottom *axisymmetry*.

Export the mesh and save it with the name *flow-past-sphere-mesh.msh*.

Open Geometry in the Outline and select Line Body. Right click on Line Body and select Suppress Body.

Close the meshing window. Right click on Mesh under Project Schematic in Ansys Workbench and select Update.

Figure 18.7 Named selections

F. Launching Ansys Fluent

8. Double click on Setup under Project Schematic in Ansys Workbench. Check the Options box Double Precision. Set the number of processes equal to the number of processor cores for your computer. To check the number of physical cores, press the Ctrl + Shift + Esc keys simultaneously to open the Task Manager. Go to the Performance tab and select CPU from the left column. You will see the number of physical cores on the bottom-right side. Close the Task Manager window. Click on the Start button in the Fluent Launcher window. Select 2D Space Axisymmetric under General on the Task Page.

Figure 18.8a) Launching Fluent

Task Page ⊠

General ⑦

Mesh

[Scale...] [Check] [Report Quality]

[Display...] [Units...]

Solver

Type
- ⦿ Pressure-Based
- ○ Density-Based

Velocity Formulation
- ⦿ Absolute
- ○ Relative

Time
- ⦿ Steady
- ○ Transient

2D Space
- ○ Planar
- ⦿ Axisymmetric
- ○ Axisymmetric Swirl

☐ Gravity

Figure 18.8b) General settings

Figure 18.8c) Geometry in Fluent

9. Double click on Models under Setup in the Outline View. Double click on Viscous Model and select Laminar Model. Click on OK to close the Viscous Model window. Double click on Boundary Conditions under Setup in the Outline View. Select the *inlet* Zone under Boundary Conditions on the Task Page. Click on Edit.... Select Components as Velocity Specification Method. Enter 0.02921469 as Axial-Velocity [m/s]. This is corresponding to Reynolds number $Re = 100$. Select Apply and Close the Velocity Inlet window.

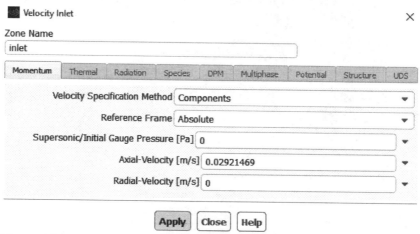

Figure 18.9 Velocity inlet settings

10. Double click on Initialization under Solution in the Outline View. Select Standard Initialization. Select Compute from inlet. Click on Initialize. Double click on Reference Values under Setup in the Outline View. Select Compute from inlet and select Initialize. Double click on Reference Values under Setup in the Outline View. Select Compute from inlet and set the reference values for Area [m²] to 0.001963496, Length [m] to 0.05 and Velocity [m/s] to 0.02921469, see Figure 18.10b). Set the Reference Zone to *solid-surface_body*.

Figure 18.10a) Solution initialization   Figure 18.10b) Reference values

Double-click on Report Definitions under Solution in the Outline View. Select New>>Force Report>>Lift... from the drop-down menu. Select *sphere* as Wall Zones. Check the boxes for Report File, Report Plot and Print to Console under Create. Enter *lift-report-def* as Name. Click on the OK button to close the Lift Report Definition window. Repeat this step for the Drag and set the Force Vector for X to 1, Y to 0 for the Drag Report Definition. Enter *drag-report-def* as Name. Close the Report Definitions window.

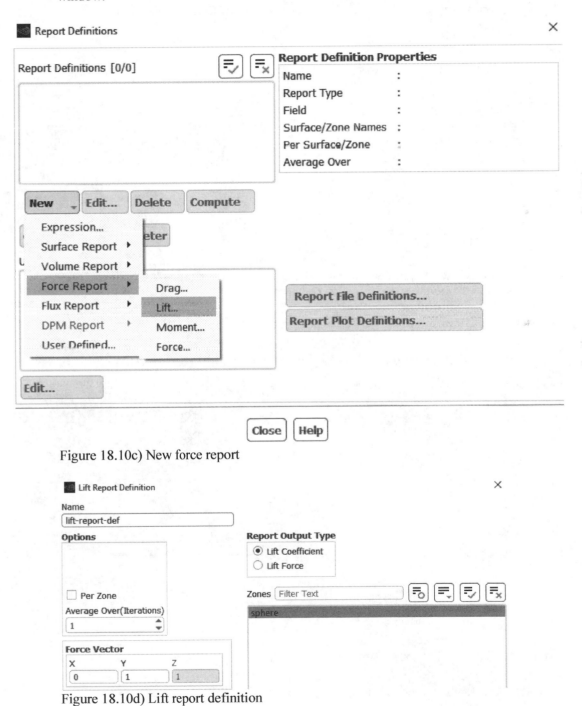

Figure 18.10c) New force report

Figure 18.10d) Lift report definition

11. Double click on Monitors and Residual under Solution in the Outline View. Check the boxes for Print to Console and Plot. Set all three Absolute Criteria to 1e-12. Click on OK to exit the Residual Monitors window.

Figure 18.11 Residual monitors

Double click on Run Calculation under Solution in the Outline View. Enter 300 for the Number of Iterations and click on Calculate.

G. Post-Processing

12. The value of the drag coefficient at $Re = 100$ for the first coarse mesh is $C_d = 1.08846$.

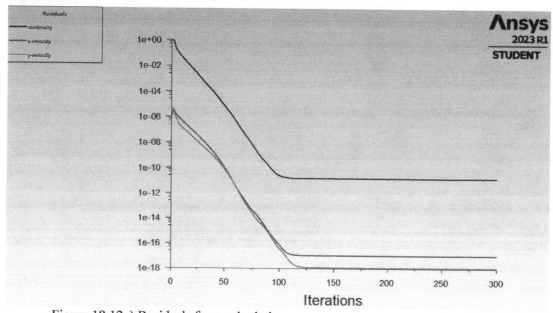

Figure 18.12a) Residuals from calculations

Double-click on Contours in Graphics under Results in the Outline View. Enter *static-pressure* as Contour Name in the Contours window. Select Contours of Pressure… and Static Pressure. Check the box Filled under Options. Deselect all Surfaces and click on Save/Display. Close the Contours window.

Figure 18.12b) Contours plot for static pressure

Double-click on Contours in Graphics under Results in the Outline View. Select Contours of Velocity and Axial Velocity. Enter *axial-velocity* as Contour Name in the Contours window. Check the box Filled under Options. Deselect all surfaces and click on Save/Display. Close the Contours window. Find the coefficient of drag by double-clicking on Reports under Results in the Outline View. Double-click on Forces under Reports. Select sphere under Wall Zones. Select Print to get the value for the total drag coefficient that you have calculated: 1.08846.

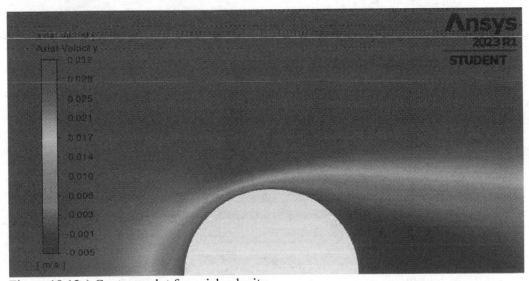

Figure 18.12c) Contours plot for axial velocity

13. Select the Domain tab in the menu. Select Adapt>>Manual…. under the Domain tab from the menu. Select Cell Registers>>New>>Region… in the Manual Mesh Adaption window and enter -0.575 for X Min [m], 1 for X Max [m], 0 for Y Min [m], and 0.625 for Y Max [m]. Select Save and Close the Region Register window. Select General Adaption Controls and set the Maximum Refinement Level to 10. Select *region_0* as Refinement Criterion in the Manual Mesh Adaption window and select Adapt. Select OK in the Information window. This increases the number of cells by a factor of 4 from 39094 to 156376 cells. Close the Manual Mesh Adaption window. Select File>>Save Project from the menu. Select File>>Export>>Case… from the menu. Save the file in the Working Directory as Files of type: CFF Case Files (*.cas.h5) with the name *Axisymmetric Coarse Mesh Flow Past a Sphere at Re = 100.*

18.13a) Adaption controls window

18.13b) Region register

18.13c) Region 0 as refinement criterion

Initialize the flow and run the calculations once again for this new mesh. Enter 300 for the Number of Iterations and click on Calculate. The value of the drag coefficient at Re = 100 for a refined mesh is C_d = 1.0955586. Continue refining the mesh further to see how the value for the drag coefficient is affected.

Number of Cells	$C_{d, Fluent}$	$C_{d, Experiment}$	% difference
39094	1.08846	1.1855	8.2
156376	1.0954523	1.1855	7.6

Table 18.1 Drag coefficient with a refined mesh at Re = 100.

H. Theory

14. The drag coefficient from experiments for a smooth sphere is found using the following curve-fit formula:

$$C_{D,Exp} = \frac{24}{Re} + \frac{6}{1+Re^{1/2}} + 0.4 \qquad 0 \leq Re \leq 200,000 \qquad (18.1)$$

We determine the Reynolds number for the flow around the sphere that is defined as

$$Re = Ud/v = 0.02921469 * 0.05 / 1.46 * 10^{-5} = 100 \qquad (18.2)$$

, where U is the magnitude of the free stream velocity, d is the diameter of the sphere, and v is the kinematic viscosity of air at room temperature. The difference between Ansys Fluent and experiments for the drag coefficient is 8.1%.

I. References

1. Clift, R., Grace, J.R. and Weber, M.E. "Bubbles, Drops and Particles", Academic Press, New York, (1978).

2. Jones, D.A. and Clarke, D.B. "Simulation of Flow Past a Sphere using the Fluent Code", Australian Government Department of Defence, Defence Science and Technology Organisation, Maritime Platforms Division, DSTO-TR-2232, (2008).

3. Roos, F.W. and Willmarth, W.W. "Some Experimental Results on Sphere and Disk Drag", AIAA Journal, **9**(2), 285-291, (1971).

4. Tabata, M. and Itakura, K., "A Precise Computation of Drag Coefficients of a Sphere", Int. J. Comput. Fluid Dynam. **9**, 303, (1998).

J. Exercises

18.1 Run Ansys Fluent Simulations for Re = 20, 40, 60, 80, 100, 120, 140, 160, 180 and 200 and compare your results with equation (18.1). Include a plot of drag coefficient versus Reynolds number in this range.

CHAPTER 19. TAYLOR-COUETTE FLOW

A. Objectives

- Using Ansys Meshing to Create the Mesh for 3D Taylor-Couette Flow
- Inserting Wall Boundary Conditions for Rotating Inner Wall
- Running Laminar Steady 3D Ansys Fluent Simulations
- Using Contour Plots for Visualizations
- Compare Results with Neutral Stability Theory

B. Problem Description

We will study the flow between two vertical cylinders where the inner wall is rotating at 1.5 rad/s. The inner and outer cylinders have radii of 30 mm and 35 mm, respectively. The height of both cylinders is 100 mm. The centrifugal instability will cause the appearance of counter-rotating vortices at low rotation speeds for the inner wall with a fixed outer wall.

C. Launching Ansys Workbench and Selecting Fluent

1. Start by launching Ansys Workbench. Double click on Fluid Flow (Fluent) under Analysis Systems in the Toolbox. Select Units>>Metric (tonne, mm, s, …) from the menu.

Figure 19.1 Selecting Fluent

D. Launching Ansys DesignModeler

2. Right click on Geometry in the Project Schematic window and select New DesignModeler Geometry. Select millimeter as the desired length unit from the menu (Units>>Millimeter) in DesignModeler. Select the ZX Plane in the Tree Outline and select Look at Face/Plane/Sketch ![icon]. Select the Sketching tab and click on Circle. Draw two circles centered at the origin. Make sure that you see a P next to the cursor before you start drawing the circles from the origin. Click on the Dimensions tab and select Radius. Click on the circles in the graphics window and enter 30 mm and 35 mm for the radii of the circles in Details View.

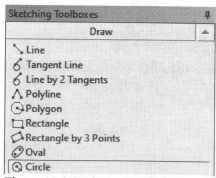

Figure 19.2a) Circle sketch tool

436

Figure 19.2b) Dimensions for circles

Select [Extrude] and apply Sketch1 under ZXPlane in the Tree Outline as Geometry in Details View. Enter 100 mm as the FD1, Depth (>0) of the extrusion. Click on Generate and select isometric view in the graphics window.

Figure 19.2c) Details for extrusion

Create a new sketch in the ZX plane. Select the ZXPlane in the Tree Outline, select New Sketch and Look At Face/Plane/Sketch . Select the new Sketch2 and select the Sketching tab and Line under Draw. Draw a line on the vertical axis through the extruded tube, see Figure 19.2d). Make sure you have the letter C at the starting point, letter V indicating a vertical line and the letter C at the end point when you complete the vertical line.

437

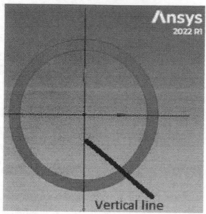

Figure 19.2d) Vertical line through tube

Select Tools>>Face Split from the menu. Turn around the extruded tube and select the face that the line goes through, see Figure 19.2e). Apply the selected face as Target Face in Details View. Select the end points of the vertical line and apply it as Tool Geometry in Details View. Click on Generate and select Face 🔲 . The face has now been split and you can select half the face individually, see Figure 19.2f).

Figure 19.2e) Selected face associated with the line

Figure 19.2f) Half the face selected after completed split

Select Tools>>Projection from the menu. Select Edges on Face as Type in Details View. Control select the two radial line segments and apply them as Edges in Details View, see Figure 19.2g). Turn around the extrusion and select the face on the opposite side. Apply this face as Target in Details View, see Figure 19.2g). Click on Generate. You have now created a total of four different flat surfaces, two at each end of the tube.

Figure 19.2g) Selected edges and target for projection

Turn the tube around to the other end and select Tools>> Face Split once again from the menu. Apply the inner cylindrical face as Target Face in Details View. Apply the point and edge as Tool Geometry as shown in Figure 19.2h). Click on Generate.

Repeat the face split once again but on the opposite side of the inner cylindrical face, see Figure 19.2i). The inner cylindrical face has now been split and you can select half the face individually, see Figure 19.2j). Close the DesignModeler window.

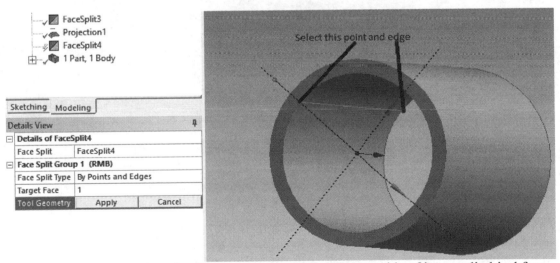

Figure 19.2h) Selected point and edge for face split on one side of inner cylindrical face

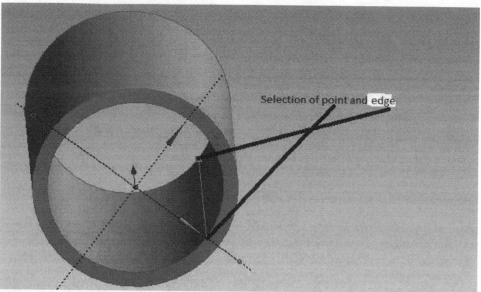

Figure 19.2i) Selected point and edge for split on opposite side of inner cylindrical face

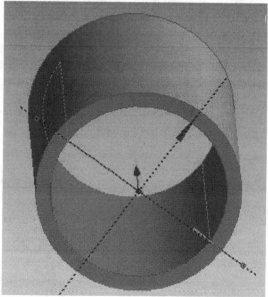

Figure 19.2j) Half the inner cylindrical face selected after completed split

E. Launching Ansys Meshing

3. Double click on Mesh under Project Schematic in Ansys Workbench. Click on Mesh in the Outline under Project. Select Home>>Tools>>Units>>Metric (mm, kg, N, …) from the menu. Select the Mesh under Project in Outline. Open Quality in Details of "Mesh" and set the Smoothing to High, see Figure 19.3a). Generate the Mesh. A tetrahedral mesh is generated.

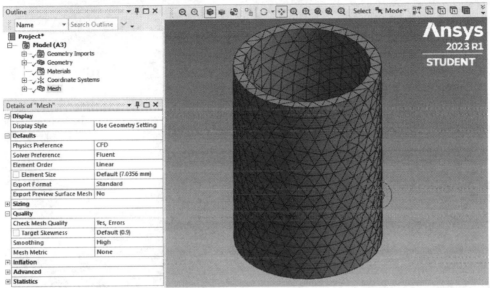

Figure 19.3a) Tetrahedral mesh

Select Mesh>>Controls>>Sizing from the menu. Select Edge , select the radial edge at the top of the extrusion and the corresponding radial line at the bottom of the tube, see Figure 19.3b). Apply the two edges as Geometry in Details of "Edge Sizing" – Sizing. Select Number of Divisions as Type under Definition. Set the Number of Divisions to 32, set Capture Curvature to No and Behavior to Hard. Select the third Bias Type and set the Bias Factor to 20, see Figure 19.3c). Repeat this sizing for the other side of the extrusion.

Figure 19.3b) Radial edges on extrusion

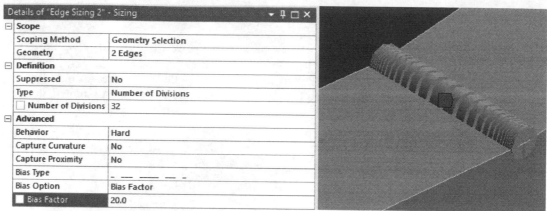

Figure 19.3c) Radial sizing with bias

Right click on Mesh in Outline and select Insert>>Method. Select the extrusion body and Apply it as Geometry under Scope in Details of "Automatic Method" – Method. Select MultiZone as Method under Definition, see Figure 19.3d).

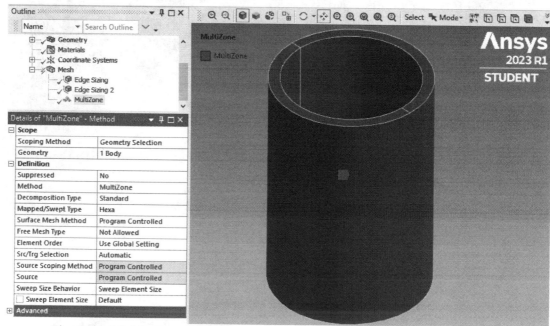

Figure 19.3d) Multizone settings

Select Mesh>>Controls>>Sizing from the menu. Select Edge, control select the eight half-circles at the top and the bottom of the extrusion and Apply the eight Edges as Geometry under Scope in Details of "Sizing" – Sizing, see Figure 19.3e). Select Number of Divisions as Type under Definition. Set the Number of Divisions to 64, set Capture Curvature to No and Behavior to Hard. Right click on Mesh in Outline and select Mesh>>Update from the menu.

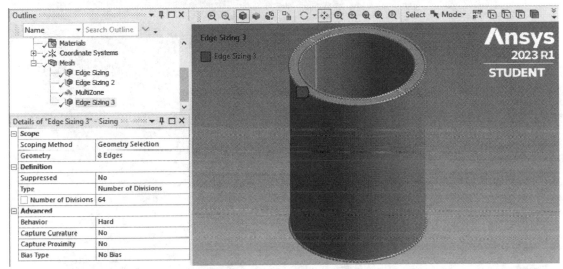

Figure 19.3e) Azimuthal sizing

Select Mesh>>Controls>>Sizing from the menu. Select Edge ⬚, control select the two inner vertical lines of the extrusion and Apply the two Edges as Geometry under Scope in Details of "Sizing" – Sizing, see Figure 19.3f). Select Number of Divisions as Type under Definition. Set the Number of Divisions to 256, set Capture Curvature to No and Behavior to Hard. Right click on Mesh in Outline and select Mesh>>Update from menu.

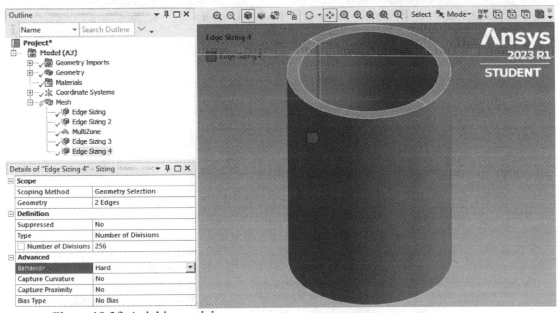

Figure 19.3f) Axial inner sizing

Open Statistics in Details of "Mesh" to see the total number of Nodes 1,085,568 and total number of Elements 1,048,576.

Figure 19.3g) Finished mesh with 1,048,576 elements

Right click in the graphics window and select Cursor Mode>>Face. Select the outer cylindrical face, right click and select Create Named Selection. Enter the name *wall_outer* and click on OK to close the Selection Name window. Repeat this for the two inner cylindrical faces (control-select the two faces) with the name *wall_inner*, the bottom two faces named *wall_bottom* and the top two faces named *wall_top*.

Figure 19.3b) Details for named selections

Select File>>Export...>>Mesh>>FLUENT Input File>>Export from the menu and save the mesh in the *Working Directory* with the name *cylinder-mesh-taylor-couette-flow.msh*. Select File>>Save Project from the menu and save the project with the name *Taylor-Couette.wbpj*. Close the Meshing window and right-click on Mesh under Project Schematic in Ansys Workbench and select Update.

F. Launching Ansys Fluent

4. Double click on Setup under Project Schematic in Ansys Workbench. Check the Options box Double Precision. Increase the Number of Solver Processes to the same number as the number of computer cores. Click on Start.

Figure 19.4a) Launching Fluent

Click on Check and Scale… under Mesh in General on the Task Page to verify the Domain Extents. Close the Scale Mesh window. Select Transient as Time under General on the Task Page. Select Relative as Velocity Formulation. Double click on Models and Viscous (SST k-omega) under Setup in the Outline View. Select Laminar as Viscous Model and click OK to close the window.

Figure 19.4b) Domain extents for Taylor-Couette flow mesh

Figure 19.4c) General settings

Double click on Materials under Setup in the Outline View. Click on Create/Edit for Fluid under Materials on the Task Page. Click on Fluent Database in the Create/Edit Materials window. Scroll down and select *water-liquid (h2o<l>)* as the Fluent Fluid Material. Click on Copy and Close the Fluent Database Materials window. Click on Change/Create and Close the Create/Edit Materials window.

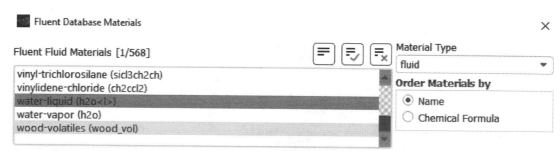

Figure 19.4d) Selection of water-liquid as fluid

5. Open Cell Zone Conditions and Fluid under Setup in the Outline View. Double click on *solid (fluid, id=2)* under Cell Zone Conditions. Select *water-liquid* as Material Name. Check the box for Frame Motion and enter Speed (rad/s) -1.5 as Rotational Velocity. Set the Rotation-Axis Direction as X 0, Y 1, Z 0. Click Apply and Close the Fluid window.

Figure 19.5a) Selection of water-liquid as fluid and setting for rotational velocity

Double click on Boundary Conditions. Select *wall_inner* under Zone in Boundary Conditions on the Task Page. Click on the Edit... button for wall Type. Select *Moving Wall* under Wall Motion. Check Absolute and Rotational under Motion and enter 1.5 as Speed (rad/s). Set the Rotation-Axis Direction as X 0, Y 1, Z 0, see Figure 19.5b). Click Apply and Close the window.

Repeat this step for the *wall_outer* Zone, use Absolute and Rotational Motion, enter 0 as Speed (rad/s) and set the Rotation-Axis Direction as X 0, Y 1, Z 0, see Figure 19.5c). Do the same for the top and bottom walls, see Figure 19.5d) and Figure 19.5e).

Figure 19.5b) Details of wall motion for inner wall.

447

Figure 19.5c) Details of wall motion for outer wall.

Figure 19.5d) Details of wall motion for top wall.

Figure 19.5e) Details of wall motion for bottom wall.

6. Double click on Methods under Solution in the Outline View. Set the Pressure-Velocity Coupling Scheme to PISO and set Momentum under Spatial Discretization to Third-Order MUSCL, see Figure 19.6a).

Figure 19.6a) Solution methods

Double click on Monitors and Residual under Solution in the Outline View. Set the Absolute Criteria for all Residuals to 0.0001. Select OK to close the Residual Monitors window.

Figure 19.6b) Residual monitors

Double click on Initialization under Solution in the Outline View and select Hybrid Initialization. Click on Initialize. Select File>>Write>>Case... from the menu. Save the file as Files of type: CFF Case Files (*.cas.h5) with the name *taylor-couette-flow.cas.h5*. Double click on Run Calculation under Solution in the Outline View. Enter 250 for the Number of Time Steps, 0.1 for Time Step Size [s] and 80 for Max Iterations/Time Step. Click on Calculate.

Figure 19.6c) Solution initialization

Figure 19.6d) Run calculations

G. Post-Processing

7. The residuals are shown in Figure 19.7a). Double-click on Contours in Graphics under Results in the Outline View. Enter *axial-velocity* as Contour Name in the Contours window. Select Contours of Velocity… and Axial Velocity. Select New Surface>>Plane…. Select YZ Plane as Method and select Create. Check the box Filled under Options. Uncheck Node Values, Boundary Values and Global Range under Options. Select Colormap Options… and select Fixed as Font Behavior under Font. Set the Font Size to 12 and select Apply. Close the Colormap window. Select *wall_outer* under Surfaces and click on Save/Display. Select the red X coordinate axis in the graphics window. Deselect *wall_outer* and select *plane-4* and click on Save/Display. Close the Contours window. Repeat this step for radial and tangential velocities.

Figure 19.7a) Residuals from calculations

Figure 19.7b) Contours settings

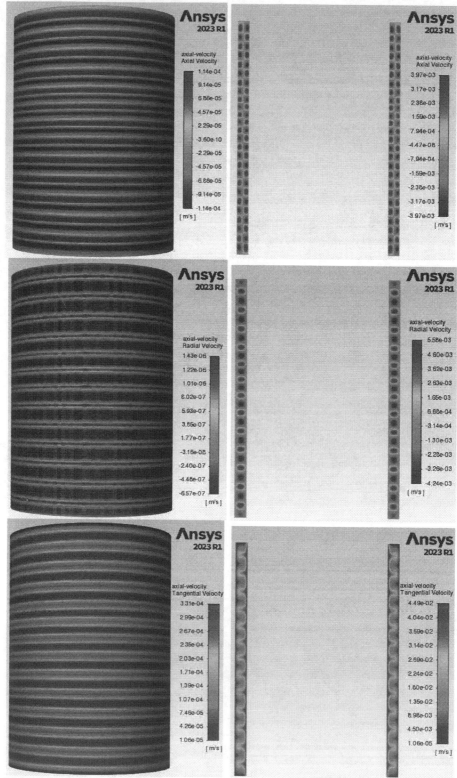

Figure 19.7c) Contours of axial, radial & tangential velocities in front & YZ plane views.

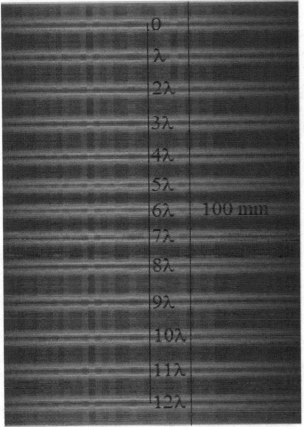

Figure 19.7d) Determining average spanwise wave length using radial velocity contours.

H. Theory

8. The instability of the flow between two vertical rotating cylinders is governed by the so-called Taylor number Ta.

$$Ta = \frac{4\Omega_i^2 d^4}{\nu^2} \qquad (19.1)$$

where Ω_i is the rotation rate of the inner cylinder, d is the distance between the cylinders and ν is the kinematic viscosity of the fluid. Below the critical $Ta_{crit} = 3430$ for a non-rotating outer cylinder, the flow is stable but instabilities will develop above this Taylor number. The non-dimensional wave number α of the centrifugal instability is determined by

$$\alpha = \frac{2\pi d}{\lambda} \qquad (19.2)$$

where λ is the spanwise wave length of the instability. The critical wave numbers $\alpha_{crit} = 3.12$ in the narrow gap limit: $\eta \to 1$. The radius ratio is defined as $\eta = r_i/r_o$ where r_i and r_o is the radius of the inner and outer cylinders, respectively. From figure 19.7c), the wave number can be determined to be

$$\alpha = \frac{2\pi \cdot 0.005}{0.007545} = 4.164 \qquad (19.3)$$

In order to determine the wave-length that is used to determine the wave number in equation (19.3), you can measure the total length in the vertical spanwise direction for twelve red or twelve blue successive bands as shown in figure 19.7d). Next, you take this vertical length and divide by twelve to get the average wave-length in the vertical spanwise direction. To get the scale for figure 19.7d), use the total height of the cylinder in the vertical direction in figure 19.7d) that corresponds to the 100 mm height of the cylinder.

The Taylor number in the Ansys Fluent simulations is

$$Ta = \frac{4 \cdot 1.5^2 \cdot 0.005^4}{(1.004 \cdot 10^{-6})^2} = 5580 \qquad (19.4)$$

The neutral stability curve can be given to the first approximation, see figure 19.8.

$$Ta = \frac{2(\pi^2 + \alpha^2)^3}{(1+\mu)\alpha^2 \left\{ 1 - \frac{16\alpha\pi^2 \cosh^2\left(\frac{\alpha}{2}\right)}{\left[(\pi^2 + \alpha^2)^2 (\sinh \alpha + \alpha)\right]} \right\}} \qquad (19.5)$$

where $\mu = \Omega_o / \Omega_i$ and Ω_o is the rotation rate of the outer cylinder.

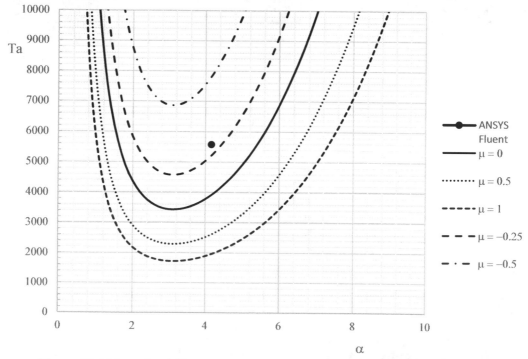

Figure 19.8 Neutral stability curves for Taylor-Couette flow. The filled circle represents result from Ansys Fluent. The radius ratio is $\eta = \frac{6}{7}$ and $\mu = 0$ in Ansys Fluent simulations and $\eta = 1$ for the stability curves.

CHAPTER 19. TAYLOR COUETTE FLOW

I. References

1. Chandrasekhar, S., Hydrodynamic and Hydromagnetic Stability, Dover, 1981.
2. Koschmieder, E.L., Benard Cells and Taylor Vortices, Cambridge, 1993.
3. Matsson, J.E, On the Integration of Fluid Flows and Fabrication, *Proceedings of the 2002 American Society for Engineering Education Annual Conference & Exposition.*, Montreal, Canada, (2002).
4. Taylor, G.I., Stability of a Viscous Liquid Contained between Two Rotating Cylinders, *Phil. Trans. Roy. Soc.* London Ser. A, Vol. **223**, (1923).

J. Exercises

19.1 Run the calculations for the flow in the Taylor-Couette apparatus with only the inner cylinder rotating $\mu = 0$ for Taylor numbers $Ta = 10000$, 20000 and 30000 and compare the spanwise wave numbers with the one determined in this chapter for $Ta = 5580$. Include your results in figure 19.8 for comparison with the neutral stability curve.

19.2 Run the calculations for the flow in the Taylor-Couette apparatus with both cylinders rotating $\mu = 1$ and for Taylor numbers $Ta = 3000$, 5000 and 10000 and determine the spanwise wave numbers. Include your results in a graph and compare with the neutral stability curve corresponding to $\mu = -1/2$, see equation (19.5).

K. Notes

CHAPTER 20. DEAN FLOW IN A CURVED CHANNEL

A. Objectives

- Using Ansys Meshing to Create the Mesh for 3D Dean Flow in a Curved Channel
- Inserting Wall Boundary Conditions
- Running Laminar Steady 3D Ansys Fluent Simulations
- Using Contour Plots for Visualizations
- Compare Results with Neutral Stability Theory

B. Problem Description

Dean flow appears in a curved tube and can also be shown in a narrow gap curved channel. We will therefore study the flow between two curved walls where the inner wall has a radius of 395 mm and the radius of the outer wall is 405 mm. The height of the curved channel is 100 mm. The centrifugal instability will cause the appearance of counter-rotating vortices at low flow rates.

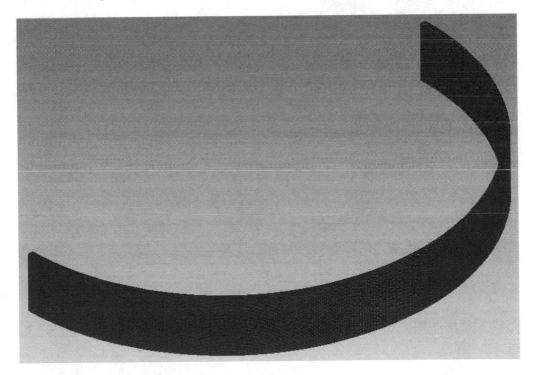

C. Launching Ansys Workbench and Selecting Fluent

1. Start by launching Ansys Workbench. Double click on Fluid Flow (Fluent) under Analysis Systems in the Toolbox. Select Units>>Metric (tonne, mm, s, …) from the menu.

Figure 20.1 Selecting Fluent

D. Launching Ansys DesignModeler

2. Right click on Geometry in the Project Schematic window and select New DesignModeler Geometry. Select millimeter as the desired length unit from the menu (Units>>Millimeter) in DesignModeler. Select the ZX Plane in the Tree Outline and select Look at Face/Plane/Sketch ![icon].

   Select the Sketching tab and Arc by Center. Draw two 180 degrees half-circle arcs centered at the origin. Make sure you have the letter *P* when you move the cursor to the origin to define the center of the arc and to start drawing the half circle from the negative horizontal axis with letter *C* showing up to the positive horizontal axis with the letter *C* to finish the arc.

   Click on the Dimensions tab and select Radius. Click on the two arcs in the graphics window and enter 395 mm and 405 mm for the radii of the half-circles in Details View. Select the Line sketch tool and draw two horizontal lines that connects the ends of the arcs.

Figure 20.2a) Arc by center sketch tool

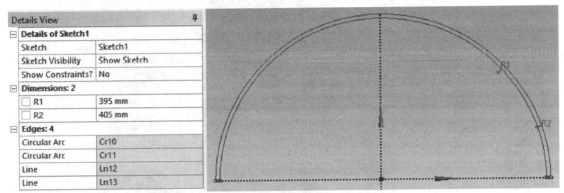

Figure 20.2b) Dimensions for half-circles

Select **Extrude** and apply Sketch1 under ZXPlane in the Tree Outline as Geometry in Details View. Enter 100 mm as the FD1, Depth (>0) of the extrusion. Click on Generate and select isometric view in the graphics window. Close the DesignModeler.

Figure 20.2c) Details for extrusion of curved channel

E. Launching Ansys Meshing

3. Double click on Mesh under Project Schematic in Ansys Workbench. Select Mesh in the Outline under Project in the Meshing window. Select Home>>Tools>>Units>>Metric (mm, kg, N, …) from the menu. Select the Mesh under Project in Outline.

   Open Quality in Details of "Mesh" and set the Smoothing to High, see Figure 20.3a). Generate the Mesh.

Figure 20.3a) Coarse mesh

Select Mesh>>Controls>>Sizing from the menu. Select Edge , select the radial edge at the top of one side the curved channel and the corresponding radial line at the bottom of the same channel, see Figure 20.3b). Apply the two edges as Geometry in Details of "Edge Sizing" – Sizing. Select Number of Divisions as Type under Definition. Set the Number of Divisions to 32, set Capture Curvature to No and Behavior to Hard. Select the third Bias Type - — —— — - and set the Bias Factor to 20, see Figure 20.3c). Repeat this sizing for the other side of the curved channel.

Figure 20.3b) Radial edges on curved channel

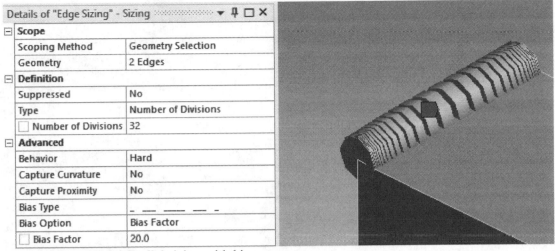

Figure 20.3c) Radial sizing with bias

Right click on Mesh in Outline and select Insert>>Method. Select the curved channel and Apply it as Geometry under Scope in Details of "Automatic Method" – Method. Select MultiZone as Method under Definition, see Figure 20.3d).

Figure 20.3d) Multizone settings for curved channel

461

Select Mesh>>Controls>>Sizing from the menu. Select Edge, control select the four half-circles at the top and the bottom of the curved channel and Apply the four Edges as Geometry under Scope in Details of "Sizing" – Sizing, see Figure 20.3e). Select Number of Divisions as Type under Definition. Set the Number of Divisions to 128, set Capture Curvature to No and Behavior to Hard. Right click on Mesh in Outline and select Mesh>>Update from the menu.

Figure 20.3e) Azimuthal sizing

Select Mesh>>Controls>>Sizing from the menu. Select Edge, control select the two inner vertical lines of the curved channel and Apply the two Edges as Geometry under Scope in Details of "Sizing" – Sizing, see Figure 20.3f). Select Number of Divisions as Type under Definition. Set the Number of Divisions to 128, set Capture Curvature to No and Behavior to Hard. Right click on Mesh in Outline and select Mesh>>Update from the menu.

Figure 20.3f) Axial inner sizing

Open Statistics in Details of "Mesh" to see the total number of Nodes 549,153 and total number of Elements 524,288.

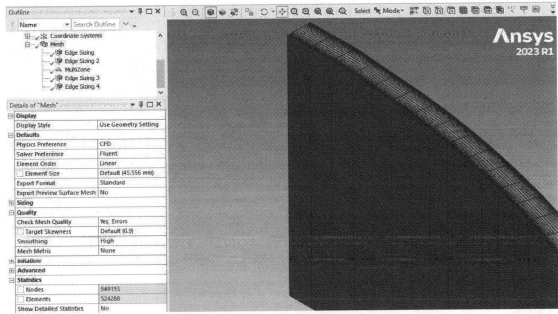

Figure 20.3g) Finished mesh with 524,288 elements

Right click in the graphics window and select Cursor Mode>>Face. Select the outer cylindrical face, right click and select Create Named Selection. Enter the name *wall_outer* and click on OK to close the Selection Name window. Do the same for the inner cylindrical wall and name this *wall_inner*. Furthermore, name the top curved flat surface as *wall top* and the bottom curved flat surface as *wall_bottom*. Select the flat inlet face and name this *inlet*. Select the flat outlet face and name this *outlet*, see Figure 20.3h).

Figure 20.3h) Details for named selections

Select File>>Export...>>Mesh>>FLUENT Input File>>Export from the menu and save the mesh in the *Working Directory* with the name *curved-channel-mesh-dean-flow.msh*. Select File>>Save Project from the menu and save the project with the name *Dean Flow in Curved Narrow Channel.wbpj*. Close the Meshing window and right-click on Mesh under Project Schematic in Ansys Workbench and select Update.

F. Launching Ansys Fluent

4. Double click on Setup under Project Schematic in Ansys Workbench. Check the Options box Double Precision. Set the Number of Solver Processes to the same number as computer cores. Click on Start.

Figure 20.4a) Launching Fluent

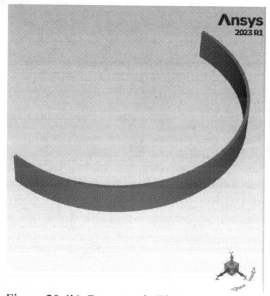

Figure 20.4b) Geometry in Fluent

Click on Check and Scale… under Mesh in General on the Task Page to verify the
Domain Extents. Close the Scale Mesh window. Set Time as Transient. Check the box
for Gravity. Enter -9.81 as Gravitational Acceleration in the Y direction. Double click on
Models and Viscous (SST k-omega) under Setup in the Outline View. Select Laminar as
Viscous Model and click OK to close the window.

Scale Mesh

Domain Extents

Xmin [m]	-2.73188e-24	Xmax [m]	0.405
Ymin [m]	0	Ymax [m]	0.1
Zmin [m]	-0.405	Zmax [m]	0.405

Figure 20.4c) Domain extents for curved channel flow mesh

Figure 20.4d) General settings including transient time and gravity

Double click on Materials under Setup in the Outline View. Click on Create/Edit for Fluid under Materials on the Task Page. Click on Fluent Database in the Create/Edit Materials window. Scroll down and select *water-liquid (h2o<l>)* as the Fluent Fluid Material. Click on Copy and Close the Fluent Database Materials window. Click on Change/Create and Close the Create/Edit Materials window.

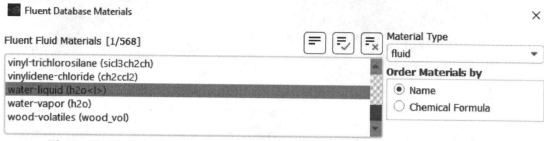

Figure 20.4e) Selection of water-liquid as fluid

5. Open Cell Zone Conditions and Fluid under Setup in the Outline View. Double click on *solid (fluid, id=2)* under Cell Zone Conditions. Select *water-liquid* as Material Name and enter *water-liquid* as the Zone Name. Select Apply and Close the Fluid window.

Figure 20.5a) Selection of water-liquid as fluid

Double click on Boundary Conditions. Select *inlet* under Zone in Boundary Conditions on the Task Page. Click on the Edit… button for velocity-inlet Type. Select *Components* as Velocity Specification Method. Select Local Cylindrical (Radial, Tangential, Axial) as Coordinate System. Enter 0.1 as Tangential-Velocity [m/s]. Enter X 0, Y 1 and Z 0 as Axis Direction, see figure 20.5b). Select Apply and Close the Velocity Inlet window.

Figure 20.5b) Details for velocity inlet of curved narrow channel

6. Double click on Methods under Solution in the Outline View. Use the settings as shown in Figure 20.6a).

Figure 20.6a) Solution methods

Double click on Monitors and Residual under Solution in the Outline View. Set the Absolute Criterial for all Residuals to 0.0001, see Figure 20.6b). Select OK to close the Residual Monitors window.

Figure 20.6b) Residual monitors

Double click on Initialization under Solution in the Outline View and select Hybrid Initialization. Click on Initialize. Select File>>Save Project from the menu. Select File>>Export>>Case… from the menu. Save the file as Files of type: CFF Case Files (*.cas.h5) with the name *dean-flow.cas.h5.* Double click on Run Calculation under Solution in the Outline View. Enter 250 for the Number of Iterations, 0.1 as Time Step Size [s] and 80 for Max Iterations/Time Step. Click on Calculate.

Figure 20.6c) Solution initialization

G. Post-Processing

7. The residuals are shown in Figure 20.7a).

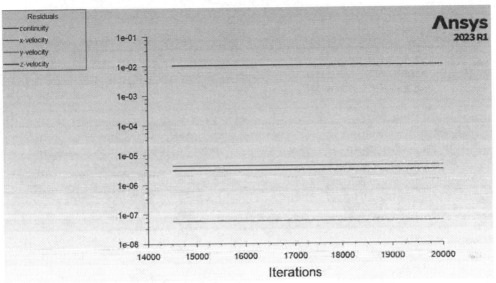

Figure 20.7a) Residuals from calculations

Double-click on Contours in Graphics under Results in the Outline View. Enter *spanwise-y-velocity* as Contour Name in the Contours window. Select Contours of Velocity... and Y Velocity. Select New Surface>>Plane.... Select *XY Plane* as Method, *xy-plane* as New Surface Name and select Create. Close the Plane Surface window. Check the box Filled under Options. Uncheck Node Values, Boundary Values and Global Range under Options. Select Colormap Options... and select Fixed as Font Behavior under Font. Set the Font Size to 10 and select Apply. Close the Colormap window. Select *wall_outer* under Surfaces and click on Save/Display, see Figure 20.7b). Close the Contours window.

Figure 20.7b) Contours settings

Select Set View From -X Direction from the left of the graphics window, see Figures 20.7c) and 20.7d). Manipulate the light properties by unchecking the box for Lighting under Display from the View tab in the menu, see Figure 20.7e).

Figure 20.7c) Setting the view   Figure 20.7d) Selecting -X direction view

Figure 20.7e) Display settings

Figure 20.7f) Y-velocity contours from inside of curved channel, flow from right to left

Figure 20.7g) Determination of spanwise wave length for curved channel flow

In Figure 20.7g), you can see six distinct streaks in the middle of the channel. The streaks indicate the presence of streamwise counter rotating vortices with a wavelength in the spanwise Y direction that can be determined from the figure.

H. Theory

8. The centrifugal instability for the flow in a curved channel is determined by the Dean number De.

$$De = Re\sqrt{\gamma} = \frac{U_b d}{v}\sqrt{\frac{d}{R}} \qquad (20.1)$$

, where U_b is the bulk flow velocity in the curved channel, d is the channel width, R its radius of curvature at the channel centerline and v is the kinematic viscosity of the fluid. Below the critical $De_{crit} = 36$, the flow is stable but instabilities will develop above this Dean number[1,2]. The non-dimensional wave number α of the centrifugal instability is determined by

$$\alpha = \frac{2\pi d}{\lambda} \qquad (20.2)$$

where λ is the spanwise wave length of the instability. The critical wave numbers $\alpha_{crit} = 3.96$. From figure 20.7c), the wave number can be determined to be

$$\alpha = \frac{2\pi \cdot 0.010}{0.00985} = 6.38 \qquad (20.3)$$

In order to determine the wave length that is used to determine the wave number in equation (20.3), you can measure the total length in the vertical spanwise direction for three red or three blue successive bands in figure 20.7d). Next, you take this vertical length and divide by three to get the average wave length in the vertical spanwise direction. To get the scale for figure 20.7d), use the total height of the curved channel in the vertical direction in figure 20.7d) that corresponds to 100 mm.

The Dean number in the Ansys Fluent simulations is

$$De = Re\sqrt{\gamma} = \frac{0.1*0.01}{10^{-6}}\sqrt{\frac{0.01}{0.4}} = 158 \qquad (20.4)$$

The neutral stability curve to the first approximation[4], see Figure 20.8.

$$De = \frac{1}{6\alpha}\left[\frac{(\pi^2+\alpha^2)(4\pi^2+\alpha^2)}{(A-B-C)(D+E-F)}\right]^{1/4} \qquad (20.5)$$

where

$$A = \frac{2(640-48\pi^2)}{27\pi^4(\alpha^2+\pi^2)^2} + \frac{64(9\pi^2-56)}{27\pi^2(\alpha^2+\pi^2)^3} - \frac{256(\alpha^2-5\pi^2)}{9\pi^2(\alpha^2+\pi^2)^4} \qquad (20.6)$$

$$B = \frac{64\pi^2\left[16\alpha\pi^6+24\alpha\pi^4\left(\alpha^2+3\right)+\alpha^3\pi^2\left(9\alpha^2-38\right)+\alpha^5\left(\alpha^2-2\right)\right]}{\left(\alpha^2+4\pi^2\right)^3\left(\alpha^2+\pi^2\right)^4(\sinh\alpha+\alpha)}$$ (20.7)

$$C = \frac{64\pi^2\left\{\left(24\pi^4-34\alpha^2\pi^2-22\alpha^4\right)\alpha\cosh\alpha+\left[16\pi^4\left(\pi^2+3\right)+4\alpha^2\pi^2\left(8\pi^2-21\right)+\alpha^4\left(19\pi^2+36\right)+3\alpha^6\right]\sinh\alpha\right\}}{\left(\alpha^2+4\pi^2\right)^3\left(\alpha^2+\pi^2\right)^4(\sinh\alpha+\alpha)}$$ (20.8)

$$D = \frac{2\left(640-48\pi^2\right)}{27\pi^4\left(\alpha^2+4\pi^2\right)^2} - \frac{64\left(9\pi^2-104\right)}{27\pi^2\left(\alpha^2+4\pi^2\right)^3} - \frac{256\left(\alpha^2-20\pi^2\right)}{9\pi^2\left(\alpha^2+4\pi^2\right)^4}$$ (20.9)

$$E = \frac{64\pi^2\left[4\alpha\pi^6+9\alpha\pi^4\left(\alpha^2+8\right)+2\alpha^3\pi^2\left(3\alpha^2+59\right)+\alpha^5\left(\alpha^2-2\right)\right]}{\left(\alpha^2+4\pi^2\right)^4\left(\alpha^2+\pi^2\right)^3(\sinh\alpha-\alpha)}$$ (20.10)

$$F = \frac{64\alpha\pi^2\left\{\left(24\pi^4-46\alpha^2\pi^2-22\alpha^4\right)\alpha\cosh\alpha+\left[4\pi^4\left(\pi^2+12\right)+\alpha^2\pi^2\left(17\pi^2-156\right)+4\alpha^4\left(4\pi^2+9\right)+3\alpha^6\right]\sinh\alpha\right\}}{\left(\alpha^2+4\pi^2\right)^4\left(\alpha^2+\pi^2\right)^3(\sinh\alpha-\alpha)}$$ (20.11)

Figure 20.8 Neutral stability curve for Dean flow in a curved channel. The filled circle represents result from Ansys Fluent.

I. References

1. Chandrasekhar, S., Hydrodynamic and Hydromagnetic Stability, Dover, 1981.
2. Dean, W.R., Fluid motion in a curved channel. *Proc. Roy. Soc.* A**121**, 402, 1928.
3. Matsson, J.E. and Alfredsson, P.H., Curvature- and rotation induced instabilities in channel flow. *J. Fluid Mech.* **210**, 537, (1990).
4. Reid, W.H., On the stability of viscous flow in a curved channel. Proc. Roy. Soc. A, **244**, 1237, (1958).

J. Exercises

20.1 Run the calculations for the flow in the curved channel but use the following tangential velocities: 0.03, 0.06, 0.09 and 0.12 m/s and determine the Dean number and the spanwise wave number for each tangential velocity.

20.2 Run the calculations for the flow in the curved channel as shown in this chapter for different mesh sizes using Element Size 1.6, 1.8, 2.0 and 2.2 mm. Determine the spanwise wave number for each case and compare with the critical value.

K. Notes

CHAPTER 21. ROTATING CHANNEL FLOW

A. Objectives

- Using Ansys Meshing to Create the Mesh for 3D Plane Channel Flow
- Inserting Boundary Conditions
- Running Laminar Steady 3D Ansys Fluent Simulations
- Using Contour Plots for Visualizations
- Compare Results with Neutral Stability Theory

B. Problem Description

In this chapter we will study the appearance of streamwise vortices for the flow in a narrow gap rotating plane channel with rectangular cross section. The height of the rectangular cross section is 100 mm and the channel width is 10 mm. The length of the channel is 1000 mm. The Coriolis force due to spanwise system rotation will cause the appearance of counter-rotating vortices and transition to turbulent flow at much lower Reynolds numbers as compared with the non-rotating plane channel flow case.

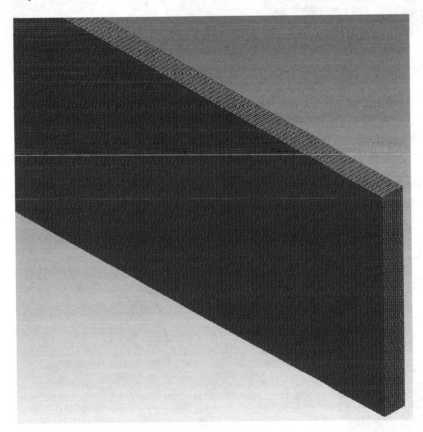

C. Launching Ansys Workbench and Selecting Fluent

1. Start by launching Ansys Workbench. Double click on Fluid Flow (Fluent) under Analysis Systems in the Toolbox. Select Units>>Metric (tonne, mm, s, …) from the menu.

Figure 21.1 Selecting Fluent

D. Launching Ansys DesignModeler

2. Right click on Geometry in the Project Schematic window and select New DesignModeler Geometry. Select millimeter as the desired length unit from the menu (Units>>Millimeter) in DesignModeler. Select the YZPlane in the Tree Outline and

   select Look at Face/Plane/Sketch ⬚. Select the Sketching tab and click on Rectangle under Draw in Sketching Toolboxes. Draw a rectangle from the origin. Make sure you have the letter *P* before you start drawing the rectangle from the origin. Click on the Dimensions tab. Click on the vertical and horizontal sides of the rectangle and enter 10 mm and 100 mm for the vertical and horizontal dimensions in Details View.

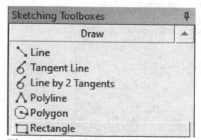

Figure 21.2a) Rectangle sketch tool

476

Figure 21.2b) Dimensions for rectangle

Select 🔲 Extrude and apply Sketch1 under YZPlane in the Tree Outline as Geometry in Details View. Enter 1000 mm as the FD1, Depth (>0) of the extrusion. Click on Generate and select isometric view in the graphics window. Close the DesignModeler.

Figure 21.2c) Details for extrusion of straight channel

477

E. Launching Ansys Meshing

3. Double click on Mesh under Project Schematic in Ansys Workbench. Select Mesh in the Outline under Project. Select Home>>Tools>>Units>>Metric (mm, kg, N, …) from the menu. Open Quality in Details of "Mesh" and set the Smoothing to High, see Figure 21.3a). Generate the Mesh.

Figure 21.3a) Coarse mesh for straight channel

Select Mesh>>Controls>>Sizing from the menu. Select Edge , select the edge at the top of the straight channel and the corresponding line at the bottom of the same channel, see Figure 21.3b). Apply the two edges as Geometry in Details of "Edge Sizing" – Sizing. Select Number of Divisions as Type under Definition. Set the Number of Divisions to 32, set Capture Curvature to No and Behavior to Hard. Select the third Bias Type - — ——— — - and set the Bias Factor to 20, see Figure 21.3c). Repeat this sizing for the other side of the straight channel.

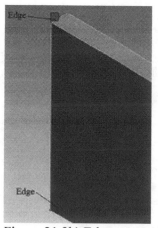

Figure 21.3b) Edges on straight channel

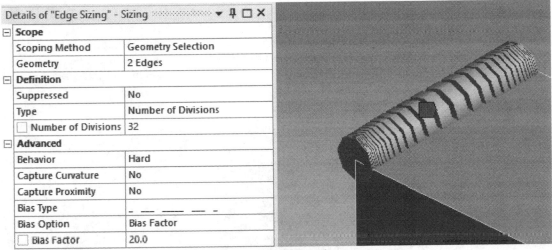

Figure 21.3c) Sizing with bias for straight channel

Right click on Mesh in Outline and select Insert>>Method. Select the straight channel and Apply it as Geometry under Scope in Details of "Automatic Method" – Method. Select MultiZone as Method under Definition, see Figure 21.3d).

Figure 21.3d) Multizone settings for straight channel

Select Mesh>>Controls>>Sizing from the menu. Select Edge, control select the four long straight edges at the top and the bottom of the channel and Apply the four Edges as Geometry under Scope in Details of "Sizing" – Sizing, see Figure 20.3e). Select Number of Divisions as Type under Definition. Set the Number of Divisions to 128, set Capture Curvature to No and Behavior to Hard. Right click on Mesh in Outline and select Mesh>>Update from the menu.

Figure 21.3e) Streamwise sizing for straight channel

Select Mesh>>Controls>>Sizing from the menu. Select Edge, control select the two vertical lines on the straight channel and Apply the two Edges as Geometry under Scope in Details of "Sizing" – Sizing, see Figure 21.3f). Select Number of Divisions as Type under Definition. Set the Number of Divisions to 128, set Capture Curvature to No and Behavior to Hard. Right click on Mesh in Outline and select Mesh>>Update.

Figure 21.3f) Spanwise sizing for straight channel

Open Statistics in Details of "Mesh" to see the total number of Nodes 549,153 and total number of Elements 524,288.

Figure 21.3g) Finished mesh with 524,288 elements for straight channel

Right click in the graphics window and select Cursor Mode>>Face. Select the larger face, right click and select Create Named Selection. Enter the name *wall_right* and click on OK to close the Selection Name window. Do the same for the other larger wall and name this *wall_left*. Furthermore, name the top face as *wall_top* and the bottom face as *wall_bottom*. Select the inlet face and name this *inlet*. Select the outlet face and name this *outlet*, see Figure 21.3h).

Select File>>Export…>>Mesh>>FLUENT Input File>>Export from the menu and save the mesh in the *Working Directory* with the name *straight-channel-mesh.msh*. Select File>>Save Project from the menu and save the project with the name *Straight Rotating Channel Flow.wbpj*. Close the Meshing window and right-click on Mesh under Project Schematic in Ansys Workbench and select Update.

Figure 21.3h) Details for named selections for straight channel

F. Launching Ansys Fluent

4. Double click on Setup under Project Schematic in Ansys Workbench. Check the Options box for Double Precision. Set the Number of Parallel Solver Processes to the same number as computer cores. Click on Start.

Figure 21.4a) Launching Fluent

Figure 21.4b) Geometry in Ansys Fluent

Click on Check and Scale… under Mesh in General on the Task Page to verify the Domain Extents. Close the Scale Mesh window. Select Time Transient under General on the Task Page. Check the box for Gravity. Enter -9.81 as Gravitational Acceleration in the Y direction.

Figure 21.4c) Domain extents for rotating channel flow mesh

Figure 21.4d) General settings including gravity and transient time

Double click on Models and Viscous under Setup in the Outline View. Set the Viscous Model to Laminar and click OK to close the window.

Figure 21.4d) Viscous model

Double click on Materials under Setup in the Outline View. Click on Create/Edit for Fluid under Materials on the Task Page. Click on Fluent Database in the Create/Edit Materials window. Scroll down and select *water-liquid (h2o<l>)* as the Fluent Fluid Material. Click on Copy and Close the Fluent Database Materials window. Click on Change/Create and Close the Create/Edit Materials window.

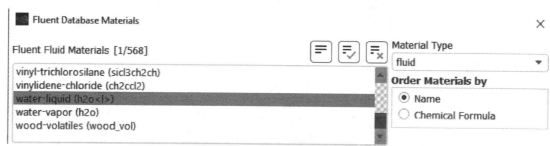

Figure 21.4e) Selection of water-liquid as fluid

5. Open Cell Zone Conditions and Fluid under Setup in the Outline View. Double click on *solid (fluid, id=2)* under Cell Zone Conditions. Select *water-liquid* as Material Name and enter *water-liquid* as the Zone Name. Check the box for Frame Motion. Set the Rotation-Axis Origin to X [m] 0, Y [m] 0 and Z [m] 0.005. Set the Rotation-Axis Direction to X 0, Y -1, and Z 0. Set the Rotational Velocity Speed [rad/s] to 0.5. Select Apply and Close the Fluid window.

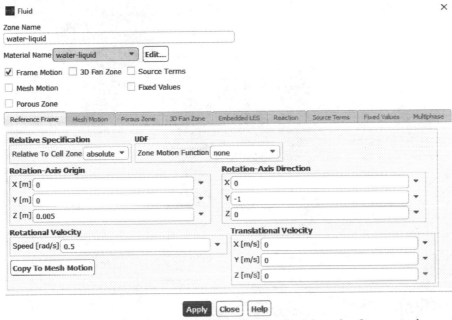

Figure 21.5a) Selection of water-liquid as fluid and settings for frame motion

Double click on Boundary Conditions. Select *inlet* under Zone in Boundary Conditions on the Task Page. Click on the Edit… button for velocity-inlet Type. Select *Components* as Velocity Specification Method. Enter 0.05 as X-Velocity [m/s]. Select Apply and Close the Velocity Inlet window.

Figure 21.5b) Details of velocity inlet

Select *outlet* under Zone in Boundary Conditions on the Task Page. Click on the Edit… button for *pressure-outlet* Type. Select *From Neighboring Cell* as Backflow Direction Specification Method. Select Apply and Close the Pressure Outlet window.

Figure 21.5c) Details of pressure outlet

Double click on Methods under Solution in the Outline View. Use the settings as shown in Figure 21.5d).

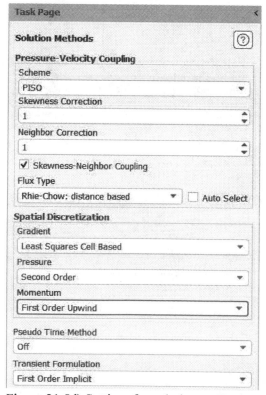

Figure 21.5d) Settings for solution methods

Double click on Monitors and Residual under Solution in the Outline View. Set the Absolute Criteria to 0.0001 for all equations and click OK to close the window.

Figure 21.5e) Settings for residual monitors

6. Double click on Initialization under Solution in the Outline View and select Hybrid Initialization. Click on Initialize.

    Select File>>Save Project from the menu. Select File>>Export>>Case… from the menu. Save the file as Files of type: CFF Case Files (*.cas.h5) with the name *rotating-channel-flow.cas.h5*.

    Double click on Run Calculation under Solution in the Outline View. Enter the settings as shown in Figure 21.6b) and click on Calculate.

Figure 21.6a) Initialization   Figure 21.6b) Running calculations

G. Post-Processing

7. The residuals are shown in Figure 21.7a).

Figure 21.7a) Residuals from calculations

Double-click on Contours in Graphics under Results in the Outline View. Enter *spanwise-y-velocity* as Contour Name in the Contours window. Select Contours of Velocity… and Y Velocity. Check the box Filled under Options. Uncheck Node Values, Boundary Values and Global Range under Options. Select Colormap Options… and select Top as Colormap Alignment. Select Fixed as Font Behavior under Font. Set the Font Size to 10 and select Apply. Close the Colormap window. Select *wall_left* under Surfaces and click on Save/Display, see Figure 21.7b). Close the Contours window.

Figure 21.7b) Contours settings for Y Velocity

Select Set View From +Z Direction, see Figures 21.7c) and 21.7d). Manipulate the light properties by unchecking the box for Lighting under Display from View in the menu, see Figure 21.7e).

Figure 21.7c) Selecting set view

Figure 21.7d) Selecting +Z direction view

Figure 21.7e) Display settings

Figure 21.7f) Contours of Y Velocity from the inside of the rotating channel flow

Figure 21.7g) Determination of spanwise wave length for rotating channel flow

489

In Figure 21.7g), you can see at least five distinct streaks in the middle of the channel. The streaks indicate the presence of streamwise counter rotating vortices with a wavelength in the spanwise Y direction that can be determined from the figure.

Double-click on Contours in Graphics under Results in the Outline View. Enter *streamwise-x-velocity* as Contour Name in the Contours window. Select Contours of Velocity… and X Velocity. Check the box Filled under Options. Uncheck Node Values, Boundary Values and Global Range under Options. Select Colormap Options… and select Fixed as Font Behavior under Font. Set the Font Size to 12 and select Apply. Close the Colormap window.

Select New Surface>>Plane…. Select *YZ Plane* as Method, enter 0.25 as X [m] and *yz-plane-x=0.25m* as New Surface Name and select Create. Close the Plane Surface window. Repeat this step and create *yz-plane-x=0.5m* and *yz-plane-x=0.75m*.

Select *inlet, yz-plane-x=0.25m, yz-plane-x=0.5m, yz-plane-x=0.75m* and *outlet* under Surfaces and click on Save/Display, see Figure 21.7h). Close the Contours window. Select Set View From +X Direction.

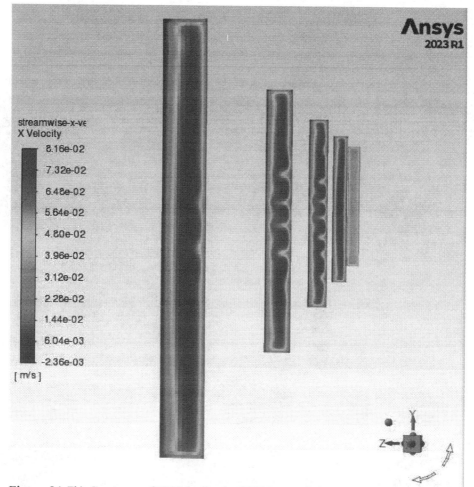

Figure 21.7h) Contours of X Velocity in YZ Plane at inlet, $x = 0.25$ m, $x = 0.5$ m, $x = 0.75$m and at the outlet of rotating plane channel flow.

H. Theory

8. The Coriolis instability for the flow in a plane channel with system rotation is governed by the Reynolds number Re and the rotation number Ro.

$$Re = \frac{U_b d}{\nu} \tag{21.1}$$

$$Ro = \frac{\Omega d}{U_b} \tag{21.2}$$

where U_b is the bulk flow velocity in the channel, d is the channel width, Ω is the rotation rate and ν is the kinematic viscosity of the fluid. The critical Reynolds number Re_{crit} is 5770 caused by Tollmien-Schlichting waves without rotation. Roll cell instabilities will develop below this Reynolds number due to the Coriolis force. The non-dimensional wave number α of the Coriolis instability is determined by

$$\alpha = \frac{2\pi d}{\lambda} \tag{21.3}$$

where λ is the spanwise wave-length of the instability. From Figure 21.7d), the wave number can be determined to be

$$\alpha = \frac{2\pi \cdot 0.01}{0.01017} = 6.18 \tag{21.4}$$

In order to determine the wave-length that is used to determine the wave number in equation (21.3), you can measure the total length in the vertical spanwise direction for six successive streaks in Figures 21.7e) – g). Next, you take this vertical length and divide by five to get the average wave-length in the vertical spanwise direction. To get the scale for figures 21.7e) – g), use the total height of the channel in the vertical direction in Figures 21.7e) – g) that corresponds to 100 mm.

The Reynolds and rotation numbers in the Ansys Fluent simulations is

$$Re = \frac{U_b d}{\nu} = \frac{0.05*0.01}{10^{-6}} = 1000 \tag{21.5}$$

$$Ro = \frac{\Omega d}{U_b} = \frac{0.5*0.01}{0.05} = 0.1 \tag{21.6}$$

I. References

1. Alfredsson, P.H. and Persson, H., Instabilities in channel flow with system rotation. *J. Fluid Mech.*, **202**, 543, (1989).
2. Chandrasekhar, S., Hydrodynamic and Hydromagnetic Stability, Dover, 1981.
3. Hart, J.E., Instability and secondary motion in a rotating channel flow. *J. Fluid Mech.*, **45**, 341, (1971).
4. Lezius, D.K. and Johnston, J.P., The structure and stability of turbulent boundary layers in rotating channel flow. *J. Fluid Mech.*, **77**, 153, (1976).
5. Speziale, C.G. and Thangam, S., Numerical study of secondary flows and roll-cell instabilities in rotating channel flow. *J. Fluid Mech.*, **130**, 377, (1983).

J. Exercises

1. Run the calculations for the flow in the rotating channel in this chapter but use the following bulk flow velocities: 0.03, 0.06, 0.09 and 0.12 m/s and determine the Reynolds number, rotation number and the spanwise wave number for each velocity.

2. Run the calculations for the flow in the rotating channel as shown in this chapter at $Re = 500$ for different rotation numbers $Ro = 0.2, 0.3, 0.4$ and 0.5. Determine the spanwise wave number for each case and compare with the critical value.

CHAPTER 22. COMPRESSIBLE FLOW PAST A BULLET

A. Objectives

- Using ICEM CFD to Create the Mesh for 2D Axisymmetric Flow Past a Bullet
- Inserting Boundary Conditions
- Running Turbulent Steady 2D Axisymmetric ANSYS Fluent Simulations
- Using Contour Plots for Visualizations
- Compare Drag Coefficient Results with Experiments

B. Problem Description

In this chapter we will study the flow around the Berger 7 mm 180 grain VLD bullet. We will be using a parabolic mesh for the compressible airflow at different Mach number past the bullet. Results from simulation drag coefficients will be compared with experiments for the bullet. ICEM CFD is not available in Ansys Student. Alternatively, you may move directly to step 21 and read the finished mesh directly in Ansys Fluent standalone mode.

C. Launching ICEM CFD

1. Start by launching ICEM CFD. Select File>>Save Project As… from the menu. Save the project with the name *BergerBullet7mm180grainVLD*.prj. Select Settings>>Selection from the menu. Uncheck the box for Auto Pick Mode. Select Settings>>Model/Units from the menu and check the box for Meters as Length Units. Click on Apply and OK. Select File>>Replay Script>>Replay Control from the menu. The Replay Control window opens.

Figure 22.1a) Length units

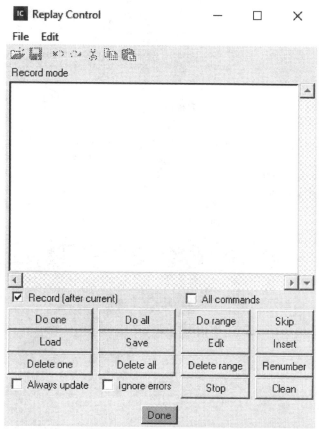

Figure 22.1b) Replay control window

D. Importing Geometries

2. Select File>>Imported Geometry>>Formatted point data from the menu. Select the file "BergerBullet7mm180grainVLD.dat" and open the file. This file can be downloaded from sdcpublications.com. Set the name for the Curve Part to BULLET. Check the boxes for Import Points, Import Curves and Import Surfaces. Click on Apply. Save the project.

Figure 22.2a) Import of formatted input point data for bullet

Figure 22.2b) Imported curve for Berger 7mm 180grain VLD

3. Select File>>Imported Geometry>>Formatted point data from the menu. Select the file "farfield.dat" and open the file. This file can be downloaded from sdcpublications.com. Check the boxes for Import Points, Import Curves and Import Surfaces. Enter the name GEOM for Point Part, FARFIELD for Curve Part and the name FLUID for Surface Part. Click on Apply. Select Merge when you get the message Geometry Already Exists. Save the project.

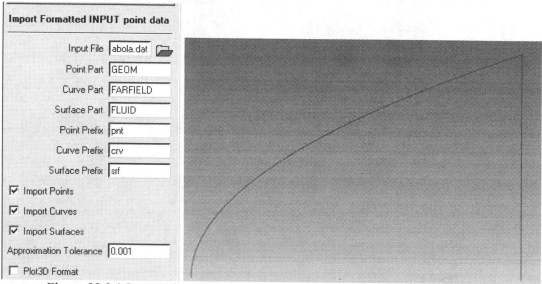

Figure 22.3a) Imported curve for mesh   Figure 22.3b) Imported curve for mesh

Check the box for Points located under Model and Geometry in the tree to the left of the graphics window. Select the Geometry tab below the menu and click on Create/Modify Curve ⟋. Uncheck the box for Inherit Part and name the Part AXIS. Select From Points ⟋ and click on the lower left end point for the parabola in the graphics window. Zoom in on the bullet using Box Zoom ⬚. Next, select ⟋ and click on the lower front end-point of the bullet and click on the mouse scroll wheel. Select ⟋ and select the lower front end-point of the bullet once again. Select the lower rear end point of the bullet and click on the mouse scroll wheel. Select ⟋ and select the lower rear end point of the bullet once again and zoom out to Fit Window ⬚ and select the lower end point of the vertical line to the right. Click on the mouse scroll wheel. Click on Apply and OK. Uncheck the box for Points under Model and Geometry. This creates three line-segments for the horizontal AXIS.

Figure 22.3c) Create Point

Figure 22.3d) Create curve

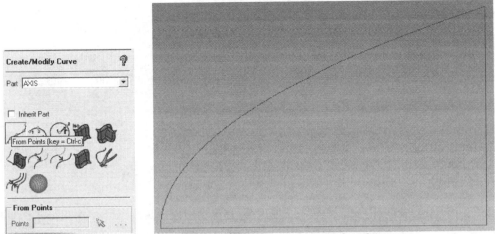

Figure 22.3e) Create curve from points   Figure 22.3f) Completed geometry

E. Creating Blocks

4. Select the Blocking tab under the menu and select Create Block. Select FLUID as Part under Create Block 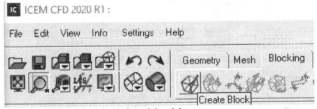. Select 2D Planar as Type under Initialize Blocks. Click on Apply. Save the project.

Figure 22.4a) Initializing blocking

Figure 22.4b) Settings for blocking   Figure 22.4c) Initial block

5. Right-click on Curves under Model and Geometry and select Show Curve Names. Check the box for Vertices under Model and Blocking. Right click on Vertices and select Numbers. Select the Blocking tab under the menu and select Associate . Select Associate Edge to Curve under Edit Association for Blocking Associations.

Select Edge(s) and select the right vertical edge of the block and click on the mouse scroll wheel. Select Curve(s) and select the right vertical curve of the mesh and click on the mouse scroll wheel. Click on Apply.

Select Edge(s) and select the lower horizontal edge of the block and click on the mouse scroll wheel. Select Curve(s) and select the three parts of the horizontal AXIS curve of the mesh and click on the mouse scroll wheel. You may need to zoom in to select the short middle section. Click on Apply.

Select Edge(s) and select the remaining two edges of the block (one vertical and one horizontal edge) and click on the mouse scroll wheel. Select Curve(s) and select the remaining parabolic curve of the mesh and click on the mouse scroll wheel. The edges of the block have now turned green. To check this, uncheck and check the box for Curves under Model and Geometry. Save the project.

Figure 22.5a) Selecting associate

Figure 22.5b) Selecting associate edge to curve

Figure 22.5c) Block and mesh after completed association

498

6. Select Snap Project Vertices under Edit Association for Blocking Associations. Click on Apply. Save the project.

Figure 22.6a) Snap project vertices   Figure 22.6b) Completed snapping of vertices

F. Setting Max Mesh Size

7. Select the Mesh tab under the menu and select Part Mesh Setup . Set Maximum size for AXIS and FARFIELD to 0.1 and Maximum size for BULLET to 0.02. Click on Apply and Dismiss. **Select the Blocking tab under the menu and select Pre-Mesh Params** **. Click on Apply.** Check the box for Pre-Mesh located under Model and Blocking in the tree to the left of the graphics window. Answer Yes when you get the question to recompute. Save the project.

Figure 22.7a) Part mesh setup

IC Part Mesh Setup

Part △	Prism	Hexa-core	Maximum size
AXIS	☐		0.1
BULLET	☐		0.02
FARFIELD	☐		0.1

Figure 22.7b) Entering maximum sizes

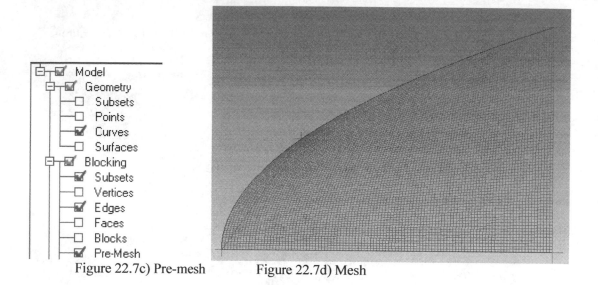

Figure 22.7c) Pre-mesh   Figure 22.7d) Mesh

G. Splitting Blocks

8. Uncheck the Pre-Mesh box and check the Blocks box under Model and Blocking in the tree next to the graphics window. Select the Blocking tab under the menu and select Split

   Block . Select Ogrid block under Split Block. Select Block(s) under Ogrid Block. Select block number 4 in the graphics window and click on the mouse scroll wheel. Click on Apply. Save the project.

Figure 22.8a) Tree including pre-mesh and blocks

Figure 22.8b) Selecting split block

Figure 22.8c) Selecting ogrid block

Figure 22.8d) Selecting block

Figure 22.8e) Block number 4 before and after selection

H. Deleting Blocks

9. Select the Blocking tab under the menu and select Delete Block . Check the box for Delete permanently. Select Blocks and select block **11** as shown in Figure 22.9b). Click on the mouse scroll wheel. Click on Apply and OK. Save the project.

Figure 22.9a) Deleting blocks

Figure 22.9b) Before and after deletion of block 11

CHAPTER 22. COMPRESSIBLE FLOW PAST A BULLET

I. Blocking Associations

10. Uncheck Blocks under Model and Blocking in the tree to the left of the graphics window. Select the Blocking tab under the menu and select Associate ⬡. Select Associate Edge to Curve ⬦ under Edit Association for Blocking Associations. Check the box for Project vertices. Select Edge(s) ⬦ and select edges 1 - 4 as shown in Figure 22.10b). Click on the mouse scroll wheel. Zoom in on the bullet using Box Zoom ⬦. Select Curve(s) ⬦ and select all curves of the bullet except the lower horizontal edge that belongs to axis, see Figure 22.10d). Click on the mouse scroll wheel. You can also use the mouse scroll wheel to zoom in and out if needed. **Select Apply Snap Project Vertices ⬦ under Edit Association for Blocking Associations. Click on Apply. Save the project.**

Figure 22.10a) Project vertices   Figure 22.10b) Selection of edges 1-4

Figure 22.10d) Selected edges for bullet

Figure 22.10d) After completion of the associations

503

J. Moving Vertices

11. Select the Blocking tab under the menu and select Move Vertex . Select Vertex under Move Vertex and select and move the rightmost vertex to the lower right corner of the bullet as shown in Figure 22.11c). In the same way, move the left vertex to the lower left corner of the bullet. Finally, move the second vertex on the right-hand side a little bit to the right to the next corner as shown in Figure 22.11e). Click on Dismiss. Zoom in on the vertices to make sure they are lined up exactly in the correct corner locations. If not, repeat this step. Save the project.

Figure 22.11a) Move vertex

Figure 22.11b) Move Vertices

Figure 22.11c) After completed movement of the rightmost vertex

Figure 22.11d) After completed movement of a left vertex

Figure 22.11e) After completed movement of all vertices

12. Select the Blocking tab under the menu and select Edit Edge . Select Method as
Automatic Linear under Split Edge. Select Edge and select line 5 as shown in Figure
22.12c). Click on the mouse scroll wheel and select Apply. Save the project.

Figure 22.12a) Edit edge   Figure 22.12b) Automatic Spline split edge meth

Figure 22.12c) Selection of line 5 for split edge

Figure 22.12d) After completed automatic linear split of line 5

Select Linear as Method under Split Edge and zoom in on corner **6** as shown in Figure 22.12d). Select Edge and select the green line segment next to the red corner point, see Figure 22.12f). Click on Apply and Dismiss. Zoom in on the different corners to make sure they are lined up exactly in the correct corner locations. If not, repeat this step. Save the project.

Figure 22.12e) Selection of linear split edge method

Figure 22.12f) Corner 6 before linear split

Figure 22.12g) Corner 6 after linear split

13. Check Blocks under Model and Blocking in the tree to the left of the graphics window.

    Select the Blocking tab under the menu and select Delete Block . Check the box for Delete permanently. Select Blocks and select block **4** as shown in Figure 22.13b). Click on the mouse scroll wheel. Click on Apply. Uncheck Blocks under Model and Blocking in the tree to the left of the graphics window. Save the project.

Figure 22.13a) Deleting blocks

Figure 22.13b) Selection of block 4 for deletion

K. Computing Wall Spacing

14. Go to the website https://www.pointwise.com/yplus/. Enter the values as shown in the left column as Input. Click on "Compute Wall Spacing". Notice the value for wall spacing 4.1015095901749715e-7 m.

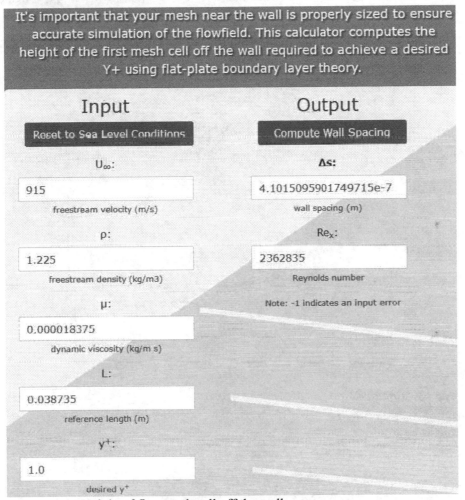

Figure 22.14 Height of first mesh cell off the wall

L. Pre-Mesh Parameters

15. Select the Blocking tab under the menu and select Pre-Mesh Params . Select Edge Params I as Meshing Parameter under Pre-Mesh Params. Click on Fit Window 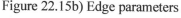 under the menu, see Figure 22.15c). Select Edge and select edge **6** as shown in Figure 22.15d). Enter 60 as the number of Nodes. Enter 4.1e-7 as Spacing 2. Set Ratio 2 to 1.2. Check the box for Copy Parameters. Click on Apply. Check the Pre-Mesh box under Model and Blocking in the tree next to the graphics window. Answer Yes when you get the question to recompute. Save the project.

Figure 22.15a) Pre-mesh parameters

Figure 22.15b) Edge parameters

Figure 22.15c) Fit window

Figure 22.15d) Lines for edge parameters

Figure 22.15e) Pre-mesh parameters   Figure 22.15f) Edge settings

Figure 22.15g) Incorrect mesh

Select the Blocking tab under the menu and select Associate ⬡. Select Associate Edge to Curve 🖐 under Edit Association for Blocking Associations. Check the box for

Project vertices. Select Edge(s) and select the horizontal edges 7 and 8 as shown in Figure 22.15g). Click on the mouse scroll wheel. Select Curve(s) and select curves 7 and 8 once again. Click on the mouse scroll wheel and select Apply. Answer Yes to the question to recompute. Save the project.

Figure 22.15g) Updated mesh after association with axis

M. Matching Edges

16. Select the Blocking tab under the menu and select Pre-Mesh Params . Select Match Edges as Meshing Parameter under Pre-Mesh Params. Zoom in around the front of the bullet, see Figure 22.16b). Select Reference Edge and select the front vertical edge of the bullet as Reference Edge. Select Target Edge(s) and select the inclined top edge of the bullet as the target edge. Click on the mouse scroll wheel. Select Edge Params as Meshing Parameter under Pre-Mesh Params. Select the top edge of the bullet once again. Enter 1.2 as Ratio 1. Click on Apply. Uncheck and check the Pre-Mesh box under Model and Blocking in the tree next to the graphics window. Answer Yes when you get the question to recompute. Repeat this step to match the edges at the rear of the bullet. Save the project.

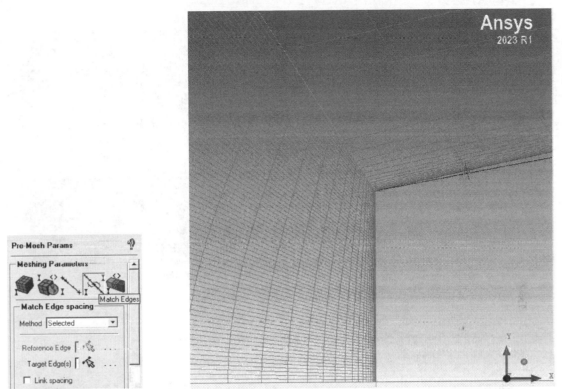

Figure 22.16a) Matching edges   Figure 22.16b) Mesh at the front end of the bullet

N. Pre-Mesh Quality

17. Select the Blocking tab under the menu and select Pre-Mesh Quality Histograms .
Select Aspect Ratio as Criterion and click on Apply. The aspect ratio is in between 1.04
and 935. One way to lower the maximum aspect ratio is to have a finer mesh. However,
since the mesh is very fine close to the surface of the bullet, it will require a much finer
mesh to substantially lower the aspect ratio. Select Quality as the Criterion and click on
Apply. The Quality is in between 0.775 and 1 which is good since the minimum quality is
not close to zero. Save the project.

Figure 22.17a) Pre-mesh quality

Figure 22.17b) Aspect ratio criterion

Figure 22.17c) Aspect ratio values

Figure 22.17d) Quality values

O. Converting to Unstructured Mesh and Selection of Solver

18. Right click on Pre-Mesh under Model and Blocking in the tree next to graphics window. Select Convert to Unstruct Mesh. Select the Output Mesh tab under the menu. Click on

    Select solver [icon]. Select Ansys Fluent as Output Solver under Solver Setup. Check the box for Set As Default. Click on Apply. Save the project.

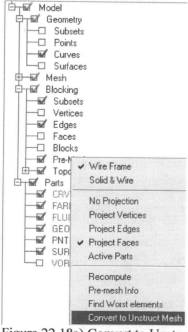

Figure 22.18a) Convert to Unstruct Mesh

P. Boundary Conditions

19. Select the Output Mesh tab under the menu and select Boundary conditions . Click on plus sign next to Edges and FARFIELD. Select Create new under FARFIELD and select *pressure far-field* as boundary condition type. Click Okay in the selection window.

    Click on plus signs next to surfaces, ? Mixed/unknown and FLUID. Select Create new under FLUID and select *fluid* as boundary condition type. Click Okay in the selection window.

    Click on the plus sign next to Edges and AXIS. Select Create new under AXIS and select *axis*. Click Okay in the selection window.

    Click on the plus sign next to Edges and BULLET. Select Create new under BULLET and select *wall*. Click Okay in the selection window.

    Click on Accept. Select File Save Project.

Figure 22.19a) Selection of boundary conditions

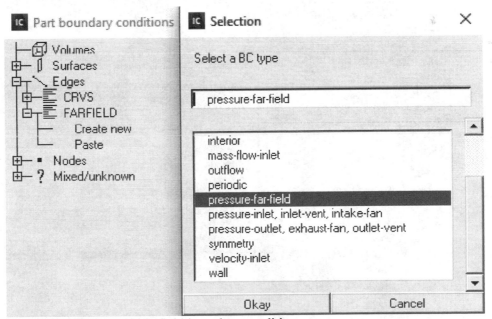

Figure 22.19b) Pressure farfield boundary condition

Figure 22.19c) Fluid boundary condition

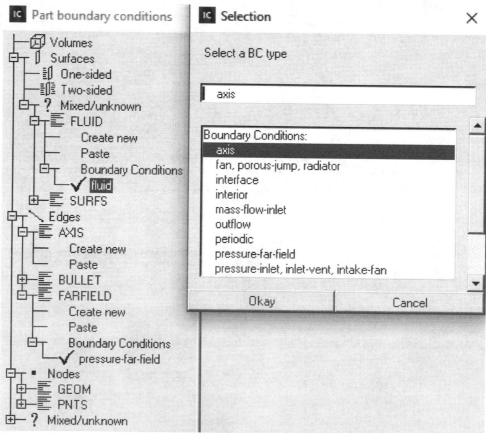

Figure 22.19d) Axis boundary condition

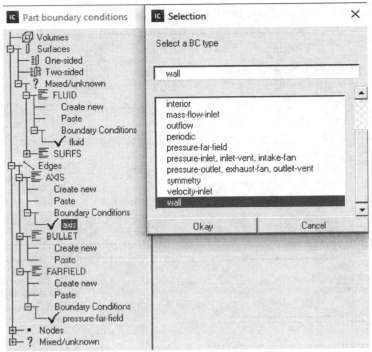

Figure 22.19e) Wall boundary condition

Q. Saving the Mesh File

20. Select the Output Mesh tab under the menu and select Write input . Answer Yes to the question Save current project first? Select *BergerBullet7mm180grainVLD.uns* and click on Open. Select 2D as Grid dimension. Enter the name for the Output file as *./BergerBullet7mm180grainVLD*, see Figure 22.20b). Click on Done and save the project.

Figure 22.20a) Write input   Figure 22.20b) Ansys Fluent

R. Starting Ansys Workbench and Setting Up Ansys Fluent

21. Start by launching Ansys Workbench. Double click on Fluent under Analysis Systems in the Toolbox. Select Units>>Metric (kg, m, s, …) from the menu. Right click on Mesh in the Project Schematic window and select Import Mesh File…>>Browse…. Select *BergerBullet7mm180grainVLD.msh* and click on Open. File>>Save from the menu and save the project with the name *Berger Bullet.wbpj*. Double click on Setup in Project Schematic. Select Double Precision under Solver Options and set the number of Solver Processes under Parallel (Local Machine) to the same number as the number of computer cores. Click on Start.

Figure 22.1a) Starting Fluent in Ansys Workbench

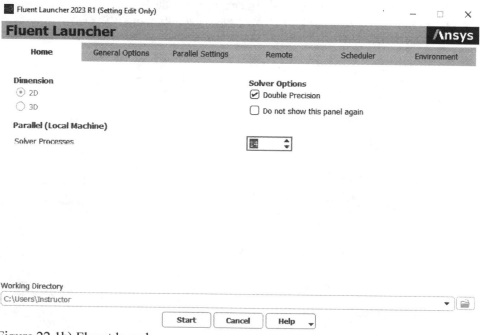

Figure 22.1b) Fluent launcher

516

Figure 22.1c) Mesh in Ansys Fluent

22. Double click on General under Setup in Outline View. Set 2D Space to *Axisymmetric*. Double click on Models and Viscous (SST k-omega) under Setup in Outline View. Set Model to *Spalart-Allmaras (1 eqn)*. Select *Strain/Vorticity-Based* under Spalart-Allmaras Production. Click OK to close the Viscous Model window.

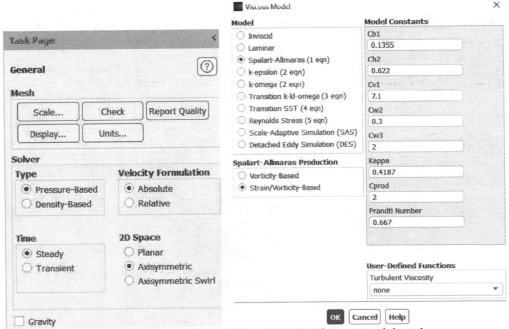

Figure 22.2a) General settings Figure 22.2b) Viscous model settings

23. Double click on Materials, Fluid and air under Setup in the Outline View. Select *ideal-gas* for Density and select *sutherland* for Viscosity. Select OK to close the Sutherland Law window. Click on Change/Create and Close the Create/Edit Materials window.

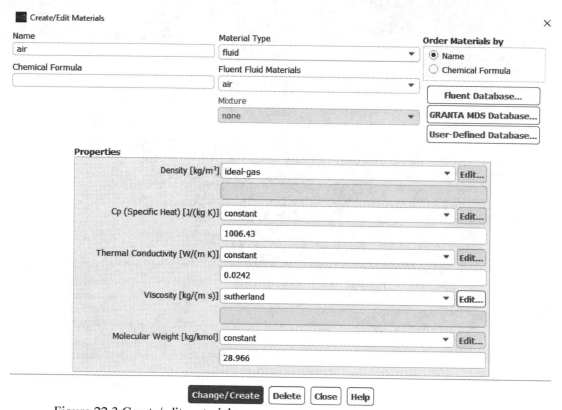

Figure 22.3 Create/edit materials

24. Double click on Boundary Conditions under Setup in the Outline View. Select *farfield* as Zone under Boundary Conditions on the Task Page. Click on Edit. Set Mach number to *2.68*. Select Apply and Close the Pressure Far-Field window.

Figure 22.4 Pressure far-field

25. Double click on Methods under Solution in the Outline View. Select the settings as shown in Figure 22.5a). Double click on Controls under Solution in the Outline View. Set *Density* to 0.5 and *Modified Turbulent Viscosity* to 0.9 under Pseudo Time Explicit Relaxation Factors.

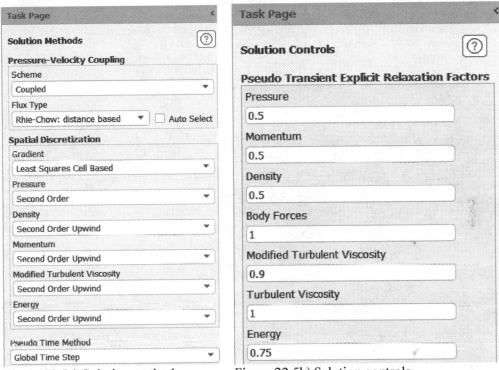

Figure 22.5a) Solution methods   Figure 22.5b) Solution controls

26. Double click on Monitors and Residual under Solution in the Outline View. Set all Residuals to 1e-06. Click OK to close the Residual Monitors window.

Figure 22.6 Residual monitors

27. Double click on Reference Values under Setup in the Outline View. Select Compute from *farfield* under Reference Values on the Task Page. Set the Area (m2) to 0.0000408894 and set Length (m) to 0.038735, see Figure 22.7a).

Double click on Report Definitions under Solution in the Outline View. Select New>>Force Report>>Drag in the Report Definitions window. Check the boxes for Report File, Report Plot and Print to Console under Create. Select *bullet* under Zones. Click OK to close the Drag Report Definition window. Close the Report Definitions window.

Figure 22.7a) Reference values   Figure 22.7b) Report definitions

Figure 22.7c) Drag report definitions

28. Double click on Initialization under Solution in the Outline View. Select Hybrid Initialization and click on Initialize. Move the cursor to the Console and press *enter* on the keyboard to get the command prompt.

   a) Enter *solve/initialize/set-fmg-initialization* and press enter 3 times.
   b) Enter *100* as number of cycles on level 1 and press enter 2 times.
   c) Enter *100* as number of cycles on level 2 and press enter 8 times.
   d) Answer *yes* to the question *enable FMG verbose?* and press enter.
   e) Enter *solve/initialize/fmg-initialization* and press enter.
   f) Answer *yes* to the question *Enable FMG initialization?*

```
> solve/initialize/set-fmg-initialization


Customize your FMG initialization:
  set the number of multigrid levels [5]

  set FMG parameters on levels ..

    residual reduction on level 1 is:  [0.001]
    number of cycles on level 1 is:  [10] 100


    residual reduction on level 2 is:  [0.001]
    number of cycles on level 2 is:  [50] 100


    residual reduction on level 3 is:  [0.001]
    number of cycles on level 3 is:  [100]


    residual reduction on level 4 is:  [0.001]
    number of cycles on level 4 is:  [500]


    residual reduction on level 5 [coarsest grid] is:   [0.001]
    number of cycles on level 5 is:  [600]


 Number of FMG (and FAS  geometric multigrid) levels: 5
* FMG customization summary:
*    residual reduction on level 0 [finest grid] is: 0.001
*    number of cycles on level 0 is: 1
*    residual reduction on level 1 is: 0.001
*    number of cycles on level 1 is: 100
*    residual reduction on level 2 is: 0.001
*    number of cycles on level 2 is: 100
*    residual reduction on level 3 is: 0.001
*    number of cycles on level 3 is: 100
*    residual reduction on level 4 is: 0.001
*    number of cycles on level 4 is: 500
*    residual reduction on level 5 [coarsest grid] is: 0.001
*    number of cycles on level 5 is: 500
* FMG customization complete



  set FMG courant-number [0.75]

  enable FMG verbose? [no] yes

> solve/initialize/fmg-initialization
 Enable FMG initialization? [no] yes
```

Figure 22.8 Full multigrid initialization

29. Double click on Run Calculation under Solution in the Outline View. Enter 400 as Number of Iterations. Click on Calculate.

Figure 22.9a) Running calculations

Figure 22.9b) Scaled residuals

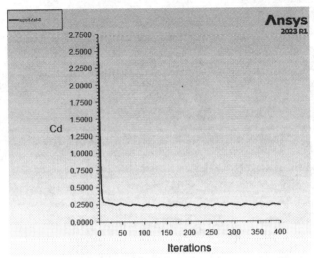

Figure 22.9c) Drag coefficient

30. Double click on Graphics and Contours under Results in the Outline View. Select Contours of Pressure and Static Pressure. Deselect all surfaces and use the settings and check boxes as shown in Figure 22.10a). Select Colormap Options…. Set Font Behavior to Fixed and Font Size to 11. Choose Top Colormap Alignment. Select Apply and Close the Colormap window. Enter the name *static-pressure-field* as Contour Name. Click on Save/Display. Zoom in on the bullet.

Create another contour plot and use the same setting but change to Contours of Velocity and Mach Number. Select Colormap Options…. Set Type to float under Number Format. Enter the name *mach-number-field* as Contour Name. Click on Save/Display. Zoom in on the bullet.

Figure 22.10a) Settings for pressure contour plot

Figure 22.10b) Pressure at the tip of bullet

Figure 22.10c) Mach number at tip of bullet

Figure 22.10d) Mach number around bullet

Double click on Reports and Forces under Results in the Outline View. Select *bullet* under Wall Zones and click on Print in the force Reports window. The value for the drag coefficient 0.242 at Ma = 2.68 has a 1.1 % difference in comparison with the experimental drag coefficient value 0.235.

```
Forces - Direction Vector (1 0 0)
                         Forces [N]                              Coefficients
Zone                   Pressure      Viscous      Total          Pressure      Viscous      Total
bullet                 4.0589581     0.83490247   4.8938606      0.19499655    0.04010958   0.23510613
--------------------------------------------------------------------------------------------------------------
Net                    4.0589581     0.83490247   4.8938606      0.19499655    0.04010958   0.23510613
```

Figure 22.10e) Forces and coefficients at Mach number 2.68

31. You will need to create the case and scheme files to complete a batch job. Select File>>Export>>Case & Data… from the menu. Save the case and data files in the *Working Directory* folder with the name ***BergerBullet7mm180grainVLD.cas.h5***. Open Notebook and create and save the scheme file with the name ***berger_bulletIII.scm*** for the batch job. The scheme file is shown in Figure 22.11a) and is available for download from *sdcpublications.com*. Select File>>Read>>Scheme… from the menu in Ansys Fluent. Open the scheme file to start the batch job. Tables 22.1, 22.2 and Figures 22.11b), 22.11c) are showing the results from the batch job in comparison with experiments, G7 drag profile and G7 ballistic coefficient.

```
(Do ((x 0.700 (+ x 0.025))) ((> x 1.150))
(begin
    (Ti-menu-load-string (format #f "file read-case BergerBullet7mm180grainVLD.cas.h5"))
    (Ti-menu-load-string (format #f "define boundary-conditions pressure-far-field farfield n 0 n ~a n 300 n 1 n 0 n n y n 10" x))
    (Ti-menu-load-string (format #f "report reference-values compute pressure-far-field farfield"))
    (Ti-menu-load-string (format #f "solve initialize hyb-initialization yes"))
    (Ti-menu-load-string (format #f "solve/initialize/set-fmg-initialization 5 0.001 100 0.001 100 0.001 100 0.001 500 0.001 500 0.75 y"))
    (Ti-menu-load-string (format #f "solve/initialize/fmg-initialization y"))
    (Ti-menu-load-string (format #f "solve iterate 400"))
    (Ti-menu-load-string (format #f "report forces wall-forces y 1 0 y Berger_bullet_drag_coefficient_data_Ma=~a" x))
    (Ti-menu-load-string (format #f "display objects display static-pressure-field"))
    (Ti-menu-load-string (format #f "display save-picture Berger_bullet_static_pressure_contour_plot_Ma=~a.jpg yes" x))
    (Ti-menu-load-string (format #f "display objects display mach-number-field"))
    (Ti-menu-load-string (format #f "display save-picture Berger_bullet_Mach_number_contour_plot_Ma=~a.jpg yes" x))
    (Ti-menu-load-string (format #f "plot residuals y y y y y"))
    (Ti-menu-load-string (format #f "display save-picture Berger_bullet_residuals_plot_Ma=~a.jpg yes" x))
    (Ti-menu-load-string (format #f "plot file report-def-0-rfile.out"))
    (Ti-menu-load-string (format #f "display save-picture Berger_bullet_drag_coefficient_plot_Ma=~a.jpg yes" x))
))
(Do ((x 1.150 (+ x 0.100))) ((> x 2.000))
(begin
    (Ti-menu-load-string (format #f "file read-case BergerBullet7mm180grainVLD.cas.h5"))
    (Ti-menu-load-string (format #f "define boundary-conditions pressure-far-field farfield n 0 n ~a n 300 n 1 n 0 n n y n 10" x))
    (Ti-menu-load-string (format #f "report reference-values compute pressure-far-field farfield"))
    (Ti-menu-load-string (format #f "solve initialize hyb-initialization yes"))
    (Ti-menu-load-string (format #f "solve/initialize/set-fmg-initialization 5 0.001 100 0.001 100 0.001 100 0.001 500 0.001 500 0.75 y"))
    (Ti-menu-load-string (format #f "solve/initialize/fmg-initialization y"))
    (Ti-menu-load-string (format #f "solve iterate 400"))
    (Ti-menu-load-string (format #f "report forces wall-forces y 1 0 y Berger_bullet_drag_coefficient_data_Ma=~a" x))
    (Ti-menu-load-string (format #f "display objects display static-pressure-field"))
    (Ti-menu-load-string (format #f "display save-picture Berger_bullet_static_pressure_contour_plot_Ma=~a.jpg yes" x))
    (Ti-menu-load-string (format #f "display objects display mach-number-field"))
    (Ti-menu-load-string (format #f "display save-picture Berger_bullet_Mach_number_contour_plot_Ma=~a.jpg yes" x))
    (Ti-menu-load-string (format #f "plot residuals y y y y y"))
    (Ti-menu-load-string (format #f "display save-picture Berger_bullet_residuals_plot_Ma=~a.jpg yes" x))
    (Ti-menu-load-string (format #f "plot file report-def-0-rfile.out"))
    (Ti-menu-load-string (format #f "display save-picture Berger_bullet_drag_coefficient_plot_Ma=~a.jpg yes" x))
))
```

Figure 22.11a) Scheme file for batch job

Figure 22.11b) Comparison between ANSYS Fluent, experiments[4] and G7 Drag Profile

Ma	V (m/s)	$C_{d,Fluent}$	$C_{d,G7}$	$i_{G7,Fluent}$	$BC_{G7,Fluent}$ (lb/in^2)
0.725	251.9535	0.1222	0.1207	1.0124	0.3156
0.75	260.6416	0.1230	0.1215	1.0121	0.3157
0.775	269.3296	0.1242	0.1226	1.0133	0.3154
0.8	278.0177	0.1237	0.1242	0.9958	0.3209
0.825	286.7057	0.1283	0.1266	1.0138	0.3152
0.85	295.3938	0.1257	0.1306	0.9624	0.3320
0.875	304.0818	0.1338	0.1368	0.9782	0.3267
0.9	312.7699	0.1384	0.1464	0.9454	0.3380
0.925	321.4579	0.1529	0.166	0.9213	0.3468
0.95	330.146	0.2016	0.2054	0.9816	0.3255
0.975	338.834	0.2237	0.2993	0.7473	0.4276
1	347.5221	0.3494	0.3803	0.9187	0.3478
1.025	356.2101	0.3688	0.4015	0.9185	0.3479
1.05	364.8982	0.3725	0.4043	0.9213	0.3468
1.075	373.5862	0.3721	0.4034	0.9223	0.3465
1.1	382.2743	0.3708	0.4014	0.9238	0.3459
1.125	390.9623	0.3692	0.3987	0.9261	0.3450
1.15	399.6504	0.3678	0.3955	0.9299	0.3436
1.2	417.0265	0.3644	0.3884	0.9381	0.3406
1.25	434.4026	0.3609	0.381	0.9471	0.3374
1.3	451.7787	0.3569	0.3732	0.9564	0.3341
1.35	469.1548	0.3523	0.3657	0.9633	0.3317
1.4	486.5309	0.3480	0.358	0.9720	0.3287
1.5	521.2831	0.3378	0.344	0.9820	0.3254
1.55	538.6592	0.3332	0.3376	0.9869	0.3238
1.6	556.0353	0.3286	0.3315	0.9913	0.3224
1.65	573.4114	0.3239	0.326	0.9935	0.3216
1.7	590.7875	0.3187	0.3209	0.9932	0.3217
1.75	608.1636	0.3137	0.316	0.9927	0.3219
1.8	625.5398	0.3093	0.3117	0.9924	0.3220
1.85	642.9159	0.3052	0.3078	0.9917	0.3222
1.9	660.292	0.3002	0.3042	0.9870	0.3238
1.95	677.6681	0.2945	0.301	0.9784	0.3266
2	695.0442	0.2889	0.298	0.9696	0.3296

Table 22.1 Drag coefficient, form factor and ballistic coefficient from Ansys Fluent in comparison with G7 model.

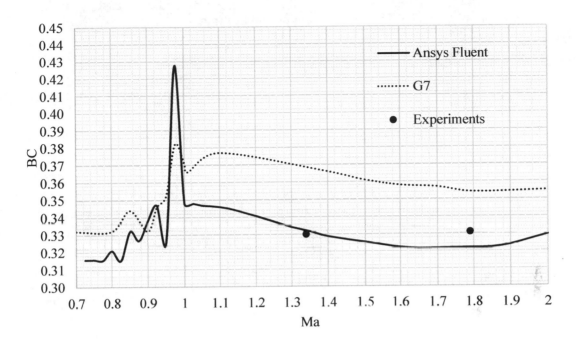

Figure 22.11c) Comparison between Ansys Fluent, experiments and G7 ballistic coefficients BC

S. Theory

32. The drag coefficient for a bullet is defined as

$$C_d = \frac{F_d}{\frac{1}{2}\rho V^2 A} \qquad (22.1)$$

where F_d (N) is the drag force on the bullet, $\rho = 1.1767$ (kg/m³) is air density, V (m/s) is the velocity of the bullet and A is frontal area of the bullet. The Mach number Ma is defined as the velocity of the bullet divided by the speed of sound $a = 347.5221$ (m/s) in the air

$$Ma = \frac{V}{a} \qquad (22.2)$$

where the speed of sound depends on the absolute temperature of the air $T = 300$ (K), the gas constant $R = 287.5514$ (J/kg-K) and γ is the adiabatic index.

$$a = \sqrt{\gamma R T} \qquad (22.3)$$

The ballistic coefficient BC (lb/m²) for the bullet is determined from

$$BC = \frac{m}{i d^2} \qquad (22.4)$$

527

where i is the form factor, $m = 0.02571$(lb) is the bullet mass and $d = 0.2837$ (in) is the diameter of the bullet. The form factor is determined from

$$i = \frac{c_d}{c_{d,GX}} \tag{22.4}$$

where $c_{d,GX}$ is the drag coefficient from the GX drag profile as shown in Figure 22.11b) and Table 22.1 for drag profile G7.

T. References

1. ANSYS Fluent Tutorial Guide, Release 18.0, January 2017.
2. Beech, D., Aerodynamic Drag Measurement and Modeling for Small Arms – Improving on Ballistic Coefficients, Applied Ballistics, 2019.
3. Litz, B., Ballistic Coefficient Testing of the Sierra .308 155 grain Matchking PALMA Bullet., Applied Ballistics, ABDOC107.2, (2007).
4. Litz, B., Berger's 7 mm VLD Bullets Part 1: Properties and Test Results, Applied Ballistics, ABDOC109.1, (2008).
5. Muruganantham, V.R. and Babin, T., Numerical Investigation of hybrid blend design target bullets, MATEC Web of Conferences **172**, 01006 (2018).

U. Exercises

22.1 Run the simulations in this chapter for the Berger 7mm 168 VLD bullet. Bullet dimensions in inches are given in Reference 4. Use the file "BergerBullet7mm168grainVLD.dat" to read the geometry for the bullet. This file is available for download at sdcpublications.com. Fill out the missing information in Table 22.3. Draw the G7 drag profile and include comparison with experiments and Ansys Fluent results in the same graph. The G7 drag profile is also available for download at sdcpublications.com.

Ma	V (m/s)	C_d, Exp.	C_d, Fluent	Percent Diff.
0.89	308.9115	------		-----
1.12	388.7425	------		-----
1.34	465.1027	0.335		
1.79	621.2939	0.297		
2.23	774.0142	0.270		
2.69	930.2054	0.249		

Table 22.2 Drag coefficient from Ansys Fluent in comparison with experiments for Berger 7mm 168 grain VLD

22.2 Run the simulations in this chapter for the Sierra .30 cal 155 grain PALMA MatchKing. Bullet dimensions in inches are given in Reference 3. Use the file "SierraBullet.30cal155grainPALMAMatchKing.dat" to read the geometry for the bullet. This file is available for download at sdcpublications.com. Fill out the missing information in Table 22.4. Draw the G7 drag profile and include comparison with experiments and Ansys Fluent results in the same graph. The G7 drag profile is also available for download at sdcpublications.com.

Ma	V (m/s)	C_d, Exp.	C_d, Fluent	Percent Diff.
0.89	308.9115	------		-----
1.12	388.7425	------		-----
1.34	465.1027	0.406		
1.79	621.2939	0.348		
2.23	774.0142	0.309		
2.69	930.2054	0.280		

Table 22.3 Drag coefficient from Ansys Fluent in comparison with experiments for Sierra .30 cal 155 grain PALMA MatchKing

V. Notes

CHAPTER 23. VERTICAL AXIS WIND TURBINE FLOW

A. Objectives

- Using Dynamic Mesh and 6DOF Solver
- Using Rotating Mesh
- Running Turbulent Transient 3D Ansys Fluent Simulations for VAWT
- Using Contour Plots for Visualizations
- Comparison of Rotor RPM with Experiments

B. Problem Description

In this chapter we will study the airflow that causes the drag based helical Savonius rotor VAWT to spin. We will be using a combination of fixed and rotating mesh domains for the incompressible airflow. Results for rotor RPM will be compared with experiments. We will create a mesh with less than 512 000 elements so that the Ansys Fluent student version can be used. The performance of this rotor has been studied by Zakaria and Ibrahim[1-2] using the commercial code AcuSolve and compared with experiments. The helical rotor has180 degree sweep angle, height over diameter ratio of 2, an overlap ratio of 0.242, and end plates with a diameter that is 10% larger than the rotor diameter of 0.5 m. The helical rotor was modeled in SOLIDWORKS.

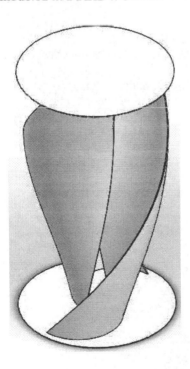

1. Start Fluent 2023 R1 in standalone mode. Read the Case File named Helical Savonius Turbine Flow II.cas.h5. This and other files can be downloaded from *sdcpublications.com*. Check the box for Double Precision under Solver Options in the Fluent Launcher. Set the number of processes equal to the number of processor cores for your computer. To check the number of physical cores, press the Ctrl + Shift + Esc keys simultaneously to open the Task Manager. Go to the Performance tab and select CPU from the left column. You will see the number of physical cores on the bottom-right side. Close the Task Manager window. Click on the Start button in the Fluent Launcher window.

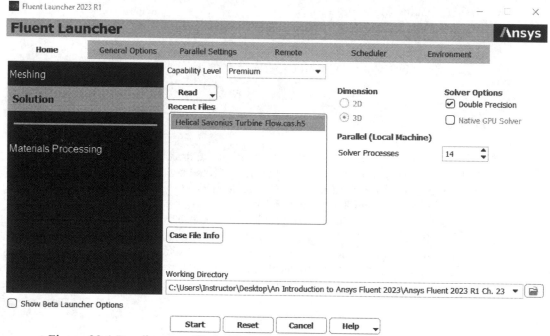

Figure 23.1 Reading case file in Fluent Launcher

2. Double click on Initialization under Solution in the Outline View. Select Hybrid Initialization and click on Initialize. Double click on Run Calculation. Set the Time Step Size (s) to 0.01 and set the Number of Time Steps to 2000. Set Max Iterations/Time Step to 50. Click on Calculate.

Figure 23.2a) Run calculation settings

Figure 23.2b) shows the velocity magnitude at the edge of the rotor end plate during the first 20 seconds. The velocity magnitude levels off at around 7.1386 m/s corresponding to rotor tip speed $v = 6.4896$ m/s and a rotor rpm of 248. This value is 19 % higher than the corresponding value of 208 rpm as listed by Zakaria and Ibrahim[1] in their experiments. The rotor tip speed $v = 6.4896$ m/s translates to a tip speed ratio of 1.30 since the free stream velocity is 5 m/s.

Figure 23.2b) Velocity magnitude at rotor end plate edge during the first 20 seconds

When calculations are complete, click on Copy Screenshot of Active Window to Clipboard in the graphics window, see Figure 23.18c). The velocity magnitude at the edge of the rotor end plate plot can be pasted into a Word document.

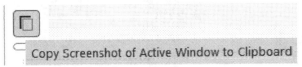

Figure 23.2c) Copying screenshot

Start Excel and open the file *end-plate-tip-velocity-magnitude-rfile.out*. Select Next> in Step 1 of the Text Import Wizard. Check the box for Space under Delimiters and select Next> in Step 2. Select Finish in Step 3. Copy the last 101 values ($t = 19 - 20$ s) from columns C and B and paste the data in columns F and G with the heading *t (s)* for column F and heading *Velocity (m/s)* for column G. Determine the average value for the velocities in column G. The average value is 7.1335 m/s.

F	G	H	I
t (s)	*Velocity* (m/s)		
19	7.150156498	*Average* (m/s) =	7.1335
19.01	7.17373848	*d* (m) =	0.5
19.02	7.177240372	*D* (m) =	0.55
19.03	7.138011932	*v* (m/s) = *Average*$*d/D$ =	6.4850
19.04	7.178740501	*N* (rpm) = $60*v/(\pi*d)$ =	247.7083
19.05	7.116882324	*N-exp.* (rpm) =	208
19.06	7.018544197	% *diff.* =	19
19.07	7.142709732	*V* (m/s) =	5
19.08	7.131412983	*TSR* = *v/V* =	1.2970
19.09	7.092211246		
19.1	7.159035683		
19.11	7.139612675		
19.12	7.134111404		
19.13	7.178254128		
19.14	7.158715248		
19.15	6.898990631		
19.16	7.175828934		
19.17	7.128138542		
19.18	7.013719559		
19.19	7.151053429		
19.2	7.096494198		
19.21	7.078816891		
19.22	7.133673668		
19.23	7.15787077		
19.24	7.121403217		
19.25	7.161549568		
19.26	7.201757908		
19.27	7.100908279		
19.28	7.156680584		
19.29	7.185548782		
19.3	7.028903008		

Table 23.1 Evaluation of data for *end-plate-tip-velocity-magnitude-rfile.out*

Open Graphics and Contours under Results in the Outline View. Double click on Velocity Magnitude under Contours. Uncheck Auto Range under Options in the Contours window. Set Max to 8 and select Colormap Options…. Set Font Behavior to Fixed and Font Size to 12 under Font in the Colormap window. Set Type to general and Precision to 2 under Number Format. Select Apply and Close the Colormap window. Select Save/Display and Close the Contours window. Figure 23.2d) shows the contours of velocity magnitude for the helical Savonius rotor after 20 seconds. It is seen that the velocity magnitude increases as expected further away from the rotation axis up to a maximum value of around 7.2 m/s at the edge of the circular end plate.

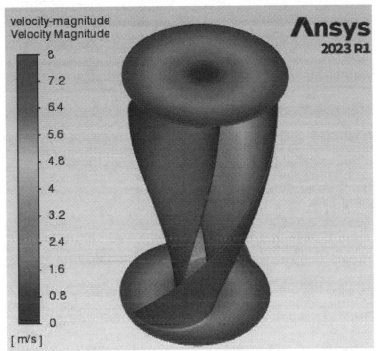

Figure 23.2d) Contours of velocity magnitude for the rotor after 20 seconds

In Figure 23.2e) we see the time signal of the drag and lift coefficients on the rotor during one second. The average value of the corresponding lift force 10.65 N is 14% higher than the average value of the drag force at 9.33 N. The corresponding average values for the drag and lift coefficients are 1.22 and 1.39, respectively.

Start Excel and open the files *downwind-drag-force-coefficient-rfile.out* and *side-force-coefficient-rfile.out*. Select Next> in Step 1 of the Text Import Wizard. Check the box for Space under Delimiters and select Next> in Step 2. Select Finish in Step 3.

Copy the last 101 values ($t = 19 - 20$ s) from columns C and B from both files and paste the data in columns A to D in a new Excel file with heading t (s) for column A, heading C_{Fx} for column B, t (s) for column C and heading C_{Fz} for column D. The period $T = 0.12$ s in Figure 23.2e) corresponds to a 180 degrees rotation of the rotor. The average value for the drag coefficient is 1.2181 and the average value for the lift coefficient is 1.3904.

A	B	C	D	E	F
t (s)	C_{Fx}	t (s)	C_{Fz}		
19	1.34698	19	1.370353	ρ (kg/m^3) =	1.225
19.01	1.321055	19.01	1.310615	V (m/s) =	5
19.02	1.260063	19.02	1.27354	A_s (m^2)=	0.5
19.03	1.194177	19.03	1.277438	$C_{Fx\text{-}ave}$ =	1.2181
19.04	1.13433	19.04	1.317141	$F_{x\text{-}ave}$ (N) =	9.3260
19.05	1.101226	19.05	1.371847	$C_{Fz\text{-}ave}$ =	1.3904
19.06	1.102258	19.06	1.42809	$F_{z\text{-}ave}$ (N) =	10.6454
19.07	1.131916	19.07	1.473198		
19.08	1.179718	19.08	1.502256		
19.09	1.240956	19.09	1.50352		
19.1	1.298699	19.1	1.479369		
19.11	1.337706	19.11	1.428145		
19.12	1.34872	19.12	1.363352		
19.13	1.311919	19.13	1.305617		
19.14	1.249186	19.14	1.271303		
19.15	1.180473	19.15	1.279072		
19.16	1.122971	19.16	1.322039		
19.17	1.094958	19.17	1.37797		
19.18	1.101778	19.18	1.431612		
19.19	1.132596	19.19	1.475916		
19.2	1.179901	19.2	1.50106		

Table 23.2 Evaluation of data for drag and lift coefficients

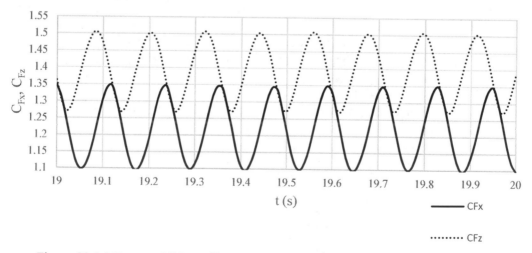

Figure 23.2e) Drag and lift coefficients on rotor during one second

C. Theory

3. In defining the geometry we follow the approach by Rogowski and Maronski[3]. The cross sectional geometry of the rotor is shown in Figure 23.3.

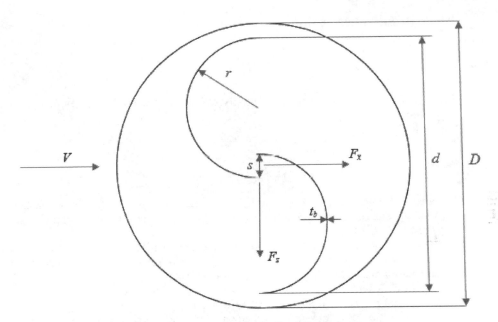

Figure 23.3 Cross sectional geometry for the helical Savonius rotor

The bucket gap width s is 0.121 m and the bucket thickness t_b is 3 mm. The bucket radius r is 0.15525 m and the rotor diameter $d = 4r - s = 0.5$ m. The overlap ratio is $s/d =$ 0.242. The twist angle of the helical rotor is 180 degrees. The end plate diameter D is 0.55 m and the end plate thickness t_e is 2 mm. The free stream velocity is V (m/s). The downwind drag force is F_x (N) and the side force or lift force is F_z (N). The rotor swept area is defined as

$$A_s = (4r - s)h \tag{23.1}$$

where $h = 1$ m is the height of the rotor. We define the rotor tip speed v (m/s) and the tip speed ratio TSR as

$$v = \omega d/2 \tag{23.2}$$

$$TSR = \frac{\omega d}{2V} = \frac{v}{V} \tag{23.3}$$

where ω (rad/s) is the rotational speed of the turbine. The turbine rotational speed N (rpm) is then given by

$$N = \frac{30\omega}{\pi} = \frac{60v}{\pi d} \tag{23.4}$$

The drag and lift coefficients can be defined as

$$C_{F_x} = \frac{F_x}{\frac{1}{2}\rho V^2 A_s} \qquad (23.5)$$

$$C_{F_z} = \frac{F_z}{\frac{1}{2}\rho V^2 A_s} \qquad (23.6)$$

where ρ (kg/m^3) is air density. We can also define the torque coefficient

$$C_T = \frac{T}{\frac{1}{2}\rho V^2 A_s \frac{d}{2}} \qquad (23.7)$$

where T (Nm) is the turbine torque. Finally, we define the power coefficient as

$$C_P = C_T * TSR \qquad (23.8)$$

D. References

1. Zakaria, A. and Ibrahim, M.S.N., "Analysis of Savonius Rotor Performance Operating at Low Wind Speeds Using Numerical Study.", International Journal of Engineering & Technology, **7**, 1549-1552, 2018.
2. Zakaria, A. and Ibrahim, M.S.N., "Numerical Performance Evaluation of Savonius Rotors by Flow-Driven and Sliding-Mesh Approaches.", International Journal of Advanced Trends in Computer Science and Engineering, **8**, No. 1, 2019.
3. Rogowski, K and Maronski, R, "CFD Computation of the Savonius rotor.", Journal of Theoretical and Applied Mechanics, **53**, 37-45, 2015.
4. Blackwell, B.F., Sheldahl, R.E. and Feltz, L.V.,"Wind Tunnel Performance Data for Two- and Three-Bucket Savonius Rotors, Sandia Laboratories, Report SAND76-0131, 1977.

E. Exercises

23.1 Run the simulations in this chapter for the Helical Savonius Turbine at the different free stream velocities as shown in Table 23.3. Fill out the table and plot drag and lift coefficients versus tip speed ratio. Also, plot N (rpm) from simulations and experiments versus free stream velocity V.

V (m/s)	N (rpm)	TSR	F_x (N)	C_{F_x}	F_z (N)	C_{F_z}	N (rpm)-Exp.	% diff.	Δt (s)	N_t
2							83		0.01	2000
3							124		0.01	2000
4							170		0.01	2000
5	248	1.298	9.322	1.218	10.643	1.390	208	19	0.01	2000
6							230		0.01	2000

Table 23.3 Turbine rotational speed from Ansys Fluent in comparison with experiments[1].

23.2 Run the simulations in this chapter for the Helical Savonius Turbine at the different aspect ratios h/d as shown in Table 23.4. Fill out the table and plot drag and lift coefficients versus tip speed ratio. Also, plot N (rpm) from simulations versus free aspect ratio h/d.

h/d	V (m/s)	N (rpm)	TSR	F_x (N)	C_{F_x}	F_z (N)	C_{F_z}	Δt (s)	N_t
2	5	248	1.298	9.322	1.218	10.643	1.390	0.01	2000
8	5							0.01	2000
32	5							0.01	2000

Table 23.4 Input data and results from Ansys Fluent simulations

h/d	A_s (m²)	h (m)	d (m)	D (m)	end-plate edge-point (m)	s (m)	Reference Length (m)
2	0.5	1	0.5	0.55	0.275	0.121	0.25
8	0.5	2	0.25	0.275	0.1375	0.0605	0.125
32	0.5	4	0.125	0.1375	0.06875	0.03025	0.0625

Table 23.5 Input data for Ansys Fluent simulations

h/d	No. Nodes	No. Elements	Orthogonal Quality Min	Growth Rate	Rotor Mesh Size (mm)
2	120506	506532	0.10131	1.2	30
8					
32					

Table 23.6 Input data and results from Ansys Meshing

h/d	Center of Mass Y (m)	Moment of Inertia (kg-m2)	Mass (kg)
2	0.50000169	0.13735168	4.01319583
8	1.00000073	0.02699234	3.22839079
32			

Table 23.7 Input data for Ansys Fluent simulations

F. Notes

CHAPTER 24. CIRCULAR HYDRAULIC JUMP

A. Objectives

- Using Ansys Fluent to Study the Flow in a Circular Hydraulic Jump
- Inserting Boundary Conditions
- Using Volume of Fluid Model for Multiphase Flow with Surface Tension
- Running Laminar 2D Axisymmetric Ansys Fluent Simulations
- Using Contour Plots for Visualizations of Volume Fraction
- Using Excel for Free Surface Plots

B. Problem Description

We will study the circular hydraulic jump for laminar flow on a flat plate. The water exits the circular tube nozzle with radius R_n and velocity U_0 and spreads out over the horizontal flat plate. The flow enters the hydraulic jump at radius R_j with a water height H after the jump.

C. Launching Ansys Workbench and Selecting Fluent

1. Start by launching Ansys Workbench. Launch Fluid Flow (Fluent) that is available under Analysis Systems in Ansys Workbench. Select Geometry under Project Schematic, right click and select Properties. Select 2D Analysis Type under Advanced Geometry Options. Right click on Geometry in Project Schematic and select New DesignModeler Geometry to start DesignModeler.

Figure 24.1 Launching Ansys Fluent and selecting 2D analysis type

D. Launching Ansys DesignModeler

2. Select **Units>>Millimeter** from the menu in DesignModeler. Select the XYPlane in the Tree Outline. Select Look At Face/Plane/Sketch [icon]. Select the Sketching tab and Polyline. Draw the shape with the dimensions as shown in Figure 24.2. Make sure that the letter *P* is showing when you move the cursor over the origin of the coordinate system and before you start drawing the polyline.

Select Concept>>Surfaces from Sketches from the menu. Select Sketch 1 under XY Plane in the Tree Outline. Apply the sketch as a Base Object in Details View. Click on Generate and close DesignModeler.

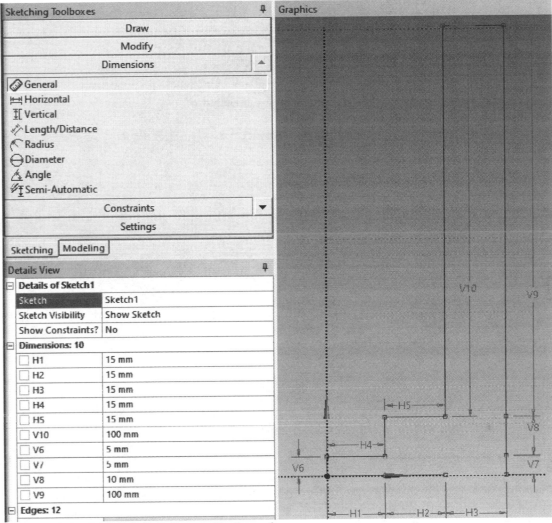

Figure 24.2 Polyline shape with dimensions

E. Launching Ansys Meshing

3. Double click on Mesh under Project Schematic in Ansys Workbench. Right-click on Mesh under Project and Model (A3) in the Meshing window and select Update. Select Mesh>>Controls>Face Meshing from the menu. Click on the polyline shape in the graphics window and Apply it as Geometry in Details of Face Meshing.

Figure 24.3a) Launching meshing window

Figure 24.3b) Selecting face meshing

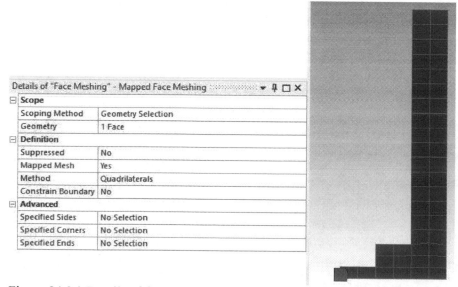

Figure 24.3c) Details of face meshing

Select Mesh>>Controls>>Sizing from the menu. Select the Edge tool [icon], control click on the two long vertical edges and Apply them as Geometry in Details of Sizing. Select Number of Divisions as Type and enter 225. Set Capture Curvature to No and select Hard as Behavior, see Figure 24.3d).

Details of "Edge Sizing" - Sizing

Scope	
Scoping Method	Geometry Selection
Geometry	2 Edges
Definition	
Suppressed	No
Type	Number of Divisions
Number of Divisions	225
Advanced	
Behavior	Hard
Capture Curvature	No
Capture Proximity	No
Bias Type	No Bias

Figure 24.3d) Details for sizing of long vertical edges

Repeat this step but control select the two right most horizontal edges (one located at the top and the other at the bottom), enter 70 as the Number of Divisions, set Capture Curvature to No and select Hard as Behavior, see Figure 24.3e). Select the first Bias Type and set the Bias Factor to 40. Select the horizontal line at the top and apply the line next to Reverse Bias.

Details of "Edge Sizing 2" - Sizing

Scope	
Scoping Method	Geometry Selection
Geometry	2 Edges
Definition	
Suppressed	No
Type	Number of Divisions
Number of Divisions	70
Advanced	
Behavior	Hard
Capture Curvature	No
Capture Proximity	No
Bias Type	___ __ _ -
Bias Option	Bias Factor
Bias Factor	40.0
Reverse Bias	1 Edge

Figure 24.3e) Details for sizing of right horizontal edges.

Repeat this step but control select the two middle horizontal edges, enter 60 as the Number of Divisions, set Capture Curvature to No and select Hard as Behavior, see Figure 24.3f).

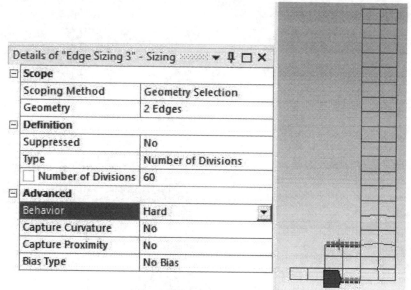

Figure 24.3f) Details for sizing of middle horizontal edges

Repeat this step but control select the two left horizontal edges, enter 25 as the Number of Divisions, set Capture Curvature to No and select Hard as Behavior, see Figure 24.3g).

Figure 24.3g) Details for sizing of left horizontal edges

Repeat this step but control select the two bottom vertical edges, enter 40 as the Number of Divisions, set Capture Curvature to No and select Hard as Behavior, see Figure 24.3h). Select the first Bias Type and set the Bias Factor to 5. Select the vertical line to the left and apply the line as Reverse Bias.

Details of "Edge Sizing 5" - Sizing ▼ 廿 □ ✕	
⊟ **Scope**	
Scoping Method	Geometry Selection
Geometry	2 Edges
⊟ **Definition**	
Suppressed	No
Type	Number of Divisions
☐ Number of Divisions	40
⊟ **Advanced**	
Behavior	Hard
Capture Curvature	No
Capture Proximity	No
Bias Type	____ __ _ _
Bias Option	Bias Factor
☐ Bias Factor	5.0
Reverse Bias	1 Edge

Figure 24.3h) Details for sizing of bottom vertical edges

Repeat this step but control select the two middle vertical edges, enter 40 as the Number of Divisions, set Capture Curvature to No and select Hard as Behavior, see Figure 24.3i). Select the second Bias Type and set the Bias Factor to 2. Right-click on Mesh under Project and Model (A3) in the Meshing window and select Update. The finished mesh is shown in Figure 24.3j). The mesh has 27,611 Nodes and 27,150 Elements.

Details of "Edge Sizing 6" - Sizing ▼ 廿 □ ✕	
⊟ **Scope**	
Scoping Method	Geometry Selection
Geometry	2 Edges
⊟ **Definition**	
Suppressed	No
Type	Number of Divisions
☐ Number of Divisions	40
⊟ **Advanced**	
Behavior	Hard
Capture Curvature	No
Capture Proximity	No
Bias Type	_ _ __ ___
Bias Option	Bias Factor
☐ Bias Factor	2.0
Reverse Bias	No Selection

Figure 24.3i) Details for sizing of middle vertical edges

Figure 24.3j) Finished mesh

Select the three lower horizontal edges, right click and select Create Named Selection. Enter the name *axis* and click OK to close the window. Control-select the three right most vertical edges and the left most horizontal edge that was not selected as an axis edge, right click and select Create Named Selection. Enter the name *wall*. Name the left most vertical edge as *inlet*. Finally, name the remaining four edges as *outlet*.

Figure 24.3k) Named selections

Select File>>Export...>>Mesh>>FLUENT Input File>>Export from the menu. Save the mesh with the name *circular-hydraulic-jump.msh*. Select File>>Save Project from the menu and save the project with the name *Circular Hydraulic Jump.wbpj*. Close the meshing window. Right click on Mesh in the Project Schematic and select Update.

F. Launching Ansys Fluent

4. Double click on Setup under Project Schematic in Ansys Workbench. Check the box for Double Precision. Set the number of Solver Processes equal to the number of processor cores for your computer. To check the number of physical cores, press the Ctrl + Shift + Esc keys simultaneously to open the Task Manager. Go to the Performance tab and select CPU from the left column. You'll see the number of physical cores on the bottom-right side. Close the Task Manager window. Click on the Start button in the Fluent Launcher window.

Figure 24.4a) Fluent launcher

Figure 24.4b) Circular hydraulic jump with mesh in Ansys Fluent

5. Select Axisymmetric for 2D Space in the General Solver settings on the Task Page in Ansys Fluent. Select Transient for Time in the General Solver settings.

    Open Models under Setup in the Outline View and double click on Viscous (SST k-omega). Select Laminar Model and click on OK to close the Viscous Model window. Double click on Multiphase. Select the Volume of Fluid Model and check the box for Implicit Body Force under Body Force Formulation. Click on Apply and Close the Multiphase Model window.

> **Why did we select Axisymmetric and Transient for Time?** This problem is axisymmetric. We choose Transient since we are interested in studying the shape of the free surface, hydraulic jump and how it deforms over time.

> **Why did we select Volume of Fluid as Multiphase model and Implicit Body Force?** The Volume of Fluid model is used when we have two or more immiscible fluids where the position of the interface between the fluids is of interest. The applications of the VOF Model includes free-surface flows, stratified flows and the tracking of liquid-gas interfaces in general. The VOF model is limited to the pressure based solver and doesn't allow for void regions without a fluid. Implicit Body Force is used when we have large body forces such as is the case when we have gravity and surface tension in multiphase flows. Checking the box for Implicit Body Force will make the solution more robust with better convergence.

Figure 24.5a) General settings

Figure 24.5b) Multiphase model

Figure 24.5c) Selecting volume of fluid model

6. Next, we double click on Materials under Setup in the Outline View. Select Fluid under Materials on the Task Page and click on Create/Edit…. Click on Fluent Database… in the Create/Edit Materials window. Scroll down in the Fluent Fluid Material section and select *water-liquid (h2o<l>)*. Click Copy at the bottom of the Fluent Database Materials window. Close the two windows.

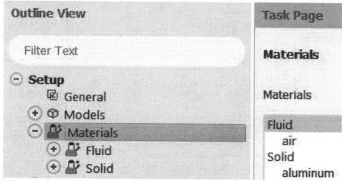

Figure 24.6a) Selecting fluid materials

Figure 24.6b) Selecting Fluent database

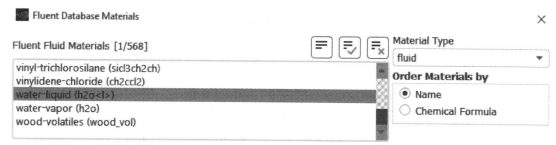

Figure 24.6c) Selection of water-liquid as fluid material

7. Select the Physics tab in the menu and select Models>>Multiphase Model. Select the Phases tab at the top of the Multiphase Model window. Select phase-1-Primary Phase and enter *air* as name for the Primary Phase. Select phase-2-Secondary Phase, select *water-liquid* as the Phase Material and enter *water* as the Name for the Secondary Phase. Click on Apply in the Multiphase Model window. Click on the Phase Interaction tab at the top of the Multiphase Model window. Select constant for Surface Tension Coefficients [N/m] and enter 0.07286 as the value. Click on Apply and Close the Multiphase Model window.

Figure 24.7a) Setting up the physics

Figure 24.7b) Editing phase-1-Primary Phase

Figure 24.7c) Name for phase-1   Figure 24.7d) Name for phase-2

Figure 24.7e) Entering value for surface tension between water and air

8. Select the Physics tab in the menu and select Operating Conditions… under Solver. Check the box for Gravity and enter 9.81 m/s² for Gravitational Acceleration in the positive X direction. Select user-input as Operating Density Method. Enter 1.225 as Operating Density [kg/m³]. Set the Reference Pressure Location for X [m] to 0.0375 and for Y [m] to 0.115. Click on the OK button to close the window.

Figure 24.8a) Selecting operating conditions

Figure 24.8b) Including gravity and specified operating density

9. Double click on Boundary Conditions under Setup in the Outline View. Select *inlet* under Zone in Boundary Conditions on the Task Page. Click on the Edit… button. Select Components as Velocity Specification Method. Set Axial-Velocity [m/s] to 0.4. Select Apply and Close the window.

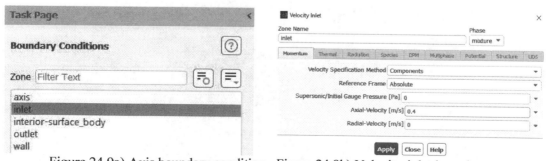

Figure 24.9a) Axis boundary condition   Figure 24.9b) Velocity-inlet boundary

10. For *inlet*, select water from the Phase drop-down menu under Boundary Conditions on the Task Page. Click on the Edit... button, make sure that the value for the Volume Fraction is 1 and click Apply and Close the window.

Figure 24.10 Details for pressure-inlet boundary condition

11. Double click on Methods under Solution in the Outline View and choose Body Force Weighted for Pressure and First Order Upwind for Momentum under Spatial Discretization on the Task Page.

Figure 24.11 Solution methods

12. Double click on Initialization under Solution in the Tree and select Compute from *inlet* on the Task Page. Set water Volume Fraction to 0. Click on Initialize.

Figure 24.12 Solution initialization

13. Select the *Domain* tab in the menu and select Adapt>>Manual…. Select Cell Registers>>New>>Region from the Manual Mesh Adaption window. Set X Max [m] to 0.015 and Y Max (m) to 0.005. Click on Display Options…. Select red as Color and no Symbol. Check the box for Filled and uncheck the boxes for Wireframe and Marker under Options. Click on OK to close the Cell Register Display Options window. Click on Save/Display in the Region Register window. Close the windows.

Figure 24.13a) Domain adaption

Figure 24.13b) Adaption controls

Figure 24.13c) Settings for region register

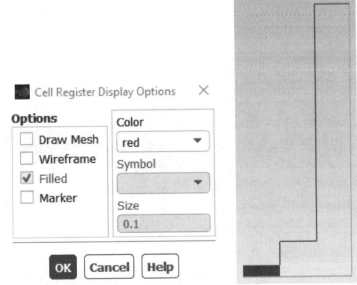

Figure 24.13d) Cell register display   Figure 24.13e) Filled inlet section

Click on Copy Screenshot of Active Window to Clipboard, see Figure 24.13f). The contour plot can be pasted into a Word document.

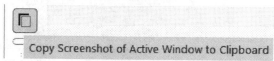

Figure 24.13f) Copying screenshot

14. Click on the Patch... button on the Solution Initialization Task Page. Select *water* as Phase in the Patch window and select Volume Fraction as Variable. Select region_0 as Registers to Patch and set the Value to 1. Click on the Patch button. Select Mixture as Phase and choose Radial Velocity as Variable. Set the Value [m/s] to 0 and click on the Patch button. Close the Patch window.

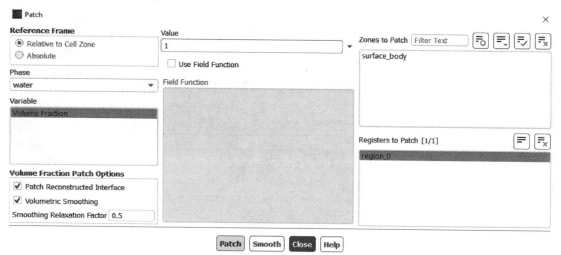

Figure 24.14a) Patching settings for water

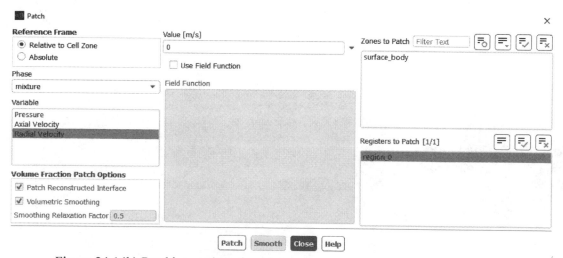

Figure 24.14b) Patching settings for mixture

15. Double click on Graphics and Contours under Results in the Outline View. Select Contours of Phases… and Volume fraction. Select *water* as the Phase. Deselect all Surfaces. Click on Colormap Options. Select general as Type and 0 for Precision under Number Format in the Colormap window. Set Colormap Alignment to Top. Select Apply and Close the Colormap window. Click on Save/Display and Close the Contours window.

Figure 24.15a) Contours window

Select the View tab in the menu and click on Views…. Select *axis* as Mirror Plane and click on Apply. Zoom out and translate the view if needed so that the entire model is visible in the graphics window. Click on the Camera… button in the Views window. Use your left mouse button to rotate the dial clockwise until the model rotates 90 degrees clockwise and appears upright. Close the Camera Parameters window. Click on the Save button under Actions in the Views window and Close the Views window. Click on Copy Screenshot of Active Window to Clipboard, see Figure 8.13f). The water volume fraction contour plot can be pasted into a Word document.

Figure 24.15b) Displaying views

Figure 24.15c) Camera parameters window

Figure 24.15d) Views window

Figure 24.15e) Water volume fraction contour plot

16. Select File>>Export>>Case... from the menu. Save the Case File with the name *Circular Hydraulic Jump.cas.h5*. Double click on Run Calculation under Solution in the Outline View and set the Time Step Size to 0.0001 s. Set the Number of Time Steps to 20000. Click on the Calculate button on the Task Page. Click OK in the Information window when the calculation is complete. Click on Copy Screenshot of Active Window to Clipboard, see Figure 24.13f). The Scaled Residuals plot can be pasted into a Word document.

Figure 24.16a) Calculation settings

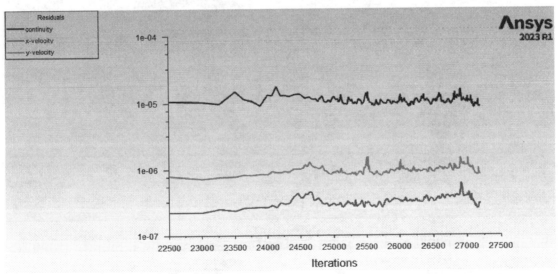

Figure 24.16b) Scaled residuals

G. Post-Processing

Double click on Graphics and Contours under Results in the Outline View. Select Contours of Phases… and Volume Fraction. Select *water* as the Phase. Deselect all Surfaces. Click on Save/Display. Click on Copy Screenshot of Active Window to Clipboard, see Figure 8.13f). The water volume fraction contour plot can be pasted into a Word document.

Figure 24.16c) Contours of water volume fraction at $t = 2$ s

Double click on Graphics and Contours under Results in the Outline View. Select Contours of Velocity… and Velocity Magnitude. Select *mixture* as the Phase. Deselect all Surfaces. Click on Colormap Options. Select general as Type and 0 for Precision under Number Format in the Colormap window. Set Font Behavior to Fixed and Font Size to 12. Set Colormap Alignment to Top. Select Apply and Close the Colormap window. Click on Save/Display and Close the Contours window. Repeat this step for Axial Velocity and Radial Velocity for the mixture.

Figure 24.16d) Contours of velocity magnitude at $t = 2$ s

Figure 24.16e) Contours of axial velocity at $t = 2$ s

562

Figure 24.16f) Contours of radial velocity at $t = 2$ s

Double click on Plots and XY Plot under Results in the Outline View. Select Velocity and Velocity Magnitude for Y Axis Function and Direction Vector for X Axis Function.

Select New Surface >> Line/Rake from the drop–down menu. Enter the following end points: x0 [m] = 0, y0 [m] = 0 and x1 [m] = 0.045, y1 [m] = 0. Select Create and Close the window.

Figure 24.16g) Line/rake surface settings

Select *line-6* under the Surfaces section in the Solution XY Plot window. Click on Curves…, select Curve #0 and select the first available pattern with no (blank) symbol. Click on Apply and select Curve # 1, select the second available pattern and no symbol as shown in Figure 24.16h).

Curves - Solution XY Plot   ✕

| Curve # | **Line Style** | | **Marker Style** | |

Curve #
1 ▲▼

Line Style

Pattern
- -   ▼

Sample

Color
red   ▼

Weight
2

Marker Style

Symbol
                               ▼

Color
red   ▼

Size
0.3

Apply Close Help

Figure 24.16h) Curve settings

Click on Apply and Close the Curves-Solution XY Plot window. Click on Axes… in the Solution XY Plot window. Select X Axis, set Type to float, set Precision under Number Format to 3 and click on Apply. Select Y Axis, set Type to general, set Precision under Number Format to 2 and click on Apply. Close the Axes – Solution XY Plot window. Click on Save/Plot in the Solution XY Plot window. Click on Copy Screenshot of Active Window to Clipboard in the graphics window, see Figure 24.13f). The XY plot can be pasted into a Word document.

Axes - Solution XY Plot   ✕

Axis

◉ X
○ Y

Label

Number Format

Type
float   ▼

Precision
3 ▲▼

Major Rules

Color
foreground   ▼

Weight
1

Options

☐ Log
☑ Auto Range
☐ Major Rules
☐ Minor Rules

Range

Minimum
0

Maximum
0

Minor Rules

Color
dark gray   ▼

Weight
1

Apply Close Help

Figure 24.16i) X-axes settings

Figure 24.16j) Y-axes settings

Figure 24.16k) Axial velocity variation along centerline (based on *x* coordinate)

Select File>>Export>>Solution Data… from the menu. Select ASCII as File Type, Node under Location and Space as Delimiter. Select *line-6* under Surfaces. Select Axial Velocity under Quantities. Click on Write and save the ASCII File in the Working Directory with the name *centerline axial velocity data*. Close the Export window.

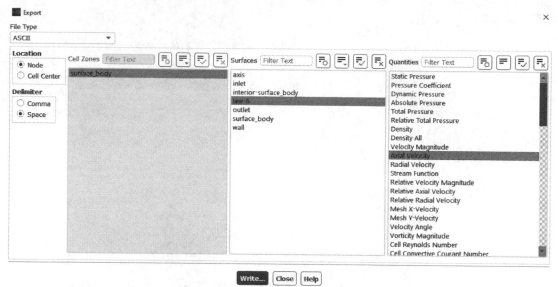

Figure 24.16l) Export settings for axial velocity data

Open the saved file in Excel. Click on Next in the Text Import Wizard – Step 1 of 3. Click on Next in Text Import Wizard – Step 2 of 3. Click on Finish in Text Import Wizard – Step 3 of 3. Plot the axial velocity along the centerline. The definition of the two coordinate systems (x, y) and (r, z) is

$$z = Z - x = 0.045 - x, r = y \qquad (24.1)$$

Label the D column as U (m/s). Enter the equation =0.045-B2 in cell $E2$ and label this column z (m). Drag cell $E2$ down the whole column.

Figure 24.16m) Axial velocity variation along centerline (based on z coordinate)

x-coordinate	U (m/s)	z (m)	x-coordinate	U (m/s)	z (m)
0.00E+00	**0.400000**	**0.045000**	2.10E-02	0.546649	0.024000
6.00E-04	0.404913	0.044400	2.13E-02	0.550967	0.023750
1.20E-03	0.405686	0.043800	2.15E-02	0.555275	0.023500
1.80E-03	0.407256	0.043200	2.18E-02	0.559572	0.023250
2.40E-03	0.409259	0.042600	2.20E-02	0.563855	0.023000
3.00E-03	0.411339	0.042000	2.23E-02	0.568122	0.022750
3.60E-03	0.413380	0.041400	2.25E-02	0.572372	0.022500
4.20E-03	0.415350	0.040800	2.28E-02	0.576603	0.022250
4.80E-03	0.417236	0.040200	2.30E-02	0.580815	0.022000
5.40E-03	0.419032	0.039600	2.33E-02	0.585007	0.021750
6.00E-03	0.420740	0.039000	2.35E-02	0.589178	0.021500
6.60E-03	0.422369	0.038400	2.38E-02	0.593327	0.021250
7.20E-03	0.423933	0.037800	2.40E-02	0.597455	0.021000
7.80E-03	0.425449	0.037200	2.43E-02	0.601561	0.020750
8.40E-03	0.426936	0.036600	2.45E-02	0.605645	0.020500
9.00E-03	0.428421	0.036000	2.48E-02	0.609708	0.020250
9.60E-03	0.429933	0.035400	2.50E-02	0.613750	0.020000
1.02E-02	0.431512	0.034800	2.53E-02	0.617771	0.019750
1.08E-02	0.433211	0.034200	2.55E-02	0.621771	0.019500
1.14E-02	0.435098	0.033600	2.58E-02	0.625752	0.019250
1.20E-02	0.437260	0.033000	2.60E-02	0.629714	0.019000
1.26E-02	0.439805	0.032400	2.63E-02	0.633658	0.018750
1.32E-02	0.442865	0.031800	2.65E-02	0.637585	0.018500
1.38E-02	0.446575	0.031200	2.68E-02	0.641495	0.018250
1.44E-02	0.451506	0.030600	2.70E-02	0.645390	0.018000
1.50E-02	**0.456531**	**0.030000**	2.73E-02	0.649271	0.017750
1.53E-02	0.458859	0.029750	2.75E-02	0.653139	0.017500
1.55E-02	0.461650	0.029500	2.78E-02	0.656993	0.017250
1.58E-02	0.464398	0.029250	2.80E-02	0.660836	0.017000
1.60E-02	0.467291	0.029000	2.83E-02	0.664666	0.016750
1.63E-02	0.470377	0.028750	2.85E-02	0.668483	0.016500
1.65E-02	0.473631	0.028500	2.88E-02	0.672287	0.016250
1.68E-02	0.477030	0.028250	2.90E-02	0.676076	0.016000
1.70E-02	0.480559	0.028000	2.93E-02	0.679847	0.015750
1.73E-02	0.484210	0.027750	2.95E-02	0.683596	0.015500
1.75E-02	0.487972	0.027500	2.98E-02	0.686878	0.015250
1.78E-02	0.491834	0.027250	3.00E-02	0.690270	0.015000
1.80E-02	0.495785	0.027000	3.08E-02	0.702131	0.014200
1.83E-02	0.499813	0.026750	3.16E-02	0.712863	0.013442
1.85E-02	0.503907	0.026500	3.23E-02	0.722935	0.012723
1.88E-02	0.508058	0.026250	3.30E-02	0.731832	0.012042
1.90E-02	0.512256	0.026000	3.36E-02	0.739755	0.011396
1.93E-02	0.516493	0.025750	3.42E-02	0.746885	0.010784
1.95E-02	0.520760	0.025500	3.48E-02	0.753231	0.010204
1.98E-02	0.525050	0.025250	3.53E-02	0.758720	0.009653
2.00E-02	0.529357	0.025000	3.59E-02	0.763262	0.009132
2.03E-02	0.533676	0.024750	3.64E-02	0.766762	0.008638
2.05E-02	0.538000	0.024500	3.68E-02	0.769109	0.008169
2.08E-02	0.542326	0.024250	**3.73E-02**	**0.770178**	**0.007725**

Figure 24.16n) Excel data for axial velocity variation along centerline

We find from the data that the axial velocity increases in the 15 mm long pipe section from $U_{in} = 0.4$ m/s at $z = 0.045$ m to $U_0 = 0.456531$ m/s at $z = Z = 0.03$ m. The axial velocity increases to a maximum velocity 0.770178 m/s at $z = 0.007725$ m.

17. Select File>>Export>>Data... from the menu. Save the data file in the *Working Directory* folder with the name *CHJ-t=2s.dat.h5*. Select the Results tab from the menu and select Surface>>Create>>Iso-Surface.... Select Surface of Constant Phases... and Volume fraction. Select *air* as the Phase and slide the Iso-Value to 0.5. Select *surface_body* under From Zones and *surface_body* under From Surface. Click on Create.

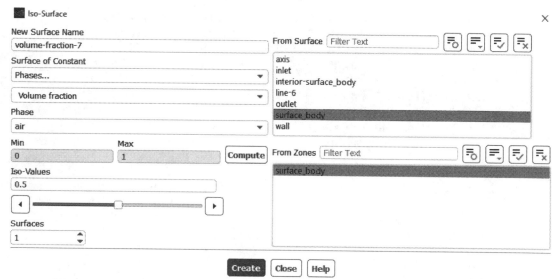

Figure 24.17a) Iso-surface settings for air phase

Select *water* as the Phase and set Iso-Values to 0.5. Select *surface_body* under From Zones and *surface_body* under From Surface. Enter *chj-free-surface-t-2s* as New Surface Name and click on Create. Close the window.

Select File>>Export>>Solution Data... from the menu. Select ASCII as File Type, Node under Location and Space as Delimiter. Select *chj-free-surface-t-2s* under Surfaces. Select Volume fraction (water) under Quantities. Click on Write... and save the ASCII File in the Working Directory with the name *chj-free-surface-coordinates-t-2s*. Close the Export window.

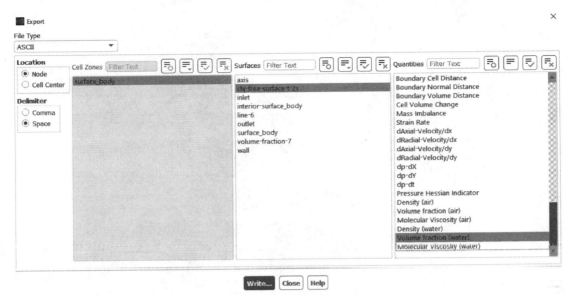

Figure 24.17b) Export settings for free surface data

Open the saved file in Excel. Click on Next in the Text Import Wizard – Step 1 of 3. Click on Next in Text Import Wizard – Step 2 of 3. Click on Finish in Text Import Wizard – Step 3 of 3. Plot the free surface. The two coordinate systems (x, y) and (r, z) are related as shown in Equation (24.1). Copy the y-coordinate column C in the Excel file and paste the data to column F. Label this column r (m). Enter the equation $-0.045-B2$ in cell $G2$ and label this column z (m). Drag cell $G2$ down the whole column.

	A	B	C	D	E	F	G
1	nodenumber	x-coordinate	y-coordinate	axial-velocity	water-vof	r (m)	z (m)
2	1	3.85E-02	3.58E-03	0.742877	5.00E-01	0.003582	0.006526
3	2	3.88E-02	3.58E-03	0.753201	5.00E-01	0.003583	0.006167
4	3	3.81E-02	3.58E-03	0.746355	5.00E-01	0.003584	0.006905
5	4	3.92E-02	3.59E-03	0.766794	5.00E-01	0.003586	0.005827
6	5	3.79E-02	3.59E-03	0.746783	5.00E-01	0.003586	0.007139
7	6	3.92E-02	3.59E-03	0.766882	5.00E-01	0.003586	0.005822
8	7	3.77E-02	3.59E-03	0.746458	5.00E-01	0.003588	0.007304
9	8	3.95E-02	3.59E-03	0.769086	5.00E-01	0.003593	0.005505
10	9	3.73E-02	3.59E-03	0.738518	5.00E-01	0.003595	0.007725
11	10	3.98E-02	3.60E-03	0.768710	5.00E-01	0.003602	0.005200

Figure 24.17c) Excel data for dimensional free surface elevation

Figure 24.17d) Free surface elevation on horizontal plate versus radial distance from axis.

Copy the r (m) column F in the Excel file and paste it to column H. Label this column r *(m)*. Enter the equation =(G4-G2)/(H4-H2) in cell $I3$ and label this column dz/dr. Drag cell $I3$ down the whole column.

	A	B	C	D	E	F	G	H	I	J
1	nodenumber	x-coordinate	y-coordinate	axial-velocity	water-vof	r (m)	z (m)	r (m)	dz/dr	
2	1	3.85E-02	3.58E-03	0.742877	5.00E-01	0.003582	0.006526	0.003582		
3	2	3.88E-02	3.58E-03	0.753201	5.00E-01	0.003583	0.006167	0.003583	=(G4-G2)/(H4-H2)	
4	3	3.81E-02	3.58E-03	0.746355	5.00E-01	0.003584	0.006905	0.003584	-97.4988	
5	4	3.92E-02	3.59E-03	0.766794	5.00E-01	0.003586	0.005827	0.003586	91.31308	
6	5	3.79E-02	3.59E-03	0.746783	5.00E-01	0.003586	0.007139	0.003586	-49.1739	
7	6	3.92E-02	3.59E-03	0.766882	5.00E-01	0.003586	0.005822	0.003586	92.00863	
8	7	3.77E-02	3.59E-03	0.746458	5.00E-01	0.003588	0.007304	0.003588	-50.409	
9	8	3.95E-02	3.59E-03	0.769086	5.00E-01	0.003593	0.005505	0.003593	64.31533	
10	9	3.73E-02	3.59E-03	0.738518	5.00E-01	0.003595	0.007725	0.003595	-31.43	
11	10	3.98E-02	3.60E-03	0.768710	5.00E-01	0.003602	0.005200	0.003602	52.21972	

Figure 24.17e) Excel data for gradient dz/dr.

We define the location of the jump Rj based on the radial location where the gradient dz/dr (as determined using the central finite difference formula in the Excel file for the derivative) has a maximum as seen in Figure 24.17f). The location of the hydraulic jump is determined to be at $Rj = 0.069576$ m.

z (m)	r (m)	dz/dr
0.000652	0.063000	0.018698
0.000658	**0.063294**	**0.02**
0.000661	0.063444	0.020666
0.000671	0.063889	0.022909
0.000681	0.064333	0.025462
0.000693	0.064778	0.027147
0.000697	0.064890	0.028389
0.000706	0.065222	0.029281
0.000719	0.065667	0.033221
0.000736	0.066111	0.039283
0.000754	0.066556	0.042231
0.000755	0.066568	0.045222
0.000774	0.067000	0.048809
0.000798	0.067444	0.055224
0.000816	0.067760	0.060756
0.000825	0.067889	0.064531
0.000853	0.068333	0.068619
0.000881	0.068712	0.074938
0.000887	0.068778	0.078283
0.000921	0.069222	0.07878
0.000950	**0.069576**	**0.08253**
0.000958	0.069667	0.077487
0.000991	0.070111	0.071702
0.001021	0.070556	0.068493
0.001022	0.070559	0.053847
0.001045	0.071000	0.046766
0.001063	0.071444	0.034727
0.001076	0.071889	0.025521
0.001084	**0.072266**	**0.02**
0.001086	0.072333	0.019022

Figure 24.17f) Excel data for gradient dz/dr.

The two different radii corresponding to before the jump $R_h = 0.063294$ mm and after the jump $R_H = 0.072266$ mm are shown in Figures 24.17d) and 24.17g). R_h and R_H were both determined from interpolated values at $dz/dr = 0.02$, see Figure 24.17f). From R_H and R_h it is possible to find $H = 0.001084$ m and $\Delta H = 0.000426$ m. H is the z coordinate corresponding to $r = R_H$. ΔH is the difference in z coordinates between $r = R_H$ and $r = R_h$, see Figure 24.17d).

Figure 24.17g) Surface slope versus radial distance from axis.

We can find the minimum radius $a = 0.003582$ m for the vertical jet from the data as shown in Figures 24.17c) and 24.17h). The corresponding vertical location is $z_a = 0.006526$ m with an axial velocity $U = 0.742877$ m/s.

Figure 24.17h) Free surface elevation for jet versus radial distance from axis.

In Figures 24.17i) and 24.17j) is the Mathematica code and the theoretical data points shown.

```
RN = 0.005; g = 9.81; ρ = 998.2; σ = 0.07286; Z = 0.030; μ = 0.001003; v = μ / ρ; U0 = 0.456531;
H = 0.001084; dH = 0.000426; Fr = U0^2 / (g * RN); We = ρ * U0^2 * RN / σ; i = 0; Clear[r]; Clear[U];
Do[{eq1 = NSolve[a / RN - (1 + 2 * (Z - z) / (Fr * RN) + 2 * (1 - (RN / a)) / We) ^ (-1 / 4) == 0, a];
   i = i + 1;
   zz[i] = z;
   r[i] = a //. eq1; U[i] = U0 * (RN / a) ^2 //. eq1;
   s = NDSolve[{y'[x] == y[x] × Cos[x + y[x]], y[0] == 1}, y[x], {x, 0, 30}];
   f[x_] := Evaluate[y[x] /. s];
   Print[i, r[i][[1]], zz[i], U[i][[1]]]}, {z, 0.03, 0, -0.001}];
m = Table[{r[i][[1]], zz[i], U[i][[1]]}, {i, 1, 31, 1}];
m = Prepend[m, {"r (m)", "z (m)", "U (m/s)"}];
m // TableForm
Export["mfile.xls", m]
```

Figure 24.17i) Mathematica code used to generate the data in Figure 24.17j).

r (m)	z (m)	U (m/s)
0.005	0.03	0.456531
0.00489225	0.029	0.476862
0.00479501	0.028	0.496399
0.00470658	0.027	0.515228
0.00462562	0.026	0.533421
0.00455109	0.025	0.551036
0.00448211	0.024	0.568126
0.004418	0.023	0.584734
0.00435817	0.022	0.6009
0.00430213	0.021	0.616656
0.00424948	0.02	0.632032
0.00419986	0.019	0.647053
0.00415298	0.018	0.661745
0.00410858	0.017	0.676126
0.00406642	0.016	0.690216
0.00402633	0.015	0.704031
0.00398811	0.014	0.717588
0.00395163	0.013	0.7309
0.00391674	0.012	0.74398
0.00388332	0.011	0.756839
0.00385127	0.01	0.769488
0.00382049	0.009	0.781938
0.00379089	0.008	0.794197
0.00376239	0.007	0.806273
0.00373493	0.006	0.818176
0.00370842	0.005	0.829911
0.00368283	0.004	0.841486
0.00365809	0.003	0.852907
0.00363415	0.002	0.86418
0.00361097	0.001	0.875311
0.00358851	0.	0.886305

Figure 24.17j) Data values for theoretical curves.

H. Theory

18. Following Bush and Aristoff (2003), we define the jet Reynolds number for this flow

$$Re = \frac{Q}{a\nu} \tag{24.2}$$

where Q (m³/s) is volume flow rate, a (m) is the minimum radius of the jet and ν (m²/s) is kinematic viscosity of the liquid.

Figure 24.18 Geometry and shape of free surface for circular hydraulic jump.

The minimum jet radius a (m) and corresponding maximum axial jet speed U_m (m/s) can be determined from

$$a = \frac{R_n}{\left\{1+\frac{2}{U_0^2}[gZ+\frac{\sigma}{\rho R_n a}(a-R_n)]\right\}^{1/4}} \tag{24.3}$$

$$U_m = U_0\left(\frac{R_n}{a}\right)^2 \tag{24.4}$$

where U_0 (m/s) is the jet velocity corresponding to nozzle radius R_n (m), g (m/s²) is acceleration due to gravity, Z (m) is the height of the nozzle exit above the horizontal plate, σ (N/m) is surface tension and ρ (kg/m³) is liquid density.

The free jet radius and axial velocity at any vertical location can be determined from

$$r(z) = \frac{R_n}{\left[1+\frac{2(Z-z)}{FrR_n}+\frac{2}{We}\left(1-\frac{R_n}{r(z)}\right)\right]^{1/4}} \tag{24.5}$$

$$U(z) = U_0\left(\frac{R_n}{r(z)}\right)^2 \tag{24.6}$$

$$Fr = \frac{U_0^2}{gR_n} \tag{24.7}$$

$$We = \frac{\rho U_0^2 R_n}{\sigma} \tag{24.8}$$

where Fr is the Froude number and We is the Weber number. The jump radius R_j (m) is given by the following equations

$$R_j g \left(\frac{aH}{Q}\right)^2 \left(1 + \frac{2}{Bo}\right) + \frac{1}{2R_j H}\left(\frac{a}{\pi}\right)^2 = 0.10132 - 0.1297\frac{1}{Re^{1/2}}\left(\frac{R_j}{a}\right)^{3/2} ; r(z) < r_v \tag{24.9}$$

$$R_j g \left(\frac{aH}{Q}\right)^2 \left(1 + \frac{2}{Bo}\right) + \frac{1}{2R_j H}\left(\frac{a}{\pi}\right)^2 = \frac{0.01676}{0.1826 + \frac{1}{Re}\left(\frac{R_j}{a}\right)^3} ; r(z) > r_v \tag{24.10}$$

$$Bo = \frac{\rho g R_j \Delta H}{\sigma} \tag{24.11}$$

$$r_v = 0.315 a Re^{1/3} \tag{24.12}$$

where Bo is Bond number, H (m) is the height of the liquid after the jump and ΔH (m) is the increase in liquid height through the jump. Table 24.1 shows a summary of numerical values used for the different variables in this chapter.

	Definition	Theory	Ansys Fluent	Percent Diff.
	Coordinates and Axial Velocity			
z	vertical coordinate defined from wall			
r	radial coordinate defined from centerline			
$r(z)$	free jet radius			
$U(z)$	axial velocity along centerline			
	Nozzle Geometry and Nozzle Velocity			
R_n (m)	nozzle radius	0.005	0.005	
Z (m)	distance from nozzle exit to wall	0.03	0.03	
U_{in} (m/s)	nozzle inlet axial velocity	0.4	0.4	
U_0 (m/s)	jet axial velocity @ nozzle exit	0.456531	0.456531	
	Jet Geometry, Max. Axial Velocity & Flow Rate			
a (m)	minimum radius of jet	0.003749	0.003582	4.5
z_a (m)	distance from minimum jet radius to wall	0.006526	0.006526	
U_m (m/s)	max axial jet velocity at minimum radius of jet	0.811915	0.742877	8.5
Q (m³/s)	volume flow rate	3.586E-05	2.994E-05	16.5

	Non-Dimensional Parameters and Flow Rate			
Bo	Bond number	3.74	3.98	6.5
Fr	Froude number	4.25	4.25	
Re	Reynolds number	9518	8320	12.6
We	Weber number	14.28	14.28	
	Hydraulic Jump Location and Thickness			
r_v (m)	radial distance to minimum liquid thickness	0.0250293	0.0323333	29.2
R_h (m)	radial location before hydraulic jump	0.063294	0.063294	
R_j (m)	radial location of hydraulic jump	0.0653327	0.069576	6.5
R_j / a	non-dimensional radial location of hydraulic jump	17.4250	19.4238	11.5
R_H (m)	radial location after hydraulic jump	0.072266	0.072266	
H	height of liquid after hydraulic jump	0.001084	0.001084	
ΔH	increase in liquid height through hydraulic jump	0.000426	0.000426	
	Fluid Properties			
σ (N/m)	surface tension	0.07286	0.07286	
ρ (kg/m³)	liquid density	998.2	998.2	
μ (kg/m-s)	liquid dynamic viscosity	0.001003	0.001003	
ν (m²/s)	kinematic viscosity of liquid	1.005E-06	1.005E-06	

Table 24.1 Numerical values from theory and simulations

I. References

1. Bush, J.W. M and Aristoff, J.M., "The influence of surface tension on the circular hydraulic jump", *Journal of Fluid Mechanics,* **489**, 229-238, (2003).
2. Bush, J.W. M, Aristoff, J.M. and Hosoi, A.E., "An experimental investigation of the stability of the circular hydraulic jump", *Journal of Fluid Mechanics,* **558**, 33-52, (2006).
3. Lokman Hosain, M.D.,"CFD Simulation of Jet Cooling and Implementation of Flow Solvers in GPU",*Master of Science Thesis*, Stockholm, Sweden, (2013).
4. Passandideh-Fard, M., Teymourtash, A.R. and Khavari, M., "Numerical Study of Circular Hydraulic Jump Using Volume-of-Fluid Method", *J. Fluids Engineering*, **133**, (2011).
5. Raghav, G.H, "Numerical analysis of hydraulic jump by an impinging jet.", *Master of Technology Thesis*, Rourkela, (2014).

J. Exercises

24.1 Use Ansys Fluent to run the simulation for different inlet velocities U_{in} = 0.2, 0.6 and 0.8 m/s and compare results in Table 24.2. Plot a graph with Reynolds number Re on the horizontal axis and R_j/a on the vertical axis based on your data in Table 24.2.

	Ansys Fluent	Ansys Fluent	Ansys Fluent	Ansys Fluent
a (m)		0.003582		
Bo		3.98		
Q (m³/s)		0.00002994		
Re		8320		
R_j (m)		0.069576		
R_j/a		19.4238		
r_v (m)		0.0323333		
Um (m/s)		0.742877		
Fr		4.25		
H		0.001084		
ΔH		0.000426		
R_n (m)	0.005	0.005	0.005	0.005
U_{in} (m/s)	0.2	0.4	0.6	0.8
U_0 (m/s)		0.456531		
We		14.28		
Z (m)	0.030	0.030	0.030	0.030
σ (N/m)	0.07286	0.07286	0.07286	0.07286
ρ (kg/m³)	998.2	998.2	998.2	998.2
v (m²/s)	0.00000100481	0.00000100481	0.00000100481	0.00000100481

Table 24.2 Numerical values for Exercise 24.1

K. Notes

CHAPTER 25. OPTIMIZED ELBOW

A. Objectives

- Using Ansys Fluent to Study the Flow in an Optimized Elbow
- Inserting Boundary Conditions
- Running Ansys Fluent Simulations
- Using Contour Plots for Visualizations of Velocity and Pressure Fields

B. Problem Description

We will study the pressure drop in a 90° elbow for an optimized design and compare the pressure drop with a standard design.

C. Launching Ansys Fluent

1. Start by launching Ansys Fluent in standalone mode.

Figure 25.1 Launching ANSYS Fluent

2. Check the box for Double Precision. Select Meshing in the upper left portion of the Fluent Launcher window. Set the number of Meshing Processes equal to the number of processor cores for your computer. The number of Meshing Processes is limited to four if you are using Ansys Student. To check the number of physical cores, press the Ctrl + Shift + Esc keys simultaneously to open the Task Manager. Go to the Performance tab and select CPU from the left column. You'll see the number of physical cores on the bottom-right side. Close the Task Manager window. Click on the Start button in the Fluent Launcher window.

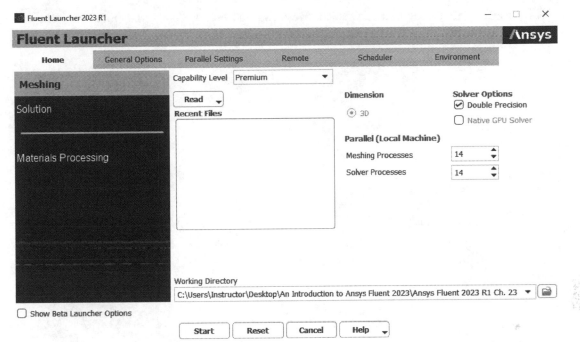

Figure 25.2 Fluent launcher window

D. Launching Ansys Fluent Meshing

3. Select Workflow Type as Watertight Geometry.

Figure 25.3 Selection of Watertight Geometry

4. Select Import Geometry and select **Mesh** as File Format. Select *in* as File Units and *cm* as Units. Browse for the File Name and open the file *optimized_elbow.msh.h5*. Select Import Geometry.

Figure 25.4 Imported geometry

5. Right click on Add Local Sizing and select Update. Right click on Generate the Surface Mesh and select Update.

Figure 25.5 Generated surface mesh

6. Select Describe Geometry. As Geometry Type, select "The geometry consists of only fluid regions with no voids". Right click on Describe Geometry and select Update. Select Update Boundaries, right click and select update. Select Update Regions, right click and select update.

Figure 25.6a) Describe geometry

Figure 25.6b) Update boundaries

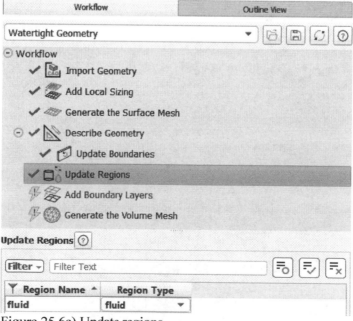

Figure 25.6c) Update regions

7. Select Add Boundary Layers. Set the Offset Method Type to aspect-ratio, Number of Layers to 6 and First Aspect Ratio to 12. Select Add Boundary Layers.

Figure 25.7 Adding boundary layers

8. Select Generate the Volume Mesh. Right click on Generate the Volume Mesh and select Update.

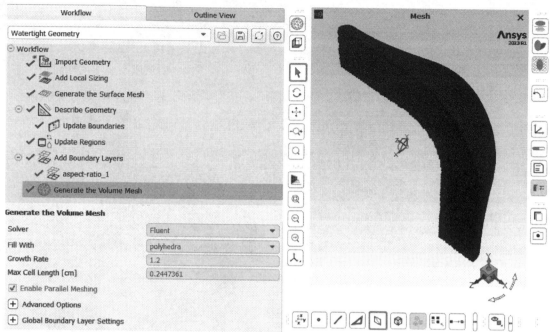

Figure 25.8 Generated volume mesh

E. Ansys Fluent Solution

9. Select Mesh>>Check from the menu. Select Switch to Solution in the upper left corner. Answer Yes to the question that you get. Double click on General under Setup in the Outline View. Click on Scale under Mesh on the Task Page.

Figure 25.9a) Switching to solution

Figure 25.9b) General Settings

Select Specify Scaling Factors under Scaling. Set the Scaling Factors in the X, Y, Z directions to **0.875**. Click on Scale. The Domain Extent will change. Close the Scale Mesh window.

Figure 25.9c) Scale Mesh

Check the Gravity box under General on the Task Page. Set the Y [m/s²] value to **-9.81**.

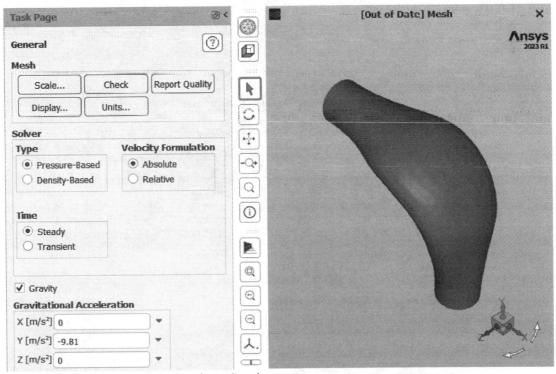

Figure 25.9d) Gravitational acceleration

10. Open Models under Setup in the Outline View and double click on Viscous (SST k-omega). Select k-epsilon (2 eqn), Realizable k-epsilon Model and Standard Wall Functions as Near-Wall Treatment and set the Model Constants and User-Defined Functions as listed in Figure 25.10a). Select OK to close the Viscous Model window.

Figure 25.10a) Viscous model

Open Materials and Fluid under Setup in the Outline View and double click on *air*. Set Density [kg/m³] to **1.190** and Viscosity [kg/(m s)] to **1.8286e-05** to correspond to values in experiments. Select Change/Create and Close the Create/Edit Materials window.

Figure 25.10b) Create/Edit Materials

11. Open Boundary Conditions under Setup in the Outline View and double click on Inlet. Set the Velocity Magnitude [m/s] to **25.99**. Select Apply and Close the Velocity Inlet window.

Figure 25.11a) Velocity inlet

Open Boundary Conditions under Setup in the Outline View and double click on Wall. Set the Roughness Height [m] under Sand-Grain Roughness to **0.00023**. Set the Roughness Constant to **0.5**. Select Apply and Close the Velocity Inlet window.

Figure 25.11b) Wall

Double click on Reference Values under Setup in the Outline View. Select Compute from inlet under Reference Values on the Task Page. Set the Area [m²] to **0.001552** and the Length [m] to **0.04445**. This is the inlet diameter. Select fluid as Reference Zone.

Figure 25.11b) Velocity inlet

12. Double click on Methods under Solution in the Outline View. Set Turbulent Kinetic Energy and Turbulent Dissipation Rate to Second Order Upwind. Set Gradient under Spatial Discretization to Least Squares Cell Based. Select Rhie-Chow: distance based Flux Type.

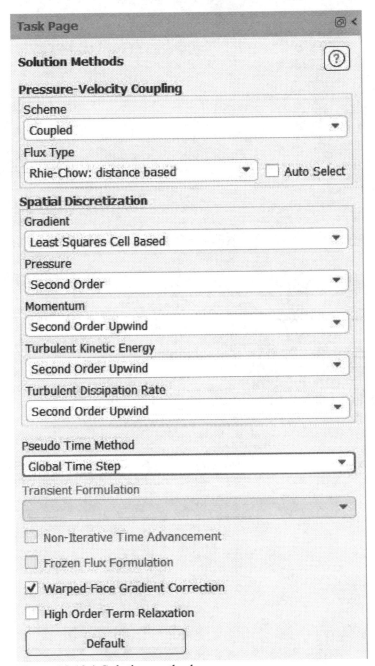

Figure 25.12a) Solution methods

Open Monitors under Solution in the Outline View and double click on Residual under Monitors. Set the Absolute Criteria for all Equations to **1e-04**. Select OK to close the Residual Monitors window.

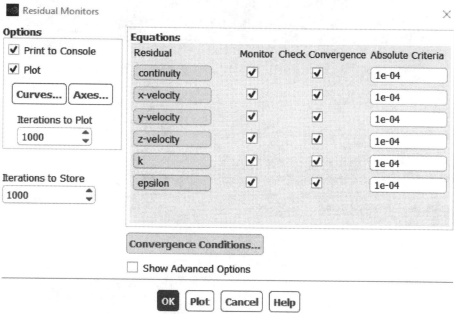

Figure 25.12b) Residual monitors

F. Initialization and Calculations

13. Double click on Initialization under Solution in the Outline View. Select Hybrid Initialization and select Initialize. Double click on Run Calculation under Solution in the Outline View. Set the Number of Iterations to 500 under Parameter. Select Calculate to start the calculations.

Figure 26.13a) Solution initialization

592

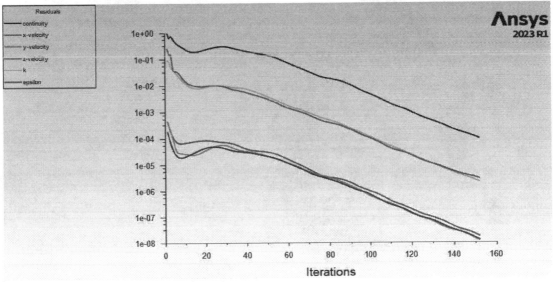

Figure 25.13b) Run calculations

Figure 25.13c) Residuals

G. Post-processing

14. Double click on Graphics under Results in the Outline View. Right click on Contours and select New…. Select Contours of Pressure and Static Pressure. Select *wall* under Surfaces. Select Colormap Options…, set Font Behavior to Fixed and set Font Size under Font to 10. Select Apply and Close the Colormap window. Click on Save/Display and Close the Contours window.

Figure 25.14a) Contours window

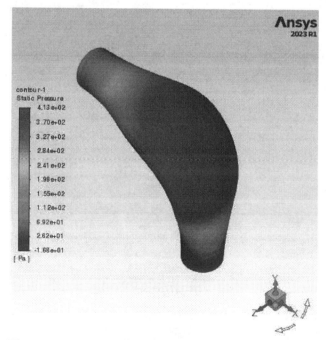

Figure 25.14b) Contours of static pressure

15. Right click on Surfaces under Results in the Outline View. Select New Plane.... Select XY Plane as Method. Select Create and Close the Plane Surface window.

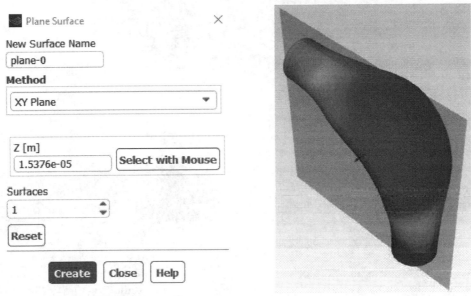

Figure 25.15 Plane surface

16. Double click on Graphics under Results in the Outline View. Right click on Contours and select New.... Select Contours of Velocity and Velocity Magnitude. Select *plane-0* under Surfaces. Select Colormap Options..., set Font Behavior to Fixed and set Font Size under Font to 10. Select Apply and Close the Colormap window. Click on Save/Display and Close the Contours window. Select the blue Z axis for the coordinate system in the graphics window to view the contour plot in the *XY* plane.

Figure 25.16a) Contours window

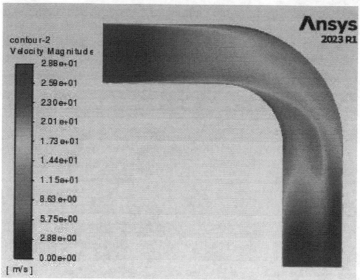

Figure 25.16b) Contours of velocity magnitude

17. Double click on Reports and Surface Integrals under Results in the Outline View. Select Area-Weighted Average as Report Type. Select Pressure... and Static Pressure as Field Variable. Select *inlet* and *outlet* as Surfaces. Select Compute and Close the Surface Integrals window. Write down the inlet pressure from the console.

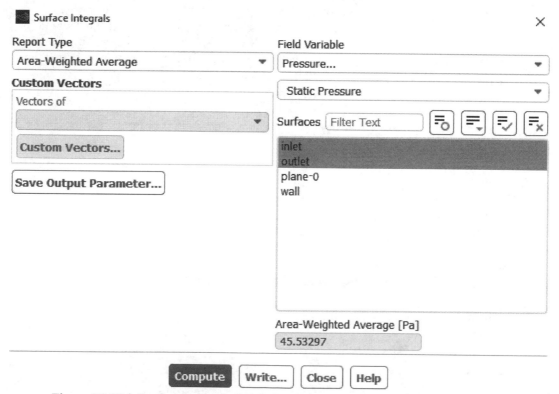

Figure 25.17a) Surface integrals for static pressure

```
            Area-Weighted Average
               Static Pressure                        [Pa]
    --------------------------------    -------------------
                             inlet              91.065965
                            outlet                      0
    --------------------------------    -------------------
                               Net              45.532972
```

Figure 25.17b) Results from console for static pressure

18. Double click on Reports and Surface Integrals under Results in the Outline View. Select Area-Weighted Average as Report Type. Select Pressure… and Pressure Coefficient as Field Variable. Select *inlet* and *outlet* as Surfaces. Select Compute and Close the Surface Integrals window. Write down the inlet pressure coefficient from the console.

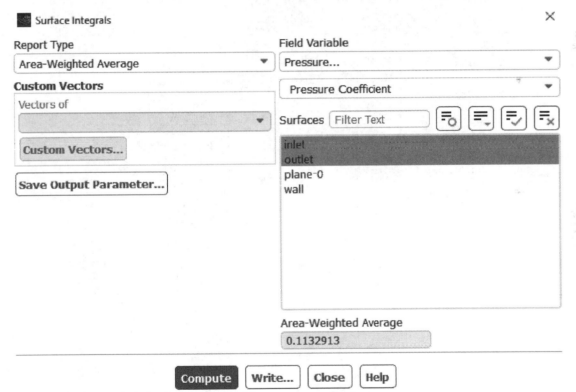

Figure 25.18a) Surface integrals for pressure coefficient

```
           Area-Weighted Average
             Pressure Coefficient
    ------------------------------    -------------------
                           inlet             0.22658261
                          outlet                      0
    ------------------------------    -------------------
                             Net             0.11329128
```

Figure 25.18b) Results from console for pressure coefficient

597

H. Batch Jobs

19. You will create the case and scheme files to complete a batch job. Select File>>Write>>Case & Data… from the menu. Save the case and data files in the *Working Directory* folder with the name ***optimized_elbow.cas.h5***.

Go back to **step 1** in this chapter and repeat all the steps but instead use the ***standard elbow***. In **step 4**, you will open the file ***standard_elbow.msh.h5***. When you arrive at this **step 19**, select File>>Write>>Case & Data… from the menu. Save the case and data files with the name ***standard_elbow.cas.h5***.

Open Notebook and create and save the scheme file with the name ***elbows_2023.scm*** for the batch job. The scheme file is shown in the Appendix at the end of this chapter and is available for download at *sdcpublications.com*. Select File>>Read>>Scheme… from the menu in Ansys Fluent. Open the scheme file to start the batch job.

U (m/s)	Re	Optimized $\Delta p_{Exp.}$ (Pa)	Optimized $K_{Exp.}$	U (m/s)	Re	Optimized Δp_{Fluent}(Pa)	Standard Δp_{Fluent}(Pa)	Optimized K_{Fluent}	Standard K_{Fluent}	Ratio K
38.47	111305	199.3	0.2263	40	115719	227.7746	380.0536	0.2392	0.3992	0.5993
38.10	110226	195.9	0.2268	39	112826	215.9873	360.4433	0.2386	0.3982	0.5992
37.57	108685	191.0	0.2274	38	109933	204.5144	341.3591	0.2380	0.3973	0.5991
36.86	106642	184.1	0.2277	37	107040	193.3628	322.7963	0.2374	0.3962	0.5990
35.69	103252	172.9	0.2281	36	104147	182.5372	304.7633	0.2367	0.3952	0.5989
33.88	98018	155.3	0.2273	35	101254	172.0214	287.2499	0.2360	0.3941	0.5989
31.65	91562	135.1	0.2266	34	98361	161.8058	270.2771	0.2352	0.3929	0.5987
30.19	87329	123.5	0.2278	33	95468	151.8842	253.8201	0.2344	0.3917	0.5984
28.62	82785	110.5	0.2268	32	92575	142.1485	237.9049	0.2333	0.3904	0.5975
25.99	75200	91.2	0.2268	31	89682	132.6727	222.5235	0.2320	0.3891	0.5962
24.55	71024	81.6	0.2275	30	86789	123.6546	207.6538	0.2309	0.3877	0.5955
23.77	68763	76.6	0.2278	29	83896	115.0255	193.3321	0.2298	0.3863	0.5950
21.78	63017	64.5	0.2284	28	81003	106.7446	179.5244	0.2288	0.3848	0.5946
20.62	59652	57.3	0.2265	27	78110	98.7485	166.2698	0.2276	0.3833	0.5939
19.10	55255	49.4	0.2276	26	75217	91.1430	153.5396	0.2266	0.3817	0.5936
18.37	53154	45.6	0.2270	25	72324	83.8325	141.3595	0.2254	0.3801	0.5930
16.65	48167	37.6	0.2279	24	69431	76.8620	129.6612	0.2242	0.3783	0.5928
15.59	45115	33.3	0.2301	23	66538	70.0828	118.3418	0.2226	0.3759	0.5922
14.32	41414	27.9	0.2288	22	63645	63.6991	107.6195	0.2212	0.3737	0.5919
13.43	38843	24.5	0.2284	21	60752	57.6351	97.3187	0.2196	0.3708	0.5922
12.41	35891	21.4	0.2337	20	57859	51.8871	87.4599	0.2180	0.3674	0.5933
10.89	31497	16.4	0.2325	19	54966	46.3135	78.1932	0.2156	0.3640	0.5923
9.69	28030	12.5	0.2238	18	52073	41.1502	69.5052	0.2134	0.3605	0.5920
8.94	25865	9.9	0.2081	17	49180	36.1393	61.3950	0.2101	0.3570	0.5886
8.25	23861	7.9	0.1952	16	46287	31.4019	53.4615	0.2061	0.3509	0.5874
6.73	19458	5.2	0.1932	15	43395	26.3251	45.5875	0.1966	0.3405	0.5775
5.35	15464	3.3	0.1941	14	40502	22.9012	36.6400	0.1964	0.3142	0.6250
4.59	13280	2.6	0.2074	13	37609	20.1257	31.1377	0.2001	0.3096	0.6463
3.51	10151	1.5	0.2075	12	34716	17.2332	26.0112	0.2011	0.3036	0.6625
				11	31823	15.2587	22.8237	0.2119	0.3170	0.6685
				10	28930	13.4566	19.8166	0.2261	0.3330	0.6791
				9	26037	11.2440	16.6564	0.2333	0.3456	0.6751
				8	23144	9.2072	13.7285	0.2418	0.3605	0.6707
				7	20251	7.3503	11.0373	0.2521	0.3785	0.6659
				6	17358	5.6744	8.5885	0.2649	0.4009	0.6607
				5	14465	4.1854	6.3906	0.2813	0.4296	0.6549
				4	11572	2.8893	4.4543	0.3035	0.4678	0.6487
				3	8678.9	1.7972	2.7964	0.3356	0.5222	0.6427

Table 25.1 Experimental results and Ansys Fluent data for $T = 21.67$ C, $p_{atm} = 100,700$ Pa, $\rho = 1.19$ kg/m³, $\mu = 1.8286 \cdot 10^{-5}$ kg/(m s), $D = 0.04445$ m, sand-grain roughness height 0.00023 m and roughness constant 0.5.

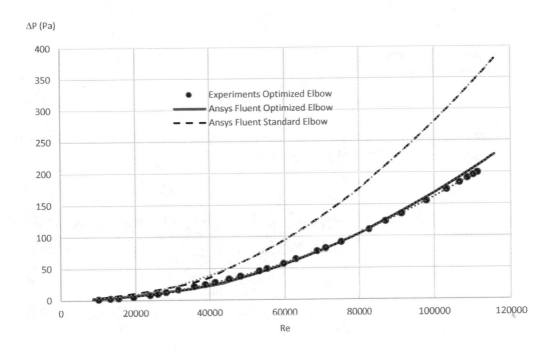

Figure 25.19a) Pressure drop versus Reynolds number for optimized elbow experiments

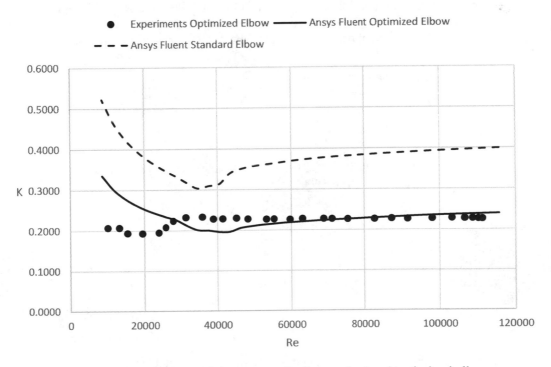

Figure 25.19b) Pressure coefficient versus *Re* for standard and optimized elbows

I. Theory

The temperature dependence for the dynamic viscosity of air can be described using the following equation

$$\mu = aT^b \tag{26.1}$$

where the constants are $a = 2.791 \cdot 10^{-7}$ and $b = 0.7355$, and T (K) is the temperature of the air. The density of air is given by the ideal gas law expressed as

$$\rho = \frac{p_{atm}}{RT} \tag{26.2}$$

where p_{atm} is the atmospheric pressure and $R = 287$ (J/(kg K)) is the gas constant. The Reynolds number is defined as

$$Re = \frac{UD\rho}{\mu} \tag{26.3}$$

where U (m/s) is the average velocity and $D = 0.04445$ (m) is the inner diameter of the pipe. The pressure coefficient can be expressed as

$$K = \frac{\Delta p}{\frac{1}{2}\rho U^2} \tag{26.4}$$

where Δp (Pa) is the pressure drop from the inlet to the outlet of the elbow. The outlet of the elbow will be at atmospheric pressure and the inlet of the elbow will be at a higher pressure than the atmospheric pressure. The percentage difference between experimental results and Ansys Fluent results for the pressure coefficient can be determined as

$$Percent\ Error = 100 \cdot \frac{|K_{Exp.} - K_{Fluent}|}{K_{Fluent}} \tag{26.5}$$

J. References

1. Baukal, C.E., Gershtein, V.Y., Li, X.," Computational Fluid Dynamics in Industrial Combustion.", CRC Press (2001).

K. Exercises

25.1 Run the calculations shown in this chapter for different surface height roughness and compare with the results in this chapter.

25.2 Run the calculations shown in this chapter for different Absolute Criteria for all Equations and compare with the results in this chapter.

25.3 Run the calculations shown in this chapter for different mesh sizes and compare with the results in this chapter.

CHAPTER 25. OPTIMIZED ELBOW

25.4 Run the Ansys Fluent calculations shown in this chapter for different velocities U (m/s) and compare with static pressure and pressure coefficient results as shown in Table 25.2. Determine the percent error between experiments and Ansys Fluent for the pressure coefficient K.

	U (m/s)	Re	$\Delta p_{Exp.}$ (Pa)	$K_{Exp.}$	$K_{Ansys\ Fluent}$	% Error
Student 1	38.47	111305	199.3	0.2263		
Student 2	38.10	110226	195.9	0.2268		
Student 3	37.57	108685	191.0	0.2274		
Student 4	36.86	106642	184.1	0.2277		
Student 5	35.69	103252	172.9	0.2281		
Student 6	33.88	98018	155.3	0.2273		
Student 7	31.65	91562	135.1	0.2266		
Student 8	30.19	87329	123.5	0.2278		
Student 9	28.62	82785	110.5	0.2268		
Student 10	25.99	75200	91.2	0.2268		
Student 11	24.55	71024	81.6	0.2275		
Student 12	23.77	68763	76.6	0.2278		
Student 13	21.78	63017	64.5	0.2284		
Student 14	20.62	59652	57.3	0.2265		
Student 15	19.10	55255	49.4	0.2276		
Student 16	18.37	53154	45.6	0.2270		
Student 17	16.65	48167	37.6	0.2279		
Student 18	15.59	45115	33.3	0.2301		
Student 19	14.32	41414	27.9	0.2288		
Student 20	13.43	38843	24.5	0.2284		

Table 25.2 Data for mean velocity, Reynolds number, pressure drop, pressure coefficient and percentage error

L. Appendix

This is the scheme file for the batch job shown in this chapter. The file is available for download at *sdcpublications.com*.

```
(Do ((y 2 (- y 1))) ((< y 1))
    (begin
        (Do ((x 40 (- x 1))) ((< x 1))
            (begin
                (if (> y 1)
                    (Ti-menu-load-string (format #f "file read-case optimized_elbow.cas.h5 ok"))
                    (Ti-menu-load-string (format #f "file read-case standard_elbow.cas.h5 ok")))
                (if (> x 10)
                    (begin
                        (Ti-menu-load-string (format #f "define models viscous ke-realizable yes"))
                        (Ti-menu-load-string (format #f "define boundary-conditions velocity-inlet inlet n n y y n ~a n 0 n n y 5 10" x))
                        (Ti-menu-load-string (format #f "define boundary-conditions wall wall n n n 0.00023 n 0.5"))
                        (Ti-menu-load-string (format #f "report reference-values velocity ~a " x))
                        (Ti-menu-load-string (format #f "solve set discretization-scheme mom 1"))
                        (Ti-menu-load-string (format #f "solve monitors residual convergence-criteria 0.0001 0.0001 0.0001 0.0001")))
                    (begin
                        (Ti-menu-load-string (format #f "define models viscous laminar yes"))
                        (Ti-menu-load-string (format #f "define boundary-conditions velocity-inlet inlet n n y y n ~a n 0" x))
                        (Ti-menu-load-string (format #f "define boundary-conditions wall wall n n n 0.00023 n 0.5"))
                        (Ti-menu-load-string (format #f "report reference-values velocity ~a " x))
                        (Ti-menu-load-string (format #f "solve set discretization-scheme mom 0"))
                        (Ti-menu-load-string (format #f "solve monitors residual convergence-criteria 0.00001 0.00001 0.00001 0.00001"))))
                (Ti-menu-load-string (format #f "solve initialize hyb-initialization yes"))
                (Ti-menu-load-string (format #f "solve iterate 300"))
                (if (> y 1)
                    (begin
                        (Ti-menu-load-string (format #f "report s-i a-w-a inlet outlet () p y optimized_elbow_static_pressure_data_U=~amps" x))
                        (Ti-menu-load-string (format #f "report s-i a-w-a inlet outlet () p-c y optimized_elbow_pressure_coeff__data_U=~amps" x)))
                    (begin
                        (Ti-menu-load-string (format #f "report s-i a-w-a inlet outlet () p y standard_elbow_static_pressure_data_U=~amps" x))
                        (Ti-menu-load-string (format #f "report s-i a-w-a inlet outlet () p-c y standard_elbow_pressure_coeff__data_U=~amps" x))))
                (Ti-menu-load-string (format #f "display objects display contour-1"))
                (Ti-menu-load-string (format #f "display views restore-view isometric"))
                (if (> y 1)
                    (Ti-menu-load-string (format #f "display save-picture optimized_elbow_static_pressure_contour_plot_U=~amps.jpg" x))
                    (Ti-menu-load-string (format #f "display save-picture standard_elbow_static_pressure_contour_plot_U=~amps.jpg" x)))
                (Ti-menu-load-string (format #f "display objects display contour-2"))
                (Ti-menu-load-string (format #f "display views restore-view front"))
                (if (> y 1)
                    (Ti-menu-load-string (format #f "display save-picture optimized_elbow_velocity_magnitude_contour_plot_U=~amps.jpg" x))
                    (Ti-menu-load-string (format #f "display save-picture standard_elbow_velocity_magnitude_contour_plot_U=~amps.jpg" x)))
                (Ti-menu-load-string (format #f "plot residuals y y y y y y"))
                (if (> y 1)
                    (Ti-menu-load-string (format #f "display save-picture optimized_elbow_residuals_plot_U=~amps.jpg" x))
                    (Ti-menu-load-string (format #f "display save-picture standard_elbow_residuals_plot_U=~amps.jpg" x)))))))
```

Figure 25.19d) Scheme file *elbows_2023* for standard and optimized elbows

CHAPTER 26. GOLF BALL

A. Objectives

- Using Ansys Fluent to Study the Flow around a Volvik ProBismuth Golf Ball
- Loading the *stl* file for the Volvik ProBismuth Golf Ball
- Using Fault-Tolerant Meshing in Ansys Fluent Meshing
- Inserting Boundary Conditions
- Running Ansys Fluent Simulations
- Using Contour Plots for Visualizations of Velocity and Pressure Fields

B. Problem Description

We will study the flow around a Volvik ProBismuth golf ball. This golf ball has an octahedron pattern of dimples and the number of dimples is 446.

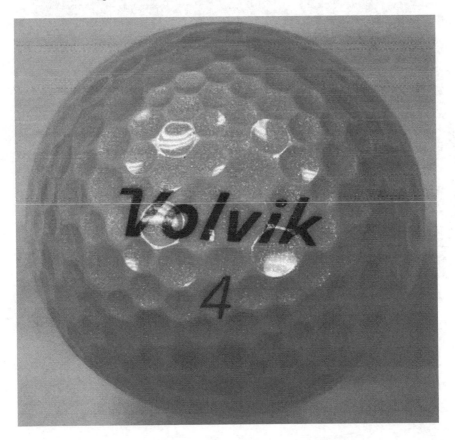

C. Launching Ansys Fluent

1. Start by launching Ansys Fluent in standalone mode.

Figure 26.1 Launching ANSYS Fluent

2. Check the box for Double Precision. Select Meshing in the upper left portion of the Fluent Launcher window. Set the number of Meshing and Solver Processes equal to the number of processor cores for your computer. To check the number of physical cores, press the Ctrl + Shift + Esc keys simultaneously to open the Task Manager. Go to the Performance tab and select CPU from the left column. You'll see the number of physical cores on the bottom-right side. Close the Task Manager window. Click on the Start button in the Fluent Launcher window.

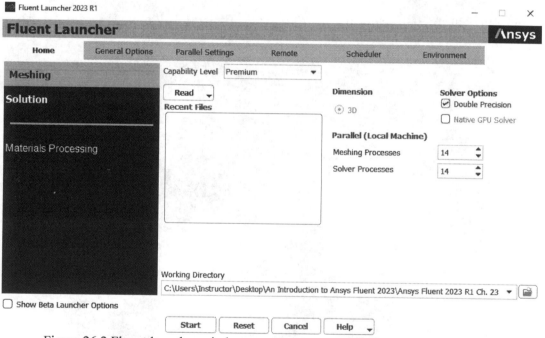

Figure 26.2 Fluent launcher window

D. Launching Ansys Fluent Meshing

3. Select Workflow Type as Fault-Tolerant Meshing.

Figure 26.3 Selection of Fault-tolerant Meshing

4. Select Import CAD and Part Management. Select *mm* as Display Unit. Browse for the CAD File. Select Other CAD Formats (*.jt *.plmxml *.stl, …) as Files of type. Open the file ***Volvik Pro Bismuth 18 scans.stl***. Select *mm* as File Unit. Select Custom as Create Meshing Objects. Select Load and check the box next to CAD Model. Right-click on Meshing Model and select Create Object. Right-click on Object and select Rename. Change the name to *golf-ball*. Drag and drop the part "Volvik Pro Bismuth 18 scans" from CAD Model to *golf-ball* under Meshing Model. Select Create Meshing Objects.

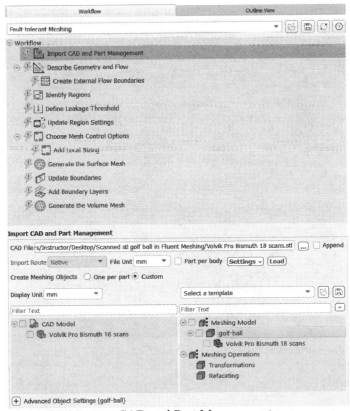

Figure 26.4 Import CAD and Part Management

5. Select External flow around object as Flow Type under Describe Geometry and Flow. Select Yes for Add an enclosure? and Yes for Add local refinement regions? Open Advanced Options and select Yes for Identify regions? and Yes for Close leakages? Select Describe Geometry and Flow.

Figure 26.5 Describe geometry and flow

6. Choose the settings as shown in Figure 26.6 including Box Parameters to Create External Flow Boundaries. Set X Min, X Max to 2,4, Y Min, Y Max to 2,2 and Z Min, Z Max to 2,2. Select the *golf_ball* or *smooth_ball* in the second case and select Create External Flow Boundaries in the lower left corner.

Figure 26.6 Create external flow boundaries

7. Choose the settings as shown in Figure 26.7 to Create Local Refinement Regions. Select Offset Box as Type and set Wake Levels to 2. Select the *golf_ball* or *smooth_ball* and select Create Local Refinement Regions.

Figure 26.7 Create local refinement regions

8. Select Identify Regions in the Workflow. Choose the settings as shown in Figure 26.8. Select *tunnel* and select Identify Region in the lower left corner.

Figure 26.8 Identify regions

9. Select Define Leakage Threshold in the Workflow. Choose the settings as shown in Figure 26.9. Select Yes for Would you like to close any leakages for your region(s)?

Select *fluid-region-1* and select Preview Leakages. Select Set View From +X Direction in the graphics window. Select Clipping Planes under the menu and check the box for Insert Clipping Planes. Select Define Leakage Threshold in the lower left corner.

Figure 26.9 Define leakage threshold

10. Select Update Region Settings in the Workflow. Choose the settings as shown in Figure 26.10 to Update Region Settings. Select Update Regions in the lower left corner.

Figure 26.10 Update region settings

11. Choose the settings as shown in Figure 26.11 to Choose Mesh Control Settings. Select Choose Options in the lower left corner.

Figure 26.11 Choose mesh control options

12. Choose the settings as shown in Figure 26.12 to Add Local Sizing. Select Add Local Sizing under Choose Mesh Control Options in the Workflow. Select *soft* as Size Functions. Enter 0.4 as Maximum Size [mm]. Select Scope To Faces and select *golf_ball* or *smooth_ball*. Select Add Local Sizing in the lower left corner.

Figure 26.12 Choose mesh control options

13. Select Generate the Surface Mesh in the Workflow. Choose the settings as shown in Figure 26.13 to Generate the Surface Mesh. Open Advanced Options and select Yes for Auto Assign Zone Types? Set the Global Minimum [mm] to 0.2. Select Generate the Surface Mesh in the lower left corner. Skip the next Step 14 for the *smooth_ball*.

Figure 26.13 Generate the surface mesh

14. Right-click on Generate the Surface Mesh and select Insert Next Task>>Identify Deviated Surfaces. Choose the settings as shown in Figure 26.14 and select Deviation>>Compute. Select Deviation>>Draw Contour. Select Identify Deviated Faces in the lower left corner.

Figure 26.14 Identify Deviated Faces

15. Select Update Boundaries in the workflow. Choose the settings as shown in Figure 26.15. Select Update Boundaries in the lower left corner.

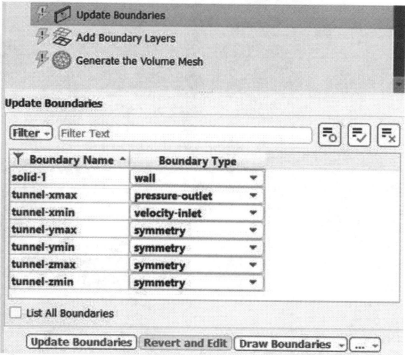

Figure 26.15 Update boundaries

16. Select Add Boundary Layers in the Workflow. Choose the settings as shown in Figure 26.16. Select Add Boundary Layers in the lower left corner.

Figure 26.16 Add boundary layers

17. Select Generate the Volume Mesh in the Workflow. Choose the settings as shown in Figure 26.17a). Select Generate Volume Mesh in the lower left corner. Use the Clipping Planes to study the mesh. Select Switch to Solution in the upper left corner. Answer yes to the question that you get.

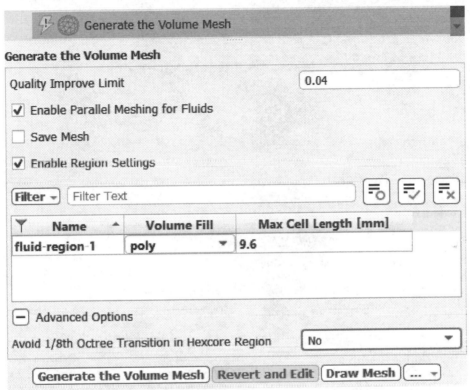

Figure 26.17a) Generate the Volume Mesh

Figure 26.17b) Finished mesh

E. Launching Ansys Fluent

18. Double click on General under Setup in the Outline View. Select Display… under Mesh on the Task Page. Select the settings as shown in Figure 26.18a) and click on Display.

Figure 26.18a) Mesh display settings

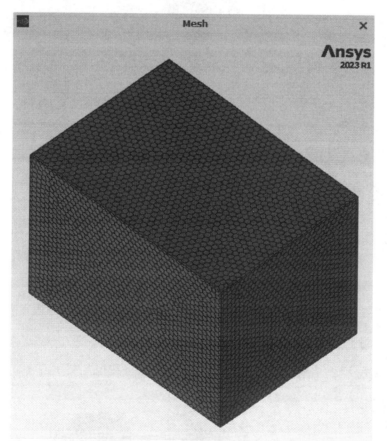

Figure 26.18b) Mesh in Ansys Fluent

19. Double click on Models under Setup in the Outline View. Double click on Viscous (SST k-omega) under Models. Select OK to close the window

Figure 26.19 Viscous model in Ansys Fluent

20. Double click on Boundary Conditions and Inlet under Setup in the Outline View. Set the Velocity Magnitude [m/s] to 30. Select Apply and Close the Velocity Inlet window.

Figure 26.20 Velocity inlet settings in Ansys Fluent

21. Double click on Methods under Solution in the Outline View. Select SIMPLE as the Scheme for the Pressure-Velocity Coupling. Select the other settings as shown in Figure 26.21.

Figure 26.21 Solution methods

22. Double click on Projected Areas under Results and Reports in the Outline View. Select *solid-1* as the Surface and enter 0.000001 as Min Feature Size [m], select X as Projection Direction and click on Compute. The computed frontal Area [m²] is 0.00144027. Close the window.

Figure 26.22 Final scaled residuals

23. Select the Physics tab from the menu and double click on Reference Values under Solver. Select Compute from *solid-1* and enter 0.00144027as Area [m²]. Enter 30 as Velocity [m/s]. Select fluid-region-1 as Reference Zone.

Figure 26.23 Reference values

24. Double-click on Report Definitions under Solution in the Outline View. Select New>>Force Report>>Drag… from the drop-down menu. Select *solid-1* under Zones. Check the box for Print to Console under Create. Select the OK button to close the Drag Report Definition window. Close the Report Definitions window.

Figure 26.24 Drag report definition

25. Double click on Initialization under Solution in the Outline View. Select Hybrid Initializations and click on Initialize. Select File>>Write>>Case and enter the name *golf_ball_flow.cas.h5*

Figure 26.25 Solution initialization in Ansys Fluent

26. Double click on Run Calculation under Solution in the Outline View. Set the Number of Iterations under Parameters to 70. Select Calculate under Solution Advancement.

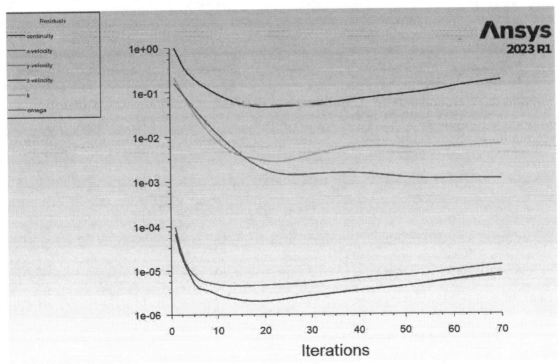

Figure 26.26 Residuals for the first 70 iterations

27. Double click on General under Setup in the Outline View. Select Transient under Time on the Task Page.

Figure 26.27 Transient simulation setting

28. Double click on Controls under Solution in the Outline View. Set the Under-Relaxation Factor for Momentum to 0.6.

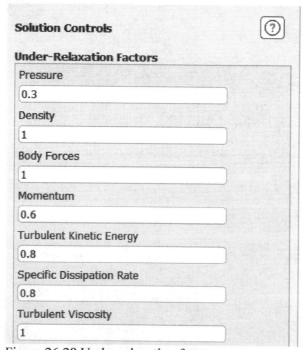

Figure 26.28 Under-relaxation factors

29. Double click on Methods under Solution in the Outline View. Set the Turbulent Kinetic Energy and Specific Dissipation Rate under Spatial Discretization to Second Order Upwind.

Figure 26.29 Solution method settings

30. Double click on Run Calculation. Set the Time Step Size [s] to 0.001. Set the Number of Time Steps to 200 and the Max Iterations/Time Step to 50. Select Calculate under Solution Advancement. Select Yes to the question that you get.

Figure 26.30a) Run calculation settings

Figure 26.30b) Final scaled residuals

F. Post-Processing

Figure 26.30c) Drag coefficient versus flow time

31. Double click on Reports under Results in the Outline View and double click on Forces under Reports. Select the *solid-1* zone and select the Print button.

Console						
	Forces [N]			Coefficients		
Zone	Pressure	Viscous	Total	Pressure	Viscous	Total
solid-1	0.25337662	0.017453829	0.27083045	0.3191347	0.02198357	0.34111827
Net	0.25337662	0.017453829	0.27083045	0.3191347	0.02198357	0.34111827

Figure 26.31 Drag coefficient from console

32. Double click on Graphics and Contours under Results in the Outline View. Select the settings as shown in Figure 26.32a) and click on Colormap Options…. Set Type to float and Precision to 0 under Number Format. Set Font Behavior to Fixed and Font Size to 12 under Font. Click on Save/Display.

Figure 26.32a) Contours window settings for static pressure

627

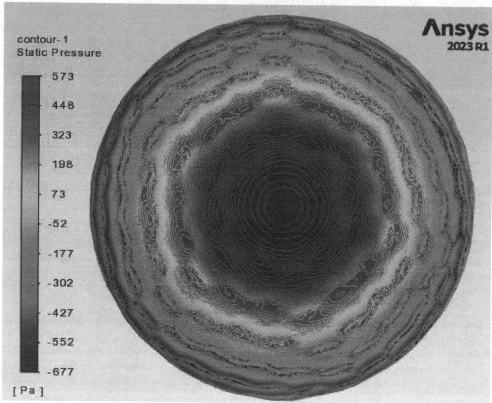

Figure 26.32b) Contours of static pressure for golf ball

33. Select Domain from the menu and select Create>>Plane…. Under Surface. Choose the settings as shown in Figure 26.33 and select Create. Close the Plane Surface window.

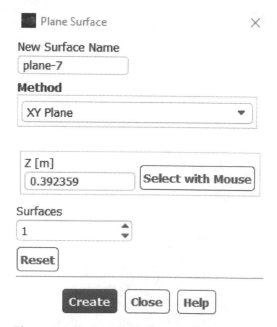

Figure 26.33 Creating a new plane

34. Double-click on Graphics and Pathlines under Results in the Outline View. Select the settings as shown in Figure 26.34a). Select *plane-7* and *solid-1* under Release from Surfaces and select Save/Display.

Figure 26.34a) Pathline settings

Figure 26.34b) Pathlines around golf ball at 30 m/s.

G. Batch Jobs

35. Double click on Monitors and Residual under Solution in the Outline View. Set the Absolute Criteria for all residuals to 0.0001. Select OK to close the Residual Monitors window.

You will need to create the case and scheme files to complete a batch job. Select File>>Write>>Case & Data... from the menu. Save the case and data files in the *Working Directory* folder with the name ***volvik_golf_ball_flow.cas.h5***.

Open Notebook and create and save the scheme file with the name ***volvik_golf_ball_scheme_file.scm*** for the batch job. The scheme file is shown in Figure 26.35a) and is available for download at *sdcpublications.com*. Select File>>Read>>Scheme... from the menu in Ansys Fluent. Open the scheme file to start the batch job.

Table 26.1 and Figure 26.35b) show the results from the batch job in comparison with the two models for smooth sphere (eq. 26.2) and golf ball (eq. 26.3) as described in the theory section. Figure 26.35b) also include results from experimental testing of different golf balls, see Jenkins *et al.* [1].

```
(Do ((x 100 (- x 5))) ((< x 5))
(begin
  (Ti-menu-load-string (format #f "file read-case volvik_golf_ball_flow.cas.h5"))
  (Ti-menu-load-string (format #f "define boundary-conditions velocity-inlet tunnel-xmin y y n ~a n 0 y n 1 n 0 n 0 n n y 5 10" x))
  (Ti-menu-load-string (format #f "report reference-values velocity ~a" x))
  (Ti-menu-load-string (format #f "solve set discretization-scheme k 0"))
  (Ti-menu-load-string (format #f "solve set discretization-scheme omega 0"))
  (Ti-menu-load-string (format #f "solve set under-relaxation mom 0.7"))
  (Ti-menu-load-string (format #f "solve set time-step 0.001"))
  (Ti-menu-load-string (format #f "solve initialize hyb-initialization yes"))
  (Ti-menu-load-string (format #f "solve dual-time-iterate 1 70"))
  (Ti-menu-load-string (format #f "solve set discretization-scheme k 1"))
  (Ti-menu-load-string (format #f "solve set discretization-scheme omega 1"))
  (Ti-menu-load-string (format #f "solve set under-relaxation mom 0.6"))
  (Ti-menu-load-string (format #f "solve set time-step 0.001"))
  (Ti-menu-load-string (format #f "solve dual-time-iterate 200 50"))
  (Ti-menu-load-string (format #f "report forces wall-forces y 1 0 0 y volvik_golf_ball_drag_coefficient_data_U=~amps" x))
  (Ti-menu-load-string (format #f "display objects display contour-1"))
  (Ti-menu-load-string (format #f "display views restore-view left"))
  (Ti-menu-load-string (format #f "display save-picture volvik_golf_ball_static_pressure_contour_plot_U=~amps.jpg" x))
  (Ti-menu-load-string (format #f "display objects display pathlines-1"))
  (Ti-menu-load-string (format #f "display views restore-view front"))
  (Ti-menu-load-string (format #f "display save-picture volvik_golf_ball_path_lines_U=~amps.jpg" x))
  (Ti-menu-load-string (format #f "plot residuals y y y y y y"))
  (Ti-menu-load-string (format #f "display save-picture volvik_golf_ball_residuals_plot_U=~amps.jpg" x))
  (Ti-menu-load-string (format #f "plot file report-def-0-rfile.out"))
  (Ti-menu-load-string (format #f "display save-picture volvik_golf_ball_drag_coefficient_plot_U=~amps.jpg" x))))
```

Figure 26.35a) Scheme file for golf ball batch job

U (m/s)	Re	Cd, eq. 26.3	Cd, Fluent
5	14669	0.454181	0.44010757
10	29337	0.484592	0.39133841
15	44006	0.424748	0.37945525
20	58675	0.245559	0.36400321
25	73344	0.234798	0.35033352
30	88012	0.233924	0.33997192
35	102681	0.236257	0.33095548
40	117350	0.238722	0.32280035
45	132019	0.241107	0.31556523
50	146687	0.243356	0.3086426
55	161356	0.245461	0.30216134
60	176025	0.247429	0.2961533
65	190694	0.249270	0.2897973
70	205362	0.250997	0.28470818
75	220031	0.252621	0.28016918
80	234700	0.254153	0.276302
85	249368	0.255602	0.27298529
90	264037	0.256975	0.26998978
95	278706	0.258281	0.26713053
100	293375	0.259525	0.26073923

Table 26.1 Comparison between Ansys Fluent and model for drag crisis (equation 26.2)

Figure 26.35b) Drag coefficient versus Reynolds number for smooth sphere (thin curve, model eq. 26.2), golf ball (thick curve, model eq. 26.3), □ (Ansys Fluent for Volvik Pro Bismuth), x (Tournament Plus, Exp.), o (Titlest, Exp.), ∇ (Callaway, Exp.), ◊ (TaylorMade, Exp.)

H. Theory

1. The drag coefficient from experiments for a smooth sphere has been found in a large Reynolds number range using the following curve fit-formula from Morrison [1]:

$$C_d = \frac{24}{Re} + \frac{2.6\left(\frac{Re}{5}\right)}{1+\left(\frac{Re}{5}\right)^{1.52}} + \frac{0.411\left(\frac{Re}{2.63\times10^5}\right)^{-7.94}}{1+\left(\frac{Re}{2.63\times10^5}\right)^{-8}} + \frac{0.25\left(\frac{Re}{10^6}\right)}{1+\left(\frac{Re}{10^6}\right)} \qquad 0.1 < Re < 10^6 \qquad (26.1)$$

The model for the drag crisis for a smooth sphere based on Achenbach [3] can be determined as

$$C_d = \frac{280}{10^3} - \frac{159}{10^3}atan\left[\frac{3}{50}\left(\frac{Re}{10^3} - 340\right)\right] + \frac{1}{30}log\left(\frac{Re}{11\cdot10^4}\right)[1 + \tanh(11\cdot10^4 - Re)] +$$
$$+ \frac{1}{11}log\left(\frac{Re}{42\cdot10^4}\right)[1 + \tanh(Re - 42\cdot10^4)] - \frac{1}{28}log\left(\frac{Re}{18\cdot10^5}\right)[1 + \tanh(Re - 18\cdot10^5)]$$
$$2\cdot10^4 < Re < 4\cdot10^6 \qquad (26.2)$$

The model for the drag crisis for a golf ball based on Bearman and Harvey [2] is

$$C_d = \frac{369}{10^3} - \frac{183}{2\cdot10^3}atan\left[\frac{7}{20}\left(\frac{Re}{10^3} - 46\right)\right] + \frac{1}{34}log\left(\frac{Re}{8\cdot10^4}\right)[1 + \tanh(Re - 8\cdot10^4)] +$$
$$+ \frac{1}{16}log\left(\frac{Re}{37\cdot10^3}\right)[1 + \tanh(37\cdot10^3 - Re)] \qquad 2\cdot10^4 < Re < 3\cdot10^5 \qquad (26.3)$$

We define the Reynolds number as

$$Re = \frac{UD\rho}{\mu} \qquad (26.4)$$

where U is free stream velocity, D is sphere or golf ball diameter, ρ is density and μ is dynamic viscosity. The drag coefficient is defined as

$$C_d = \frac{F_d}{\frac{1}{2}\rho U^2 A} \qquad (26.5)$$

where F_d is drag force and A is frontal area of the sphere or golf ball.

I. References

1. P. E. Jenkins, J. Arellano, M. Ross and M. Snell, "Drag Coefficients of Golf Balls", *World Journal of Mechanics*, **8**, 236-241, 2018.
2. F. A. Morrison, "An Introduction to Fluid Mechanics", *Cambridge University Press*, 2013.
3. E. Achenbach, "Experiments on the flow past spheres at very high Reynolds numbers", *J. Fluid Mech.*, **54** (3), 565-575, 1972.
4. P.W. Bearman and J.K. Harvey, "Golf Ball Aerodynamics", *Aeronautical Quarterly*, **27**, 112-122, 1976.
5. M. Geier, A. Pasquali and M. Schonherr, "Parametrization of the cumulant lattice Boltzmann method for fourth order accurate diffusion Part II: application to flow around a sphere at drag crisis", Journal of Computational Physics, July 2017.

6. S. M. Nakhostin,"Investigation of transitional turbulence models to predict drag crisis for flow over spheres and cylinder", Master's Thesis, University of Stavanger, 2019.
7. J. Hart,"Comparison of turbulence modeling approaches to the simulation of a dimpled sphere", Procedia Engineering, 147, 68-73, 2016.

J. Exercise

26.1. Run Ansys Fluent calculations shown in this chapter for different velocities U (m/s) and compare your static pressures, pathlines and drag coefficient results as shown in Table 26.2. Determine the percent error between the golf ball model as shown in eq. 26.3 and Ansys Fluent results for the drag coefficient.

Student	U (m/s)	Re	Cd, eq. 26.3	Cd, Fluent	Percent Difference
A	7.5				
B	12.5				
C	17.5				
D	22.5				
E	27.5				
F	32.5				
G	37.5				
H	42.5				
I	52.5				
J	57.5				
K	62.5				
L	67.5				
M	72.5				
N	77.5				
O	82.5				
P	87.5				
Q	92.5				
R	97.5				

Table 26.2 Free stream velocity, Reynolds number, drag coefficients and percentage error for 18 different velocities

26.2. Run Ansys Fluent calculations shown in this chapter for the golf balls as shown in Table 26.3 and 26.4. Compare the static pressure, pathlines and drag coefficient results with this chapter results. Determine the percent error between the golf ball model as shown in eq. 26.3 and Ansys Fluent drag coefficients for the different commercial golf balls at a free stream velocity of $U = 28$ m/s.

Manufacturer	Model	Design	Number of Dimples
Callaway	ERC SOFT	Hex	332
Mizuno	RD 566	D micro dimple	566
Volvik	ProBismuth	Octahedron	446
Wilson Staff	50 ELITE	--------	302

Table 26.3 Model, design and number of dimples for different commercial golf balls

Manufacturer	A (m²)	Re	Cd, eq. 26.3	Cd, Fluent	Percent Difference
Callaway					
Mizuno					
Volvik	0.0014423734				
Wilson Staff					

Table 26.4 Frontal area, Reynolds number, golf ball model drag coefficient, Ansys Fluent drag coefficient and percentage error for different commercial golf balls at $U = 28$ m/s.

K. Notes

CHAPTER 27. MODEL CAR: AUDI R8

A. Objectives

- Using Ansys Fluent to Study the Flow around an Audi R8
- Loading the *stl* file for the Audi R8
- Using Fault-Tolerant Meshing in Ansys Fluent Meshing
- Inserting Boundary Conditions
- Running Ansys Fluent Simulations
- Using Contour Plots for Visualizations of Velocity and Pressure Fields

B. Problem Description

We will study the flow around an Audi R8 model car.

C. Launching Ansys Fluent

1. Start by launching Ansys Fluent in standalone mode.

Figure 27.1 Launching ANSYS Fluent

2. Check the box for Double Precision. Select Meshing in the upper left portion of the Fluent Launcher window. Set the number of Meshing and Solver Processes equal to the number of processor cores for your computer. To check the number of physical cores, press the Ctrl + Shift + Esc keys simultaneously to open the Task Manager. Go to the Performance tab and select CPU from the left column. You'll see the number of physical cores on the bottom-right side. Close the Task Manager window. Click on the Start button in the Fluent Launcher window.

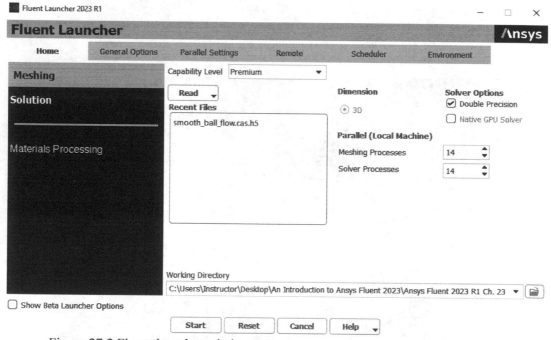

Figure 27.2 Fluent launcher window

D. Ansys Fluent Meshing

3. Select Workflow Type as Fault-Tolerant Meshing.

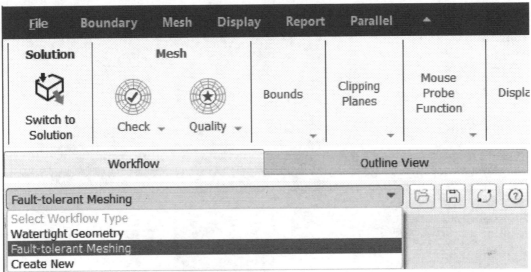

Figure 27.3 Selection of Fault-Tolerant Meshing

4. Select Import CAD and Part Management. Select *mm* as Display Unit. Browse for the CAD File. Select Other CAD Formats (*.jt *.plmxml *.stl) as Files of type. Open the file *Audi R8 Scan.stl*. Select *mm* as File Unit. Select Custom as Create Meshing Objects. Select Load and check the box next to CAD Model. Right-click on Meshing Model and select Create Object. Right-click on Object and select Rename. Change the name to *Audi-R8*. Drag and drop the part *Audi R8 Scan* from CAD Model to *Audi-R8* under Meshing Model. Select Create Meshing Objects.

Figure 27.4 Import CAD and Part Management

5. Select External flow around object as Flow Type under Describe Geometry and Flow. Select Yes for Add an enclosure? and Yes for Add local refinement regions? Open Advanced Options and select Yes for Identify regions?, Yes for Close leakages? and Yes for Extract Edge Features. Select Describe Geometry and Flow.

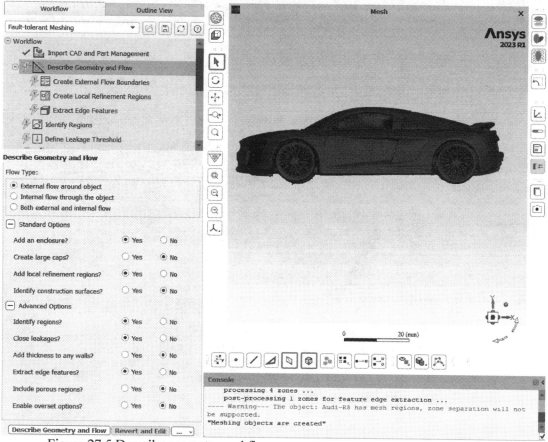

Figure 27.5 Describe geometry and flow

6. Choose the settings as shown in Figure 27.6a) to Create External Flow Boundaries. Select Create External Flow Boundaries in the lower left corner.

Figure 27.6a) Create external flow boundaries

Figure 27.6b) External flow boundaries

7. Choose the settings as shown in Figure 27.7a) to Create Local Refinement Regions. Select Create Local Refinement Regions in the lower left corner.

Figure 27.7a) Create local refinement regions

Figure 27.7b) Local refinement region

Select Extract Edge Features in the Workflow. Choose the settings as shown in Figure 26.7c) to Extract Edge Features. Select Extract Edge Features in the lower left corner.

Figure 27.7c) Create edge features

Figure 27.7d) Edge features where tires intersect the ground

8. Select Identify Regions in the Workflow. Choose the settings as shown in Figure 27.8 to Identify Regions. Select Identify Region in the lower left corner.

Figure 27.8 Identify regions

9. Select Define Leakage Threshold in the Workflow. Set Would you like to close any leakages for your region(s)? to Yes. Choose the settings as shown in Figure 27.9 to Define Leakage Threshold. Set Maximum Leakage Size [mm] to 0.5. Select Clipping Planes in the menu. Check the box for Insert Clipping Planes. Select *fluid-region-1* and select Preview Leakages. Select Set View From -X Direction in the graphics window. Select Define Leakage Threshold in the lower left corner.

Figure 27.9 Define leakage threshold

10. Select Update Region Settings in the Workflow. Choose the settings as shown in Figure 27.10 to Update Region Settings. Select Update Regions in the lower left corner.

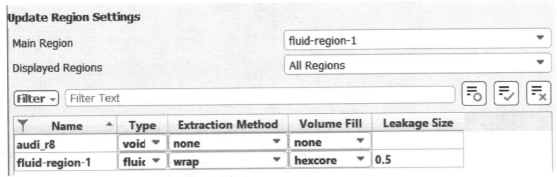

Update Region Settings

Main Region	fluid-region-1
Displayed Regions	All Regions

Name	Type	Extraction Method	Volume Fill	Leakage Size
audi_r8	void	none	none	
fluid-region-1	fluid	wrap	hexcore	0.5

Figure 27.10 Update region settings

11. Choose the settings as shown in Figure 27.11 to Choose Mesh Control Settings. Select Choose Options in the lower left corner.

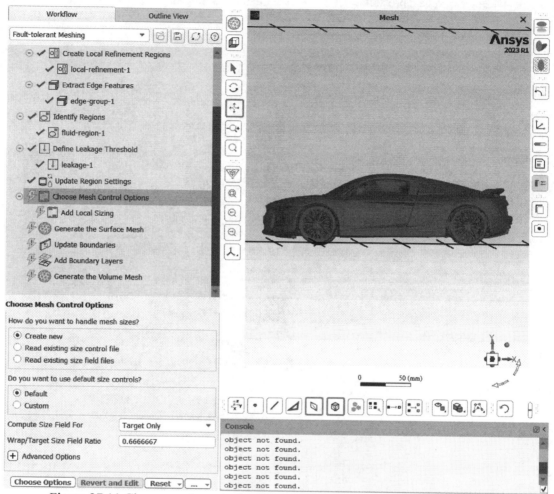

Figure 27.11 Choose mesh control options

12. Select default-curvature under Add Local Sizing in the Workflow. Select Revert and Edit. Choose the settings as shown in Figure 27.12a) to decrease the Minimum Size to 1. Select Update.

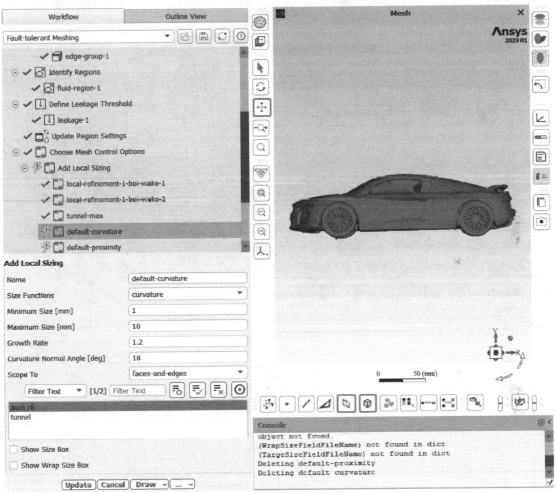

Figure 27.12a) Settings for default-curvature

Select default-proximity under Add Local Sizing in the Workflow. Choose the settings as shown in Figure 27.12b) to decrease the Minimum Size to 1. Select Update.

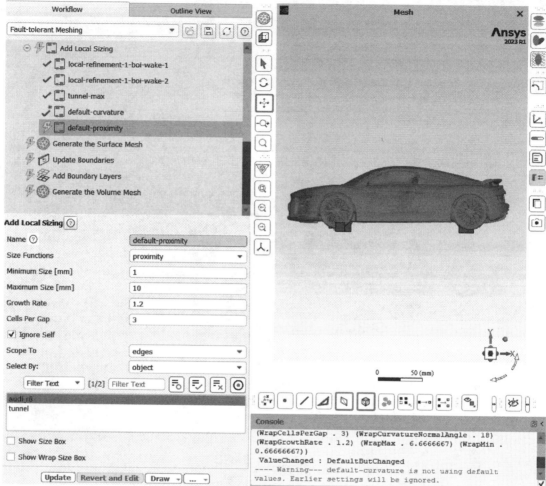

Figure 27.12b) Settings for default-proximity

Select Add Local Sizing in the Workflow. Choose the settings as shown in Figure 27.12c) to add a soft size control. Select Add Local Sizing in the lower left corner.

Figure 27.12c) Adding a soft size control

13. Select Generate the Surface Mesh in Workflow. Choose the settings as shown in Figure 27.13a) to Generate the Surface Mesh. Select Generate the Surface Mesh in the lower left corner.

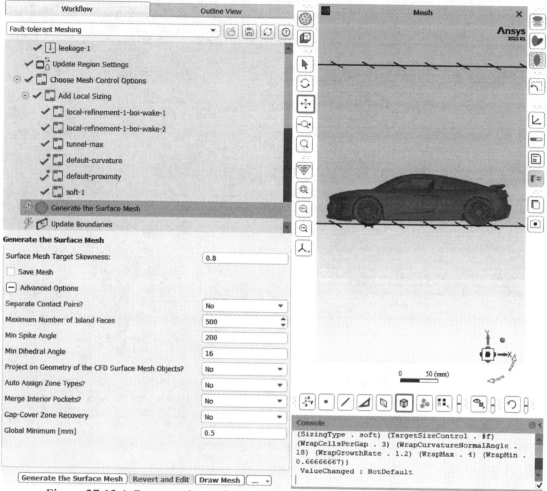

Figure 27.13a) Generate the surface mesh

Select Clipping Planes and check the box for Insert Clipping Planes. Select Limit in Y. Drag the slider to the left and right to adjust the clipping plane.

Figure 27.13b) Clipping planes

Figure 27.13c) Finished surface mesh

14. Select Update Boundaries in the Workflow. Select Revert and Edit. Choose the settings as shown in Figure 27.14. Select Update Boundaries in the lower left corner.

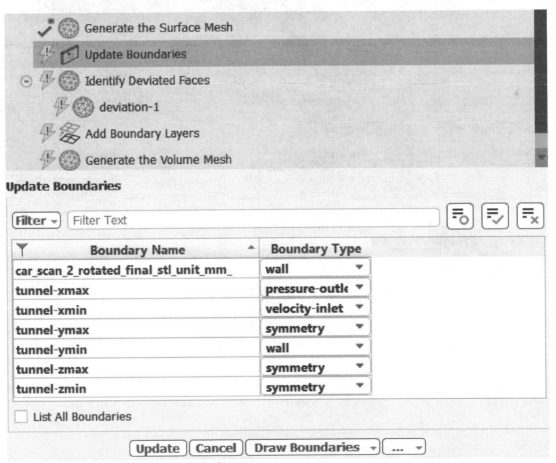

Figure 27.14 Update boundaries

15. Right-click on Generate the Surface Mesh and select Insert Next Task>>Identify Deviated Faces. Select Identify Deviated Faces in the Workflow. Choose the settings as shown in Figure 27.15 including Overlay with Geometry and select Deviation>>Compute. Select Deviation>>Draw Contour. Select Identify Deviated Faces.

Figure 26.15 Identify Deviated Faces

16. Select Add Boundary Layers in the Workflow. Choose the settings as shown in Figure 27.16 including setting the Number of Layers to 10. Select Add Boundary Layers in the lower left corner.

Figure 26.16 Add boundary layers

650

17. Select Generate the Volume Mesh in the Workflow. Choose the settings as shown in Figure 27.17 including setting the Max Cell Length to 4. This is completed by right-clicking on Max Cell Length [mm] and selecting Set Max Cell Length [mm] and entering 4 in the blank region. Select Generate Volume Mesh in the lower left corner.

Figure 26.17a) Generate the Volume Mesh

Select Clipping Planes and check the box for Insert Clipping Planes and Draw Cell Layer. Select Limit in Z. Drag the slider to the left and right to adjust the clipping plane.

Figure 27.17b) Finished mesh

Select Mesh >> Check from the menu. Select File >> Write >> Mesh… from the menu. Save the mesh with the name *Audi_R8.msh.h5*. Select File >> Write >> Case… from the menu. Save the cas file with the name *Audi_R8.msh.cas.h5*. Select Solution >> Switch to Solution in the upper left corner. Answer yes to the question that you get.

```
Console

Domain extents.
   x-coordinate: min = -6.408524e+02, max = 1.077853e+03.
   y-coordinate: min = 8.288524e+00, max = 2.299090e+02.
   z-coordinate: min = 3.526365e+02, max = 9.078843e+02.
Volume statistics.
   minimum volume: 1.951937e-04.
   maximum volume: 3.961081e+04.
     total volume: 1.999195e+08.
Face area statistics.
    minimum face area: 1.103575e-04.
    maximum face area: 1.937598e+03.
    average face area: 8.470460e+00.
Checking number of nodes per edge.
Checking number of nodes per face.
Checking number of nodes per cell.
Checking number of faces/neighbors per cell.
Checking cell faces/neighbors.
Checking isolated cells.
Checking face handedness.
Checking periodic face pairs.
Checking face children.
Checking face zone boundary conditions.
Checking for invalid node coordinates.
Checking poly cells.
Checking zones.
Checking neighborhood.
Checking interfaces.
Checking modified centroid.
Checking non-positive or too small area.
Done.
```

Figure 27.17c) Console output from checking the mesh

E. Ansys Fluent

18. Double click on General under Setup in the Outline View. Select Display… under Mesh on the Task Page. Select the settings as shown in Figure 27.18a) and click on Display. Close the Mesh Display window.

Figure 27.18a) Mesh display settings

Figure 27.18b) Surface mesh in Ansys Fluent

19. Double click on Models under Setup in the Outline View. Double click on Viscous (SST k-omega) under Models. Select OK to close the window

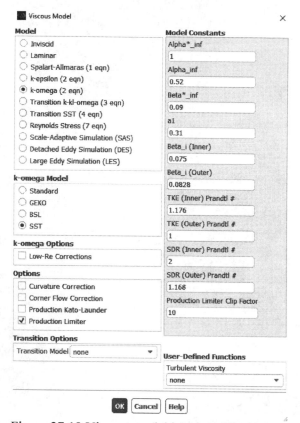

Figure 27.19 Viscous model in Ansys Fluent

20. Double click on Boundary Conditions and Inlet under Setup in the Outline View. Set the Velocity Magnitude [m/s] to 30. Select Apply and Close the Velocity Inlet window.

Figure 27.20 Velocity inlet settings in Ansys Fluent

21. Double click on Methods under Solution in the Outline View. Select SIMPLE as the Scheme for the Pressure-Velocity Coupling. Select the other settings as shown in Figure 27.21.

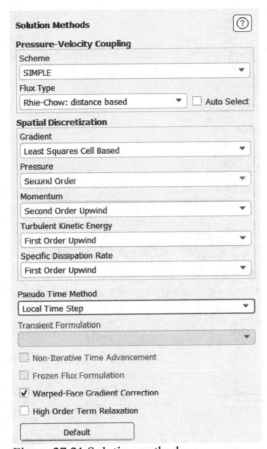

Figure 27.21 Solution methods

22. Double click on Projected Areas under Results and Reports in the Outline View. Select *wrap-origin-car_scan_2_rotated_final_stl_unit_mm_* as the Surface and enter 0.000005 as Min Feature Size [m], select X as Projection Direction and click on Compute. The computed frontal Area [m²] is 0.00616434. Close the window.

Figure 27.22 Final scaled residuals

23. Select the Physics tab from the menu and select Reference Values under Solver. Select Compute from *wrap-origin-car_scan_2_rotated_final_stl_unit_mm_* and enter 0.00616434 as Area [m²]. Enter 30 as Velocity [m/s]. Select *prism-cells-32366* as Reference Zone.

Figure 27.23 Reference values

24. Double-click on Report Definitions under Solution in the Outline View. Select New>>Force Report>>Drag… from the drop-down menu. Select *wrap-origin-car_scan_2_rotated_final_stl_unit_mm_* under Zones. Check the box for Print to Console under Create. Select the OK button to close the Drag Report Definition window. Close the Report Definitions window.

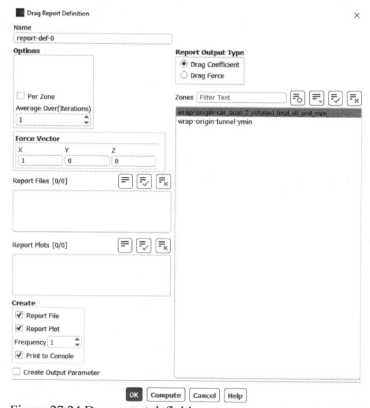

Figure 27.24 Drag report definition

25. Double click on Initialization under Solution in the Outline View. Select Hybrid Initializations and click on Initialize. Select File>>Write>>Case and enter the name *Audi_R8_Flow.cas.h5*

Figure 27.25 Solution initialization in Ansys Fluent

26. Double click on Run Calculation under Solution in the Outline View. Set the Number of Iterations under Parameters to 70. Select Calculate.

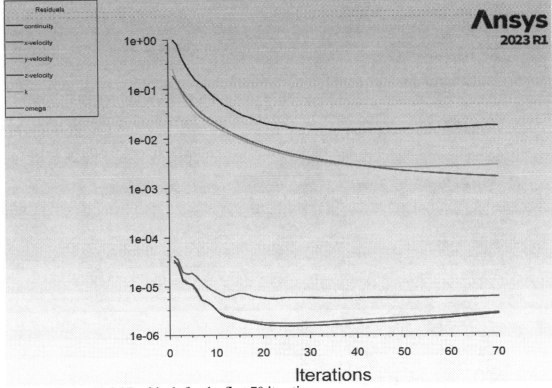

Figure 27.26 Residuals for the first 70 iterations

27. Double click on General under Setup in the Outline View. Select Transient under Time on the Task Page.

Figure 27.27 Transient simulation setting

28. Double click on Controls under Solution in the Outline View. Set the Under-Relaxation Factor for Momentum to 0.6.

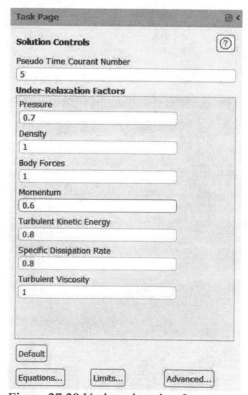

Figure 27.28 Under-relaxation factors

29. Double click on Methods under Solution in the Outline View. Set the Turbulent Kinetic Energy and Specific Dissipation Rate under Spatial Discretization to Second Order Upwind. Double click on Monitors and Residual. Set the Absolute Criteria for all Residuals to 0.0001. Select OK to close the Residual Monitor window.

Figure 27.29 Solution method settings

30. Double click on Run Calculation. Set the Time Step Size [s] to 0.001. Set the Number of Time Steps to 200 and the Max Iterations/Time Step to 50. Select Calculate under Solution Advancement. Create a new report file by answering yes in the window that appears.

Figure 27.30a) Run calculation settings

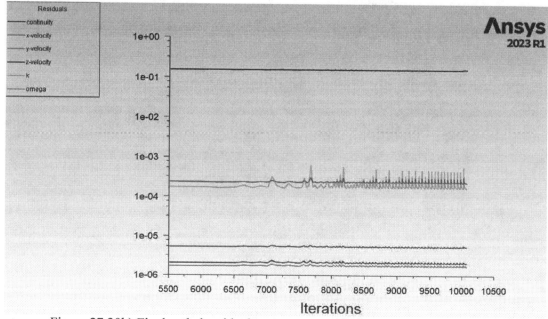

Figure 27.30b) Final scaled residuals

F. Post-Processing

31. Double click on Monitors and Report Plots under Solution in the Outline View and double click on *report-def-0-rplot* under Report Plots. Select Axes… at the bottom of the window. Select X Axis and set Precision to 2 under Number Format. Uncheck Auto Range under Options, set Minimum to 0 and Maximum to 0.2 under Range. Select Apply and select Y Axis. Set Precision for Y Axis to 2 under Number Format and select Apply. Close the Axes – Report Plots window. Select Plot and OK in the Edit Report Plot window.

Figure 27.31a)) Drag coefficient versus flow time

Double click on Reports under Results in the Outline View and double click on Forces under Reports. Select the *wrap-origin-car_scan_2_rotated_final_stl_unit_mm_* as Wall Zones and select the Print button. The drag coefficient is 0.377. This can be compared with the value 0.36 as listed in technical specifications, a 5 % difference.

Console							
	Forces [N]			Coefficients			
Zone	Pressure	Viscous	Total	Pressure	Viscous	Total	
wrap-origin-car_scan_2_rotated_final_stl_unit_mm_	1.1599371	0.12158689	1.281524	0.3413495	0.035780927	0.37713043	
Net	1.1599371	0.12158689	1.281524	0.3413495	0.035780927	0.37713043	

Figure 27.31b) Drag coefficient from console

32. Double click on Graphics and Contours under Results in the Outline View. Select the settings as shown in Figure 27.32a) and click on Save/Display.

Figure 27.32a) Contours window settings for static pressure

Figure 27.32b) Contours of static pressure for Audi R8.

33. Select File>>Export to CFD Post… from the menu. Select ASCII as Format and check the box for Open CFD-Post. Select *prism-cells-32366* under Cell Zones and highlight all Surfaces. Select Static Pressure, Pressure Coefficient, Dynamic Pressure, Absolute Pressure, Total Pressure, Velocity Magnitude, X Velocity, Y Velocity, Z Velocity, Vorticity Magnitude and Skin Friction Coefficient as Quantities. Select Write… and save the file with the name *Audi_R8_Flow_Data*.

Figure 27.33 Export settings to CFD-Post

34. Check the boxes for *wrap origin car_scan_2_rotated_final_stl_unit_mm_* and *wrap origin tunnel ymin* under Cases, *Audi_R8_Flow_Data* and *prism cells 32366* in the Outline. Select Location>>Line from under the menu. Select OK in the Insert Line window. Set the locations for the Two Points in Details of Line 1. Set Point 1 to 96, 20, 630 and Point 2 to 96, 75, 630. Select Sample as Line Type and set the number of Samples to 50. Select Apply.

Figure 27.34a) Line settings

663

Select Location>>Plane from under the menu. Select OK in the Insert Plane window. Select All Domains as Domains in Details of Plane 1. Select XY Plane as Method and 630 [mm] for Z under Definition in Details of Plane 1. Select Type as None under Plane Bounds in Details of Plane 1. Select Slice as Plane Type. Select Apply.

Figure 27.34b) Plane settings

Select the Streamline icon from under the menu. Select OK in the Insert Streamline window. Select Surface Streamline as Type in Details of Streamline 1. Select Plane 1 as Surfaces under Definition. Select Start From Locations, Line 1 for Locations, Max Number of Points as Reduction and set Max Points to 100. Select Velocity as Variable and Forward and Backward as Direction. Select Apply in the bottom left corner.

Figure 27.34c) Streamline settings

Double click on Default Legend View 1 under User Locations and Plots. Select the Definition tab in Details of Default Legend View 1. Select Horizontal. Select Location X Justification as Center and Y Justification as Top. Select the Appearance tab, select Fixed Precision and set Precision to 0 under Text Parameters. Uncheck the box for Line 1 under User Locations and Plots. Select Apply at the bottom left corner in Details of Default Legend View 1.

Figure 27.34d) Details of Default Legend View 1

Figure 27.34e) Streamlines around Audi R8.

G. Theory

35. The drag force from aerodynamic simulations of the 2016 Audi R8 LMS has been found in a large velocity range, see [1].

$$F_d = \alpha V^2 + \beta V \qquad\qquad 0 < V < 70 \ m/s \qquad\qquad (27.1)$$

, where V (m/s) is velocity and the two constants are $\alpha = 0.5338 \ (\frac{Ns^2}{m^2})$ and $\beta = 0.0922 \ (\frac{Ns}{m})$. We define the Reynolds number Re as

$$Re = \frac{VL\rho}{\mu} \qquad\qquad 0 < Re < 2.2 \cdot 10^7 \qquad\qquad (27.2)$$

, where $L = 4.58$ (m) is the length of the car, $\rho = 1.225 \ (\frac{kg}{m^3})$ is the density of air, and $\mu = 1.79 \cdot 10^{-5} \ (\frac{kg}{ms})$ is the dynamic viscosity of air. The drag coefficient can be determined

$$C_d = \frac{2}{A}\left(\frac{\alpha}{\rho} + \frac{\beta L}{\mu}\frac{1}{Re}\right) \qquad\qquad (27.3)$$

, where $A = 2.08$ m2 is the frontal area of the car. The drag curve is shown in Figure 27.35 and the Ansys Fluent data point has been included for reference. The drag curve approaches the value 0.419 asymptotically as the Reynolds number is increasing.

$$C_d = 2\frac{\alpha}{\rho A} = 0.419, \ Re \rightarrow \infty \qquad\qquad (27.4)$$

Figure 27.35 Drag curve for 2016 Audi R8 in comparison with Ansys Fluent result.

H. References

1. 2016 Audi R8 LMS Aerodynamic Simulation Report, AirShaper sample report, 2019.

I. Exercise

27.1. Run Ansys Fluent calculations shown in this chapter for different velocities V (m/s) and compare your static pressures, streamlines and drag coefficient results as shown in Table 27.1. Determine the percent error between the golf ball model as shown in eq. 27.3 and Ansys Fluent results for the drag coefficient.

Student	V (m/s)	Re	Cd, eq. 27.3	Cd, Fluent	Percent Difference
A	7.5				
B	12.5				
C	17.5				
D	22.5				
E	27.5				
F	32.5				
G	37.5				
H	42.5				
I	52.5				
J	57.5				
K	62.5				
L	67.5				
M	72.5				
N	77.5				
O	82.5				
P	87.5				
Q	92.5				
R	97.5				

Table 27.1 Free stream velocity, Reynolds number, drag coefficients and percentage error for 18 different velocities

27.2. Run Ansys Fluent calculations shown in this chapter for the Audi R8 but use water as the fluid in the simulation. Compare the static pressure, streamlines and drag coefficient results with this chapter's results.

J. Notes

CHAPTER 28. DROP IMPACT ON FLUID LAYER

A. Objectives

- Using Ansys Fluent to Study the Impact of a Drop on a Layer of Fluid
- Inserting Boundary Conditions
- Mesh Refinement in Ansys Fluent
- Running Ansys Fluent Simulations
- Using Contour Plots and Animations for Visualizations of Water Phase

B. Problem Description

We will study the impact of a water droplet into a thin layer of water on a horizontal surface.

C. Launching Ansys Workbench and Selecting Fluent

1. Start by launching Ansys Workbench. Launch Fluid Flow (Fluent) that is available under Analysis Systems in Ansys Workbench. Select Geometry under Project Schematic, right click and select Properties. Select 2D Analysis Type under Advanced Geometry Options. Right click on Geometry in Project Schematic and select New DesignModeler Geometry to start DesignModeler.

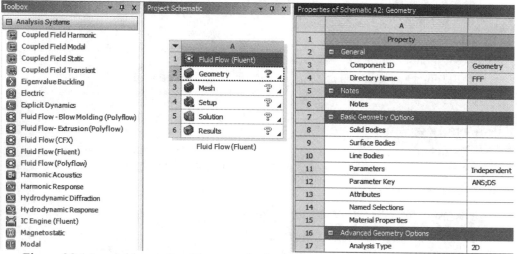

Figure 28.1 Launching Ansys Fluent and selecting 2D analysis type

D. Launching Ansys DesignModeler

2. Select **Units>>Millimeter** from the menu in DesignModeler. Select the XYPlane in the Tree Outline. Select Look At Face/Plane/Sketch ⬛. Select the Sketching tab and Rectangle. Draw a square from the origin with the dimensions 50 x 50 mm. Make sure that the letter P is showing when you move the cursor over the origin of the coordinate system and before you start drawing the rectangle.

Select Concept>>Surfaces from Sketches from the menu. Select Sketch 1 under XY Plane in the Tree Outline. Apply the sketch as a Base Object in Details View. Click on Generate and close DesignModeler.

Figure 28.2 Square with dimensions

E. Launching Ansys Meshing

3. Double click on Mesh under Project Schematic in Ansys Workbench. Right-click on Mesh under Project and Model (A3) in the Meshing window and select Update. Set the Element Size to 0.6 mm under Defaults in Details of "Mesh". Select Mesh>>Controls>Face Meshing from the menu. Select the square in the graphics window and Apply it as Geometry in Details of Face Meshing. Select Quadrilaterals as Method under Definition in Details of "Face Meshing" – Mapped Face Meshing.

Figure 28.3a) Launching meshing window

Figure 28.3b) Selecting face meshing

Right-click on Mesh under Project and Model (A3) in the Meshing window and select Update. The finished mesh is shown in Figure 28.3c). The mesh has 7056 Nodes and 6889 Elements.

Figure 28.3c) Finished mesh

Select the lower horizontal edge, right click and select Create Named Selection. Enter the name *axis* and click OK to close the window. Select the left vertical edge, right click and select Create Named Selection. Enter the name *wall-bottom*. Name the right most vertical edge as *outlet*. Finally, name the top horizontal edge as *wall-side*.

Figure 28.3d) Named selections

Select File>>Export…>>Mesh>>FLUENT Input File>>Export from the menu. Save the mesh with the name *drop-impact.msh*. Select File>>Save Project from the menu and save the project with the name *Drop Impact on Fluid Layer.wbpj*. Close the meshing window. Right click on Mesh in the Project Schematic and select Update.

F. Launching Ansys Fluent

4. Double click on Setup under Project Schematic in Ansys Workbench. Check the box for Double Precision. Set the number of parallel solver processes equal to the number of processor cores for your computer. To check the number of physical cores, press the Ctrl + Shift + Esc keys simultaneously to open the Task Manager. Go to the Performance tab and select CPU from the left column. You'll see the number of physical cores on the bottom-right side. Close the Task Manager window. Click on the Start button in the Fluent Launcher window.

Fluent Launcher 2023 R1 (Setting Edit Only)

Fluent Launcher /\nsys

| Home | General Options | Parallel Settings | Remote | Scheduler | Environment |

Dimension
- ( ) 2D
- ( ) 3D

Parallel (Local Machine)

Solver Processes   14

Solver Options
- [✓] Double Precision
- [ ] Do not show this panel again

Working Directory
C:\Users\Instructor

Start   Cancel   Help ▾

Figure 28.4a) Fluent launcher

Figure 28.4b) Mesh in Ansys Fluent

Select Domain>>Adapt>>Manual… from the menu. Select Cell Registers>>New>>Region…. Set X Max [m] to 0.015 and Y Max [m] to 0.012. Enter *refinedmeshregion* as name. Select Save and Close the window. Select *refinedmeshregion* as Refinement Criterion in the Manual Mesh Adaption window and click on General Adaption Controls. Set Maximum Refinement Level to 8 and click OK to close the General Adaptions Controls window. Select Adapt three times in the Manual Mesh Adaption window and select OK in the Information window. Close the Manual Mesh Adaption window.

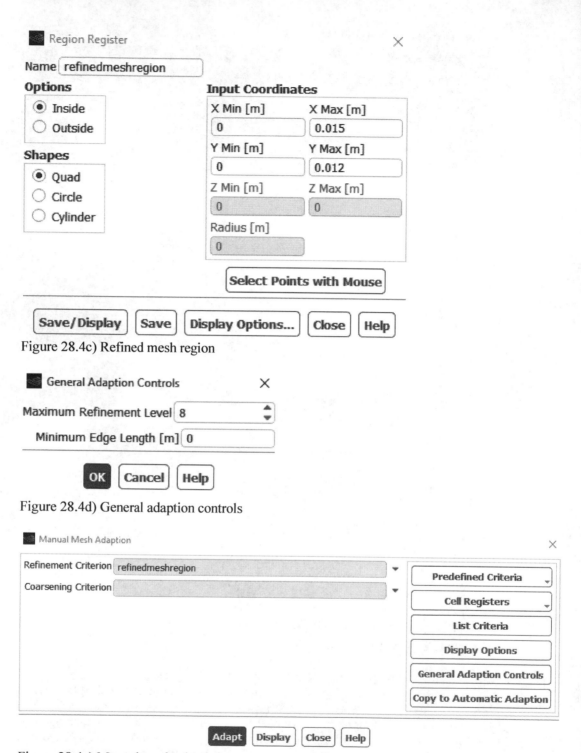

Figure 28.4c) Refined mesh region

Figure 28.4d) General adaption controls

Figure 28.4e) Manual mesh adaption

Select Display… under Mesh on the Task Page. Check the boxes for Edges and Faces, select all Surfaces and click on Display. Close the Mesh Display window.

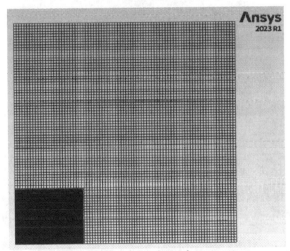

Figure 28.4f) Mesh display

Figure 28.4g) Refined mesh region

Figure 28.4h) Details of refined mesh

5. Select Axisymmetric for 2D Space in the General Solver settings on the Task Page in Ansys Fluent. Select Transient for Time in the General Solver settings. Check the Gravity bow and enter -9.81 for X [m/s²].

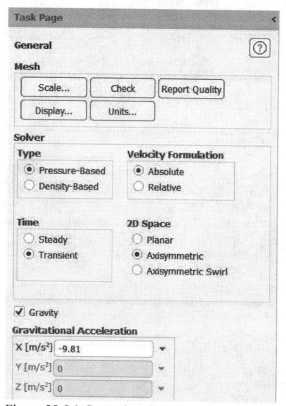

Figure 28.5a) General settings

Open Models under Setup in the Outline View and double click on Viscous (SST k-omega). Select Laminar Model and click on OK to close the Viscous Model window. Double click on Multiphase (Off). Select the Volume of Fluid Model and check the box for Implicit Body Force under Body Force Formulation. Set the Number of Eulerian Phases to 3. Click on Apply and Close the Multiphase Model window.

> **Why did we select Axisymmetric and Transient for Time?** This problem is axisymmetric. We choose Transient since we are interested in studying the falling drop hitting the fluid layer over time.

> **Why did we select Volume of Fluid as Multiphase model and Implicit Body Force?** The Volume of Fluid model is used when we have two or more immiscible fluids where the position of the interface between the fluids is of interest. The applications of the VOF Model includes free-surface flows, stratified flows and the tracking of liquid-gas interfaces in general. The VOF model is limited to the pressure-based solver and doesn't allow for void regions without a fluid. Implicit Body Force is used when we have large body forces such as is the case when we have gravity and surface tension in multiphase flows. Checking the box for Implicit Body Force will make the solution more robust with better convergence.

Figure 28.5b) Multiphase model

Figure 28.5c) Selecting volume of fluid model

6. Next, we double click on Materials under Setup in the Outline View. Select Fluid under Materials on the Task Page and click on Create/Edit…. Click on Fluent Database… in the Create/Edit Materials window. Scroll down in the Fluent Fluid Material section and select *water-liquid (h2o<l>)*. Click Copy at the bottom of the Fluent Database Materials window and Close the same window. Set Density [kg/m³] to 997.6 and Viscosity [kg/(m s)] to 0.0009395. Select Change/ Create and Close the Create/Edit Materials window.

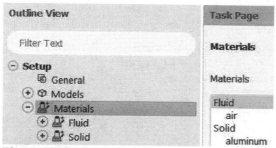

Figure 28.6a) Selecting fluid materials

Figure 28.6b) Selecting Fluent database

Figure 28.6c) Selection of water-liquid as fluid material

Figure 28.6d) Property settings for water

CHAPTER 28. DROP IMPACT ON FLUID LAYER

Select air under Materials and Fluid on the Task Page and click on Create/Edit…. Set Density [kg/m³] to 1.2041 and Viscosity [kg/(m s)] to 1.8e-05. Select Change/ Create and Close the Create/Edit Materials window.

Figure 28.6e) Property settings for air

7. Select the Physics tab in the menu and select Models>>Multiphase Model. Select the Phases tab at the top of the Multiphase Model window. Select *air* as Phase Material for phase-1-Primary Phase and enter *air* as name for the Primary Phase and select Apply.

Select phase-2-Secondary Phase, select *water-liquid* as the Phase Material and enter *film* as the Name for the Secondary Phase. Click on Apply in the Multiphase Model window.

Select phase-3-Secondary Phase, select *water-liquid* as the Phase Material and enter *drop* as the Name for the phase. Click on Apply in the Multiphase Model window.

Click on the Phase Interaction tab at the top of the Multiphase Model window. For Phase Pairs *air film*, select constant for Surface Tension Coefficients [N/m] and enter 0.072496 as the value. Check the box for Surface Tension Force Modeling and select Continuum Surface Stress under Model. Check the box for Wall Adhesion and select Apply.

Select the *air drop* Phase Pairs and set the Surface Tension Coefficient [N/m] to the same constant value 0.072496 and select Apply.

Select the *film drop* Phase Pairs and set the Surface Tension Coefficient [N/m] to the same constant value 0.072496, select Apply and Close the Multiphase Model window.

CHAPTER 28. DROP IMPACT ON FLUID LAYER

Figure 28.7a) Setting up the physics

Figure 28.7b) Editing the phases

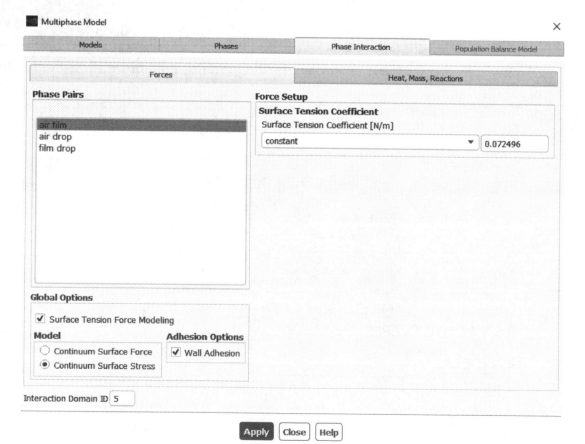

Figure 28.7c) Multiphase model settings

680

8. Double click on Boundary Conditions under Setup in the Outline View. Select *wall-bottom* under Zone in Boundary Conditions on the Task Page. Click on the Edit... button. Set the Contact Angles [deg] for *film air*, *drop air* and *drop film* to 2. Select Apply and Close the window.

Figure 24.8 Bottom wall boundary condition

9. Double click on Methods under Solution in the Outline View and choose PISO as scheme under Pressure-Velocity Coupling. Double click on Monitors and Residual under Solution in the Outline View. Set Absolute Criteria for all Equations to 1e-05.

Figure 28.9a) Solution methods

Figure 28.9b) Residual monitors

10. Double click on Initialization under Solution in the Tree and select Hybrid Initialization under Initialization Methods. Click on Initialize.

Figure 28.10 Solution initialization

11. Select the *Domain* tab in the menu and select Adapt>>Manual…. Select Cell Registers>>New>>Region from the Manual Mesh Adaption window. Set X Max [m] to 0.000725 and Y Max [m] to 0.05. Enter *filmliquid* as Name. Click on Display Options…. Select green as Color and no Symbol. Check the box for Filled and uncheck the boxes for Wireframe and Marker under Options. Click on OK to close the Cell Register Display Options window. Click on Save/Display in the Region Register window. Close the windows.

Figure 28.11a) Domain adaption

Figure 28.11b) Adaption controls

Region Register ✕

Name [filmliquid]

Options
- ⊙ Inside
- ○ Outside

Shapes
- ⊙ Quad
- ○ Circle
- ○ Cylinder

Input Coordinates

X Min [m] [0] X Max [m] [0.000725]

Y Min [m] [0] Y Max [m] [0.05]

Z Min [m] [0] Z Max [m] [0]

Radius [m] [0]

[Select Points with Mouse]

[Save/Display] [Save] [Display Options...] [Close] [Help]

Figure 28.11c) Settings for film liquid region

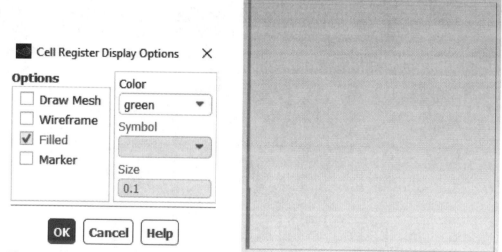

Cell Register Display Options ✕

Options
- ☐ Draw Mesh
- ☐ Wireframe
- ☑ Filled
- ☐ Marker

Color [green ▼]

Symbol [▼]

Size [0.1]

[OK] [Cancel] [Help]

Figure 28.11d) Film cell register display   Figure 24.11e) Filled liquid film to the left

Click on Copy Screenshot of Active Window to Clipboard, see Figure 24.11f). The contour plot can be pasted into a Word document.

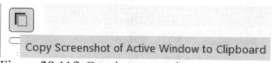

Copy Screenshot of Active Window to Clipboard

Figure 28.11f) Copying screenshot

Select the *Domain* tab in the menu and select Adapt>>Manual.... Select Cell Registers>>New>>Region from the Manual Mesh Adaption window. Select Circle under Shapes. Set X Center [m] to 0.0101 and Radius [m] to 0.003125. Enter *dropliquid* as Name. Click on Display Options.... Select red as Color and no Symbol. Check the box for Filled and uncheck the boxes for Wireframe and Marker under Options. Click on OK to close the Cell Register Display Options window. Click on Save/Display in the Region Register window. Close the windows.

Figure 28.11g) Settings for drop liquid region

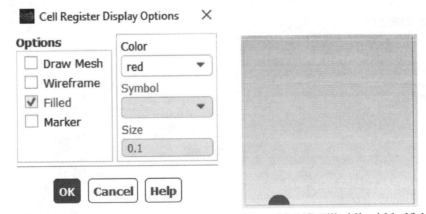

Figure 28.11h) Drop register display   Figure 28.11i) Filled liquid half drop

12. Click on the Patch... button on the Solution Initialization Task Page. Select *film* as Phase in the Patch window and select Volume Fraction as Variable. Select *filmliquid* as Registers to Patch and set the Value to 1. Click on the Patch button. Select *drop* as Phase and choose Volume Fraction as Variable. Select *dropliquid* as Registers to Patch, set the Value to 1 and click on the Patch button. Select *mixture* as Phase and choose Axial Velocity as Variable. Select *dropliquid* as Registers to Patch and set the Value [m/s] to -2.631 and click on the Patch button. Close the Patch window.

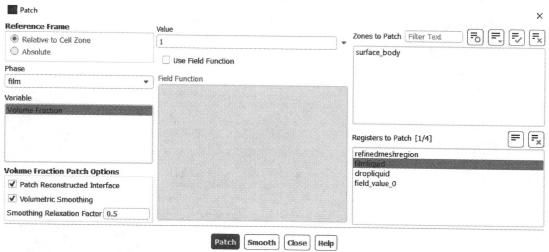

Figure 28.12a) Patching settings of volume fraction for liquid film

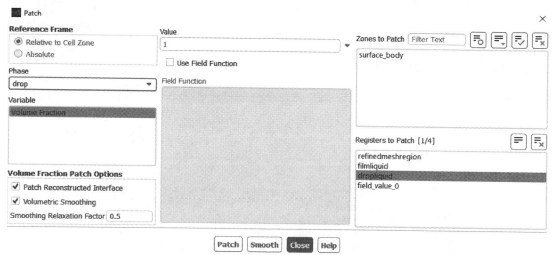

Figure 28.12b) Patching settings of volume fraction for liquid drop

Figure 28.12b) Patching settings of axial velocity for liquid drop

13. Double click on Graphics and Contours under Results in the Outline View. Select Contours of Phases… and Volume fraction for *air* as the Phase. Deselect all Surfaces. Click on Colormap Options…. Select general as Type and 0 for Precision under Number Format in the Colormap window. Set Colormap Alignment to Top. Select Apply and Close the Colormap window. Click on Save/Display and Close the Contours window.

Figure 28.13a) Contours window

Figure 28.13b) Contours window

Select the View tab in the menu and click on Views…. Select *axis* as Mirror Plane and click on Apply. Zoom out and translate the view if needed so that the entire model is visible in the graphics window. Click on the Camera… button in the Views window. Use your left mouse button to rotate the dial counter-clockwise until the model rotates 90 degrees and appears upright. Close the Camera Parameters window. Click on the Save button under Actions in the Views window and Close the Views window. Click on Copy Screenshot of Active Window to Clipboard. The water volume fraction contour plot can be pasted into a Word document.

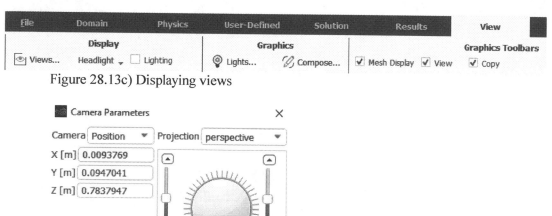

Figure 28.13c) Displaying views

Figure 28.13d) Camera parameters window

Figure 28.13e) Views window

Figure 28.13f) Air volume fraction contour plot at $t = 0$

14. Select File>>Export>>Case and Data… from the menu. Save the Case and Data File with the name *Drop Impact On Fluid Layer.cas.h5*. Double click on Run Calculation under Solution in the Outline View and set the Time Step Size to 5e-6 s. Set the Number of Time Steps to 480. Set Max Iterations/Time Step to 50. Click on the Calculate button on the Task Page. Click OK in the Information window when the calculation is complete. Click on Copy Screenshot of Active Window to Clipboard. The Scaled Residuals plot can be pasted into a Word document. Continue running the simulation with 600 time steps each time corresponding to time increments of 3 ms.

Figure 28.14a) Calculation settings

Figure 28.14b) Scaled residuals

G. Post-Processing

15. In Figure 28.15, we show the volume fraction of air at the moment when the liquid drop is getting in contact with the liquid film.

Figure 28.15a) Volume fraction of air at t = 2.4 ms, time step size 5e-6 s.

Figure 28.15b) Volume fraction of air at t = 5.4, 8.4 and 11.4 ms.

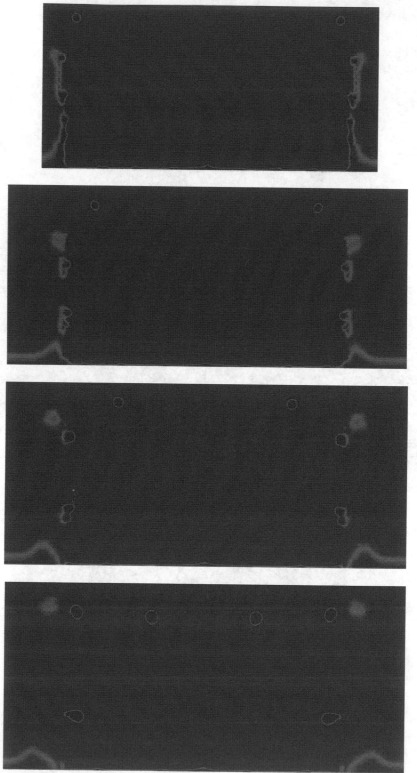

Figure 28.15c) Volume fraction of air at t = 14.4, 17.4, 20.4 and 23.4 ms.

H. Theory

16. This flow is governed by a number of non-dimensional parameters such as the Reynolds number Re, the Weber number We, the Ohnesorge number Oh, and the Froude number Fr. The Reynolds number is defined as

$$Re = \frac{U_0 D \rho}{\mu} \tag{28.1}$$

where U_0 (m/s) is the initial velocity of the liquid drop, D (m) is the drop diameter, ρ (kg/m^3) is the density of the drop liquid, and μ (kg/m s) is the dynamic viscosity of the drop liquid. The Froude number is defined by

$$Fr = \frac{U_0^2}{gD} \tag{28.2}$$

where g is acceleration due to gravity. The definition of the Weber number is the following

$$We = \frac{U_0^2 D \rho}{\sigma} \tag{28.3}$$

where σ is surface tension and the Ohnesorge number is defined as

$$Oh = \frac{\mu}{\sqrt{\sigma D \rho}} \tag{28.4}$$

We also define a non-dimensional number related to secondary splashing

$$Y = \frac{We\, Oh^{-0.4}}{2100 - 5880\left(\frac{H}{D}\right)^{1.44}} \tag{28.5}$$

where H is the liquid pool thickness.

Froude number	$Fr = 113$
Initial liquid drop velocity	$U_0 = 2.631$ m/s
Initial liquid drop diameter	$D = 0.00625$ m
Liquid film thickness	$H = 0.000725$ m
Ohnesorge number	$Oh = 0.0014$
Reynolds number	$Re = 17460$
Secondary splashing number	$Y = 3.5$
Weber number	$We = 595$

Table 28.1 Different values used in the simulation

I. References

1. Hiller N., Kuhlman J.M., Dine M., Gray D.D., "Drop Impingement on Wet and Dry Surfaces", 42nd AIAA Fluid Dynamics Conference and Exhibition, 25 – 28 June 2012, New Orleans, Louisiana.

J. Exercise

28.1 Run Ansys Fluent calculations shown in this chapter for different initial drop velocities U_0 (m/s) and compare your results.

28.2 Run Ansys Fluent calculations shown in this chapter for different liquid film thickness H (m) and compare your results.

CHAPTER 29. DEPOSITION FLOW IN ADDITIVE MANUFACTURING

A. Objectives

- Using Ansys Fluent to Study the Deposition Flow in Fused Filament Fabrication
- Inserting Boundary Conditions
- Using Volume of Fluid Model for Multiphase Flow with Surface Tension
- Running Laminar 2D and 3D Ansys Fluent Simulations
- Using Contour Plots for Visualizations of Volume Fraction

B. Problem Description

We will study the extrusion-based additive manufacturing process in Fused Filament Fabrication (FFF). The flow of ABS plastic (Acrylonitrile Butadiene Styrene) exits from a nozzle and approaches the moving bed. The flow leaves the circular exit of the extrusion head with diameter D_e and volumetric flow rate Q.

C. Launching Ansys Workbench and Selecting Fluent for the 2D Case

1. Start by launching Ansys Workbench. Launch Fluid Flow (Fluent) that is available under Analysis Systems in Ansys Workbench. Select Geometry under Project Schematic, right click and select Properties. Select 2D Analysis Type under Advanced Geometry Options. Right click on Geometry in Project Schematic and select New DesignModeler Geometry to start DesignModeler.

Figure 29.1 Launching Ansys Fluent and selecting 2D analysis type

D. Launching Ansys DesignModeler for the 2D Case

2. Select **Units>>Millimeter** from the menu in DesignModeler. Select the XYPlane in the Tree Outline. Select Look At Face/Plane/Sketch ![icon]. Zoom in in the graphics window using your mouse wheel so that the horizontal scale at the bottom of the window shows from 0 – 1.000 (mm). Select the Sketching tab and Polyline. Draw the shape with the dimensions as shown in Figure 29.2a). Make sure that the letter *P* is showing when you move the cursor over the origin of the coordinate system and before you start drawing the polyline.

Select Concept>>Surfaces from Sketches from the menu. Select Sketch 1 under XY Plane in the Tree Outline. Apply the sketch as a Base Object in Details View. Click on Generate and close DesignModeler.

Figure 29.2a) Polyline shape with dimensions

Figure 29.2b) Generated surface and computational domain

E. Launching Ansys Meshing for the 2D Case

3. Double click on Mesh under Project Schematic in Ansys Workbench. Right-click on Mesh under Project and Model (A3) in the Meshing window and select Update. Select Mesh>>Controls>Face Meshing from the menu. Click on the polyline shape in the graphics window and Apply it as Geometry in Details of Face Meshing.

Figure 29.3a) Launching meshing window

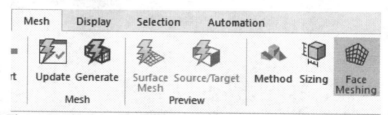

Figure 29.3b) Selecting face meshing

Figure 29.3c) Details of face meshing

Select Mesh>>Controls>>Sizing from the menu. Select the Edge tool , control click on the four horizontal edges and Apply them as Geometry in Details of Sizing. Select Home>>Tools>>Units>>Metric (mm, kg, N, s, mV, mA). Select Element Size as Type and enter 0.01 mm. Set Capture Curvature to No and select Hard as Behavior, see Figure 29.3d).

Figure 29.3d) Details for sizing of horizontal edges

Repeat this step but control select the two vertical edges, enter 0.01 mm as the Element Size and select Hard as Behavior, see Figure 29.3e). Generate the mesh that will have 8077 nodes.

Figure 29.3e) Details for sizing of vertical edges.

Figure 29.3j) Finished 2D mesh

Control-select the two inclined edges, right click and select Create Named Selection. Enter the name *wall*. Name the bottom horizontal edge as *moving bed*. Name the horizontal edge in between the inclined edges as *inlet*. Finally, name the remaining four edges as *outlet*.

Figure 29.3k) Named selections

Select File>>Export…>>Mesh>>FLUENT Input File>>Export from the menu. Save the mesh with the name *2D deposition flow.msh*. Select File>>Save Project from the menu and save the project with the name *2D Deposition Flow in Additive Manufacturing.wbpj*. Close the meshing window. Right click on Mesh in the Project Schematic and select Update.

F. Launching Ansys Fluent for the 2D Case

4. Double click on Setup under Project Schematic in Ansys Workbench. Check the box for Double Precision. Set the number of parallel processes equal to the number of processor cores for your computer. To check the number of physical cores, press the Ctrl + Shift + Esc keys simultaneously to open the Task Manager. Go to the Performance tab and select CPU from the left column. You'll see the number of physical cores on the bottom-right side. Close the Task Manager window. Click on the Start button in the Fluent Launcher window.

Figure 29.4a) Fluent 2D launcher

Figure 29.4b) Deposition 2D mesh in Ansys Fluent

5. Select Transient for Time in the General Solver settings. Open Models under Setup in the Outline View and double click on Viscous (SST k-omega). Select Laminar Model and click on OK to close the Viscous Model window. Double click on Multiphase. Select the Volume of Fluid Model. Click on Apply and Close the Multiphase Model window.

Why did we select Transient for Time? We choose Transient since we are interested in studying the shape of the free surface, hydraulic jump and how it deforms over time.

Why did we select Volume of Fluid as Multiphase model? The Volume of Fluid model is used when we have two or more immiscible fluids where the position of the interface between the fluids is of interest. The applications of the VOF Model includes free-surface flows, stratified flows and the tracking of liquid-gas interfaces in general. The VOF model is limited to the pressure based solver and doesn't allow for void regions without a fluid.

Figure 29.5a) General settings

Figure 29.5b) Viscous model

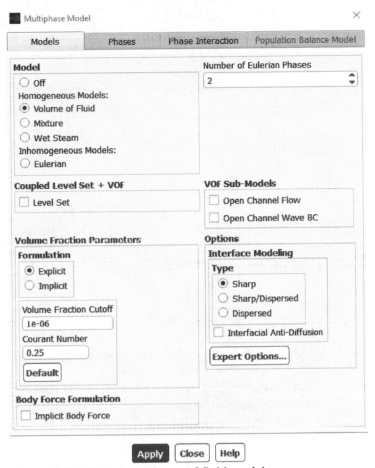

Figure 29.5c) Selecting volume of fluid model

6. Next, we double click on Materials under Setup in the Outline View. Select Fluid under Materials on the Task Page and click on Create/Edit.... Click on Fluent Database... in the Create/Edit Materials window. Scroll down in the Fluent Fluid Material section and select *water-liquid (h2o<l>)*. Click Copy at the bottom of the Fluent Database Materials window. Close the window.

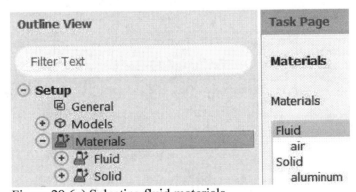

Figure 29.6a) Selecting fluid materials

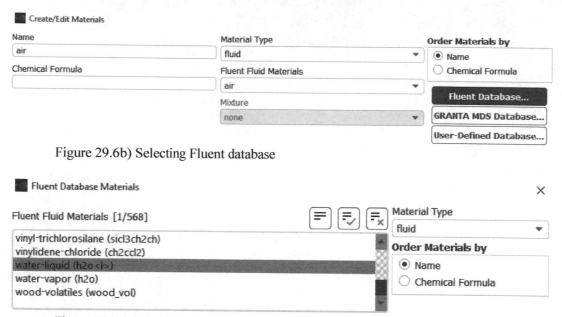

Figure 29.6b) Selecting Fluent database

Figure 29.6c) Selection of water-liquid as fluid material

Set the Name to *abs*, delete the chemical formula, set the Density [kg/m³] to 1200 and select non-newtonian-power-law as Viscosity [kg/(m s)]. Click on Edit… next to non-newtonian-power-law and set Consistency Index, k [kg s^n-2 m⁻¹] to 7040 and Power-Law Index, n to 0.507. Select OK to close the window. Answer Yes to the Question Change/Create mixture and Overwrite water-liquid? Close the Create/Edit Materials window.

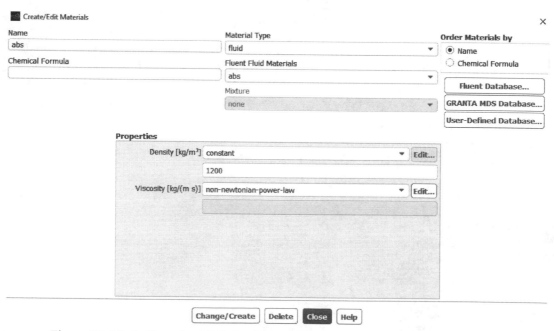

Figure 29.6d) Settings for ABS plastic material

Figure 29.6e) Non-Newtonian power law settings

7. Select the Physics tab in the menu and select Models>>Multiphase Model. Select the Phases tab at the top of the Multiphase Model window. Select phase-1-Primary Phase and enter *air* as name for the Primary Phase. Select Apply and select phase-2-Secondary Phase, select *abs* as the Phase Material and enter *abs* as the Name for the Secondary Phase. Click on Apply and select the Phase Interaction tab. Enter 0.06 as Surface Tension Coefficient [N/m], click on Apply and Close the Multiphase Model window.

Figure 29.7a) Setting up the physics

Figure 29.7b) Editing phase-1-Primary Phase

Figure 29.7c) Name for phase-1   Figure 29.7d) Name for phase-2

Figure 29.7e) Surface tension coefficient

8. Select the Physics tab in the menu and select Operating Conditions… under Solver. Check the box for Gravity and enter -9.81 m/s² for Gravitational Acceleration in the Y direction. Select user-input as Operating Density Method. Enter 1.225 as Operating Density [kg/m³]. Set the Reference Pressure Location for X [m] to 0 and for Y [m] to 0.0002. Click on the OK button to close the window.

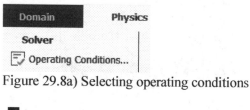

Figure 29.8a) Selecting operating conditions

Figure 29.8b) Including gravity and specified operating density

9. Double click on Boundary Conditions under Setup in the Outline View. Select *inlet* under Zone in Boundary Conditions on the Task Page. Click on the Edit... button. Select Magnitude, Normal to Boundary as Velocity Specification Method. Set Velocity Magnitude [m/s] to 0.01432. Select Apply and Close the window.

Figure 29.9 Inlet velocity magnitude

10. For *inlet*, select *abs* from the Phase drop-down menu under Boundary Conditions on the Task Page. Click on the Edit... button, make sure that the value for the Volume Fraction is 1 and click Apply and Close the window.

Figure 29.10a) Volume fraction for velocity inlet boundary condition

Double click on *moving_bed* under Boundary Conditions on the Task Page. Select Moving Wall under Wall Motion, mixture as Phase, select Components under Motion and set X-velocity [m/s] to -0.03. Select Apply and Close the Wall window.

Figure 29.10b) X-velocity for moving bed

11. Double click on Methods under Solution in the Outline View and choose First Order Upwind for Momentum under Spatial Discretization on the Task Page. Double click on Monitors and Residual under Solution in the Outline View. Set Absolute Criteria for all Residuals to 0.0001. Double click on Initialization under Solution in the Tree and select Compute from *inlet* on the Task Page. Set *abs* Volume Fraction to 0. Click on Initialize.

Figure 29.11a) Solution methods

Figure 29.11b) Solution initialization

Figure 29.11c) Residual monitors

12. Select File>>Export>>Case… from the menu. Save the Case File with the name *2D Deposition Flow.cas.h5*. Double click on Run Calculation under Solution in the Outline View and set the Time Step Size to 0.0001 s. Set the Number of Time Steps to 100 and set Max Iterations/Time Step to 60. Click on the Calculate button on the Task Page. Click OK in the Information window when the calculation is complete.

Figure 29.12a) Calculation settings for 2D flow

Figure 29.12b) Scaled residuals for 2D flow

G. Post-Processing for the 2D Case

13. Double click on Graphics and Contours under Results in the Outline View. Select Contours of Phases… and Volume Fraction. Select *abs* as the Phase. Deselect all Surfaces. Click on Save/Display. Click on Copy Screenshot of Active Window to Clipboard ⬜. The *abs* volume fraction contour plot can be pasted into a Word document.

Figure 29.13a) Contours of ABS volume fraction at *t* = 0.01 s

Double click on Graphics and Contours under Results in the Outline View. Select Contours of Velocity… and Velocity Magnitude. Select *mixture* as the Phase. Deselect all Surfaces. Click on Colormap Options. Select general as Type and 2 for Precision under Number Format in the Colormap window. Set Font Behavior to Fixed, Font Size to 12 and set Colormap Alignment to Top. Select Apply and Close the Colormap window. Click on Save/Display and Close the Contours window. Repeat this step for Y-velocity and X-velocity for the mixture. Continue calculations in increments of 0.01 s until 0.05s.

Figure 29.13b) Contours of Velocity Magnitude, Y-Velocity and X-velocity at *t* = 0.01 s

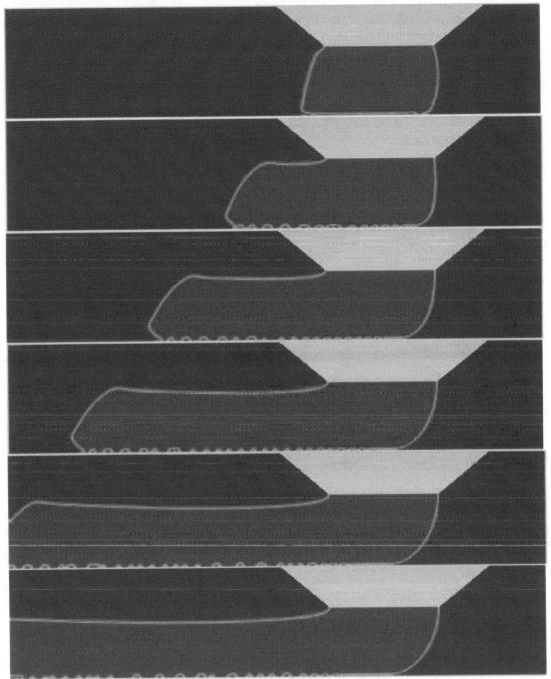

Figure 29.13c) Contours of ABS volume fraction at $t = 0.02 - 0.07$s

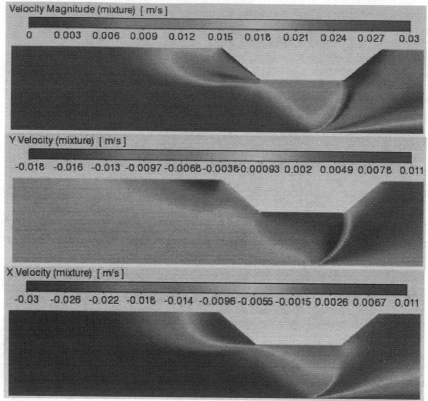

Figure 29.13d) Contours of Velocity Magnitude, Y-Velocity and X-velocity at $t = 0.07$ s

H. Launching Ansys Workbench and Selecting Fluent for the 3D Case

14. Start by launching Ansys Workbench. Launch Fluid Flow (Fluent) that is available under Analysis Systems in Ansys Workbench . Right click on Geometry in Project Schematic and select New DesignModeler Geometry to start DesignModeler.

Figure 29.14 Launching Ansys DesignModeler Geometry

I. Launching Ansys DesignModeler for the 3D Case

15. Select **Units>>Millimeter** from the menu in DesignModeler. Select the ZXPlane in the Tree Outline. Select Look At Face/Plane/Sketch 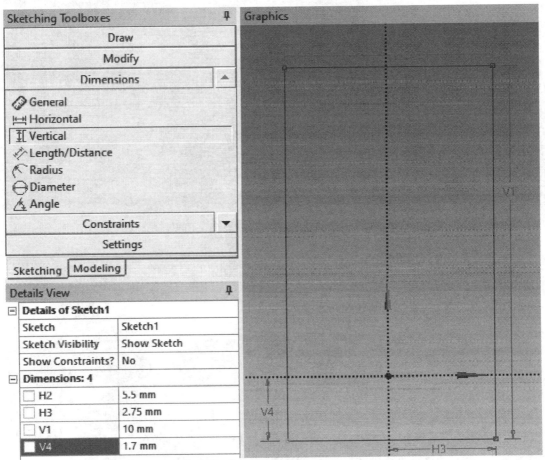. Zoom in in the graphics window using your mouse wheel so that the horizontal scale at the bottom of the window shows from 0 – 5.000 (mm). Select the Sketching tab and Rectangle. Draw the rectangle with the dimensions as shown in Figure 29.15a).

Figure 29.15a) Rectangle with dimensions

Select Extrude and Apply Sketch1 as Geometry in Details View. Set the Depth FD1 to 1.55 mm and click on Generate.

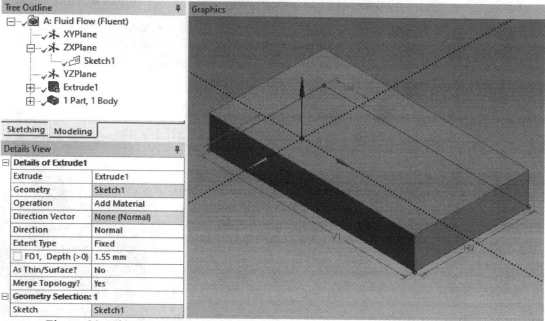

Figure 29.15b) Generated volume for computational domain

Select New Plane 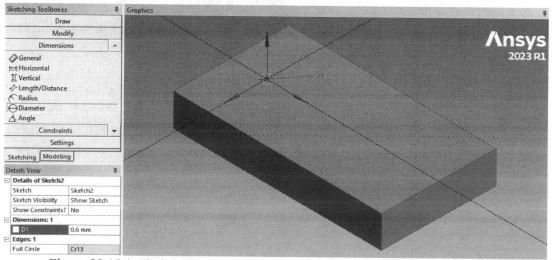 and select Offset Z as Transform 1 (RMB) in Details View. Enter 1.55 mm as FD1, Value 1 and click on Generate. Select New Sketch on Plane4 and draw a circle at the origin with a diameter of 0.6 mm.

Figure 29.15c) Circle in Sketch2 with diameter 0.6 mm

Select Extrude and Apply Sketch2 in Details View. Choose Cut Material as Operation and enter 1.3 mm as the Depth. Select Reversed Direction and click on Generate.

Figure 29.15d) Generated extruded cut in computational domain

Select Plane4 and select New Sketch ![icon] on Plane4 and draw a circle at the origin with a diameter of 0.5 mm.

Figure 29.15e) Generated Sketch3 on Plane4

Select Extrude and Apply Sketch3 in Details View. Choose Add Material as Operation and select Both-Asymmetric Direction. Set Depth FD1 to 0.5 mm, Depth FD4 to 1.3 mm and click on Generate. Close DesignModeler.

Figure 29.15f) Generated Extrude3

J. Launching Ansys Meshing for the 3D Case

16. Double click on Mesh under Project Schematic in Ansys Workbench. Select Mesh in the Outline and set the Element Size to 0.2 mm. Right-click on Mesh under Project and Model (A3) in the Meshing window and select Update. The mesh has 15987 Nodes and 79383 Elements

Figure 29.16a) Launching meshing window

Figure 29.16b) Finished 3D mesh

Figure 29.16c) Details of 3D mesh

Select the Face tool and select the circular face, right click and select Create Named Selection. Enter the name *inlet*. Name the bottom rectangular horizontal face as *moving bed*. Zoom in and control select and name the two cylindrical faces and the circular segment at the bottom of the extruded cut as *wall*. Finally, name the remaining five rectangular faces as *outlet*.

Figure 29.16d) Named selections

Select File>>Export…>>Mesh>>FLUENT Input File>>Export from the menu. Save the mesh with the name *3D deposition flow.msh*. Select File>>Save Project from the menu and save the project with the name *3D Deposition Flow in Additive Manufacturing.wbpj*. Close the meshing window. Right click on Mesh in the Project Schematic and select Update.

K. Launching Ansys Fluent for the 3D Case

17. Double click on Setup under Project Schematic in Ansys Workbench. Check the box for Double Precision. Set the number of parallel processes equal to the number of processor cores for your computer. To check the number of physical cores, press the Ctrl + Shift + Esc keys simultaneously to open the Task Manager. Go to the Performance tab and select CPU from the left column. You'll see the number of physical cores on the bottom-right side. Close the Task Manager window. Click on the Start button in the Fluent Launcher window.

Figure 29.17a) Ansys Fluent 3D launcher

Figure 29.17b) 3D computational domain in Ansys Fluent

18. Select Transient for Time in the General Solver settings. Open Models under Setup in the Outline View and double click on Viscous (SST k-omega). Select Laminar Model and click on OK to close the Viscous Model window. Double click on Multiphase. Select the Volume of Fluid Model. Click on Apply and Close the Multiphase Model window.

> **Why did we select Transient for Time?** We choose Transient since we are interested in studying the shape of the free surface, hydraulic jump and how it deforms over time.

> **Why did we select Volume of Fluid as Multiphase model?** The Volume of Fluid model is used when we have two or more immiscible fluids where the position of the interface between the fluids is of interest. The applications of the VOF Model includes free-surface flows, stratified flows and the tracking of liquid-gas interfaces in general. The VOF model is limited to the pressure based solver and doesn't allow for void regions without a fluid.

Figure 29.18a) General settings

Figure 29.18b) Viscous model

Figure 29.18c) Selecting volume of fluid model

19. Next, we double click on Materials under Setup in the Outline View. Select Fluid under Materials on the Task Page and click on Create/Edit…. Click on Fluent Database… in the Create/Edit Materials window. Scroll down in the Fluent Fluid Material section and select *water-liquid (h2o<l>)*. Click Copy at the bottom of the Fluent Database Materials window. Close the window.

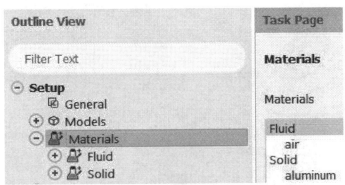

Figure 29.19a) Selecting fluid materials

Figure 29.19b) Selecting Fluent database

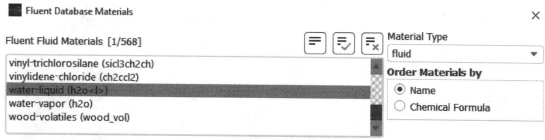

Figure 29.19c) Selection of water-liquid as fluid material

Set the Name to *abs*, delete the chemical formula, set the Density [kg/m³] to 1126 and select non-newtonian-power-law as Viscosity [kg/(m s)]. Click on Edit... next to non-newtonian-power-law and set Consistency Index, k [kg s^n-2 m⁻¹] to 499 and Power-Law Index, n to 0.7. Select OK to close the window. Answer Yes to the question to Change/Create mixture and Overwrite water-liquid. Close the Create/Edit Materials window.

Create/Edit Materials

Name	Material Type	Order Materials by
abs	fluid	● Name
Chemical Formula	Fluent Fluid Materials	○ Chemical Formula
	abs	
	Mixture	Fluent Database...
	none	GRANTA MDS Database...
		User-Defined Database...

Properties

Density [kg/m³] constant Edit...
1126

Viscosity [kg/(m s)] non-newtonian-power-law Edit...

Change/Create   Delete   Close   Help

Figure 29.19d) Settings for ABS plastic material

Figure 29.19e) Non-Newtonian power law settings

20. Select the Physics tab in the menu and select Models>>Multiphase Model. Select the Phases tab at the top of the Multiphase Model window. Select phase-1-Primary Phase and enter *air* as name for the Primary Phase. Select Apply and select phase-2-Secondary Phase, select *abs* as the Phase Material and enter *abs* as the Name for the Secondary Phase. Click on Apply and select the Phase Interaction tab. Enter 0.06 as constant for Surface Tension Coefficient [N/m], click on Apply and Close the Multiphase Model window.

Figure 29.20a) Setting up the physics

Figure 29.20b) Editing phase-1-Primary Phase

Figure 29.20c) Name for phase-1   Figure 29.20d) Name for phase-2

Figure 29.20e) Surface tension coefficient

21. Select the Physics tab in the menu and select Operating Conditions… under Solver. Check the box for Gravity and enter -9.81 m/s² for Gravitational Acceleration in the Y direction. Select user-input as Operating Density Method. Enter 1.225 as Operating Density [kg/m³]. Set the Reference Pressure Location for X [m] to 0.005 and for Y [m] to 0.00155. Click on the OK button to close the window.

Figure 29.21a) Selecting operating conditions

Figure 29.21b) Including gravity and specified operating density

22. Double click on Boundary Conditions under Setup in the Outline View. Select *inlet* under Zone in Boundary Conditions on the Task Page. Click on the Edit... button. Select Magnitude, Normal to Boundary as Velocity Specification Method. Set Velocity Magnitude [m/s] to 0.0045. Select Apply and Close the window.

Figure 29.22 Inlet velocity magnitude

23. For *inlet*, select *abs* from the Phase drop-down menu under Boundary Conditions on the Task Page. Click on the Edit... button, make sure that the value for the Volume Fraction is 1 and click Apply and Close the window.

Figure 29.23a) Volume fraction for velocity inlet boundary condition

Double click on *moving_bed* under Boundary Conditions on the Task Page. Select Moving Wall under Wall Motion, mixture as Phase, select Components under Motion and set X-velocity [m/s] to 0.0045. Select Apply and Close the Wall window.

Figure 29.23b) X-velocity for moving bed

24. Double click on Methods under Solution in the Outline View and choose First Order Upwind for Momentum under Spatial Discretization on the Task Page.

Double click on Initialization under Solution in the Tree and select Compute from *inlet* on the Task Page. Set *abs* Volume Fraction to 0. Click on Initialize.

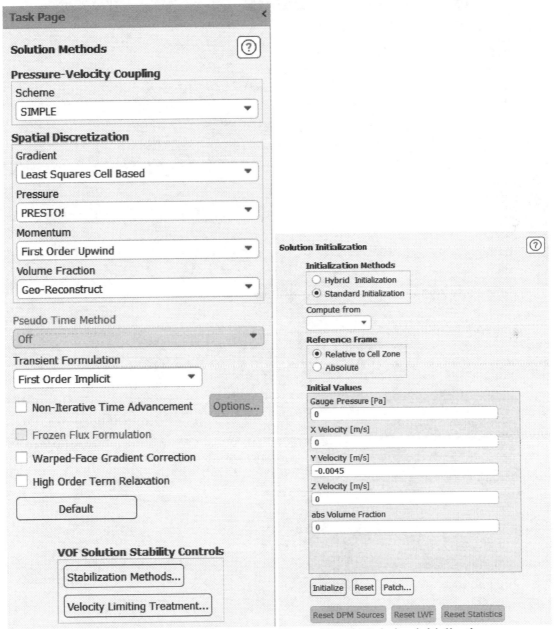

Figure 29.24a) Solution methods   Figure 29.24b) Solution initialization

25. Select File>>Export>>Case… from the menu. Save the Case File with the name *3D Deposition Flow.cas.h5*. Double click on Monitors and Residual under Solution in the Outline View. Set the Absolute Criteria for all Residuals to 1e-05. Select OK to close the Residual Monitors window. Double click on Run Calculation under Solution in the Outline View and set the Time Step Size to 0.05 s. Set the Number of Time Steps to 40 and set Max Iterations/Time Step to 100. Click on the Calculate button on the Task Page. Click OK in the Information window when the calculation is complete.

Figure 29.25a) Residual settings for 3D flow

Figure 29.25b) Calculation settings for 3D flow

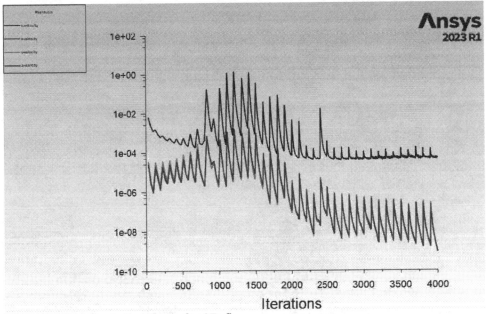

Figure 29.25c) Scaled residuals for 3D flow

L. Post-Processing for the 3D Case

26. Double click on Graphics and Contours under Results in the Outline View. Select Contours of Phases… and Volume Fraction. Select *abs* as the Phase. Deselect all Surfaces. Select Colormap Options… and set Colormap Alignment to Top. Set Font Behavior to Fixed and Font Size to 12. Select general as Type and set Precision to 0 under Number Format. Select Apply and Close the Colormap window. Select New Surface>>Plane… under Display State. Select ZX Plane and Y [m] = 0. Create a new plane and Close the Plane Surface window. Select *plane-4* under Surfaces in the Contours window and click on Save/Display. Click on Copy Screenshot of Active Window to Clipboard ⬜ . The *abs* volume fraction contour plot can be pasted into Word document.

Figure 29.26a) Contours of ABS volume fraction in ZX Plane and $y = 0$ at $t = 2$ s

729

Repeat this step but select XY Plane with Z[m] = 0 and YZ Plane with X[m] = 0.005.

Figure 29.26b) Contours of ABS volume fraction in XY Plane and $z = 0$ at $t = 2$ s

Figure 29.26c) Contours of ABS volume fraction in YZ Plane and $x = 0.005$ at $t = 2$ s

Repeat this step once again but select New Surface>>Iso-Surface… under Display State in the Contours window. Select Surface of Constant Phases… and Volume fraction. Select *abs* as Phase and click on Compute. Slide the Iso-Value to 0.5 and select *solid* From Zones. Click on Create and Close the Iso-Surface window. Select the new Surface in the Contours window and also select *inlet*, *moving_bed* and *wall* under Surfaces. Select Save/Display.

Figure 29.26d) Settings for iso-surface of constant *ABS* phase and volume fraction

Figure 29.26e) Iso-surface for *ABS* phase in isometric, front and side views at $t = 2$ s

M. Theory

27. The extrusion based additive manufacturing process is shown in Figure 29.27a).

Figure 29.27a) Extrusion based additive manufacturing[1]

We define the volume flow rate Q from the nozzle creating a slice with layer height H and layer width W on the printing bed. Examples of H and W values for the da Vinci 1.0 Pro are 0.2 mm and 0.3 mm, respectively. The feed velocity v will then be

$$v = \frac{Q}{WH} \tag{1}$$

Examples of feed velocities (print speeds) for the da Vinci printer is 60, 90 and 120 mm/s. For a no-slip condition in between the feed pinch rollers and filament, the feed velocity can alternatively be expressed as

$$v = \omega R \tag{2}$$

where ω is angular velocity and R is the radius of the rollers. The force F to push the melt through the heated liquefier is

$$F = A\Delta P = \frac{\pi d^2 \Delta P}{4} \tag{3}$$

where A is cross sectional area of the filament, d is diameter of the filament and ΔP is pressure drop through the liquefier. For the da Vinci 1.0 Pro, the filament diameter $d =$ 1.75 mm. We can now determine the necessary torque T and power P for the two roller motors.

$$T = \frac{F}{2}R = \frac{A\Delta PR}{2} = \frac{\pi d^2 \Delta PR}{8} \tag{4}$$

$$P = \omega T = \frac{\pi d^2 \Delta PQ}{8WH} \tag{5}$$

The bi-polar stepper motor Model 17PM-K142B for the da Vinci printer has a holding torque of 0.38 Nm. We can also relate critical pressure P_{cr} from Euler buckling analysis to maximum pressure drop in the liquefier

$$P_{cr} = \frac{\pi E_m}{16}\left(\frac{d}{L}\right)^2 = \Delta P_{max} \tag{6}$$

where L is the distance from the rollers to the liquefier and E_m is the modulus of elasticity of the filament.

Figure 29.27b) Buckling of the filament[1]

We assume that the viscous behavior of the melted filament in the liquefier follows the power-law viscosity model

$$\mu = K\dot{\gamma}^{n-1} \tag{7}$$

where μ is dynamic viscosity, $\dot{\gamma}$ is shear rate and K (consistency index) and n (power-law index) are constants. The temperature T dependence of viscosity is considered by introducing

$$\mu = \mu_{T_0} e^{E\left(\frac{1}{T} - \frac{1}{T_0}\right)} \tag{8}$$

where μ_{T_0} is dynamic viscosity evaluated at reference temperature T_0 and E is activation energy.

Figure 29.27c) Liquifier zones[1]

The pressure drop in the different sections of the liquefier can be modeled according to Michaeli[2]

$$\Delta P_1 = 2KL_1 v^n \left(\frac{D_1}{2}\right)^{-(n+1)} \left(\frac{1+3n}{n}\right)^n e^{E\left(\frac{1}{T}-\frac{1}{T_0}\right)} \tag{9}$$

$$\Delta P_2 = \frac{2}{3n\tan\beta/2} \left(\frac{D_1}{2}\right)^{2n} (D_2^{-3n} - D_1^{-3n}) \left(\frac{1+3n}{n}\right)^n e^{E\left(\frac{1}{T}-\frac{1}{T_0}\right)} \tag{10}$$

$$\Delta P_3 = 2KL_2 v^n \left(\frac{D_1}{2}\right)^{2n} \left(\frac{D_2}{2}\right)^{-(3n+1)} \left(\frac{1+3n}{n}\right)^n e^{E\left(\frac{1}{T}-\frac{1}{T_0}\right)} \tag{11}$$

$$\Delta P = \Delta P_1 + \Delta P_2 + \Delta P_3 \tag{12}$$

The shear rate $\dot{\gamma}$ through a circular opening of diameter D_2 can be expressed as[3,4]

$$\dot{\gamma} = \frac{8Q(1+3n)}{\pi n D_2^3} \tag{13}$$

The calculated pressure drop can now be determined as

$$\Delta P = \frac{128\mu Q(L_1+L_2)}{\pi D_2^4} \tag{14}$$

Typical values for nozzle lengths are $L_1 = 10$ mm and $L_2 = 0.8$ mm. The exit diameter is $D_2 = 0.4$ mm. The power law index is $n = 0.225$ and the constant $K = 1400$ Pa s^n.

N. References

1. Turner, B.N., Strong, R. and Gold, S.A., "A review of melt extrusion additive manufacturing processes: I. Process design and modeling", Rapid Prototyping Journal, Volume 20, No. 3, 192-204, (2014).

2. Michaeli, W.,"Extrusion dies for plastics and rubber: design and engineering computations", Hanser Verlag, Munchen, (2003).

3. Duty, C. et al., "What makes material printable? A viscoelastic model for extrusion-based 3D printing of polymers", Journal of Manufacturing Processes 35, 526 – 537, (2018).

4. Dealy, J.M, and Wissbrun, K.F., "Melt rheology and its role in plastic processing: theory and applications", New York, NY: Spring Science + Business Media, (1990).

5. Talluri, D.J., "Numerical modeling of the fiber deposition flow in extrusion-based 3D bioprinting.", M.S. Thesis, Rowan University, 2021.

6. Chen, J., "Direct Rheological Characterization of Polymer Composite Filament Used in Fused Filament Fabrication Additive Manufacturing.", M.S. Thesis, Baylor University, 2018.

7. Shadvar, N., Foroozmehr, E., Badrossamay, M., Amouhadi, I., and Dindarloo, A.S, "Computational analysis of the extrusion process of fused deposition modeling of acrylonitrile-butadiene-styrene", International Journal of Material Forming 14:121-131, 2021.

N. Exercises

29.1 Use Ansys Fluent to run the simulation for different feed velocities v = 0.004, 0.003 and 0.002 m/s and compare results for contour plots. Use the same velocity for the moving wall as the feed velocity.

	Ansys Fluent	Ansys Fluent	Ansys Fluent
Mesh Size (mm)	0.2	0.2	0.2
Max Iter. per Time Step	100	100	100
Time Step Size (s)	0.05	0.05	0.05
Number of Time Steps	45	60	90
Total Simulation Time (s)	2.25	3	4.5
g (m/s^2)	9.81	9.81	9.81
H (m)	0.00025	0.00025	0.00025
σ (N/m)	0.060	0.060	0.060
D_2 (m)	0.0005	0.0005	0.0005
v (m/s)	0.004	0.003	0.002

Table 29.1 Numerical values for Exercise 29.1

29.2 Use Ansys Fluent to run the simulation in this chapter for the same feed velocity v = 0.004 m/s but for a different height H = 0.00035 m and compare results for contour plots. Use the same velocity for the moving wall as the feed velocity.

O. Notes

CHAPTER 30. DRAPING OF PLASTIC SHEET

A. Objectives

- Using Ansys Polyflow to Study Draping Process of a Plastic Sheet
- Inserting Boundary Conditions
- Running Ansys Polyflow Simulations
- Using Contour Plots and Animations for Visualizations of Draping Process

B. Problem Description

We will in this chapter study the draping process of a plastic sheet on a positive mold. An adiabatic mold will be moving upward towards the plastic sheet that is fixed at three boundaries. Pressure will be applied to form the sheet. Only half the mold and sheet are modeled as a symmetry boundary condition will be used.

C. Launching Ansys Workbench and Selecting Polyflow

1. Start by launching Ansys Workbench. Launch Fluid Flow (Polyflow) that is available under Analysis Systems in Ansys Workbench. Right click on Geometry in Project Schematic and select New DesignModeler Geometry to start DesignModeler.

Figure 30.1 Launching Ansys Polyflow

D. Launching Ansys DesignModeler

2. Select **Units>>Meter** from the menu in DesignModeler. Select File>>Import External Geometry File…. Open the file *DrapingIII.IGS* and select Generate. This file is available for download from *sdcpublications.com*.

Figure 30.2a) Imported draping geometry in DesignModeler

Select Tools>>Mid-Surface from the menu. Set Selection Method to Automatic in Details View. Set Minimum Thresholds to 6.25e-06 m, Maximum Threshold to 0.000625 m and all three Tolerances to 6.25e-05 m. You will need to enter these values manually in the respective fields as you will not be able to copy and paste from here. Select All Bodies To Search, No Extra Trimming, select Yes for Find Face Pairs Now and select Generate.

Figure 30.2b) Creation of Mid-Surface

Right-click in graphics window and choose Selection Filter>>Edge. Control select the 12 edges as shown in Figure 30.2c). The total length of the 12 edges is 0.21569 m as shown at the bottom of the screen. This is the theoretical line length of formed part *LFP* as defined in the theory section. Right-click in the graphics window and choose Selection Filter>>Vertex. Control select the two points at the end points of the twelve-line segments as shown in Figure 30.2d). The distance between the vertices is 0.079474 m. This is the theoretical line length before forming *LBF*.

Select Tools>>Options from the menu and select Geometry under DesignModeler in the Options window. Set the Measure Selection Limit to 50 under Selection, see Figure 30.2e). Select Graphics under DesignModeler in the Options window. Select Both Sides as Highlight Selection under Graphics. Click OK to close the window.

Right-click in the graphics window and choose Selection Filter>>Face. Select Box Select from the menu, see Figure 30.2f). Box select around the mold to select the 39 faces as shown in Figure 30.2g). The total surface area for the faces is 0.037495 m². This is the theoretical surface area of formed part *SAFP* as defined in the theory section. Select Single Select from the menu and select the plane sheet above the mold. The surface area of the undeformed sheet is 0.02 m², see Figure 30.2h). Close DesignModeler.

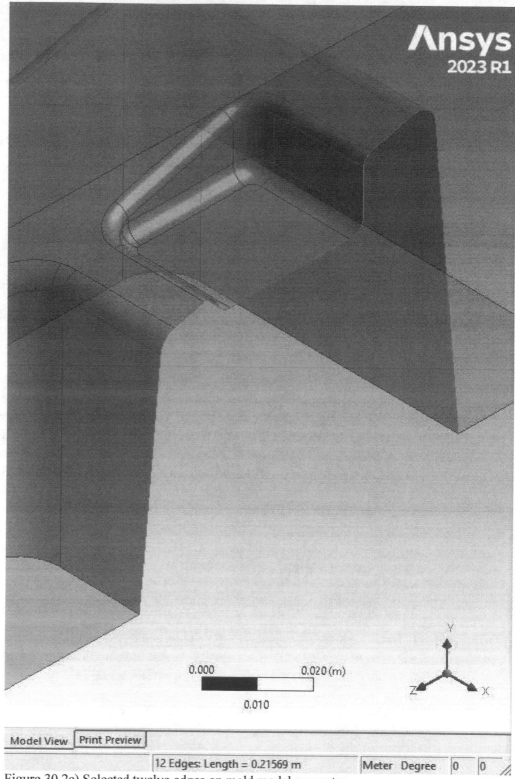

Figure 30.2c) Selected twelve edges on mold model geometry

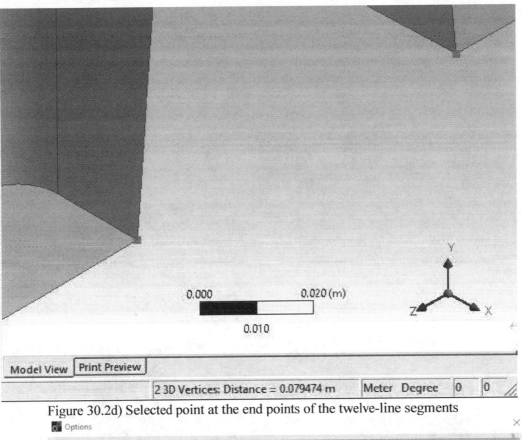

2 3D Vertices: Distance = 0.079474 m   Meter Degree   0   0

Figure 30.2d) Selected point at the end points of the twelve-line segments

Figure 30.2e) Options window

Figure 30.2f) Box select

39 Faces: Area = 0.037495 m²   Meter Degree 0 0

Figure 30.2g) Selected faces on mold model geometry

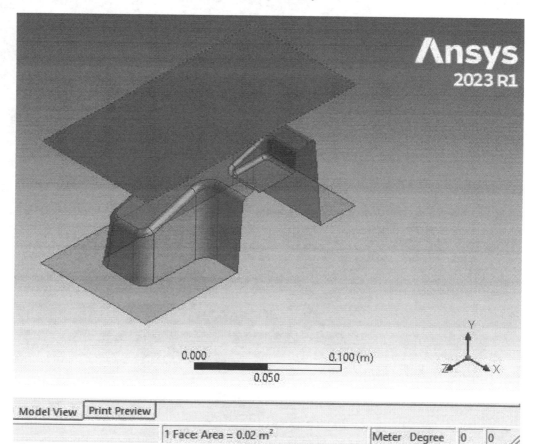

1 Face: Area = 0.02 m²   Meter Degree 0 0

Figure 30.2h) Selected face of undeformed plastic sheet

E. Launching Ansys Meshing

3. Double click on Mesh under Project Schematic in Ansys Workbench. Open Geometry under Model (A3) in the Outline. Right-click on the first part, select Rename and name the first part as Mold. Open Graphic Properties in Details of "Mold" and select red Color. Rename the second part as Film and select blue color.

    Select Mesh in the Outline. Select the Home tab in the menu and select Tools>>Units>>Metric (m, kg, N, s, V, A). Set the Element Size to 9.e-003 m under Defaults in Details of "Mesh". Right-click on Mesh under Project and Model (A3) in the Meshing window and select Update. Click on the plus sign next to Statistics in Details of "Mesh". The mesh has 6177 Nodes and 6124 Elements.

Figure 30.3a) Mesh for mold and sheet

Figure 30.3b) Subdomains and boundaries

Select the shorter edge Boundary 1 for Subdomain 1 (the flat sheet), see Figure 30.3b), right click and select Create Named Selection. Enter the name *Fixed1* and click OK to close the window. Select the longer edge Boundary 2 for Subdomain 1, right click and select Create Named Selection. Enter the name *Fixed2*. Name the other shorter edge Boundary 3 as *Fixed3*. Finally, name the other longer edge Boundary 4 as *Symmetry*, see Figure 30.3c).

Figure 30.3c) Named selections

Select File>>Export...>>Mesh>>POLYFLOW Input File>>Export from the menu. Save the mesh with the name *draping.poly*. Select File>>Save Project from the menu and save the project with the name *Draping Project*. Close the meshing window. Right click on Mesh in the Project Schematic and select Update.

F. Launching Ansys Polydata

4. Double click on Setup under Project Schematic in Ansys Workbench. Select the Mesh tab at the bottom of the middle Polydata section. Select Grid as Display option. Select the Menus tab at the bottom of the Polydata section. Select Create a new task in the Polydata section. Select F.E.M. task, Time-dependent problem and 2D shell geometry. Select Accept the current setup.

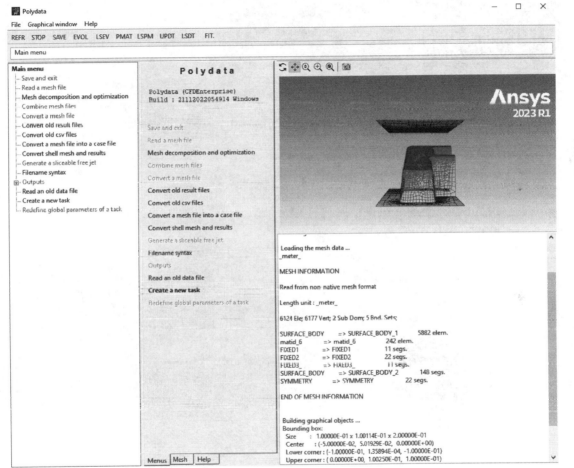

Figure 30.4a) Polydata main menu

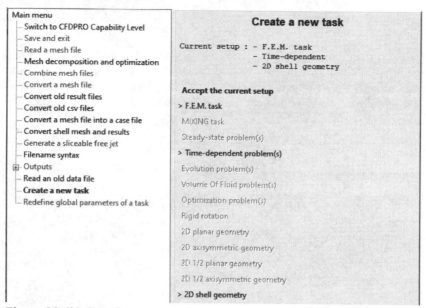

Figure 30.4b) Creating a new task

5. Select Define molds in the F.E.M. Task 1 section and select Create a new mold in the Define molds window.

Figure 30.5a) Defining a mold

Figure 30.5b) Creating a new mold

Select Adiabatic mold in the Create a new mold window. Accept the New value: Mold 1.

Figure 30.5c) Selection of an adiabatic mold

Figure 30.5d) Title for adiabatic mold

Select Domain of the mold in the Mold 1 window. Select matid_6 and select Remove in the Domain of the mold window. Select Upper level menu.

Figure 30.5e) Selecting domain for the mold

Figure 30.5f) Defining the mold

Select Contact conditions in the Mold 1 window. Select No contact along SURFACE_BODY_1 and select Modify. Select Contact in the Contact condition along SURFACE_BODY_1 window. Select Upper level menu two times.

Figure 30.5g) Selection of contact conditions for the mold

Figure 30.5h) Modifying contact

Figure 30.5i) Including contact for the mold surface

Select Mold motion in the Mold 1 window. Select Modify motion type: fixed mold. Choose 1 – Translational velocity imposed in the Specify the type of mold motion window and select OK to close the same window.

Main menu > F.E.M. Task 1 > Define molds > Mold 1

Main menu	Mold 1
Switch to CFDPRO Capability Level	
Save and exit	Adiabatic mold
Read a mesh file	
Mesh decomposition and optimization	Upper level menu
Combine mesh files	Domain of the mold
Convert a mesh file	Material data
Convert old result files	Thermal boundary conditions
Convert old csv files	
Convert a mesh file into a case file	Contact conditions
Convert shell mesh and results	Interpolation
Generate a sliceable free jet	
Filename syntax	Mold motion
Outputs	

Figure 30.5j) Selecting mold motion

Main menu > F.E.M. Task 1 > Define molds > Mold 1 > Mold motion

Main menu
- Switch to CFDPRO Capability Level
- Save and exit
- Read a mesh file
- Mesh decomposition and optimization
- Combine mesh files
- Convert a mesh file
- Convert old result files
- Convert old csv files
- Convert a mesh file into a case file
- Convert shell mesh and results

Mold motion

Actual motion of the mold is enabled
The mold is fixed (no force, no velocity imposed).

Reset to default values

Disable actual motion

Upper level menu

Modify motion type : fixed mold

Figure 30.5k) Selection of motion type

Specify the type of mold motion   ✕

0 - Fixed mold (no velocity, no force imposed)
1 - Translation velocity imposed
2 - Translation force imposed
3 - General velocity driven motion

Update type of mold motion (between 0 and 3).

Old value: 0

New value: 1

OK

Figure 30.5l) Imposing translational velocity for the mold

Select the EVOL tab in the menu and choose Modify translational velocity in the Mold motion window. Leave the value for the translational velocity-x as 0 and click OK to close the Request from Polydata window. Select Upper level menu in the Time dependence of the velocity-x window. Set translational velocity-y to 1 and click OK in the Request from Polydata window.

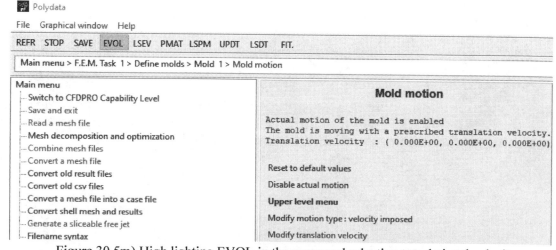

Polydata

File   Graphical window   Help

REFR   STOP   SAVE   EVOL   LSEV   PMAT   LSPM   UPDT   LSDT   FIT.

Main menu > F.E.M. Task 1 > Define molds > Mold 1 > Mold motion

Main menu
- Switch to CFDPRO Capability Level
- Save and exit
- Read a mesh file
- Mesh decomposition and optimization
- Combine mesh files
- Convert a mesh file
- Convert old result files
- Convert old csv files
- Convert a mesh file into a case file
- Convert shell mesh and results
- Generate a sliceable free jet
- Filename syntax

Mold motion

Actual motion of the mold is enabled
The mold is moving with a prescribed translation velocity.
Translation velocity : (0.000E+00, 0.000E+00, 0.000E+00)

Reset to default values

Disable actual motion

Upper level menu

Modify motion type : velocity imposed

Modify translation velocity

Figure 30.5m) High lighting EVOL in the menu and selecting translational velocity

Figure 30.5n) Translational velocity-x

Figure 30.5o) Translational velocity–y

Select f(t) = Ramp function in the Time dependence of the velocity-y window. Select Modify the value of a in the Time dependence of the velocity-y window. Enter 0.097 as the value for a. Click OK in the Request from Polydata window. Modify the values for b = 1, c = 0.103 and d = 0 in the same way. Deselect the EVOL tab in the menu.

Figure 30.5p) Selection of ramp function as transient motion

Request from Polydata   ✕

Update a.

Old value: 1

New value: 0.097

OK

Figure 30.5q) Entering the value for a

Request from Polydata   ✕

Update b.

Old value: 1

New value: 1

OK

Figure 30.5r) Modifying the value for b

Request from Polydata   ✕

Update c.

Old value: 1

New value: 0.103

OK

Figure 30.5s) Selecting the value for c

Request from Polydata   ✕

Update d.

Old value: 1

New value: 0

OK

Figure 30.5t) Editing the value for d

Select Upper level menu in the Time dependence of the velocity-y window. Set translational velocity-z to 0 and click OK in the Request from Polydata window. Select Upper level menu three times in the Mold motion window to arrive at the F.E.M. Task 1 window.

Figure 30.5u) Translational velocity–z

6. Select Create a sub-task in the F.E.M. Task 1 window. Select Shell model: Gen. Newtonian isothermal. Enter the New value: Film in the Request from Polydata window.

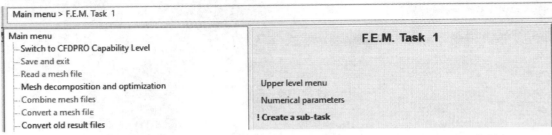

Figure 30.6a) Creating a sub-task

Shell model : **Gen. Newtonian isothermal**

Shell model : Gen. Newtonian non-isothermal

Shell model : Viscoelastic isothermal

Shell model : Viscoelastic non-isothermal

Figure 30.6b) Choosing shell model

Figure 30.6c) Updating the title for the sub-task

Select Domain of the sub-task. Select SURFACE_BODY_1 and Remove in the Domain of the sub-task window. Select Upper Level Menu.

Figure 30.6d) Selecting domain for the sub-task

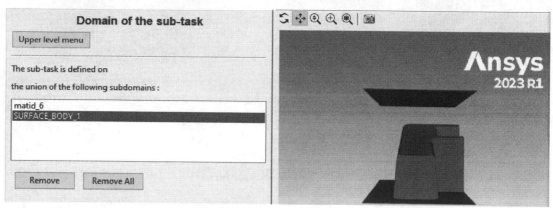

Figure 30.6e) Removing SURFACE_BODY_1

Select Flow boundary conditions and select Zero wall velocity (vn=vs=0) along SYMMETRY. Select Modify and select Plane of symmetry (fs=0,vn=0). Select normal direction along X axis. Select Upper Level Menu.

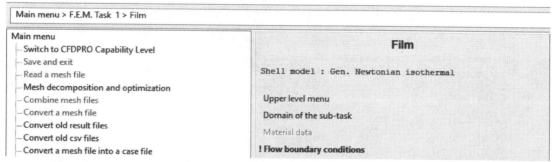

Figure 30.6f) Selecting flow boundary conditions

Figure 30.6g) Selecting symmetry boundary condition

Main menu > F.E.M. Task 1 > Film > Flow boundary conditions > Plane of symmetry along SYMMETRY

Read a mesh file
Mesh decomposition and optimization
Combine mesh files
Convert a mesh file
Convert old result files
Convert old csv files
Convert a mesh file into a case file
Convert shell mesh and results
Generate a sliceable free jet
Filename syntax
Outputs
Read an old data file
Create a new task

Flow boundary condition along SYMMETRY

Current choice : Plane of symmetry
                 No force postprocessor

Enable force postprocessor

Upper level menu

Zero wall velocity (vn=vs=0)

Slip conditions

Porous wall

> Plane of symmetry (fs=0, vn=0)

Figure 30.6h) Selecting plane of symmetry (fs=0,vn=0)

Main menu > F.E.M. Task 1 > Film > Flow boundary conditions > Plane of symmetry along SYMMETRY

Read a mesh file
Mesh decomposition and optimization
Combine mesh files
Convert a mesh file
Convert old result files
Convert old csv files
Convert a mesh file into a case file
Convert shell mesh and results
Generate a sliceable free jet
Filename syntax

Plane of symmetry along SYMMETRY

boundary condition : (fs,vn) = (0,0)
normal to plane of symmetry :
(nx,ny,nz)=   0.1000000E+01   0.0000000E+00   0.0000000E+00

Upper level menu

Switch to computed normals

> normal direction along X axis

Figure 30.6i) Selecting normal direction

Check the box for Inflation pressure imposed in the Flow boundary conditions window. Select constant in the Flow boundary conditions window. Enter 100000 Pa as the New value: and click OK. Select EVOL in the menu and select Upper level menu in the Flow boundary conditions window. Select f(t) = Ramp function. Select Modify the value of a and enter 0.1 as the New value: and click OK. Enter the values 0, 0.11 and 1.0 as b, c, and d values. Select Upper level menu and deselect EVOL in the menu. Select Upper level menu once again.

Figure 30.6j) Selecting inflation pressure

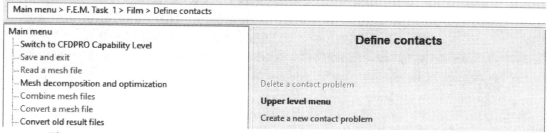

Figure 30.6k) Entered values for a – d

Select Define contacts in the Film window. Select create a new contact problem and click on Select a contact wall.

Figure 30.6l) Defining contacts

Figure 30.6m) Creating a new contact problem

Figure 30.6n) Selecting a contact wall

Select Mold 1: Contact along SURFACE_BODY_1 and click on Select in the Contact wall selection window. Select Modify slipping coefficient and enter 1e+10 as the New value. Select Modify penalty coefficient and enter 1e+10 as the New value. Select Modify penetration accuracy and enter 0.005 as the New value.

Figure 30.6o) Choosing Mold 1 as contact wall

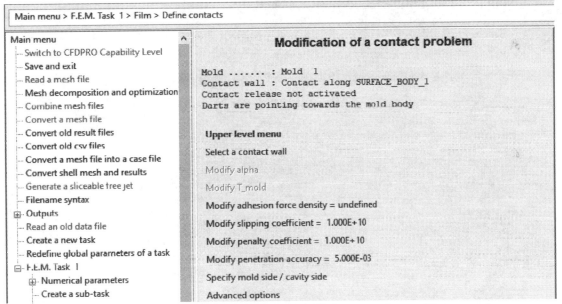

Figure 30.6p) Modified values for slipping coefficient, penalty coefficient and penetration accuracy

Select Specify mold side / cavity side in the Modification of a contact problem window. Select from the menu Graphical window>>Sizing Darts>>Size up. Answer Yes to the question "Are darts pointing towards the mold body?". Select Upper level menu two times. Select OK when you get the following message:

```
The velocity prediction must be disabled
if some contacts are defined.
>> modification automatically done
```

Figure 30.6q) Darts directed towards mold body

Select Define layers in the Film window and select Create a new layer in Define layers window. Enter film as the New value: and click OK.

Figure 30.6r) Defining layers

Figure 30.6s) Creating a new layer

Select Material data in the film window and select Shear-rate dependence of viscosity in the Material data window.

Figure 30.6t) Selection of material data

Figure 30.6u) Selection of shear-rate dependence of viscosity

Select Constant viscosity in the Shear-rate dependence of viscosity window. Select Modify fac. and enter 100000 Pa·s in the Request from Polydata window. Select Upper level menu two times.

Figure 30.6v) Modifying the constant viscosity

Select Density in the Materials data window and select Modification of density in the next window. Enter 1000 kg/m^3 as New value. Select Upper level menu. Select Inertia terms in the Materials data window and select Inertia will be taken into account in the Inertia terms window. Select Upper level menu. Select Layer permeability in the Material data window and select Modify coef. of permeability. Enter the value 5e-12. Select Upper level menu two times. Select Initial thickness in the film window and select Constant in the Initial thickness window. Enter 0.00035 m as initial thickness. Select Upper level menu four times.

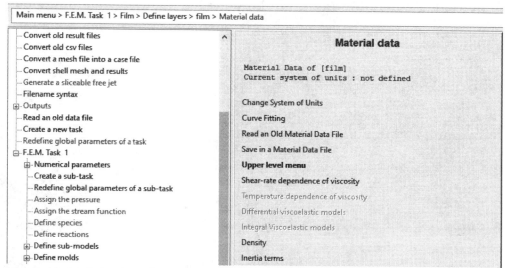

Figure 30.6x) Selection of Density in the Materials data window

Figure 30.6y) Modification of Density

Figure 30.6z) Taking into account inertia terms

7. Select Numerical Parameters in the F.E.M. Task 1 window. Select Modify the transient iterative parameters.

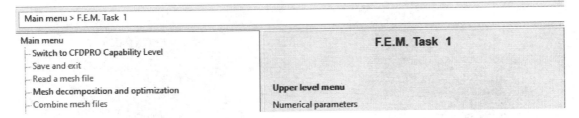

Figure 30.7a) Selecting numerical parameters

Figure 30.7b) Modifying transient iterative parameters

Select Modify the upper time limit and enter 0.6 as the value. Select Modify the initial value of the time-step and enter 0.001 as the value. Select Modify the min value of the time-step and enter 1e-6 as the value. Select Modify the max value of the time-step and enter 0.01. Select Upper level menu three times.

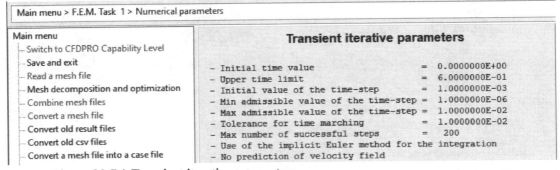

Figure 30.7c) Transient iterative parameters

8. Select Outputs in the Polydata window and select Enable CFD-Post output in the Outputs window (this may already be enabled). Select Upper level menu two times.

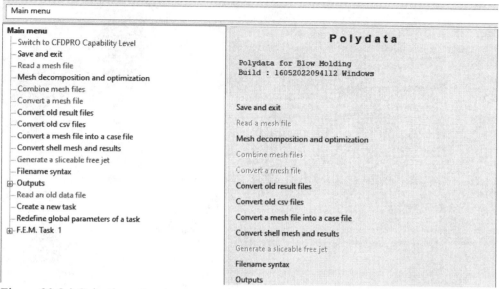

Figure 30.8a) Selection of outputs

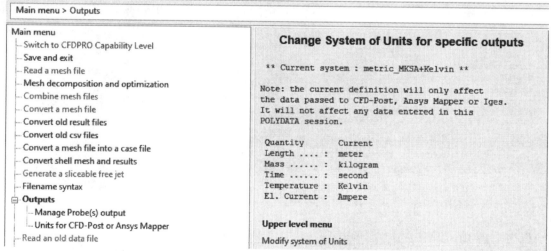

Figure 30.8b) Unit system

9. Select Save and exit in the Polydata window. Select Accept in the Field Management window and Continue.

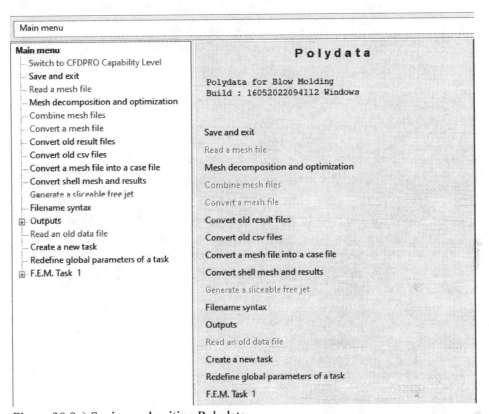

Figure 30.9a) Saving and exiting Polydata

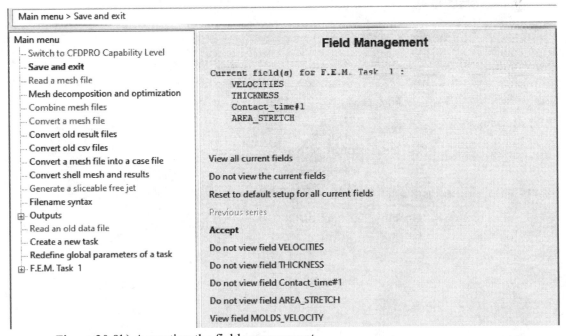

Figure 30.9b) Accepting the field management

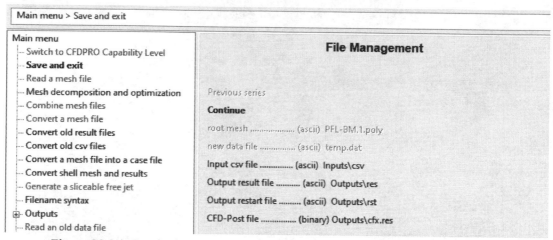

Figure 30.9c) Continuing the field management

10. Right click on Solution in the Project Schematic and select Update. Right click on Solution in the Project Schematic once again and select Listing Viewer…. Check the box for Refresh every and change the value from 10 to 5 seconds. Close the window when computations have finished.

Figure 30.10a) Updating the solution

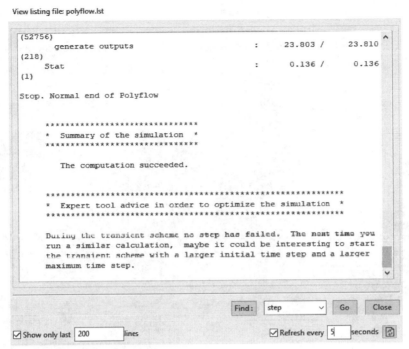

Figure 30.10b) Listing viewer

G. CFD-Post

11. Right click on Results in the Project Schematic and select Edit. Select Insert>>Contour from the menu in CFD-Post and click OK in the Insert Contour window.

Figure 30.11a) Viewing results

Figure 30.11b) Inserting a contour in CFD-Post

12. Select matid_6_surf as Locations in the Geometry Details of Contour 1. Select THICKNESS as Variable. Select User Specified as Range, enter 0.1 [mm] as Min and 0.35 [mm] as Max. Select the Render tab under Details of Contour 1 and uncheck the box for Show Contour lines.

Figure 30.12a) Details of Contour 1

Uncheck the box for Wireframe under User Locations and Plots in the Outline. Click on Apply twice at the bottom left of Details of Contour 1.

Figure 30.12b) CFD-Post Outline

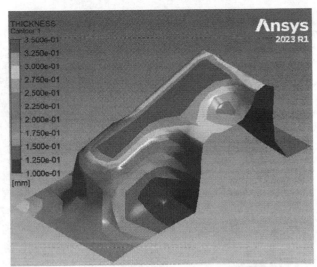

Figure 30.12c) Contours of thickness for half model plastic sheet

13. Double click on Default Transform under User Locations and Plots in the Outline. Uncheck the box for Instancing Info From Domain in the Details of Default Transform. Set the Number of Graphical Instances to 1. Select Value as Determine Angle From under Instance Definition. Set the Angle to 180 [degree] under Instance Definition. Check the box for Apply Reflection. Select YZ Plane as Method under Apply Reflection. Click on Apply at the bottom left of Details of Default Transformation.

Figure 30.13a) Details of default transformation

Figure 30.13b) Contours of thickness for full model plastic sheet

14. Select the Timestep Selector from the menu. Select Step 1 in the Timestep Selector and click Apply. Repeat this for Steps 35, 70 and 109.

Figure 30.14a) Time selector

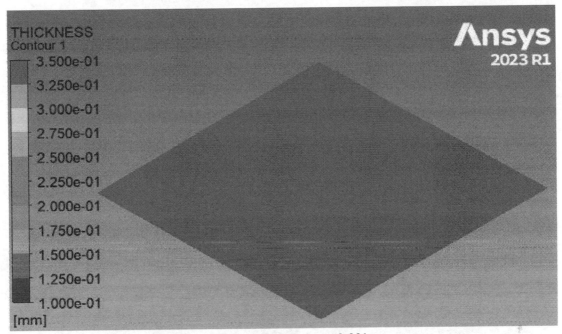

Figure 30.14b) Contours of thickness at step 1, t = 0.001 s

Figure 30.14c) Contours of thickness at step 35, t = 0.0728967 s

15. Select the Animation 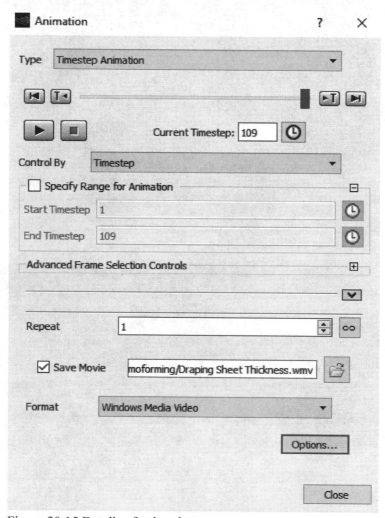 from the menu. Select Timestep Animation as Type. Check the box for Save Movie. Browse and save the file with the name *Draping Sheet Thickness.wmv*. Set Repeat to 1 and click on play. Close the Animation window after completion and close CFD-Post.

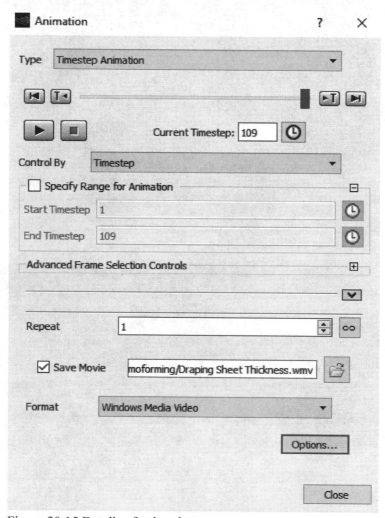

Figure 30.15 Details of animation

H. Refining the Mesh

16. Right click on Setup and select Edit… in the Workbench Project Schematic. Select F.E.M. Task 1 at the bottom of the Polydata window. Select Numerical parameters in the F.E.M. Task 1 window and select Adaptive meshing in the Numerical parameters window. Select Adaptive meshing for contacts in the Adaptive Meshing window. Select Enable all the local criteria in the Global criteria for contact window. Select Switch to 'Global Criteria' mode.

Adaptive Meshing

```
Remeshing by subdivision selected
Nstep: adaptive meshing every [Nstep] time/evolution steps
Maxdiv: maximum number of recursive subdivisions for one element
Remeshing by tri. generator not available
Mesh Conformalization: activated
Additional adaptive meshing at start: inactivated
```

Modify Nstep =   5

Modify Maxdiv =   3

Switch to Partial Triangulation 2D

Disable Mesh Conformalization

Enable additional adaptive meshing at start

Reset parameters of adaptive meshing (to default values)

Upper level menu

Adaptive meshing for contacts

Adaptive meshing for remeshings

Figure 30.16a) Adaptive meshing

Select Modify size_min and enter 0.0001 as the New value and click OK to close the Request from Polydata window. Set tolerance to 0.00001, size_max to 0.001 and dist_crit to 0.0005. Select Upper level menu.

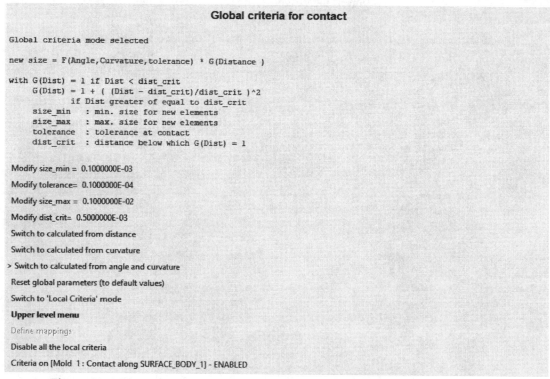

Figure 30.16b) Global criteria for contact

Select Modify Nstep in the Adaptive Meshing window, set New value: to 4 and click OK to close the window. Select Modify Maxdiv and set the value to 3. Select Upper level menu three times. Select Save and exit, select Accept in the Field Management window and select Continue in the Field Management window. Right click on Solution in the Workbench Project Schematic and select Update. Right click on solution and select Listing Viewer…. Check the box for Refresh every and set the time to 5 seconds.

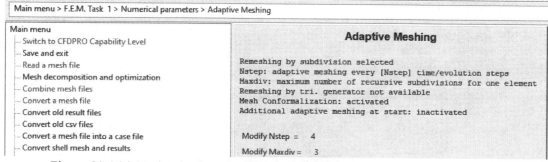

Figure 30.16c) Updated values for adaptive meshing

17. Right click on Results in the Workbench Project Schematic and select Edit…. Double click on Contour 1 under User Locations and Plots in the Outline.

Figure 30.17 Refined thickness contour plot

18. Select the Timestep Selector ⏲ from the menu. Select Step 1 in the Timestep Selector and click Apply. Repeat this for Steps 30, 45, 60, 90, 120 and 177.

Step 30, $t = 0.0493919$ s   Step 45, $t = 0.0713015$ s

Step 60, $t = 0.0958232$ s   Step 90, $t = 0.109884$ s

Step 120, $t = 0.146121$ s   Step 177, $t = 0.6$ s

Figure 30.18a) Refined thickness contour plot at different times

Select the Animation 🎞 from the menu. Select Timestep Animation as Type. Check the box for Save Movie. Browse and save the file with the name *Draping Sheet Thickness Refined Mesh.wmv*. Set Repeat to 1 and click on play. Close the Animation window after completion of the movie.

19. Select Insert>>Clip Plane from the menu. Accept the Name Clip Plane 1 and click OK to close the Insert Clip Plane window. Select YZ Plane a Method and set the X value to 0.0 [mm] in Details of Clip Plane 1. Select Apply, right click in the graphics window and select Clip Scene>>Clip Plane 1.

Figure 30.19 Inserted clip plane

20. Select Insert>>Location>>Iso Clip from the menu. Accept the Name Iso Clip 1 and click OK to close the Insert Iso Clip window. Select matid_6_surf from the drop-down menu next to Locations. Right click in the rectangular area under Visibility parameters and select New. Select THICKNESS as Variable under Visibility Parameters Properties and Visible where >= 0 [mm]. Select the Render tab under Details of Iso Clip 1. Check the box to Show Mesh Lines and click on Apply.

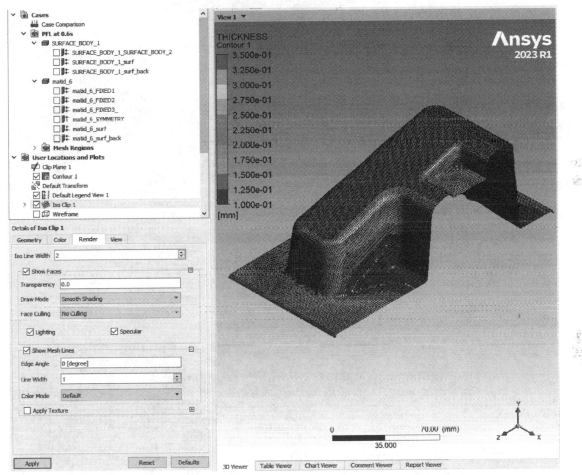

Figure 30.20 Inserted clip plane

21. Select Insert>>Location>>Point Cloud from the menu. Accept the Name Point Cloud 1 and click OK. Select matid_6_SYMMETRY as Locations under Details of Point Cloud 1. Select the Color tab and set the color to black. Select the Symbol tab and choose Ball as symbol and set the Symbol Size to 0.3. Select Apply.

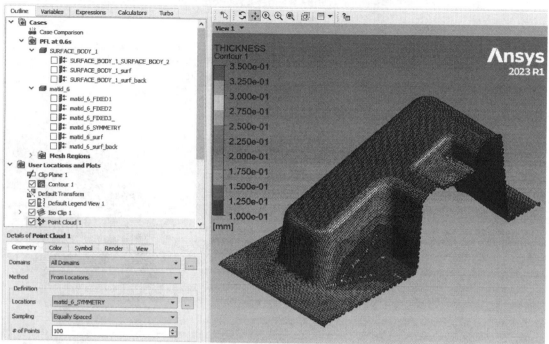

Figure 30.21a) Inserted point cloud

Figure 30.21b) Point cloud in YZ plane

22. Select File>>Export>>Export… from the menu. Select Point Cloud 1 as Locations and THICKNESS as variable. Check the boxes for Line and Face Connectivity and Node Numbers. Enter *test1* as Name Aliases. Click on Save to save the file with the name *test1.csv*.

Figure 30.22 Exporting point cloud

23. Open the file *test1.csv* in Excel. Save the file in Excel with the name *test2.xlsx*. Select Node Numbers 0 – 97 in column D and select Home>>Editing>>Sort & Filter>>Sort Smallest to Largest. Select Expand the selection in the Sort Warning window and click on Sort. Delete the 8 first rows of data including Node Numbers 90 – 97. Delete the 25 last rows of data including Node Numbers 0 – 24. Delete the five rows at the top.

Select column E, right click and select Insert. Label the new column as Distance along line [mm]. enter 0 in cell E2. Enter the following formula in cell E3: =E2+SQRT((C3-C2)^2+(D3-D2)^2). Drag cell E3 down to the end of the column.

Select column F, right click and select Insert. Copy column D to column F. Enter *LBF* [mm] in cell H1, *LFP* [mm] in cell I1 and *LDR* in cell J1. Enter =D66-D2 in cell H2, enter =E66 in cell I2 and enter =I2/H2 in cell J2. Enter 79.474 in cell H6, 215.69 in cell I6 and =I6/H6 in cell J6. Enter =100*ABS(J6-J2)/J6 in cell J9. There is a difference of 7.8 % between Ansys Polyflow simulation and theory for the Line Draw Ratio.

Select the data in columns E and G. Select Insert>>Scatter>>Scatter with Smooth Lines from the menu in Excel. Insert Axis Titles labeled Distance along Line [mm] on the horizontal axis and Thickness [mm] on the vertical axis.

	A	B	C	D	E	F	G	H	I	J
1	Node Number	X [mm]	Y [mm]	Z [mm]	Distance along line [mm]	Z [mm]	THICKNESS [mm]	*LBF* [mm]	*LFP* [mm]	*LDR*
2	89	-3.43E-06	1.03E+02	-7.13E+01	0.00E+00	-7.13E+01	8.46E-02	87.93	219.93	2.50
3	88	-2.42E-06	1.09E+02	-6.90E+01	6.466898612	-6.90E+01	9.55E-02			
4	87	-6.28E-06	1.11E+02	-6.88E+01	8.543793872	-6.88E+01	1.03E-01	Theory	Theory	Theory
5	86	-1.76E-06	1.15E+02	-6.85E+01	12.40248349	-6.85E+01	1.13E-01	*LBF* [mm]	*LFP* [mm]	*LDR*
6	85	6.23E-07	1.19E+02	-6.82E+01	16.00577258	-6.82E+01	1.20E-01	79.474	215.69	2.71
7	84	-6.76E-06	1.22E+02	-6.78E+01	19.42407714	-6.78E+01	1.25E-01			
8	83	1.10E-04	1.25E+02	-6.76E+01	22.73437058	-6.76E+01	1.29E-01			% diff.
9	82	-1.39E-06	1.30E+02	-6.72E+01	27.42231734	-6.72E+01	1.34E-01			7.8
10	81	-2.67E-06	1.34E+02	-6.68E+01	31.95122545	-6.68E+01	1.40E-01			
11	80	-8.08E-07	1.36E+02	-6.67E+01	33.39891921	-6.67E+01	1.42E-01	*SAS* [m^2]	*SAFP* [m^2]	*ADR*
12	79	-8.21E-07	1.40E+02	-6.62E+01	37.66413175	-6.62E+01	1.44E-01	0.02	0.0378674	1.8934
13	78	-3.60E-07	1.43E+02	-6.59E+01	40.44044767	-6.59E+01	1.46E-01			
14	77	-2.47E-06	1.46E+02	-6.57E+01	43.18847572	-6.57E+01	1.48E-01			
15	76	-1.42E-07	1.51E+02	-6.48E+01	48.5346674	-6.48E+01	1.50E-01			
16	75	-1.42E-07	1.54E+02	-6.46E+01	51.18743603	-6.46E+01	1.50E-01	0.02	0.037495	1.8748
17	74	-2.19E-07	1.56E+02	-6.44E+01	53.83113984	-6.44E+01	1.49E-01			
18	73	-3.47E-07	1.60E+02	-6.43E+01	57.77839503	-6.43E+01	1.49E-01			
19	72	-2.84E-06	1.63E+02	-6.43E+01	60.37648172	-6.43E+01	1.52E-01			1.0

Figure 30.23a) Data for thickness of formed plastic sheet

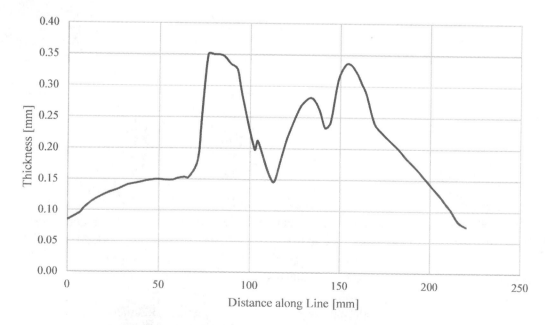

Figure 30.23b) Chart for thickness versus distance along line

Select the data in columns F and G. Select Insert>>Scatter>>Scatter with Smooth Lines from the menu in Excel. Insert Axis Titles labeled Z [mm] on the horizontal axis and Thickness [mm] on the vertical axis.

Figure 30.23c) Deformed sheet dimensions and wall thickness distribution

Select Functions Calculator ⬚ from the menu in CFD-Post. Select area as Function in the Function Calculator and choose matid_6_surf as Location. Check the box for Clear previous results on calculate. Click on Calculate. Results shows the Area on matid_6_surf as 37867.4 [mm^2].

Figure 30.23d) Function calculator

Go to the Excel file as shown in Figure 30.23a). Enter *SAS* [m^2] in cell H11, *SAFP* [m^2] in cell I11 and *ADR* in cell J11. Enter 0.02 in cell H12, enter 0.0378674 in cell I12 and enter =I12/H12 in cell J12. Enter 0.02 in cell H16, 0.037495 in cell I16 and =I16/H16 in cell J16. Enter =100*ABS(J16-J12)/J16 in cell J19. There is a difference of 1 % between Ansys Polyflow simulation and theory for the Area Draw Ratio.

I. Theory

24. The Aerial Draw Ratio *ADR* is the overall stretching of the plastic sheet. The *ADR* > 1 is determined from the Surface Area of the Formed Part *SAFP* divided by the original Surface Area of the Sheet *SAS* (or clamping frame opening) that was used to form the part[2].

$$ADR = SAFP / SAS \tag{1}$$

The maximum values of *ADR* for different plastics are shown in Table 30.1. We see from this table that the plastic Polystyrene *PS* has the highest max *ADR* of all plastics listed.

Plastic	Maximum *ADR*
ABS	5.5
Acrylic	3.4
HDPE	6.5
LDPE	6.0
PP	7.5
Polystyrene PS	8.0
PVC	4.3

Table 30.1 Maximum aerial draw ratio for different plastics

The Average Thickness Reduction *ATR* is

$$ATR = 1 / ADR \tag{2}$$

Another draw ratio is the Linear Draw Ratio *LDR*. This ratio compares the length of a straight line on the sheet before and after forming (including only the forming area).

$$LDR = LFP / LBF \tag{3}$$

where *LFP* is the line length on formed part and *LBF* is the line length before forming.

J. References

1. Klein P. "Fundamentals of Plastics Thermoforming", Synthesis Lectures on Materials Engineering, ISBN: 978-3-031
2. Li Y. "Study of thickness distribution of thermoformed medical PVC blister", IOP Conf. Series: Earth and Environmental Science **81**, 2017.
3. Modlawski M., Jaruga T. "Computer simulation of thermoforming process and its verification using rapid tooling mould", MATEC Web of Conferences **157**, 02032, 2018.
4. Sarakhun, J., Tamna S., Thammakornbunjut E. "Finite Element Analysis of Plastic Cups", International Journal of Mechanical and Production Engineering, **6** (7), 2018.

K. Exercises

Exercise 30.1. Model the plastic cup shown below and run Ansys Polyflow simulations as shown in this chapter. Include thickness contour plots at different times during the forming process and plot thickness graphs in a cross section through the center of the cup.

Figure 30.24 Dimension for plastic cup used by Sarakhun et al. [4]

CHAPTER 31. FREE FORMING OF PLASTIC SHEET

A. Objectives

- Using Ansys Polyflow to Study Free Forming of a Plastic Sheet
- Inserting Boundary Conditions
- Running Ansys Polyflow Simulations
- Using Contour Plots and Animations for Visualizations of Free Forming

B. Problem Description

We will in this chapter study the free forming process of the plastic sheet that is fixed at the circular edge. Pressure will be applied to form the sheet. Only half the sheet is modeled as a symmetry boundary condition will be used.

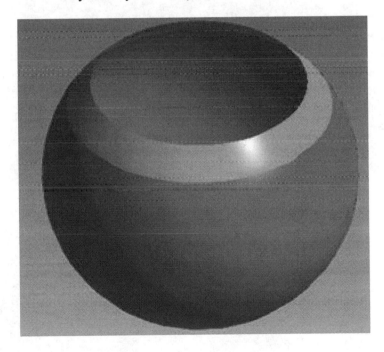

C. Launching Ansys Workbench and Selecting Polyflow

1. Start by launching Ansys Workbench. Launch Fluid Flow (Polyflow) that is available under Analysis Systems in Ansys Workbench. Right click on Geometry in Project Schematic and select New DesignModeler Geometry to start DesignModeler.

Figure 31.1 Launching Ansys Polyflow

D. Launching Ansys DesignModeler

2. Select **Units>>Meter** from the menu in DesignModeler. Right click on ZX Plane in the Tree Outline and select Look at. Select the sketching tab and Arc by Center. Create a half circle from the origin. Make sure that you have a *P* in the graphics window when you move the cursor to the origin before drawing the half circle. Draw the half circle in the third and fourth quadrants. Select Dimensions and set the radius of the half circle to 0.15 m. Use the mouse wheel to zoom in on the half circle in the graphics window. Select Draw and Line. Draw a line connecting the end points of the half circle. Select Extrude from under the menu. Select Sketch1 from under the ZX Plane in the Tree Outline. Apply the sketch in Details View, set FD1, Depth (>0) to 0.00035 m in Details View and click on Generate.

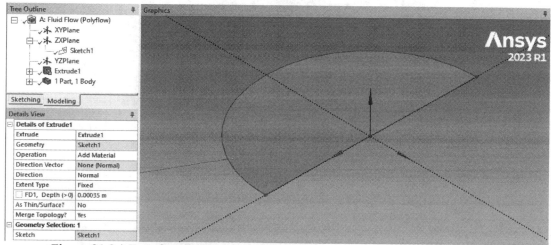

Figure 31.2a) Free forming geometry in DesignModeler

Select the Tools>>Mid-Surface from the menu. Set Selection Method to Automatic in Details View. Set Minimum Thresholds to 6.25e-06 m, Maximum Threshold to 0.000625 m and all three Tolerances to 6.25e-05 m. Select OK when you get a warning. Select All Bodies To Search and No Extra Trimming. Control select both sides of the half-circle and apply as Face Pairs in Details of Midsurf1. Close DesignModeler.

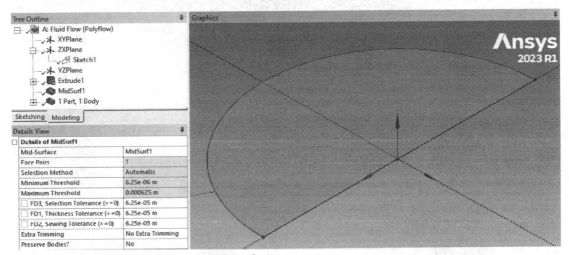

Figure 31.2b) Creation of Mid-Surface

E. Launching Ansys Meshing

3. Double click on Mesh under Project Schematic in Ansys Workbench. Open Geometry under Model (A3) in the Outline. Right-click on the Surface Body, select Rename and name the part as *Film*. Open Graphic Properties in Details of "Film" and select blue Color. Select Mesh in the Outline. Set the Element Size to 0.009 m under Defaults in Details of "Mesh". Select Meters as Export Unit. Right-click on Mesh under Project and Model (A3) in the Meshing window and select Update. The mesh has 512 Nodes and 471 Elements.

Figure 31.3a) Meshed half circle

Right click in the graphics window and select Cursor Mode>>Edge. Select the circular edge of the half-circle, right click and select Create Named Selection. Enter the name *Fixed1* and click OK to close the window. Select the straight edge, right click and select Create Named Selection. Enter the name *Symmetry*.

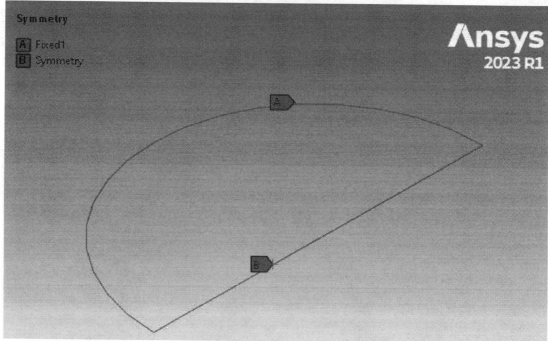

Figure 31.3b) Named selections

Select File>>Export…>>Mesh>>POLYFLOW Input File>>Export from the menu. Save the mesh with the name *free forming.poly*. Select File>>Save Project from the menu and save the project with the name *FreeForming*. Close the meshing window. Right click on Mesh in the Project Schematic and select Update.

F. Launching Ansys Polydata

4. Double click on Setup under Project Schematic in Ansys Workbench. Select the Menus tab in the middle Polydata section. Select Create a new task in the Polydata section. Select F.E.M. task, Time-dependent problem and 2D shell geometry. Select Accept the current setup in the Create a new task section.

Figure 31.4a) Polydata main menu

Figure 31.4b) Accepting the current setup

5. Select Create a sub-task in the F.E.M. Task 1 section. Select Shell model: Gen. Newtonian isothermal. Enter the New value: Film in the Request from Polydata window.

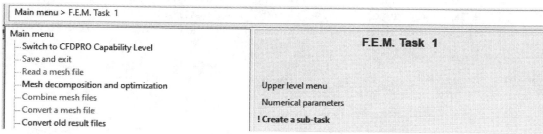

Figure 31.5a) Creating a sub-task

Shell model : Gen. Newtonian isothermal

Shell model : Gen. Newtonian non-isothermal

Shell model : Viscoelastic isothermal

Shell model : Viscoelastic non-isothermal

Figure 31.5b) Choosing shell model

Request from Polydata ✕

Update title.

Old value: F.E.M. Task 1 Sub-task 1

New value: Film

OK

Figure 31.5c) Updating the title for the sub-task

Select Domain of the sub-task and select Upper Level Menu in the next window. Select Flow boundary conditions and select Zero wall velocity (vn=vs=0) along SYMMETRY. Select Modify and select Plane of symmetry (fs=0, vn=0). Select normal direction along X axis. Select Upper Level Menu.

Main menu > F.E.M. Task 1 > Film

Main menu
— Switch to CFDPRO Capability Level
— Save and exit
— Read a mesh file
— Mesh decomposition and optimization
— Combine mesh files
— Convert a mesh file
— Convert old result files
— Convert old csv files
— Convert a mesh file into a case file

Film

Shell model : Gen. Newtonian isothermal

Upper level menu
Domain of the sub-task
Material data
! Flow boundary conditions

Figure 31.5d) Selecting flow boundary conditions

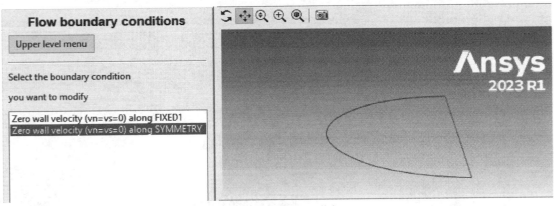

Figure 31.5e) Selecting symmetry boundary condition

Figure 31.5f) Selecting plane of symmetry (fs=0,vn=0)

Figure 31.5g) Selecting normal direction

Check the box for Inflation pressure imposed in the Flow boundary conditions section. Select constant in the Flow boundary conditions window. Enter 100000 Pa as the New value: and click OK. Select EVOL in the menu and select Upper level menu in the Flow boundary conditions window. Select f(t) = Ramp function. Select Modify the value of a and enter 0.1 as the New value: and click OK. Enter the values 0, 0.11 and 1.0 as b, c, and d values. Select Upper level menu and deselect EVOL in the menu. Select Upper level menu once again.

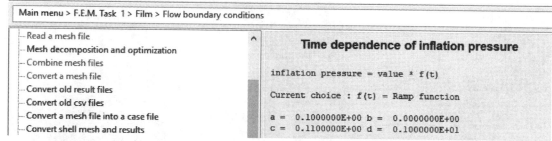

Figure 31.5h) Selecting inflation pressure

Figure 31.5i) Entered values for a – d

Select Define layers in the Film window section and select Create a new layer in Define layers window. Enter *film* as the New value: and click OK.

Figure 31.5j) Defining layers

Figure 31.5k) Creating a new layer

Select Material data in the film window and select Shear-rate dependence of viscosity in the Material data window.

Figure 31.5l) Selection of material data

Figure 31.5m) Selection of shear-rate dependence of viscosity

Select Constant viscosity in the Shear-rate dependence of viscosity window. Select Modify fac. and enter 100000 Pa·s. Select Upper level menu two times.

Figure 31.5n) Modifying the constant viscosity

Select Density in the Materials data window and select Modification of density in the next window. Enter 1000 kg/m³ as New value:. Select Upper level menu. Select Inertia terms in the Materials data window and select Inertia will be taken into account in the Inertia terms window. Select Upper level menu.

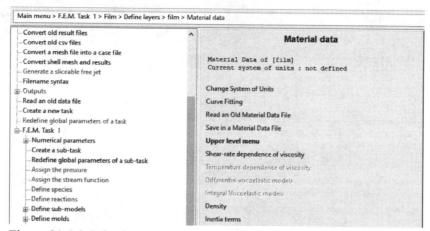

Figure 31.5o) Selection of Density in the Materials data window

Main menu > F.E.M. Task 1 > Film > Define layers > film > Material data > Density

Convert old result files	**Density**
Convert old csv files	
Convert a mesh file into a case file	density = 1.0000000E+00
Convert shell mesh and results	
Generate a sliceable free jet	
Filename syntax	**Upper level menu**
Outputs	Modification of density

Figure 31.5p) Modification of Density

Main menu > F.E.M. Task 1 > Film > Define layers > film > Material data > Inertia terms

Convert old result files	**Inertia terms**
Convert old csv files	
Convert a mesh file into a case file	Inertia will be taken into account
Convert shell mesh and results	in the momentum equations
Generate a sliceable free jet	
Filename syntax	**Upper level menu**
Outputs	> Inertia will be taken into account
Read an old data file	Inertia will be neglected
Create a new task	

Figure 31.5q) Inertia terms

Select Layer permeability in the Material data window and select Modify coef. of permeability. Enter the value 5e-12 and select OK in the Request from Polydata window. Select Upper level menu twice. Select Initial thickness in the film window and select Constant in the Initial thickness window. Enter 0.00035 m as initial thickness. Select Upper level menu four times.

Figure 31.5r) Layer permeability

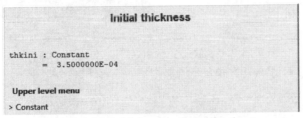

Figure 31.5s) Initial thickness of plastic sheet

6. Select Numerical Parameters in the F.E.M. Task 1 window. Select Modify the transient iterative parameters.

Figure 31.6a) Selecting numerical parameters

Figure 31.6b) Modifying transient iterative parameters

Select Modify the upper time limit and enter 0.12 as the value. Select Modify the initial value of the time-step and enter 0.001 as the value. Select Modify the min value of the time-step and enter 1e-6 as the value. Select Modify the max value of the time-step and enter 0.01. Select Upper level menu three times.

```
          Transient iterative parameters

 - Initial time value                     =  0.0000000E+00
 - Upper time limit                       =  1.2000000E-01
 - Initial value of the time-step         =  1.0000000E-03
 - Min admissible value of the time-step  =  1.0000000E-06
 - Max admissible value of the time-step  =  1.0000000E-02
 - Tolerance for time marching            =  1.0000000E-02
 - Max number of successful steps         =     200
 - Use of the implicit Euler method for the integration
```

Figure 31.6c) Transient iterative parameters

7. Select Outputs in the Polydata window and select Enable CFD-Post output in the Outputs window. Select Upper level menu two times.

Figure 31.7a) Selection of outputs

Figure 31.7b) Enabling CFD-Post output

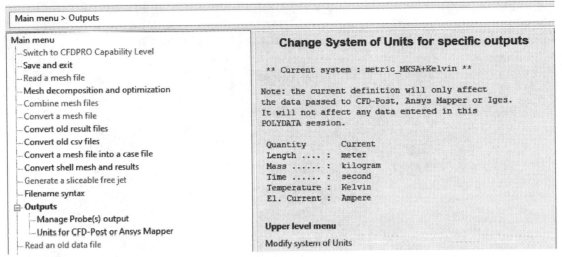

Figure 31.7c) Unit system

8.   Select Save and exit in the Polydata window. Select Accept in the Field Management window and Continue.

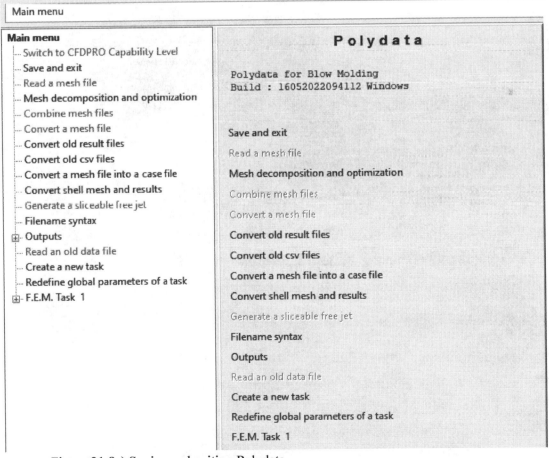

Figure 31.8a) Saving and exiting Polydata

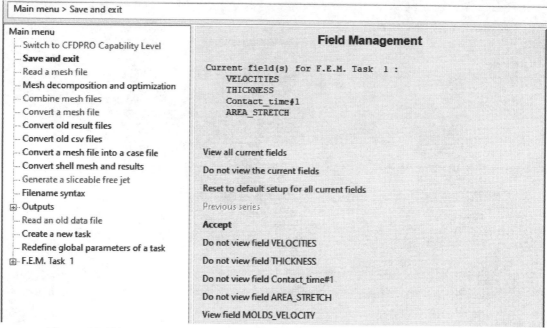

Figure 31.8b) Accepting the field management

Figure 31.8c) Continuing the field management

9. Right click on Solution in the Project Schematic and select Update. Right click on Solution in the Project Schematic once again and select Listing Viewer…. Check the box for Refresh every and change the value from 10 to 5 seconds. Close the window when computations have finished.

Figure 31.9a) Updating the solution

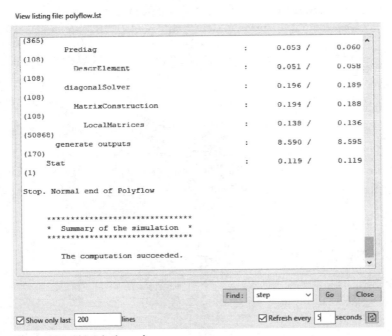

Figure 31.9b) Listing viewer

G. CFD-Post

10. Right click on Results in the Project Schematic and select Edit. Select Insert>>Contour from the menu in CFD-Post and click OK in the Insert Contour window.

Figure 31.10a) Viewing results

Figure 31.10b) Inserting a contour in CFD-Post

11. Select SURFACE_BODY_surf as Locations in the Geometry Details of Contour 1. Select THICKNESS as Variable. Select User Specified as Range and enter 0 [mm] as Min and 0.35 [mm] as Max. Select the Render tab under Details of Contour 1 and uncheck the box for Show Contour lines.

Figure 31.11a) Details of Contour 1

Uncheck the box for Wireframe under User Locations and Plots in the Outline. Click on Apply twice at the bottom left of Details of Contour 1.

Figure 31.11b) CFD-Post Outline

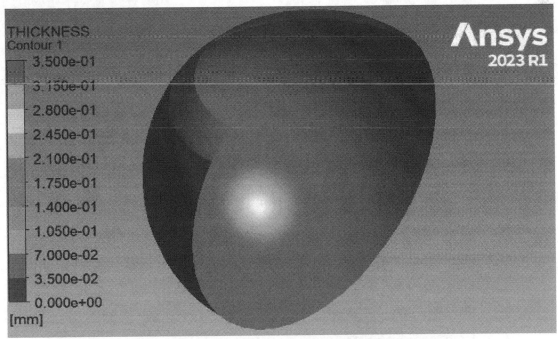

Figure 31.11c) Contours of thickness for half model plastic sheet

12. Double click on Default Transform under User Locations and Plots in the Outline. Uncheck the box for Instancing Info From Domain in the Details of Default Transform. Set the Number of Graphical Instances to 1. Select Value as Determine Angle From under Instance Definition. Set the Angle to 180 [degree] under Instance Definition. Check the box for Apply Reflection. Select YZ Plane as Method under Apply Reflection. Click on Apply at the bottom left of Details of Default Transformation.

Figure 31.12a) Details of default transformation

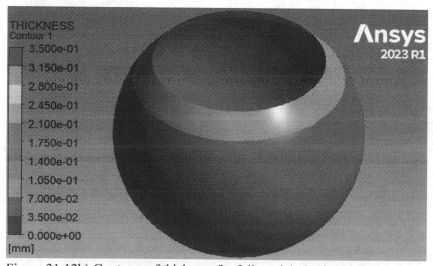

Figure 31.12b) Contours of thickness for full model plastic sheet

13. Select the Timestep Selector from the menu. Select Step 1 in the Timestep Selector and click Apply. Repeat this for Steps 30, 60, 65, 66, 67, 68, 69, 70, 71 and 85.

Figure 31.13a) Time selector

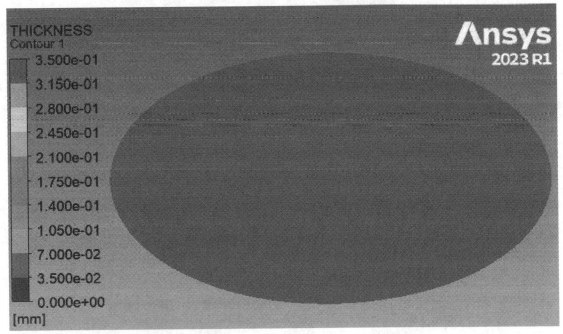

Figure 30.14b) Contours of thickness at step 1

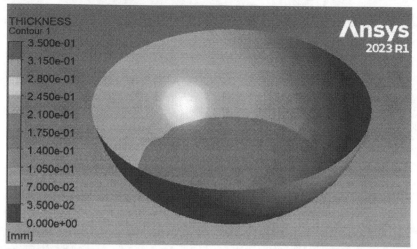

Figure 31.13c) Contours of thickness at step 65, t = 0.113081 s (lenticular shape)

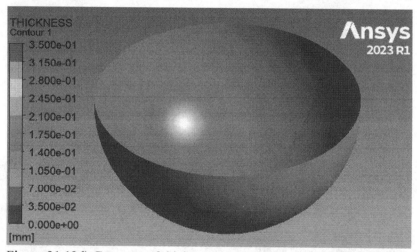

Figure 31.13d) Contours of thickness at step 68, t = 0.115268 s (hemispherical shape)

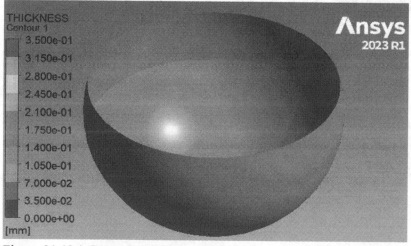

Figure 31.13e) Contours of thickness at step 71, t = 0.116872 s (bulbous shape)

14. Select the Animation ▦ from the menu. Select Timestep Animation as Type. Check the box for Save Movie. Browse and save the file with the name *Free Forming Sheet Thickness.wmv*. Set Repeat to 1 and click on play. Close the Animation window after completion, close the Timestep Selector window and close CFD-Post.

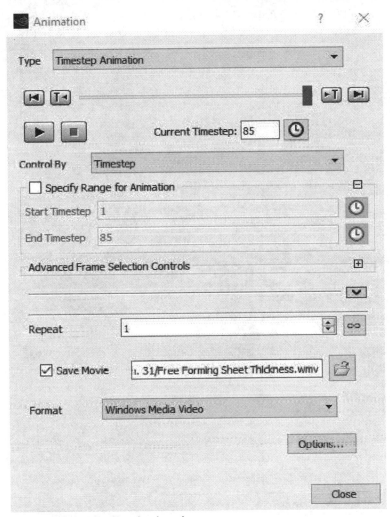

Figure 31.14 Details of animation

H. Refining the Mesh

15. Right click on Setup and select Edit… in the Workbench Project Schematic. Select F.E.M. Task 1 at the bottom of the Polydata window. Select Numerical parameters in the F.E.M. Task 1 window and select Adaptive meshing in the Numerical parameters window. Select Adaptive meshing for remeshings in the Adaptive Meshing window. Select Enable all the local criteria in the Global criteria for remeshings window.

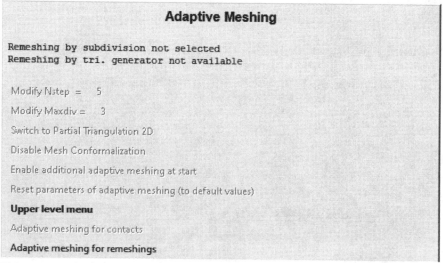

Figure 31.15a) Adaptive meshing

Select Modify size and enter 0.0001 as the New value and click OK to close the Request from Polydata window. Select Upper level menu.

```
                        Global criteria for remeshings

Global criteria mode selected

size      : size for new elements
Tquality  : if current quality of element < Tquality (element too stretched)
            then this element is selected for remeshing
      quality=1 (maximum) == element is equilateral
      quality=0 (minimum) == element is flat

Modify Tquality = 0.8000000E+00

Modify size   = 0.1000000E-03

Reset global parameters (to default values)

Switch to 'Local Criteria' mode

Manage zones for remeshing

Upper level menu

Disable all the local criteria

Criteria on [Film] - ENABLED
```

Figure 31.15b) Global criteria for remeshings

Select Modify Nstep in the Adaptive Meshing window, set New value: to 6 and click OK to close the window. Select Modify Maxdiv and set the value to 4. Select Upper level menu once.

Adaptive Meshing

```
Remeshing by subdivision selected
Nstep: adaptive meshing every [Nstep] time/evolution steps
Maxdiv: maximum number of recursive subdivisions for one element
Remeshing by tri. generator not available
Mesh Conformalization: activated
Additional adaptive meshing at start: inactivated


 Modify Nstep =   6

 Modify Maxdiv =   4
```

Figure 31.15c) Updated values for adaptive meshing

Select Modify the transient iterative parameters in the Numerical parameters section and select Modify the upper time limit in the Transient iterative parameters section. Set the New Value: to 0.1175 and select OK. Select Upper level menu once. Select No previous solution in the Numerical parameters section and select Upper level menu twice. Select Save and exit, select Accept and select Continue in the Field Management window. Right click on Solution in the Workbench Project Schematic and select Update. Right click on solution and select Listing Viewer…. Check the box for Refresh every and set the time to 5 seconds.

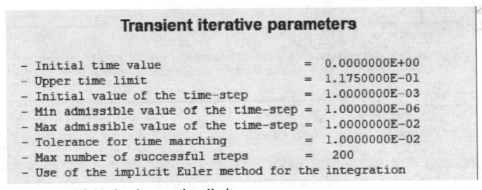

Transient iterative parameters

```
 - Initial time value                    = 0.0000000E+00
 - Upper time limit                      = 1.1750000E-01
 - Initial value of the time-step        = 1.0000000E-03
 - Min admissible value of the time-step = 1.0000000E-06
 - Max admissible value of the time-step = 1.0000000E-02
 - Tolerance for time marching           = 1.0000000E-02
 - Max number of successful steps        =   200
 - Use of the implicit Euler method for the integration
```

Figure 31.15d) Updated upper time limit

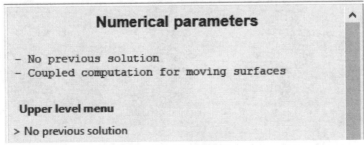

Numerical parameters

```
 - No previous solution
 - Coupled computation for moving surfaces


Upper level menu

> No previous solution
```

Figure 31.15d) Updated upper time limit

805

16. Right click on Results in the Workbench Project Schematic and select Edit…. Double click on Contour 1 under User Locations and Plots in the Outline. Select the Timestep Selector ⊙ from the menu. Select Step 1 in the Timestep Selector and click Apply. Repeat this for Steps 14, 20, 25, 30, 35, 40, 45, 50, 55 and 60.

Step 14, $t = 0.104$ s   Step 20, $t = 0.108408$ s

Step 25, $t = 0.111858$ s   Step 30 , $t = 0.114177$ s

Step 35, $t = 0.115642$ s   Step 40, $t = 0.116395$ s

Figure 31.16a) Refined thickness contour plot at different times

Select the Animation ⊞ from the menu. Select Timestep Animation as Type. Check the box for Save Movie. Browse and save the file with the name *Free Forming Thickness Refined Mesh.wmv*. Set Repeat to 1 and click on play. Close the Animation window after completion and Close CFD-Post.

17. Right click on Results in the Workbench Project Schematic and select Edit…. Double click on Contour 1 under User Locations and Plots in the Outline. Select the Timestep Selector ⏲ from the menu. Select Step 14 in the Timestep Selector, click Apply and Close the window.

    Select Insert>>Clip Plane from the menu. Accept the Name Clip Plane 1 and click OK to close the Insert Clip Plane window. Select YZ Plane as Method and set the X value to 0.0 [mm] in Details of Clip Plane 1. Select Apply, right click in the graphics window and select Clip Scene>>Clip Plane 1.

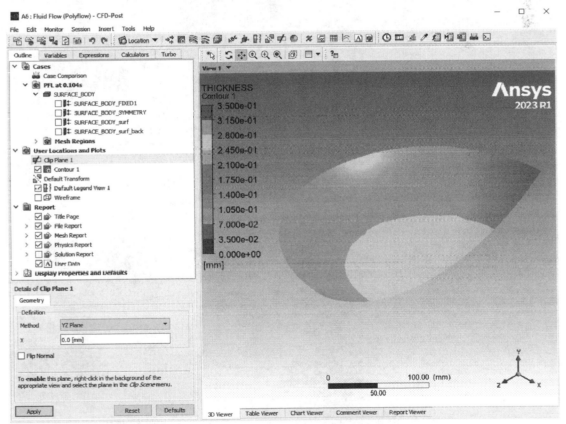

Figure 31.17 Inserted clip plane for lenticular plastic shape

18. Select Insert>>Location>>Iso Clip from the menu. Accept the Name Iso Clip 1 and click OK to close the Insert Iso Clip window. Select *SURFACE_BODY_surf* from the drop-down menu next to Locations. Right click in the rectangular area under Visibility parameters and select New. Select THICKNESS as Variable under Visibility Parameters Properties and Visible where >= 0 [mm]. Select the Render tab under Details of Iso Clip 1. Check the box to Show Mesh Lines and click on Apply. Select the Geometry tab and move the cursor on *SURFACE_BODY_surf* next to Locations in Details of Iso Clip 1.

Figure 31.18 Iso-clip with mesh shown for lenticular plastic sheet

19. Select Insert>>Location>>Point Cloud from the menu. Accept the Name Point Cloud 1 and click OK. Select *SURFACE_BODY_SYMMETRY* as Locations under Details of Point Cloud 1. Select the Color tab and set the color to black. Select the Symbol tab and choose Ball as symbol and set the Symbol Size to 0.01. Select Apply.

Figure 31.19a) Inserted point cloud for lenticular shape

Figure 31.19b) Point cloud in YZ plane for lenticular shape

20. Select File>>Export>>Export... from the menu. Select Point Cloud 1 as Locations and THICKNESS as variable. Click on Save to save the file with the name *freeforming-lenticular.csv*.

Figure 31.20 Exporting point cloud for lenticular shape

21. Open the file *freeform-lenticular.csv* in Excel. Save the file in Excel with the name *freeform-lenticular.xlsx*. Select rows 7 - 95 in column C and select Home>>Sort & Filter>>Sort Smallest to Largest. Select Expand the selection in the Sort Warning window and click on Sort. Delete rows 1 – 5 and delete column A.

Select column C, right click and select Insert. Label the new column |Y| [mm]. Enter =ABS(A2) in cell C2 and drag this cell down the C column. Label column D as *h* [mm] and column E as h_{ave} [mm]. Enter =(MAX(D1:D90)+MIN(D1:D90))/2 in cell E2. Label column F as η and enter =E2/0.35 in cell F2. Label column G as *a* [mm] and enter =C46-(C47-C46)*B46/(B47-B46) in cell G2. Label column H as α and enter =G2/150 in cell H2. Label column I as β and enter =(ATAN(C2/(B2+150))+ATAN(C90/ABS(B90-150)))/2 in cell I2. Label column J as *R* [mm] and enter =(150^2+G2^2)/(2*G2) in cell J2. Label column K as ρ and enter =J2/150 in cell K2. Finally, label column L as κ and enter =1/K2 in cell L2.

Enter *Theory* in cells F4, H4 and I4 and enter *% Diff.* in cells F8, H8 and I8. Enter η in cell F5, α in cell H5 and β in cell I5. Enter =0.5*(1+SQRT(1-L2^2)) in cell F6, enter =(1-SQRT(1-L2^2))/L2 in cell H6 and enter =(PI()/2)-ATAN(SQRT((1/L2^2)-1)) in cell I6.

Enter =100*ABS(F2-F6)/F6 in cell F9, =100*ABS(H2-H6)/H6 in cell H9 and =100*ABS(I2-I6)/I6 in cell I9. There is a maximum difference of 12 % between Ansys Polyflow simulation and theory.

	A	B	C	D	E	F	G	H	I	J	K	L
1	Y [mm]	Z [mm]	|Y| [mm]	h [mm]	have [mm]	η	a [mm]	α	β	R [mm]	ρ	κ
2	-1.7411	-147.6161	1.7411	0.2901	0.2830	0.8085	54.4969	0.3633	0.6154	233.6823	1.5579	0.6419
3	-3.2818	-145.6935	3.2818	0.2898								
4	-6.0289	-142.2071	6.0289	0.2893		Theory		Theory	Theory			
5	-7.5267	-140.1913	7.5267	0.2888		η		α	β			
6	-9.0778	-138.1074	9.0778	0.2883		0.8834		0.3633	0.6970			
7	-11.3788	-134.9371	11.3788	0.2876								
8	-13.0279	-132.5381	13.0279	0.2870		% Diff.		% Diff.	% Diff.			
9	-16.5460	-127.3326	16.5460	0.2860		8		0	12			

Figure 31.21a) Data for thickness and shape of lenticular sheet

Figure 31.21b) Scatter plot for lenticular sheet thickness *h* (mm) versus *Z* coordinate

810

Figure 31.21c) Lenticular sheet Y and Z coordinates for $X = 0$

22. Repeat steps 17 – 21 for the bulbous plastic sheet at Step 35. Right click on Results in the Workbench Project Schematic and select Edit…. Double click on Contour 1 under User Locations and Plots in the Outline. Select the Timestep Selector ⏱ from the menu. Select Step 35 in the Timestep Selector, click Apply and close the window.

Select Insert>>Clip Plane from the menu. Accept the Name Clip Plane 2 and click OK to close the Insert Clip Plane window. Select YZ Plane as Method and set the X value to 0.0 [mm] in Details of Clip Plane 2. Select Apply, right click in the graphics window and select Clip Scene>>Clip Plane 2.

Figure 31.22 Inserted clip plane for bulbous plastic shape

23. Select Insert>>Location>>Iso Clip from the menu. Accept the Name Iso Clip 2 and click OK to close the Insert Iso Clip window. Select *SURFACE_BODY_surf* from the drop-down menu next to Locations. Right click in the rectangular area under Visibility parameters and select New. Select THICKNESS as Variable under Visibility Parameters Properties and Visible where >= 0 [mm]. Select the Render tab under Details of Iso Clip 2. Check the box to Show Mesh Lines and click on Apply. Select the Geometry tab under Details of Iso Clip 2 and move the cursor on *SURFACE_BODY_surf* next to Locations.

Figure 31.23 Iso-clip with mesh shown for bulbous plastic sheet

24. Select Insert>>Location>>Point Cloud from the menu. Accept the Name Point Cloud 2 and click OK. Select *SURFACE_BODY_SYMMETRY* as Locations under Details of Point Cloud 2. Select the Color tab and set the color to black. Select the Symbol tab and choose Ball as symbol and set the Symbol Size to 0.01. Select Apply.

Figure 31.24a) Inserted point cloud for bulbous shape

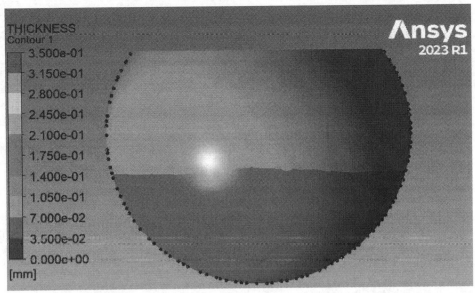

Figure 31.24b) Point cloud in YZ plane for bulbous shape

25. Select File>>Export>>Export… from the menu. Select Point Cloud 2 as Locations and THICKNESS as variable. Click on Save to save the file with the name *freeforming-bulbous.csv*.

Figure 31.25 Exporting point cloud for bulbous shape

26. Open the file *freeform-bulbous.csv* in Excel. Save the file in Excel with the name *freeform-bulbous.xlsx*. Select rows 7 - 96 in column C and select Home>>Sort & Filter>>Sort Smallest to Largest. Select Expand the selection in the Sort Warning window and click on Sort. Delete rows 1 – 5 and delete column A.

Select column C, right click and select Insert. Label the new column |Y| [mm]. Enter =ABS(A2) in cell C2 and expand this cell down the C column. Label column D as h [mm] and column E as h_{ave} [mm]. Enter =(MAX(D1:D90)+MIN(D1:D90))/2 in cell E2. Label column F as η and enter =E2/0.35 in cell F2. Label column G as a [mm] and enter =C44-(C45-C44)*B44/(B45-B44) in cell G2. Label column H as α and enter =G2/150 in cell H2. Label column I as β and enter
((PI()+ATAN(C2/(B2+150)))+(PI()-ATAN(C90/ABS(B90-150))))/2
in cell I2. Label column J as R [mm] and enter =(150^2+G2^2)/(2*G2) in cell J2. Label column K as ρ and enter =J2/150 in cell K2. Finally, label column L as κ and enter =1/K2 in cell L2.

Enter *Theory* in cells F4, H4 and I4 and enter *% Diff.* in cells F8, H8 and I8. Enter η in cell F5, α in cell H5 and β in cell I5. Enter =0.5*(1-SQRT(1-L2^2)) in cell F6, enter =(1+SQRT(1-L2^2))/L2 in cell H6 and enter =(PI()/2)+ATAN(SQRT((1/L2^2)-1)) in cell I6.

Enter =100*ABS(F2-F6)/F6 in cell F9, =100*ABS(H2-H6)/H6 in cell H9 and =100*ABS(I2-I6)/I6 in cell I9. There is a maximum difference of 13 % between Ansys Polyflow simulation and theory.

	A	B	C	D	E	F	G	H	I	J	K	L
1	Y [mm]	Z [mm]	|Y| [mm]	h [mm]	h_{ave} [mm]	η	a [mm]	α	β	R [mm]	ρ	κ
2	-102.8446	-183.1744	102.8446	0.0767	0.0766	0.2188	282.1085	1.8807	1.8837	180.9325	1.2062	0.8290
3	-97.0717	-183.1290	97.0717	0.0775								
4	-90.5118	-183.0715	90.5118	0.0786		*Theory*		*Theory*	*Theory*			
5	-113.6020	-182.9555	113.6020	0.0752		η		α	β			
6	-119.3731	-182.3901	119.3731	0.0742		0.2204		1.8807	2.1642			
7	-80.5744	-182.0055	80.5744	0.0804								
8	-73.9646	-181.0242	73.9646	0.0819		*% Diff.*		*% Diff.*	*% Diff.*			
9	-130.8596	-180.8326	130.8596	0.0725		1		0	13			

Figure 31.26a) Data for thickness and shape of bulbous sheet

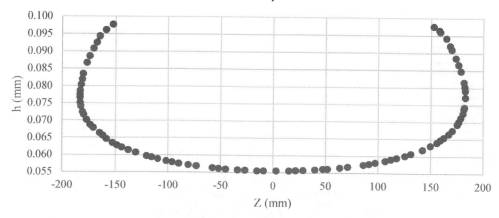

Figure 31.26b) Scatter plot for bulbous sheet thickness h (mm) versus Z coordinate

Figure 31.26c) Bulbous sheet Y and Z coordinates for $X = 0$

I. Theory

Free Forming

27. In free forming, a thin spherical plastic sheet is formed by inflation from a circular opening into the shape of a sphere[1,2], see Figure 30.27a).

Figure 31.27a) Spherical deformed plastic sheet

From the geometry in Figure 31.27a) we find for the lenticular sheet that

$$R^2 = r_0^2 + (R - a)^2 \tag{1}$$

and for the bulbous sheet

$$R^2 = r_0^2 + (a - R)^2 \tag{2}$$

Both equations (1) and (2) give the same equation for the radius R

$$R = \frac{r_0^2 + a^2}{2a} \tag{3}$$

The continuity equation in spherical coordinates $(r, \theta, \emptyset)$ for constant fluid density can be expressed as

$$\frac{1}{r^2}\frac{\partial(r^2 u_r)}{\partial r} + \frac{1}{r\sin\theta}\frac{\partial(u_\theta \sin\theta)}{\partial \theta} + \frac{1}{r\sin\theta}\frac{\partial(u_\emptyset)}{\partial \emptyset} = 0 \tag{4}$$

where $u_r, u_\theta, u_\emptyset$ are radial, azimuthal and meridional components of fluid velocity. Assuming constant density and zero velocity in the meridional and azimuthal directions, this equation becomes

$$\frac{\partial(r^2 u_r)}{\partial r} = 0 \tag{5}$$

We can solve this equation with the condition that fluid velocity at $r = R$ is $u_r(R) = -\frac{dR}{dt}$ for a lenticular sheet and $u_r(R) = \frac{dR}{dt}$ for a bulbous sheet. We get the radial velocity component

$$u_r = \frac{\left|\frac{dR}{dt}\right| R^2}{r^2} \tag{6}$$

The equation of motion in the radial direction for an incompressible fluid is

$$\rho_s \left(\frac{\partial u_r}{\partial t} + u_r \frac{\partial u_r}{\partial r} + \frac{u_\theta}{r}\frac{\partial u_r}{\partial \theta} + \frac{u_\emptyset}{r\sin\theta}\frac{\partial u_\emptyset}{\partial \emptyset} - \frac{u_\theta^2 + u_\emptyset^2}{r}\right)$$
$$= -\frac{\partial P}{\partial r} + \left(\frac{1}{r^2}\frac{\partial(r^2\varepsilon_{rr})}{\partial r} + \frac{1}{r\sin\theta}\frac{\partial(\varepsilon_{\theta r}\sin\theta)}{\partial \theta} + \frac{1}{r\sin\theta}\frac{\partial\varepsilon_{\emptyset r}}{\partial \emptyset} - \frac{\varepsilon_{\theta\theta} + \varepsilon_{\emptyset\emptyset}}{r}\right) + \rho_f g_r \tag{7}$$

Where ρ_s is sheet density, P is pressure and ε_{ij} is the viscous stress tensor.[3] This equation can be reduced to the following form for a thin sheet with inertia and gravity neglected.

$$-\frac{\partial P}{\partial r} + \left(\frac{1}{r^2}\frac{\partial(r^2\varepsilon_{rr})}{\partial r} - \frac{\varepsilon_{\theta\theta} + \varepsilon_{\emptyset\emptyset}}{r}\right) - \rho_s\left(\frac{\partial u_r}{\partial t} + u_r\frac{\partial u_r}{\partial r}\right) = 0 \tag{8}$$

We introduce the rr component of the stress tensor $\sigma_{ij} = -P\delta_{ij} + \varepsilon_{ij}$ where δ_{ij} is Kronecker delta.

$$\sigma_{rr} = -P + \varepsilon_{rr} \tag{9}$$

816

Equation (8) can be written as

$$\frac{\partial \sigma_{rr}}{\partial r} = \frac{-2\varepsilon_{rr} + \varepsilon_{\theta\theta} + \varepsilon_{\emptyset\emptyset}}{r} + \rho_s \left(\frac{\partial u_r}{\partial t} + u_r \frac{\partial u_r}{\partial r} \right) \tag{10}$$

Integration of this equation from R to $R+h$ where h is the sheet thickness gives

$$\Delta P = \int_R^{R+h} \left(\frac{2\varepsilon_{rr} - \varepsilon_{\theta\theta} - \varepsilon_{\emptyset\emptyset}}{r} - \rho_s u_r \frac{\partial u_r}{\partial r} \right) dr = \frac{h}{R} \left(2\varepsilon_{rr} - \varepsilon_{\theta\theta} - \varepsilon_{\emptyset\emptyset} - \rho_s R u_r \frac{\partial u_r}{\partial r} \right) \bigg|_{r=R} \tag{11}$$

The different components of the viscous stress tensor are

$$\varepsilon_{rr} = 2\mu \frac{\partial u_r}{\partial r}, \quad \varepsilon_{\theta\theta} = 2\mu \left(\frac{1}{r} \frac{\partial u_\theta}{\partial \theta} + \frac{u_r}{r} \right), \quad \varepsilon_{\emptyset\emptyset} = 2\mu \left(\frac{1}{r \sin\theta} \frac{\partial u_\emptyset}{\partial \emptyset} + \frac{u_r}{r} + \frac{\cot\theta \, u_\theta}{r} \right) \tag{12a}$$

where μ is sheet dynamic viscosity. Equations (12a) can be reduced to equations (13). We can alternatively rewrite equation (10) as

$$\frac{\partial \sigma_{rr}}{\partial r} = -\frac{4\mu}{r} \frac{\partial u_r}{\partial r} + 4\mu \frac{u_r}{r^2} + \rho_s \left(\frac{\partial u_r}{\partial t} + u_r \frac{\partial u_r}{\partial r} \right) = \rho_s \frac{\partial u_r}{\partial t} + \left(\rho_s u_r - \frac{4\mu}{r} \right) \frac{\partial u_r}{\partial r} + 4\mu \frac{u_r}{r^2} \tag{12b}$$

$$\varepsilon_{rr} = \frac{-4\mu \left| \frac{dR}{dt} \right| R^2}{r^3}, \quad \varepsilon_{\theta\theta} = \varepsilon_{\emptyset\emptyset} = \frac{2\mu \left| \frac{dR}{dt} \right| R^2}{r^3}, \quad u_r = \frac{\left| \frac{dR}{dt} \right| R^2}{r^2}, \quad \frac{\partial u_r}{\partial r} = -2 \frac{\left| \frac{dR}{dt} \right| R^2}{r^3} \tag{13}$$

Inserting these equations in equation (11) results in

$$\Delta P = -\frac{12\mu h \left| \frac{dR}{dt} \right|}{R^2} + \frac{2\rho_s h \left| \frac{dR}{dt} \right|^2}{R} = -2 \frac{h \left| \frac{dR}{dt} \right|}{R^2} \left(6\mu - \rho_s R \left| \frac{dR}{dt} \right| \right) \approx -\frac{12\mu h \left| \frac{dR}{dt} \right|}{R^2} \tag{14a}$$

$$\Delta P = \frac{pt}{c} \qquad 0 \le t \le c \tag{14b}$$

$$\Delta P = p \qquad c \le t \tag{14c}$$

$$\left| \frac{dR}{dt} \right| = -\frac{R^2}{12\mu h} \frac{pt}{c} \qquad 0 \le t \le c \tag{14d}$$

$$\left| \frac{dR}{dt} \right| = -\frac{R^2}{12\mu h} p \qquad t > c \tag{14e}$$

where pressure $p = 100{,}000$ Pa and $c = 0.01$ is a constant for the ramp function. Next, we transform equation (14) into non-dimensional form using non-dimensional sheet thickness $\eta = h/h_0$, non-dimensional sheet radius of curvature $\rho = R/r_0$ and non-dimensional sheet aspect ratio $\gamma = h_0/r_0$ where h_0 is initial thickness of the plastic sheet and r_0 is the radius of the opening through which the plastic is formed.

$$\left| \frac{dR}{dt} \right| = r_0 \left| \frac{d\rho}{dt} \right|, \quad R^2 = \rho^2 r_0^2 \tag{15}$$

$$\left| \frac{d\rho}{dt} \right| = \frac{\Delta P \rho^2}{12\mu\eta\gamma} - \frac{\rho_s r_0^2}{6\mu} \rho \left| \frac{d\rho}{dt} \right|^2 \tag{16}$$

Introducing non-dimensional time as $\tau = \Delta P t / 12\mu$ and non-dimensional radial velocity as $U = 12\mu u_r / \Delta P r_0$ we get from equations (6) and (16)

$$\left|\frac{d\rho}{dt}\right| = \frac{\Delta P}{12\mu}\left|\frac{d\rho}{d\tau}\right| \tag{17}$$

$$U = \left|\frac{d\rho}{d\tau}\right| = \frac{\rho^2}{\eta\gamma} - \delta\rho\left|\frac{d\rho}{d\tau}\right|^2, \quad \delta = \frac{\rho_s r_0^2 \Delta P}{72\mu^2} \tag{18a}$$

$$U = \left|\frac{d\rho}{d\tau}\right| = \frac{\rho^2}{\eta\gamma}, \quad \delta = 0 \tag{18b}$$

Variable	Symbol	Definition	Units
Sheet surface area	A		m^2
Height of sheet	a		m
Non-dimensional height of sheet	α	a/r_0	
Angle at edge of sheet	β		rad
Sheet thickness	h		m
Non-dimensional sheet aspect ratio	γ	h_0/r_0	
Non-dimensional sheet thickness	η	h/h_0	
Initial sheet thickness	h_0		m
Non-dimensional curvature	κ	r_0/R	
Pressure	P		Pa
Pressure difference	ΔP		Pa
Radial coordinate	r		m
Radius of opening	r_0		m
Radius of curvature	R		m
Non-dimensional radius of curvature	ρ	R/r_0	
Sheet density	ρ_s		kg/m^3
Azimuthal coordinate	θ		rad
Meridional coordinate	$\emptyset$		rad
Sheet dynamic viscosity	μ		kg/m-s
Viscous stress tensor	ε_{ij}		Pa
Non-dimensional viscous stress tensor	E_{ij}	$\varepsilon_{ij}/\Delta P$	
Stress tensor	σ_{ij}		Pa
Time	t		s
Non-dimensional time	τ	$\Delta Pt/12\mu$	
Radial velocity	u_r		m/s
Non-dimensional radial velocity	U	$12\mu u_r/\Delta Pr_0$	
Azimuthal velocity	u_θ		m/s
Meridional velocity	$u_\emptyset$		m/s
Sheet volume	V		m^3

Table 31.1 Dimensional and non-dimensional variables

The lenticular plastic sheet thickness can be derived from the volume of a thin spherical sheet defined as the surface area at radius $r = R$ times the sheet thickness h which equals the original undeformed volume of the sheet.

$$V = Ah = 2\pi R\left(R - \sqrt{R^2 - r_0^2}\right)h = \pi r_0^2 h_0 \tag{19a}$$

$$h = \frac{r_0^2 h_0}{2R\left(R - \sqrt{R^2 - r_0^2}\right)} \tag{19b}$$

We then get equation (19c) for the non-dimensional lenticular sheet thickness. We can alternatively express the sheet thickness in terms of sheet curvature $\kappa = 1/\rho$.

$$\eta = \frac{h}{h_0} = \frac{r_0^2}{2R\left(R - \sqrt{R^2 - r_0^2}\right)} = \frac{1}{2\rho\left(\rho - \sqrt{\rho^2 - 1}\right)} = \frac{1}{2}\left(1 + \sqrt{1 - \kappa^2}\right), \qquad \eta \geq 1/2 \tag{19c}$$

We can rewrite equations (14d,e) for a lenticular sheet

$$\left|\frac{dR}{dt}\right| = -\frac{R^2}{12\mu h}\frac{pt}{c} = -\frac{pR^3\left(R - \sqrt{R^2 - r_0^2}\right)t}{6\mu r_0^2 c} \qquad 0 \leq t \leq c \tag{19d}$$

$$\left|\frac{dR}{dt}\right| = -\frac{R^2}{12\mu h}p = -\frac{pR^3\left(R - \sqrt{R^2 - r_0^2}\right)}{6\mu r_0^2} \qquad t > c \tag{19e}$$

The bulbous sheet volume can be expressed as

$$V = Ah = 4\pi R^2 - 2\pi R\left(R - \sqrt{R^2 - r_0^2}\right)h = 2\pi R\left(R + \sqrt{R^2 - r_0^2}\right)h = \pi r_0^2 h_0 \tag{20a}$$

$$h = \frac{r_0^2 h_0}{2R\left(R + \sqrt{R^2 - r_0^2}\right)} \tag{20b}$$

and we derive equation (20b) for the non-dimensional bulbous sheet thickness.

$$\eta = \frac{h}{h_0} = \frac{r_0^2}{2R\left(R + \sqrt{R^2 - r_0^2}\right)} = \frac{1}{2\rho\left(\rho + \sqrt{\rho^2 - 1}\right)} = \frac{1}{2}\left(1 - \sqrt{1 - \kappa^2}\right), \qquad \eta \leq 1/2 \tag{20c}$$

and we can also rewrite equations (14d,e) for a bulbous sheet

$$\left|\frac{dR}{dt}\right| = -\frac{R^2}{12\mu h}\frac{pt}{c} = -\frac{pR^3\left(R + \sqrt{R^2 - r_0^2}\right)t}{6\mu r_0^2 c} \qquad 0 \leq t \leq c \tag{20d}$$

$$\left|\frac{dR}{dt}\right| = -\frac{R^2}{12\mu h}p = -\frac{pR^3\left(R + \sqrt{R^2 - r_0^2}\right)}{6\mu r_0^2} \qquad t > c \tag{20e}$$

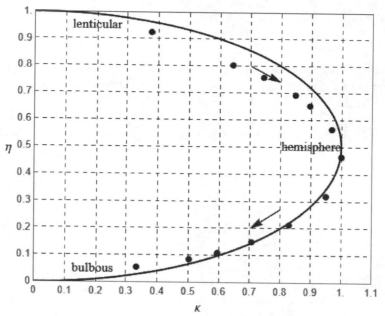

Figure 31.27b) Non-dimensional sheet thickness η versus curvature κ, • Ansys Polyflow

The non-dimensional radial velocity can be expressed for a lenticular shape as

$$U = \frac{\rho^2}{\eta\gamma} = \frac{\rho}{2\gamma\left(\rho - \sqrt{\rho^2-1}\right)} = \frac{1}{2\gamma\kappa^2}\left(1 + \sqrt{1 - \kappa^2}\right), \qquad \eta \geq 1/2 \qquad (21)$$

and for the bulbous shape we have the following expression

$$U = \frac{\rho^2}{\eta\gamma} = \frac{\rho}{2\gamma\left(\rho + \sqrt{\rho^2-1}\right)} = \frac{1}{2\gamma\kappa^2}\left(1 - \sqrt{1 - \kappa^2}\right), \qquad \eta \leq 1/2 \qquad (22)$$

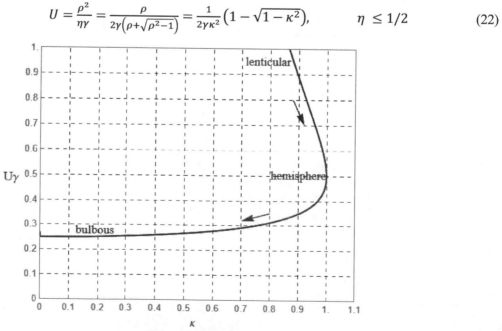

Figure 31.27c) Non-dimensional sheet thickness η versus curvature κ, • Ansys Polyflow

820

The tensile stress ratio TSR can also be determined.

$$TSR = \frac{\varepsilon_{rr}|_{r=R}}{\varepsilon_{rr}|_{r=R+h}} = \frac{(R+h)^3}{R^3} = (1+\frac{h}{R})^3 = (1+\gamma\eta\kappa)^3 \tag{23}$$

Figure 31.27d) Tensile stress ratio versus curvature κ

The non-dimensional height of the spherical plastic sheet is

$$\alpha = \rho - \sqrt{\rho^2 - 1} = \frac{1-\sqrt{1-\kappa^2}}{\kappa} \qquad \text{lenticular sheet} \quad \alpha \le 1 \tag{24}$$

$$\alpha = \rho + \sqrt{\rho^2 - 1} = \frac{1+\sqrt{1-\kappa^2}}{\kappa} \qquad \text{bulbous sheet} \quad \alpha \ge 1 \tag{25}$$

Figure 31.27e) Non-dimensional sheet height α versus curvature κ, • Ansys Polyflow

821

The angle at the edge of the spherical plastic sheet is

$$\beta = \frac{\pi}{2} - atan\frac{R-a}{r_0} = \frac{\pi}{2} - atan\sqrt{\rho^2 - 1} = \frac{\pi}{2} - atan\sqrt{\kappa^{-2} - 1} \qquad \text{lenticular sheet } \beta \leq \frac{\pi}{2} \qquad (26)$$

$$\beta = \frac{\pi}{2} + atan\frac{a-R}{r_0} = \frac{\pi}{2} + atan\sqrt{\rho^2 - 1} = \frac{\pi}{2} + atan\sqrt{\kappa^{-2} - 1} \qquad \text{bulbous sheet } \beta \geq \frac{\pi}{2} \qquad (27)$$

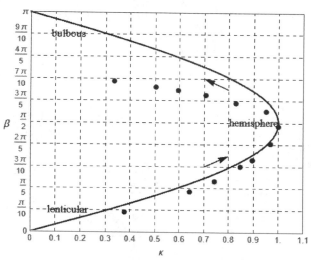

Figure 31.27f) Angle at edge of sheet β versus curvature κ, • Ansys Polyflow

We will now substitute equations (20) and (22) in equation (18) and we get the following equations for the lenticular sheet.

$$\left|\frac{d\rho}{d\tau}\right| = \frac{2\rho^3\left(\rho - \sqrt{\rho^2 - 1}\right)}{\gamma} \qquad (28)$$

$$\frac{d\tau}{d\rho} = -\frac{\gamma}{2\rho^3\left(\rho - \sqrt{\rho^2 - 1}\right)} \qquad (29)$$

$$\tau = \frac{\gamma}{4}\left[\frac{2}{\rho} + \frac{\sqrt{\rho^2 - 1}}{\rho^2} + 2atan\left(\rho - \sqrt{\rho^2 - 1}\right)\right] \qquad (30)$$

$$\tau = \frac{\gamma}{4}\left[\kappa\left(2 + \kappa\sqrt{\kappa^{-2} - 1}\right) - 2atan\left(\sqrt{\kappa^{-2} - 1} - \kappa^{-1}\right)\right] \qquad (31)$$

Transition from lenticular to bulbous sheet occurs for $\kappa = 1$ at nondimensional time

$$\tau = \frac{\gamma}{8}(4 + \pi) \qquad (32)$$

For the bulbous sheet we have the following corresponding equations

$$\left|\frac{d\rho}{d\tau}\right| = \frac{2\rho^3\left(\rho + \sqrt{\rho^2 - 1}\right)}{\gamma} \qquad (33)$$

$$\frac{d\tau}{d\rho} = \frac{\gamma}{2\rho^3\left(\rho + \sqrt{\rho^2 - 1}\right)} \qquad (34)$$

$$\tau = \frac{\gamma}{4}\left[4 - \frac{\pi}{2} - \frac{2}{\rho} + \frac{\sqrt{\rho^2-1}}{\rho^2} + 2atan\left(\frac{\sqrt{\rho^2-1}}{\rho-1}\right)\right] \tag{35}$$

$$\tau = \frac{\gamma}{4}\left[4 - \frac{\pi}{2} - 2\kappa + \kappa^2\sqrt{\kappa^{-2}-1} + 2atan\left(\frac{\sqrt{\kappa^{-2}-1}}{\kappa^{-1}-1}\right)\right] \tag{36}$$

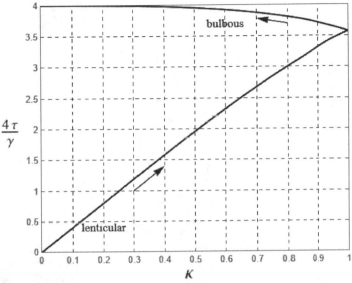

Figure 31.27g) Non-dimensional time τ versus curvature κ

The components of the viscous stress tensor in equation (13) can be written in non-dimensional form at $r = R$ using equations (15) and (17)

$$E_{rr}\Delta P = \frac{-4\mu\left|\frac{dR}{dt}\right|R^2}{R^3} = \frac{-4\mu r_0\left|\frac{d\rho}{dt}\right|}{R} = \frac{-\Delta P}{3\rho}\left|\frac{d\rho}{d\tau}\right| \tag{37}$$

$$E_{rr} = -\frac{1}{3\rho}\left|\frac{d\rho}{d\tau}\right| \tag{38}$$

$$E_{\theta\theta}\Delta P = E_{\emptyset\emptyset}\Delta P = \frac{2\mu\left|\frac{dR}{dt}\right|R^2}{R^3} = \frac{2\mu r_0\left|\frac{d\rho}{dt}\right|}{R} = \frac{\Delta P}{6\rho}\left|\frac{d\rho}{d\tau}\right| \tag{39}$$

$$E_{\theta\theta} = E_{\emptyset\emptyset} = \frac{1}{6\rho}\left|\frac{d\rho}{d\tau}\right| \tag{40}$$

Using equation (28) for a lenticular sheet we get the following components of the viscous stress tensor

$$E_{rr} = -\frac{1}{3\rho}\left|\frac{d\rho}{d\tau}\right| = -\frac{2\rho^2\left(\rho-\sqrt{\rho^2-1}\right)}{3\gamma} \tag{41}$$

$$E_{\theta\theta} = E_{\emptyset\emptyset} = \frac{1}{6\rho}\left|\frac{d\rho}{d\tau}\right| = \frac{\rho^2\left(\rho-\sqrt{\rho^2-1}\right)}{3\gamma} \tag{42}$$

and for a bulbous sheet

$$E_{rr} = -\frac{1}{3\rho}\left|\frac{d\rho}{d\tau}\right| = -\frac{2\rho^2\left(\rho+\sqrt{\rho^2-1}\right)}{3\gamma} \tag{43}$$

$$E_{\theta\theta} = E_{\emptyset\emptyset} = \frac{1}{6\rho}\left|\frac{d\rho}{d\tau}\right| = \frac{\rho^2\left(\rho+\sqrt{\rho^2-1}\right)}{3\gamma} \tag{44}$$

Transition from lenticular to bulbous sheet occurs for $\kappa = 1$ at nondimensional viscous stress $E_{\theta\theta} = E_{\emptyset\emptyset} = \frac{1}{3\gamma}$. The minimum viscous stress is $E_{\theta\theta} = E_{\emptyset\emptyset} = \frac{4}{9\sqrt{3}\gamma}$ at $\kappa = \frac{\sqrt{3}}{2}$.

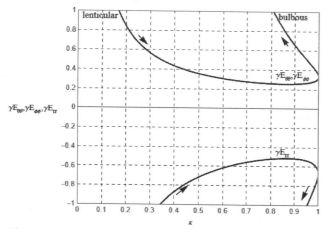

Figure 31.27h) Non-dimensional viscous stresses $\gamma E_{rr}, \gamma E_{\theta\theta}, \gamma E_{\emptyset\emptyset}$ versus curvature κ

J. References

1. Baird D.G. and Collias D.I. "Polymer Processing Principles and Design", Wiley & Sons, New York, 1998.
2. Kershner M. "Thermoforming Plastic Cups and Cones", Master's Thesis, University of Wisconsin-Madison, 2007.
3. Cengel Y.A. and Cimbala J.M. "Fluid Mechanics, Fundamentals and Applications", 4th Ed., McGraw Hill, 2018.

K. Exercises

31.1 Run the simulations in this chapter but use a negative pressure -100000 Pa instead of the positive pressure.

31.2 Run the simulations in this chapter but use a thickness of the plastic sheet of 0.00075 m instead of 0.00035 m.

31.2 Run the simulations in this chapter but use a square 0.3 m x 0.3 m geometry instead of a circular shape of the plastic sheet.

CHAPTER 32. BLOW MOLDING OF PLASTIC BOTTLE

A. Objectives

- Using Ansys Polyflow to Study Blow Molding Process of a Plastic Preform
- Inserting Boundary Conditions
- Running Ansys Polyflow Simulations
- Using Contour Plots and Animations for Visualizations of Blow Molding

B. Problem Description

We will in this chapter study the blow molding process of a plastic preform. The two mold halves will be moving towards the plastic preform. Pressure will be applied to form the preform into a bottle.

C. Launching Ansys Workbench and Selecting Polyflow

1. Start by launching Ansys Workbench. Launch Fluid Flow (Polyflow) that is available under Analysis Systems in Ansys Workbench. Right click on Geometry in Project Schematic and select New DesignModeler Geometry to start DesignModeler.

Figure 32.1 Launching Ansys Polyflow

D. Launching Ansys DesignModeler

2. Select **Units>>Meter** from the menu in DesignModeler. Select File>>Import External Geometry File…. Open the file *BlowMoldingBottleII.STEP* and select Generate. This file is available for download from *sdcpublications.com*.

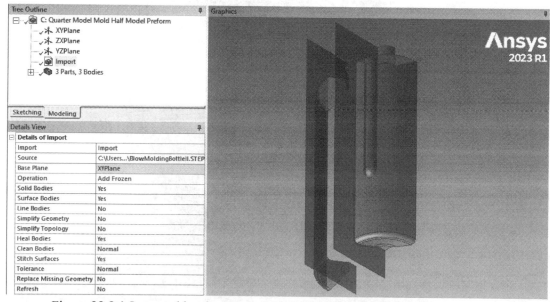

Figure 32.2a) Imported bottle and preform geometry in DesignModeler

Right-click in graphics window and choose Selection Filter>>Edge. Control select the 10 edges as shown in Figure 32.2b). The total length of the 10 edges is 0.28542 m as shown at the bottom of the screen. This is the theoretical line length of formed bottle *LFB* as defined in the theory section. Right-click in graphics window and choose Selection Filter>>Edge. Control select the 2 edges as shown in Figure 32.2b). The total length of the 2 edges is 0.131 m as shown at the bottom of the screen. This is the theoretical line length before forming *LBF*.

Select Tools>>Options from the menu and select Geometry under DesignModeler in the Options window. Set the Measure Selection Limit to 50 under Selection. Select Graphics under DesignModeler in the Options window. Select Both Sides as Highlight Selection under Graphics. Click OK to close the window.

Right-click in the graphics window and choose Selection Filter>>Face. Control select the 10 faces of the mold as shown in Figure 30.2c). The total surface area for the faces is 0.014519 m². This is the theoretical surface area of formed bottle *SAFB* as defined in the theory section. Control select the 2 faces of the preform as shown in Figure 30.2c). The surface area of the undeformed preform is 0.0027929 m². Close DesignModeler.

Figure 32.2b) Edges on mold and preform symmetry lines

Figure 32.2c) Faces on mold and preform

E. Launching Ansys Meshing

3. Double click on Mesh under Project Schematic in Ansys Workbench. Open Geometry under Model (A3) in the Outline. Right-click on the first part, select Rename and name the first part as MoldFront. Open Graphic Properties in Details of "Mold" and select blue Color. Rename the second part as MoldBack and select green color. Rename the third part as Preform and select red color.

   Select Mesh in the Outline. Select the Home tab in the menu and select Tools>>Units>>Metric (m, kg, N, s, V, A). Set the Element Size to 1.e-002 m under Defaults in Details of "Mesh". Right-click on Mesh under Project and Model (A3) in the Meshing window and select Update. Click on the plus sign next to Statistics in Details of "Mesh". The mesh has 7519 Nodes and 7317 Elements.

Figure 32.3a) Mesh for molds and preform

Right click in the graphics window and select View>>Right. Select the top half-circular edge of the red preform, right click and select Create Named Selection. Enter the name *Symmetry1* and click OK to close the window. Select the remaining four edges of the preform, right click and select Create Named Selection. Enter the name *Symmetry2* and click OK to close the window. Make sure that Geometry lists 4 Edges under Scope in Details of "Symmetry 2" when you select Symmetry2 under Named Selections in the Outline.

Figure 32.3b) Named selection

Select File>>Export…>>Mesh>>POLYFLOW Input File>>Export from the menu. Save the mesh with the name *blowmoldingbottle.poly*. Select File>>Save Project from the menu and save the project with the name *BottleBlowMoldingProject*. Close the meshing window. Right click on Mesh in the Project Schematic and select Update.

F. Launching Ansys Polydata

4. Double click on Setup under Project Schematic in Ansys Workbench. Select the Mesh tab at the bottom of the middle Polydata section. Select Grid as Display option. Select the Menus tab at the bottom of the Polydata section. Select Create a new task in the Polydata section. Select F.E.M. task, Time-dependent problem and 2D shell geometry. Select Accept the current setup.

Figure 32.4a) Polydata main menu

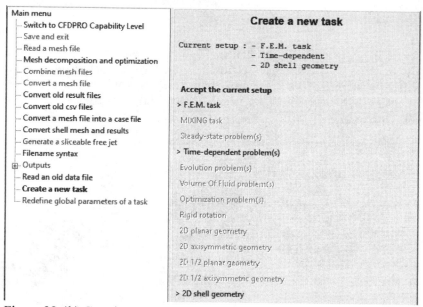

Figure 32.4b) Creating a new task

5. Select Define molds in the F.E.M. Task 1 section and select Create a new mold in the Define molds window.

Figure 32.5a) Defining a mold

Figure 32.5b) Creating a new mold

Select Mold with constant and uniform temperature in the Create a new mold window. Enter the New value: Mold-Front.

Figure 32.5c) Selection of a mold with constant and uniform temperature

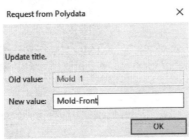

Figure 32.5d) Title for first mold

Select Domain of the mold in the Mold-Front window. Select matid_4 and select Remove in the Domain of the mold window. Select matid_6 and select Remove in the Domain of the mold window. Select Upper level menu.

Main menu > F.E.M. Task 1 > Define molds > Mold-Front

Main menu
- Save and exit
- Read a mesh file
- Mesh decomposition and optimization
- Combine mesh files
- Convert a mesh file
- Convert old result files
- Convert old csv files
- Convert a mesh file into a case file
- Convert shell mesh and results
- Generate a sliceable free jet
- Filename syntax
- Outputs
- Read an old data file

Mold-Front

Mold with constant and uniform temperature

Upper level menu

Domain of the mold

Material data

Thermal boundary conditions

Contact conditions

Interpolation

Mold motion

Figure 32.5e) Selecting domain for the mold

Figure 32.5f) Defining the mold

Select Contact conditions in the Mold-Front window. Select No contact along matid_8 and select Modify. Select Contact in the Contact condition along matid_8 window. Select Upper level menu twice.

Figure 32.5g) Selection of contact conditions for the mold

Figure 32.5h) Modifying contact

Figure 32.5i) Including contact for the mold surface

Select Mold motion in the Mold-Front window. Select Modify motion type: fixed mold. Choose 1 – Translational velocity imposed in the Specify the type of mold motion window and select OK to close the same window.

Figure 32.5j) Selecting mold motion

Mold motion

```
Actual motion of the mold is enabled
The mold is fixed (no force, no velocity imposed).


Reset to default values

Disable actual motion

Upper level menu

Modify motion type : fixed mold
```

Figure 32.5k) Selection of motion type

Specify the type of mold motion   ✕

0 - Fixed mold (no velocity, no force imposed)
1 - Translation velocity imposed
2 - Translation force imposed
3 - General velocity driven motion

Update type of mold motion (between 0 and 3).

Old value: | 0 |

New value: | 1 |

| OK |

Figure 32.5l) Imposing translational velocity for the mold

Select the EVOL tab in the menu and choose Modify translational velocity in the Mold
motion window. Leave the value for the translational velocity-x as 0 and click OK to
close the Request from Polydata window. Select Upper level menu in the Time
dependence of the velocity-x window. Set translational velocity-y to 0 and click OK in
the Request from Polydata window.

Figure 32.5m) High lighting EVOL in the menu and selecting translational velocity

Figure 32.5n) Translational velocity-x

Figure 32.5o) Translational velocity–y

Select Upper level menu in the Time dependence of the velocity-y window. Set translational velocity-z to 0.189473 and click OK in the Request from Polydata window. Select f(t) = Ramp function in the Time dependence of the velocity-z window. Select Modify the value of a in the Time dependence of the velocity-y window. Enter 0.09 as the value for a. Click OK in the Request from Polydata window. Modify the values for b = -1, c = 0.1 and d = 0 in the same way. Deselect the EVOL tab in the menu.

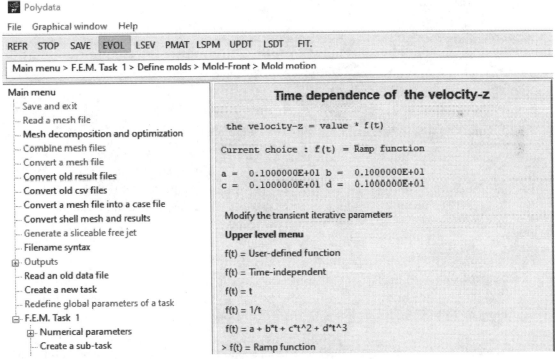

Figure 32.5p) Selection of ramp function as transient motion

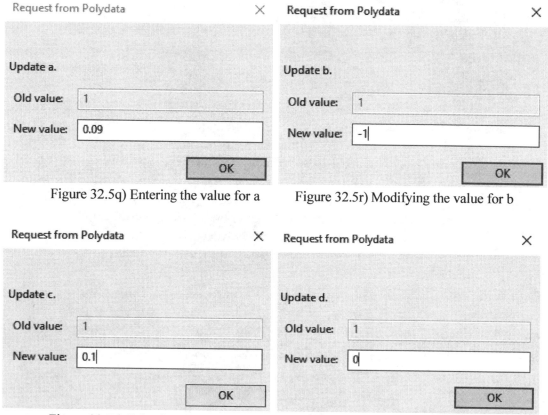

Figure 32.5q) Entering the value for a   Figure 32.5r) Modifying the value for b

Figure 32.5s) Selecting the value for c   Figure 32.5t) Editing the value for d

Select Upper level menu three times in the Mold motion window to arrive at the Define Molds window. Select Create a new mold. Answer No to the question whether you want to copy the data of an existing mold. Select Mold with constant and uniform temperature in the Create a new mold section. Enter the name Mold-Back as New value: and select OK to close the Request from Polydata window.

Select Domain of the mold in the Mold-Back section. Select matid_4 and Remove this selection in the Domain of the Mold. Select matid_8 and Remove this selection in the Domain of the Mold, see Figure 32.5u). Select Upper Level menu once.

Select Contact conditions in the Mold-Back section. Select No contact along matid_6 and select Modify. Select Contact and select Upper level menu twice.

Select Mold motion in the Mold-Back section. Select Modify motion type: fixed mold. Choose 1 – Translational velocity imposed in the Specify the type of mold motion window and select OK to close the same window.

Figure 32.5u) Defining domain of back mold half

Select the EVOL tab in the menu and choose Modify translational velocity in the Mold motion window. Leave the value for the translational velocity-x as 0 and click OK to close the Request from Polydata window. Select Upper level menu in the Time dependence of the velocity-x window. Set translational velocity-y to 0 and click OK in the Request from Polydata window.

Select Upper level menu in the Time dependence of the velocity-y window. Set translational velocity-z to 0.189473 and click OK in the Request from Polydata window. Select f(t) = Ramp function in the Time dependence of the velocity-z window. Select Modify the value of a in the Time dependence of the velocity-y window. Enter 0.09 as the value for a. Click OK in the Request from Polydata window. Modify the values for b = 1, c = 0.1 and d = 0 in the same way. Deselect the EVOL tab in the menu.

Select Upper level menu four times in the Mold motion window to arrive at the F.E.M. Task 1 window.

6.   Select Create a sub-task in the F.E.M. Task 1 window. Select Shell model: Gen. Newtonian isothermal. Enter the New value: Preform in the Request from Polydata window and select OK to close the window.

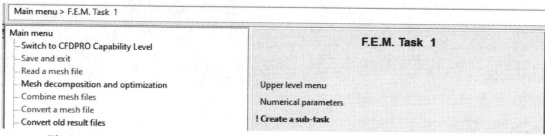

Figure 32.6a) Creating a sub-task

Figure 32.6b) Choosing shell model

Figure 32.6c) Updating the title for the sub-task

Select Domain of the sub-task. Select matid_6 and Remove in the Domain of the sub-task window. Select matid_8 and Remove in the Domain of the sub-task window. Select Upper Level Menu.

Figure 32.6d) Selecting domain for the sub-task

Figure 32.6e) Removing matid_6 and matid_8

Select Flow boundary conditions and select Zero wall velocity (vn=vs=0) along SYMMETRY1. Select Modify and select Plane of symmetry (fs=0,vn=0). Select normal direction along Y axis. Select Upper Level Menu. Select Zero wall velocity (vn=vs=0) along SYMMETRY2. Select Modify and select Plane of symmetry (fs=0,vn=0). Select normal direction along X axis. Select Upper Level Menu.

Figure 32.6f) Selecting flow boundary conditions

Figure 32.6g) Selecting symmetry boundary condition

Flow boundary condition along SYMMETRY1

Current choice : Plane of symmetry
                 No force postprocessor

Enable force postprocessor

Upper level menu

Zero wall velocity (vn=vs=0)

Slip conditions

Porous wall

> Plane of symmetry (fs=0, vn=0)

Figure 32.6h) Selecting plane of symmetry (fs=0,vn=0)

Plane of symmetry along SYMMETRY1

boundary condition : (fs,vn) = (0,0)
normal to plane of symmetry :
(nx,ny,nz)= 0.0000000E+00 0.1000000E+01 0.0000000E+00

Upper level menu

Switch to computed normals

normal direction along X axis

> normal direction along Y axis

normal direction along Z axis

arbitrary normal direction

Figure 32.6i) Selecting normal direction

Check the box for Inflation pressure imposed in the Flow boundary conditions window. Select constant in the Flow boundary conditions window. Enter -0.9e6 Pa as the New value: and click OK. Select EVOL in the menu and select Upper level menu in the Flow boundary conditions window. Select f(t) = Ramp function. Select Modify the value of a and enter 0.1 as the New value: and click OK. Enter the values 0, 0.2 and 1.0 as b, c, and d values. Select Upper level menu and deselect EVOL in the menu. Select Upper level menu once again.

Request from Polydata   ✕

Update constant.

Old value: 0

New value: -0.9e6

                                              OK

Figure 32.6j) Selecting inflation pressure

Time dependence of inflation pressure

```
inflation pressure = value * f(t)

Current choice : f(t) = Ramp function

a =  0.1000000E+00 b =  0.0000000E+00
c =  0.2000000E+00 d =  0.1000000E+01
```

Figure 32.6k) Entered values for a – d

Select Define contacts in the Preform window. Select create a new contact problem and click on Select a contact wall.

Preform

```
Shell model : Gen. Newtonian isothermal
```

Upper level menu

Domain of the sub-task

Material data

Flow boundary conditions

Global remeshing

Rigid translation

Interpolation

Define contacts

! Define layers

Figure 32.6l) Defining contacts

Define contacts

Delete a contact problem

Upper level menu

Create a new contact problem

Figure 32.6m) Creating a new contact problem

Modification of a contact problem

```
Mold and contact wall are undefined !
Contact release not activated
Darts are pointing towards the mold body
```

Upper level menu

! Select a contact wall

Figure 32.6n) Selecting a contact wall

Select Mold-Front: Contact along matid_8 and click on Select in the Contact wall selection window. Select Modify penetration accuracy and enter 0.001 as the New value.

Figure 32.6o) Choosing Mold-Front as contact wall

Figure 32.6p) Modified value for penetration accuracy

Select Specify mold side / cavity side in the Modification of a contact problem window. Answer No to the question "Are darts pointing towards the mold body?". Select from the menu Graphical window>>Sizing Darts>>Size up four times. Rotate the view to display the front mold, see Figure 32.6q).

Select Upper level menu once. Select create a new contact problem and click on Select a contact wall. Select Mold-Front: Contact along matid_6 and click on Select in the Contact wall selection window. Select Modify penetration accuracy and enter 0.001 as the New value. Select Specify mold side / cavity side in the Modification of a contact problem window. Answer No to the question "Are darts pointing towards the mold body?". Select from the menu Graphical window>>Sizing Darts>>Size down four times.

Figure 32.6q) Darts directed away from front and back mold bodies

Select Upper level menu two times. Select OK if you get the following message:

```
The velocity prediction must be disabled
if some contacts are defined.
>> modification automatically done
```

Select Define layers in the Preform window and select Create a new layer in Define layers window. Enter preform as the New value: and click OK.

Figure 32.6r) Defining layers

Figure 32.6s) Creating a new layer

Select Material data in the preform window and select Shear-rate dependence of viscosity in the Material data window.

Figure 32.6t) Selection of material data

Figure 32.6u) Selection of shear-rate dependence of viscosity

Select Constant viscosity in the Shear-rate dependence of viscosity window. Select Modify fac. and enter 6622 Pa·s in the Request from Polydata window. Select Upper level menu twice.

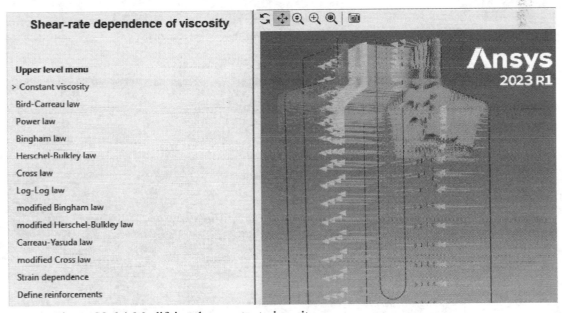

Figure 32.6v) Modifying the constant viscosity

Select Density in the Materials data window and select Modification of density in the next window. Enter 960 kg/m^3 as New value. Select Upper level menu.

Select Inertia terms in the Materials data window and select Inertia will be taken into account in the Inertia terms window. Select Upper level menu twice. Select Initial thickness in the film window and select Constant in the Initial thickness window. Enter 0.002 m as initial thickness. Select Upper level menu four times.

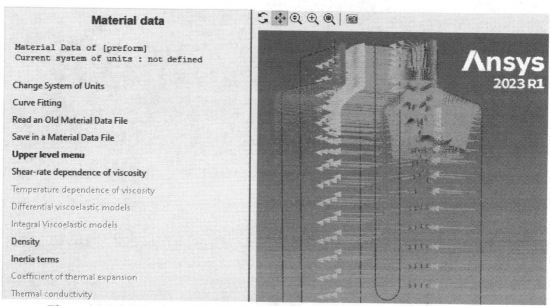

Figure 32.6x) Selection of Density in the Materials data window

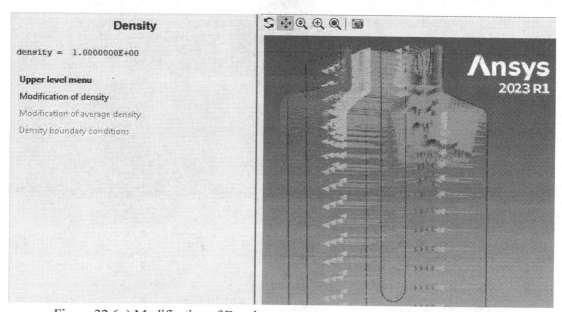

Figure 32.6y) Modification of Density

Figure 32.6z) Taking into account inertia terms

7. Select Numerical Parameters in the F.E.M. Task 1 window. Select Modify the transient iterative parameters.

Figure 32.7a) Selecting numerical parameters

Figure 32.7b) Modifying transient iterative parameters

Select Modify the min value of the time-step and enter 1e-6 as the value. Select Upper level menu three times.

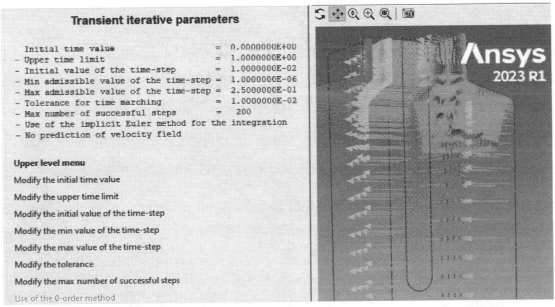

Figure 32.7c) Transient iterative parameters

8. Select Outputs in the Polydata window and select Enable CFD-Post output in the Outputs window (this may already be enabled). Select Set units for CFD-Post, Ansys Mapper or Iges. Make sure that the units are set to metric_MKSA+Kelvin. Select Upper level menu twice.

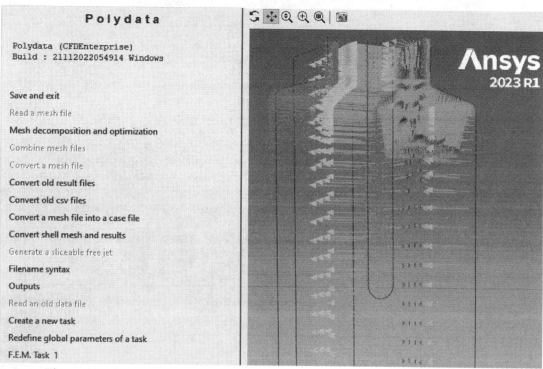

Figure 32.8a) Selection of outputs

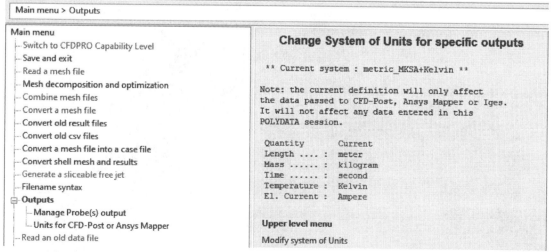

Figure 32.8b) Unit system

9. Select Save and exit in the Polydata window. Select Accept in the Field Management window and Continue.

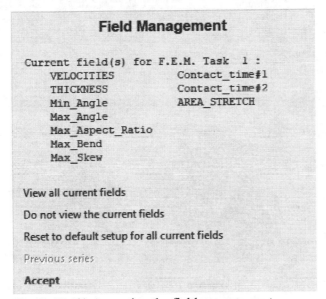

Figure 32.9a) Saving and exiting Polydata

Figure 32.9b) Accepting the field management

Figure 32.9c) Continuing the field management

10. Right click on Solution in the Project Schematic and select Update. Right click on Solution in the Project Schematic once again and select Listing Viewer…. Check the box for Refresh every and change the value from 10 to 5 seconds. Close the window when computations have finished.

Figure 32.10a) Updating the solution

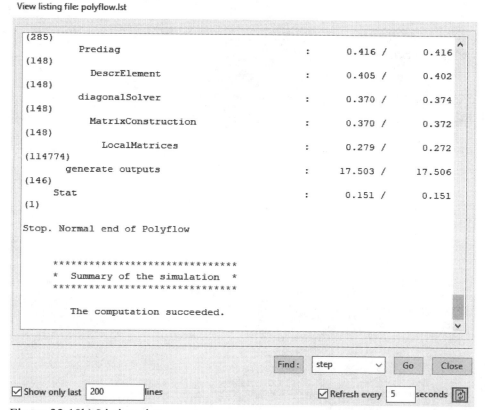

Figure 32.10b) Listing viewer

G. CFD-Post

11. Right click on Results in the Project Schematic and select Edit. Select Insert>>Contour from the menu in CFD-Post and click OK in the Insert Contour window.

Figure 32.11a) Viewing results   Figure 32.11b) Inserting a contour in CFD-Post

12. Select matid_4_surf as Locations in the Geometry Details of Contour 1. Select THICKNESS as Variable. Select Global as Range. Select the Render tab under Details of Contour 1 and uncheck the box for Show Contour lines.

Figure 32.12a) Details of Contour 1

Uncheck the box for Wireframe under User Locations and Plots in the Outline. Click on Apply twice at the bottom left of Details of Contour 1.

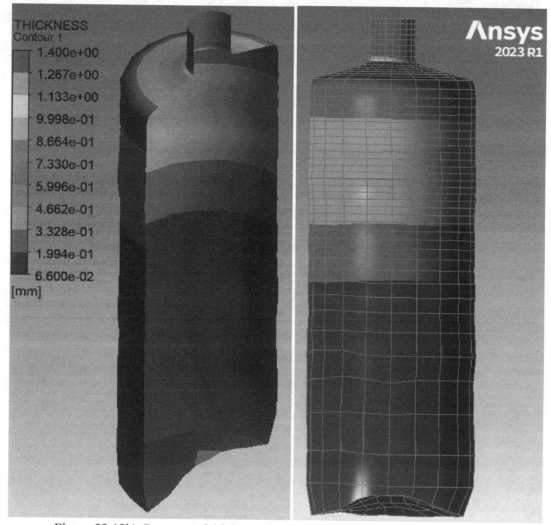

Figure 32.12b) Contours of thickness for blow molded bottle

Double click on Default Transform under User Locations and Plots in the Outline. Uncheck the box for Instancing Info From Domain in the Details of Default Transform. Set the Number of Graphical Instances to 1. Select Value as Determine Angle From under Instance Definition. Set the Angle to 180 [degree] under Instance Definition. Check the box for Apply Reflection. Select YZ Plane as Method and set X value in [mm] to -12.18 under Apply Reflection. Click on Apply at the bottom left of Details of Default Transformation.

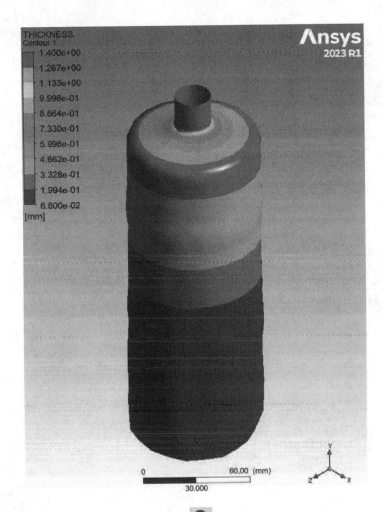

13. Select the Timestep Selector from the menu. Select Step 1 in the Timestep Selector and click Apply. Repeat this for Steps 15, 20, 25, 30, 35, 40, 45 and 73.

Figure 32.13a) Time selector

Figure 32.13b) Contours of thickness at steps 1, 15, 20, 25, 30, 35, 40 and 45, t = 0.01 s, 0.120725 s, 0.124918 s, 0.125667 s, 0.126473 s, 0.127282 s, 0.127873 s and 0.128773 s.

14. Select the Animation from the menu. Select Timestep Animation as Type. Check the box for Save Movie. Browse and save the file with the name *Blow Molding Bottle Thickness.wmv*. Set Repeat to 1 and click on play. Close the Animation window after completion and Close CFD-Post.

Figure 32.14 Details of animation

H. Refining the Mesh

15. Right click on Setup and select Edit… in the Workbench Project Schematic. Select F.E.M. Task 1 at the bottom of the Polydata window. Select Numerical parameters in the F.E.M. Task 1 window and select Adaptive meshing in the Numerical parameters window. Select Adaptive meshing for contacts in the Adaptive Meshing window. Select Enable all the local criteria in the Global criteria for contact window. Select Switch to 'Global Criteria' mode.

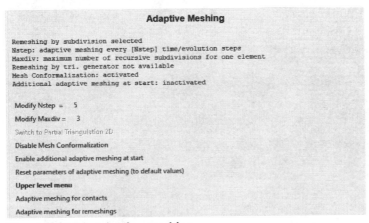

Figure 32.15a) Adaptive meshing

Select Modify size_min and enter 0.0001 as the New value and click OK to close the Request from Polydata window. Set tolerance to 0.00001, size_max to 0.001 and dist_crit to 0.0005. Select Upper level menu.

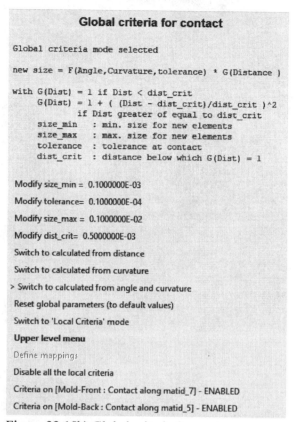

Global criteria for contact

```
Global criteria mode selected

new size = F(Angle,Curvature,tolerance) * G(Distance )

with G(Dist) = 1 if Dist < dist_crit
     G(Dist) = 1 + ( (Dist - dist_crit)/dist_crit )^2
          if Dist greater of equal to dist_crit
    size_min   : min. size for new elements
    size_max   : max. size for new elements
    tolerance  : tolerance at contact
    dist_crit  : distance below which G(Dist) = 1
```

Modify size_min = 0.1000000E-03

Modify tolerance= 0.1000000E-04

Modify size_max = 0.1000000E-02

Modify dist_crit= 0.5000000E-03

Switch to calculated from distance

Switch to calculated from curvature

> Switch to calculated from angle and curvature

Reset global parameters (to default values)

Switch to 'Local Criteria' mode

Upper level menu

Define mappings

Disable all the local criteria

Criteria on [Mold-Front : Contact along matid_7] - ENABLED

Criteria on [Mold-Back : Contact along matid_5] - ENABLED

Figure 32.15b) Global criteria for contact

Select Modify Nstep in the Adaptive Meshing window, set New value: to 4 and click OK to close the window. Select Modify Maxdiv and set the value to 3. Select Upper level menu three times. Select Save and exit, select Next Series and Accept in the Field Management window and select Continue in the Field Management window. Right click on Solution in the Workbench Project Schematic and select Update. Right click on solution and select Listing Viewer…. Check the box for Refresh every and set the time to 5 seconds.

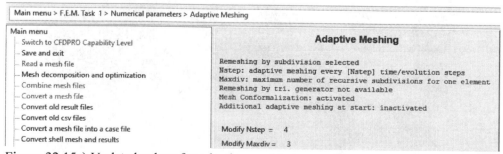

Main menu > F.E.M. Task 1 > Numerical parameters > Adaptive Meshing

Main menu
- Switch to CFDPRO Capability Level
- Save and exit
- Read a mesh file
- Mesh decomposition and optimization
- Combine mesh files
- Convert a mesh file
- Convert old result files
- Convert old csv files
- Convert a mesh file into a case file
- Convert shell mesh and results

Adaptive Meshing

```
Remeshing by subdivision selected
Nstep: adaptive meshing every [Nstep] time/evolution steps
Maxdiv: maximum number of recursive subdivisions for one element
Remeshing by tri. generator not available
Mesh Conformalization: activated
Additional adaptive meshing at start: inactivated
```

Modify Nstep =   4

Modify Maxdiv =   3

Figure 32.15c) Updated values for adaptive meshing

16. Right click on Results in the Workbench Project Schematic and select Edit…. Double click on Contour 1 under User Locations and Plots in the Outline.

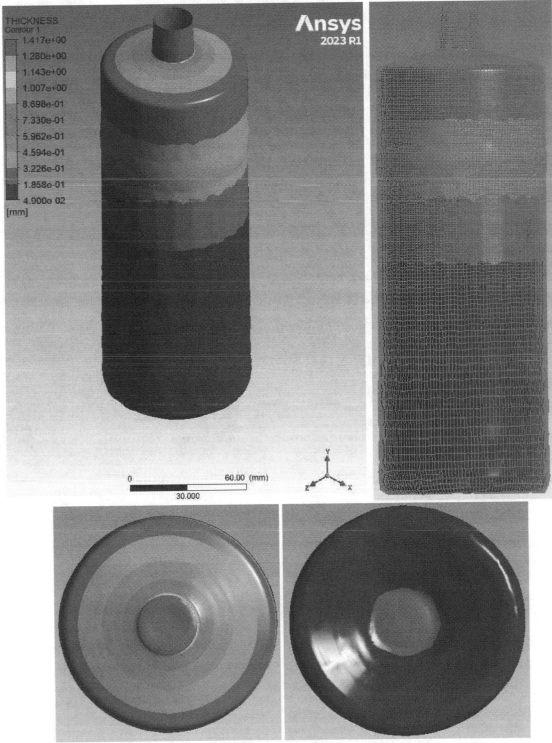

Figure 32.16 Refined mesh thickness contour plots at different views

17. Select the Animation ⬚ from the menu. Select Timestep Animation as Type. Check the box for Save Movie. Browse and save the file with the name *Blow Molding Bottle Thickness Refined Mesh.wmv*. Set Repeat to 1 and click on play. Close the Animation window after completion of the movie.

Select Insert>>Clip Plane from the menu. Accept the Name Clip Plane 1 and click OK to close the Insert Clip Plane window. Select YZ Plane a Method and set the X value in [mm] to -12.18 in Details of Clip Plane 1. Select Apply, right click in the graphics window and select Clip Scene>>Clip Plane 1.

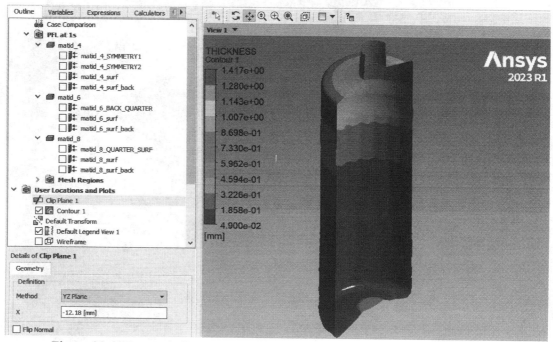

Figure 32.17 Inserted clip plane

18. Select Insert>>Location>>Iso Clip from the menu. Accept the Name Iso Clip 1 and click OK to close the Insert Iso Clip window. Select matid_4_surf from the drop-down menu next to Locations. Right click in the rectangular area under Visibility parameters and select New. Select THICKNESS as Variable under Visibility Parameters Properties and Visible where >= 0 [mm]. Select the Render tab under Details of Iso Clip 1. Check the box to Show Mesh Lines and click on Apply.

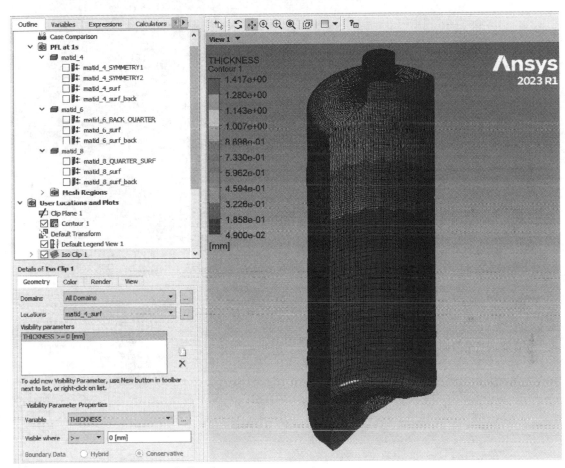

Figure 32.18 Inserted clip plane

19. Select Insert>>Location>>Point Cloud from the menu. Accept the Name Point Cloud 1 and click OK. Select matid_4_SYMMETRY2 as Locations under Details of Point Cloud 1. Set # of Points to 1000. Select the Color tab and set the color to black. Select the Symbol tab and choose Ball as symbol and set the Symbol Size to 0.15. Select Apply.

Figure 32.19a) Inserted point cloud

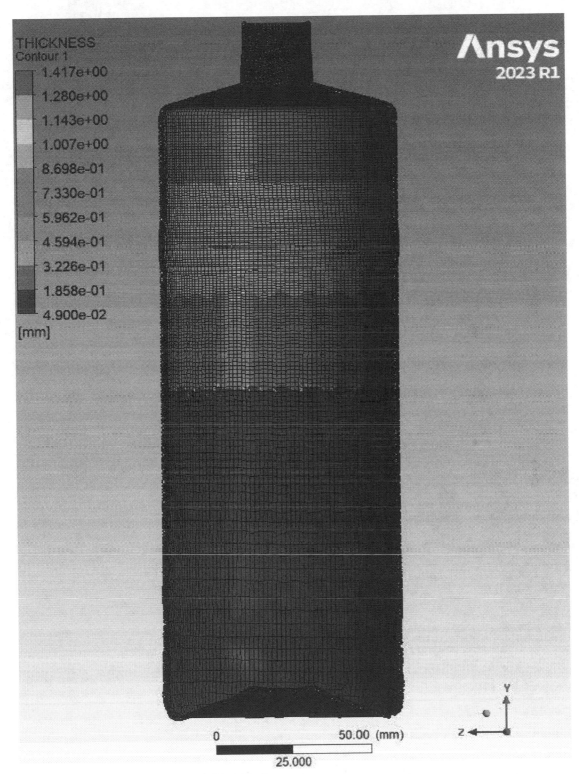

Figure 32.19b) Point cloud in YZ plane

20. Select File>>Export>>Export… from the menu. Select Point Cloud 1 as Locations and THICKNESS as variable. Check the boxes for Line and Face Connectivity and Node Numbers. Enter *bottlethicknessdata* as Name Aliases. Click on Save to save the file with the name *test1.csv*.

Figure 32.20 Exporting point cloud

21. Open the file *bottlethicknessdata.csv* in Excel. Save the file in Excel with the name *bottlethicknessdata.xlsx*. Delete the first five rows.

Insert a new column to the left of the THICKNESS [mm] column. Copy the Y [mm] data to the new column E. Select columns D and E (Z [mm] and Y [mm]) and select Insert>>Scatter>>Scatter from the menu in Excel. Insert Axis Titles labeled Z [mm] on the horizontal axis and Y [mm] on the vertical axis.

Copy the Y [mm] data to the new column G. Select the data in columns F and G (THICKNESS [mm] and Y [mm]). Select Insert>>Scatter>>Scatter from the menu in Excel. Insert Axis Titles labeled Z [mm] on the horizontal axis and Thickness [mm] on the vertical axis. Save the Excel document with the name Figures 32.21a) and 32.21b).

Figures 32.21a), b) Bottle coordinates a) and wall thickness distribution b)

Same the Excel document with the name Figure 32.21c). Select column D (Z [mm]) and select Home>>Editing>>Sort & Filter>>Sort Smallest to Largest. Select Expand the selection in the Sort Warning window and click on Sort. Delete rows 2 - 474.

Label column H as Distance along line [mm] and enter 0 in cell H2. Enter the following formula in cell H3: =H2+SQRT((C3-C2)^2+(D3-D2)^2). Drag cell H3 down to cell H20. Copy the value in cell H18 and paste the value in cell J1 (9.034692). Delete rows 2 – 18.

Select column C (Y [mm]) and select Home>>Editing>>Sort & Filter>>Sort Largest to Smallest. Select Expand the selection in the Sort Warning window and click on Sort.

Enter 0 in cell H2. Enter the following formula in cell H3: =H2+SQRT((C3-C2)^2+(D3-D2)^2). Drag cell H3 down the whole column. Copy the value in cell H408 and paste the value in cell K1 (236.017708). Delete rows 2 – 408.

Select column D (Z [mm]) and select Home>>Editing>>Sort & Filter>>Sort Smallest to Largest. Select Expand the selection in the Sort Warning window and click on Sort.

Enter 0 in cell H2. Enter the following formula in cell H3: =H2+SQRT((C3-C2)^2+(D3-D2)^2). Drag cell H3 down the whole column to cell H61. Copy the value in cell J1 and paste the value in cell H62. Copy the value in cell K1 and paste the value in cell H63. Add the values in rows H61 - H63. The total value in cell H64 is 285.295745.

Enter *LBF* [mm] in cell I1, *LFB* [mm] in cell J1 and *LDR* in cell K1. Enter 131 in cell I2, enter =H64 in cell J2 and enter =J2/I2 in cell K2. Enter 131 in cell H6, 285.42 in cell J6 and =J6/I6 in cell K6. Enter =100*ABS(J6-J2)/J6 in cell K9. There is a difference of 0.04 % between Ansys Polyflow simulation and theory for the Line Draw Ratio.

	A	B	C	D	E	F	G	H	I	J	K
1	Node Number	X [mm]	Y [mm]	Z [mm]	Y [mm]	THICKNESS [mm]	Y [mm]	Distance along line [mm]	LBF [mm]	LFB [mm]	LDR
2	270	-1.22E+01	-1.53E+02	1.03E+01	-1.53E+02	2.01E-01	-1.53E+02	0	131	285.295745	2.18
3	271	-1.22E+01	-1.53E+02	1.03E+01	-1.53E+02	2.01E-01	-1.53E+02	0.001179373			
4	267	-1.22E+01	-1.54E+02	1.12E+01	-1.54E+02	1.88E-01	-1.54E+02	0.959526041	Theory	Theory	Theory
5	264	-1.22E+01	-1.54E+02	1.20E+01	-1.54E+02	1.80E-01	-1.54E+02	1.920806788	LBF [mm]	LFB [mm]	LDR
6	261	-1.22E+01	-1.55E+02	1.29E+01	-1.55E+02	1.73E-01	-1.55E+02	2.936938428	131	285.42	2.18
7	257	-1.22E+01	-1.55E+02	1.34E+01	-1.55E+02	1.70E-01	-1.55E+02	3.437557906			
8	258	-1.22E+01	-1.55E+02	1.34E+01	-1.55E+02	1.70E-01	-1.55E+02	3.4387635			% diff.
9	254	-1.22E+01	-1.56E+02	1.47E+01	-1.56E+02	1.61E-01	-1.56E+02	5.039815938			0.04
10	255	-1.22E+01	-1.56E+02	1.47E+01	-1.56E+02	1.61E-01	-1.56E+02	5.041030838			
11	250	-1.22E+01	-1.56E+02	1.56E+01	-1.56E+02	1.60E-01	-1.56E+02	6.114213978	Theory	Theory	Theory
12	251	-1.22E+01	-1.56E+02	1.56E+01	-1.56E+02	1.60E-01	-1.56E+02	6.115423203	SAP [m^2]	SAFB [m^2]	ADR
13	247	-1.22E+01	-1.57E+02	1.66E+01	-1.57E+02	1.58E-01	-1.57E+02	7.196911782	0.0027929	0.029038	10.40
14	246	-1.22E+01	-1.57E+02	1.69E+01	-1.57E+02	1.58E-01	-1.57E+02	7.465970913			
15	243	-1.22E+01	-1.57E+02	1.76E+01	-1.57E+02	1.56E-01	-1.57E+02	8.274226558	Ansys	Ansys	Ansys
16	242	-1.22E+01	-1.57E+02	1.79E+01	-1.57E+02	1.56E-01	-1.57E+02	8.544798478	SAP [m^2]	SAFB [m^2]	ADR
17	239	-1.22E+01	-1.57E+02	1.87E+01	-1.57E+02	1.53E-01	-1.57E+02	9.354161084	0.0027929	0.0285847	10.23
18	238	-1.22E+01	-1.57E+02	1.92E+01	-1.57E+02	1.49E-01	-1.57E+02	9.891483095			
19	235	-1.22E+01	-1.58E+02	1.96E+01	-1.58E+02	1.47E-01	-1.58E+02	10.30406905			% diff.
20	234	-1.22E+01	-1.58E+02	1.99E+01	-1.58E+02	1.45E-01	-1.58E+02	10.59417088			1.59

Figure 32.21c) Data for thickness of formed plastic sheet

Select Functions Calculator ⬛ from the menu in CFD-Post. Select area as Function in the Function Calculator and choose matid_4_surf as Location. Check the box for Clear previous results on calculate. Click on Calculate. Results shows the Area on matid_4_surf as 28584.7 [mm^2].

Go to the Excel file as shown in Figure 32.21c). Enter *SAP* [m^2] in cell I12, *SAFB* [m^2] in cell J12 and *ADR* in cell K12. Enter 0.0027929 in cell I13, enter =2*0.014519 in cell J13 and enter =J13/I13 in cell K13. Enter 0.0027929 in cell I17, 0.0285847 in cell J17 and =J17/I17 in cell K17. Enter =100*ABS(J17-J13)/J17 in cell K20. There is a difference of 1.6 % between Ansys Polyflow simulation and theory for the Area Draw Ratio.

Figure 32.21b) Calculated bottle surface area

I. Theory

22. The Aerial Draw Ratio *ADR* is the overall stretching of the preform. The *ADR* > 1 is determined from the Surface Area of the Formed Bottle *SAFB* divided by the original Surface Area of the Preform *SAP* that was used to form the bottle.

$$ADR = SAFB / SAP \tag{1}$$

The maximum values of *ADR* for different plastics are shown in Table 30.1. We see from this table that the plastic Polystyrene *PS* has the highest max *ADR* of all plastics listed.

Plastic	Maximum *ADR*
ABS	5.5
Acrylic	3.4
HDPE	6.5
LDPE	6.0
PP	7.5
Polystyrene PS	8.0
PVC	4.3

Table 32.1 Maximum aerial draw ratio for different plastics

The Average Thickness Reduction *ATR* is

$$ATR = 1 / ADR$$

(2)

Another draw ratio is the Linear Draw Ratio *LDR*. This ratio compares the length of the symmetry line on the preform before and after blow molding.

$$LDR = LFB / LBF$$

(3)

where *LFB* is the line length on formed bottle and *LBF* is the line length before forming.

J. References

1. Bordival M., Schmidt F., Le Maoult Y. and Velay V., "Simulation of the Two Stages Stretch-Blow Molding Process: Infrared Heating and Blowing Modeling", NUMIFORM 2007 – 9th International Conference on Numerical Methods in Industrial Forming Processes, Porto, Portugal, June 2007.
2. Daver F. and Demirel B., "A Simulation Study of the Effect of Preform Cooling Time in Injection Stretch Blow Molding", Journal of Materials Processing Technology, 212, 2400-2405, 2012.
3. Gupta S, Uday V., Raghuwanshi A.S., Chowkshey S., Das S.N., and Suresh S, "Simulation of Blow Molding Using Ansys Polyflow", APCBEE Procedia 5, 468-473, 2013.
4. Hassan Y.R., Farahaty M.A., Ibrahim S.I. and Nassar M.A, "Design and Producing Blowing Mold Using Additive Manufacturing", International Design Journal Vol. 9: Iss. 4, Article 6, 2019.
5. Kadam A.V., Patel S. and Vajir A., "Simulation of Blow Molding of Polyethylene Bottle Using Ansys Polyflow", International Research Journal of Engineering and Technology (IRJET), Vol. 6: Iss. 4, April, 2019.
6. Lontos A. and Gregoriou A., "A Numerical Investigation of the Effect of Preform Length for the Fabrication of 1.5 Lt PET Bottle Through the Injection Stretch Blow Molding Process", MATEC Web of Conferences 188, 01021, 2018.
7. Lontos A. and Gregoriou A., "The Effect of Deformation Rate on the Wall Thickness of 1.5 LT PET bottle during ISBM (Injection Stretch Blow Molding) Process", Procedia CIRP 81, 1307-1312, 2019.

8. Nixon J. , Menary G.H. and Yan S., "Finite Element Simulations of Stretch-Blow Moulding with Experimental Validation over a Broad Process Window", Int J Mater Form 10:793-809, 2017.

9. Peplinski K. and Bielinski M., "Polyflow Software Use to Optimize the Parison Thickness in Blowing Extrusion.", Journal of Polish CIMAC, 4, Gdansk, 2009.

10. Peplinski K. and Mozer A., "Ansys-Polyflow Software Use to Select the Parison Diameter and its Thickness Distribution in Blowing Extrusion.", Journal of Polish CIMAC, 5, Gdansk, 2010.

11. Suraya S., Azman M.D., Fatchurrohman N., Jaafar A.A., Yusoff A.R., "Simulation of the Effect of Bottle Wall Thickness Distribution using Blow Moulding Technique", IOP Conf. Series: Materials Science and Engineering 114, 2016.

12. Tan Z.Q. Rosli N. and Oktaviandri M.,"Simulation of Effect of Preform Diameter in Injection Stretch Blow Molding", IOP Conf. Series: Materials Science and Engineering 319, 2018.

13. Zimmer J. and Stommel M.,"Method for the Evaluation of Stretch Blow Molding Simulations with Free Blow Trials", IOP Conf. Series: Materials Science and Engineering 48, 2013.

K. Exercises

Exercise 32.1. Run the simulations in this chapter but use a quarter model of the preform to complete the simulation.

Exercise 32.2. Run the simulations in this chapter for different initial thickness of the preform: 1, 3 and 4 mm.

L. Notes

CHAPTER 33. SUPERSONIC FLOW PAST A CONE

A. Objectives

- Create Mesh in Ansys Meshing for 2D Axisymmetric Flow Past a Cone
- Inserting Boundary Conditions
- Running Turbulent Steady 2D Axisymmetric Ansys Fluent Simulations
- Using Contour Plots for Visualizations

B. Problem Description

In this chapter we will study the flow around a cone. You will be creating the geometry in Ansys Designmodeler and the mesh in Ansys Meshing.

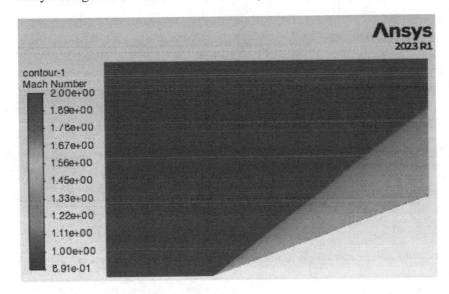

C. Starting Ansys Workbench and Setting Up Ansys Geometry and Mesh

1. Start by launching Ansys Workbench. Double click on Fluid Flow (Fluent) under Analysis Systems in the Toolbox. Right click on Geometry in Project Schematic and select Properties. Select 2D Analysis Type under Advanced Geometry Options. Right click on Geometry and select New Design Modeler Geometry. Select Units>>Meter from the menu. Select XYPlane in the Tree Outline, right click on XYPlane and select Look at. Roll the mouse wheel forward so that the horizontal scale at the bottom of the graphics window is showing 0.000 to 1.000 (m).

   Select the Sketching tab under the Tree Outline and select Draw>>Polyline. Start from the origin and make sure you have the letter *P*. Draw an inclined line upward to the right in the first quadrant and left click. Continue drawing a vertical line upward and make sure you have the letter *V* indicating a vertical line. Left click and draw a horizontal line to the left to the vertical axis. Make sure you have the letter *H* indicating a horizontal line. Left click and draw another horizontal line further to the left in the second quadrant. Left click and draw a vertical line down to the horizontal axis. Finally, left click and draw a horizontal line to the right to the origin. Right click and select Closed End.

   Select the Dimensions tab under Sketching Toolboxes and select Horizontal dimension. Left click on the two vertical sides of the geometry and 3 m in Details View as the length for the upper horizontal edge. Select General dimensions, left click on the left vertical edge of the geometry and enter 2 m as the length in Details View. Left click on the lower horizontal line and enter 1 m as the length in Details View. Select Angle and enter left click on the inclined line followed by left clicking on the horizontal axis.

Figure 33.1a) Geometry with dimensions in DesignModeler

Select Concept>>Surfaces from Sketches from the menu. Apply Sketch1 as Base Object and select Generate. Close DesignModeler.

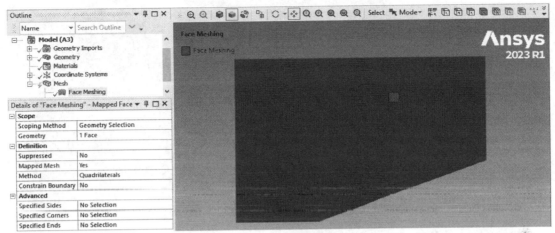

Figure 33.1b) Generated surface in DesignModeler

Double click on Mesh under Project Schematic in Workbench. Right-click on Mesh under Project in Outline and select Insert>>Face Meshing. Select Face Meshing under Mesh in Outline. Select the geometry in the graphics window and apply the geometry under Scope in Details of "Face Meshing".

Figure 33.1c) Face meshing in Ansys Meshing

Right-click on Mesh under Project in Outline and select Insert>>Sizing. Select Edge above the graphics window. Select the two horizontal edges to the left on the geometry. Apply the edges as Geometry under Scope in Details of "Edge Sizing". Select Number of Divisions as Type under Definition in Details of "Edge Sizing". Set Number of Divisions to 50, set Capture Curvature to No and set Behavior to Hard. Select the first Bias Type, set the Bias Factor to 5 and Apply the upper horizontal edge as Reverse Bias.

Figure 33.1d) First element sizing in Ansys Meshing

Insert another edge sizing but this time apply the two vertical edges. Select Number of Divisions as Type under Definition in Details of "Edge Sizing". Set Number of Divisions to 200, set Capture Curvature to No and set Behavior to Hard. Select the second Bias Type, set the Bias Factor to 25 and Apply the left vertical edge as Reverse Bias.

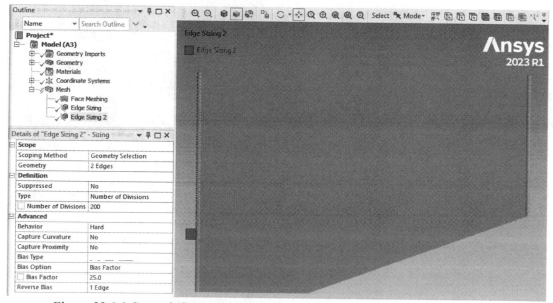

Figure 33.1e) Second element sizing in Ansys Meshing

Insert a third edge sizing but this time apply the upper horizontal edge to the right and the inclined edge. Select Number of Divisions as Type under Definition in Details of "Edge Sizing". Set Number of Divisions to 300, set Capture Curvature to No and set Behavior to Hard.

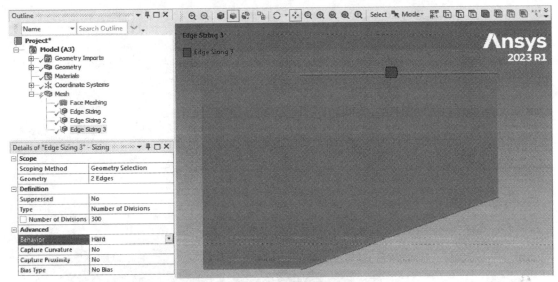

Figure 33.1f) Third element sizing in Ansys Meshing

Right click on Mesh under Project in the Outline and select Update. Click on the plus sign next to Statistic in Details of Mesh. The mesh has 70000 Elements and 70551 Nodes.

Figure 33.1g) Finished mesh in Ansys Meshing

Select Edge 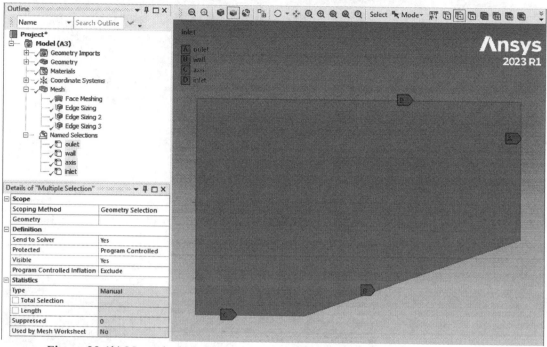 above the graphics window. Select the vertical edge to the right, right-click and select Create Named Selection. Enter *outlet* as Selection Name and click OK to close the window. Select the inclined edge and name this edge as *wall*. Select the lower horizontal edge and name this edge as *axis*. Select the remaining three edges and name these edges as *inlet*.

Figure 33.1h) Named edges in Ansys Meshing

Select File>>Export...>>Mesh>>Fluent Input File>>Export from the menu in Ansys Meshing and save the mesh with the name *cone-mesh-20-degrees.msh*. Select File>>Save Project from the menu in Ansys Meshing. Enter the name *Supersonic Flow over Cone with Half-Angle 20 degrees* and save the project. Exit Ansys Meshing. Right click on Mesh under Project Schematic in Ansys Workbench and select Update.

D. Starting Ansys Fluent

2. Double click on Setup under Project Schematic in Ansys Workbench. Select Double Precision under Solver Options. Set the number of Solver Processes under Parallel (Local Machine). Select the General Options tab in the menu and check the box for Display Mesh After Reading. Click on Start. Select the Density-Based Solver Type on the Task Page in Ansys fluent. Select Axisymmetric as 2D Space under Solver on the Task Page.

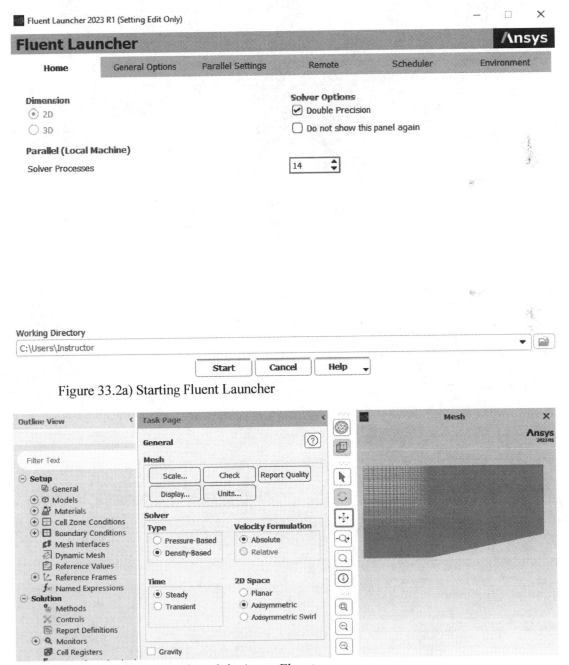

Figure 33.2a) Starting Fluent Launcher

Figure 33.2b) Imported mesh in Ansys Fluent

3. Double click on Models and Energy (Off) under Setup in the Outline View. Check the Energy Equation box and select OK to close the Energy window. Double click on Models and Viscous (SST k-omega) under Setup in Outline View. Keep the settings as shown in Figure 33.2a). Click OK to close the Viscous Model window. Select the Physics tab in the menu and select Operating Conditions…. Set the Operating Pressure to 0.

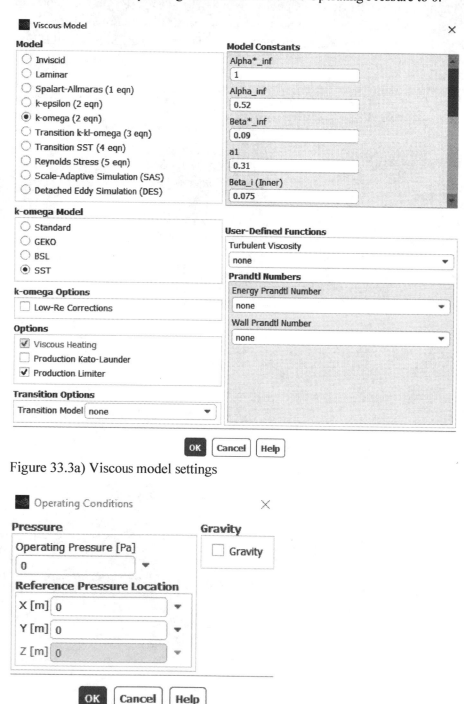

Figure 33.3a) Viscous model settings

Figure 33.3b) Setting the operating pressure

4. Double click on Materials, Fluid and air under Setup in the Outline View. Select *ideal-gas* for Density and select *sutherland* for Viscosity. Select Three Coefficient Method and OK to close the Sutherland Law window. Click on Change/Create and Close the Create/Edit Materials window.

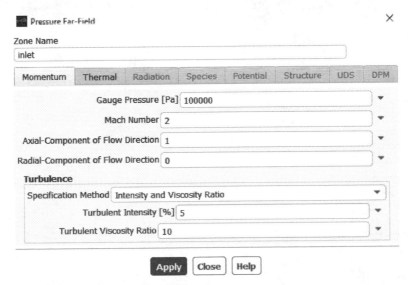

Figure 33.4 Create/edit materials

5. Double click on Boundary Conditions under Setup in the Outline View. Select *inlet* as Zone under Boundary Conditions on the Task Page. Select pressure-far-field as Type on the Task Page. Set the Gage Pressure to 100000 and set Mach number to *2*. Select the Thermal tab and set the temperature to 300 K. Select Apply and Close the Pressure Far-Field window.

Figure 33.5a) Pressure far-field

Select *outlet* as Zone under Boundary Conditions on the Task Page. Select pressure-outlet as Type on the Task Page. Set the Gage Pressure to 100000. Select the Thermal tab and set the temperature to 343.3949 K. This temperature value can be found from data at https://devenport.aoe.vt.edu/aoe3114/calc.html

For the Conical Shock Relations, set Gamma to 1.4, M1 to 2, Cone angle to 20 degrees, and click on Calculate. To find the temperature value after the conical shock, use the temperature ratio T2/T1 = 1.14464983 and calculate T2 = 343.3949 K after the shock using T1 = 300 K. Select Apply and Close the Pressure Outlet window.

Conical Shock Relations Perfect Gas, Gamma = 1.4 , angles in degrees.

INPUT: M1 = 2   Cone angle ∨ = 20   [Calculate]

M_c=	1.56672098	Cone ang.=	20	Wave ang.=	37.8266034	Shock turn ang.=	8.60032511
P_2/P_1=	1.58848925	P_{02}/P_{01}=	0.98998816	rho_2/rho_1=	1.38775126	T_2/T_1=	1.14464983
P_c/P_1=	1.91421444	P_{0c}/P_{01}=	0.98998816	rho_c/rho_1=	1.58552567	T_c/T_1=	1.20730586

Figure 33.5b) Conical shock relations

6. Double click on Methods under Solution in the Outline View. Select the settings as shown in Figure 22.5a). Select Options… next to Higher Order Term Relaxation. Select All Variables under Variables and click OK to close the Relaxation Options window.

Figure 33.6a) Solution methods

Move the cursor to the Console and press *enter* on the keyboard to get the command prompt. Type >solve and Enter in the Console. Type set and Enter. Type numerics and Enter. Set the blending factor to 0.7 and Enter. Type q and Enter.

Double click on Report Definitions under Solution in the Outline View. Select New>>Surface Report>>Area-Weighted Average... in the Report Definitions window. Select wall under surfaces in the Surface Report Definition window. Check the box for Print to Console under Create and select OK to close the window.

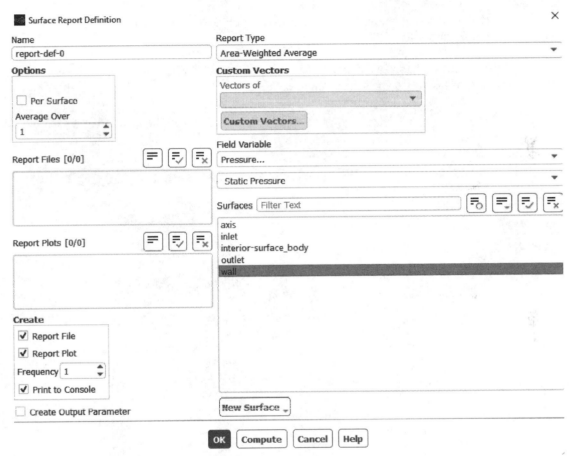

Figure 33.6b) Surface report definition

E. Initializing the Solution and Running the Calculations

7. Double click on Initialization under Solution in the Outline View. Select Hybrid Initialization and click on Initialize.

    Move the cursor to the Console and press *enter* on the keyboard to get the command prompt. Enter *solve/initialize/fmg/yes*.

    Select File>>Write>>Case & Data… from the menu. Enter the name *cone-mesh-20-deg.cas.h5* and select OK.

    Double click on Run Calculation under Solution in the Outline View. Enter 500 as Number of Iterations and select Calculate.

Figure 33.7a) Scaled residuals

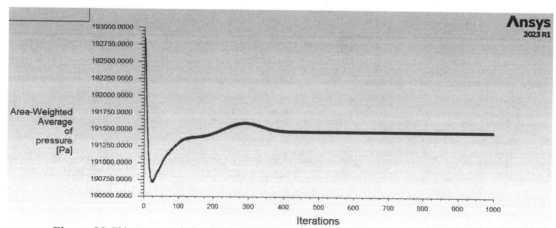

Figure 33.7b) Area-weighted average static pressure

F. Post-Processing

8. Double click on Graphics and Contours under Results in the Outline View. Select Contours of Pressure and Static Pressure. Deselect all surfaces and select Colormap Options…. Set Font Behavior to Fixed and Font Size to 12. Select Apply and Close the Colormap window. Select Save/Display and Close the Contours window. Create contours for Velocity/Mach number, Density and Static Temperature.

Figure 33.8a) Static pressure and Mach number contours

Figure 33.8b) Density and static temperature contours

Double click on Plots and XY Plot under Results in the Outline View. Check the box for Node Values under Options and uncheck all the other boxes. Select Mesh and Y-Coordinate as Y Axis Function. Select Density and Density under X Axis Function. Select outlet under Surfaces. Select Axes… and select X Axis in the Axes – Solution XY Plot window. Set Type to float and Precision to 1 under Number Format. Uncheck Auto Range and set Minimum to 1 and Maximum to 2 under Range. Click on Apply. Select Y Axis and set Type to float and Precision to 1 under Number Format. Uncheck Auto Range and set Minimum to 0 and Maximum to 2 under Range. Click on Apply. Close the window.

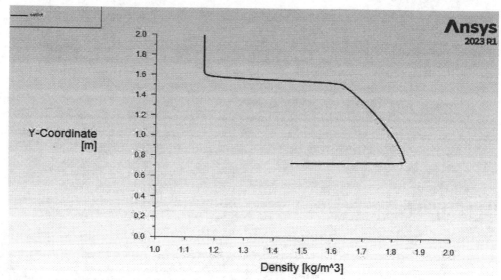

Figure 33.8c) Density variation at the outlet in Ansys Fluent

Select File>>Export>>Solution Data… from the menu. Select ASCII as File Type in the Export window. Select Node under Location, Space under Delimiter, outlet under Surfaces and Density under Quantities. Select Write… and save the file with the name *cone-mesh-20-deg-outlet-density-variation*.

Open the saved file in Excel. Select Next in steps 1 and 2 and select Finish in step 3. Copy column C to column E. Select columns D and E and insert a scatter plot for the data, see Figure 33.8d). We will now focus our attention on the four data points as shown encircled in Figure 33.8d) and delete all the other data. Copy columns D and E to columns F and G.

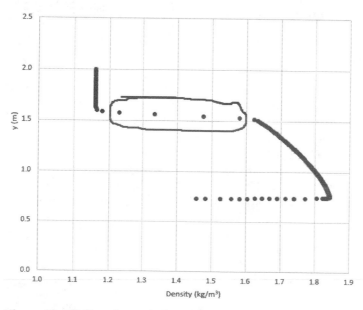

Figure 33.8d) Density variation at the outlet in Excel

Delete the first 173 data points in columns F and G (rows 2-174). After this has been completed and remaining cells been moved up in the two columns, clear rows 6 – 29 so that only four data points are left. Take the average of the remaining four data points in column G2 – G5 which we define as the y-coordinate for the shock wave at the outlet. The average y-coordinate is 1.5526 m as shown in cell G6. We use this value as the location for the shock wave and we can now determine the shock angle in cell H6 as =180*ATAN(G6/2)/PI(). The shock angle is 37.8220069 degrees. This value can be compared with the wave angle 37.8266034 degrees that we determined in Figure 33.5b). The difference is only 0.012 %.

	A	B	C	D	E	F	G	H	I	J
1	nodenumber	x-coordinate	y-coordinate	density	y-coordinate	density	y-coordinate	Shock Angle	Theory	Percent Diff.
2	1	2.00E+00	7.28E-01	1.46E+00	7.28E-01	1.58E+00	1.5314			
3	2	2.00E+00	7.29E-01	1.48E+00	7.29E-01	1.47E+00	1.5454			
4	3	2.00E+00	7.30E-01	1.53E+00	7.30E-01	1.33E+00	1.5596			
5	4	2.00E+00	7.31E-01	1.56E+00	7.31E-01	1.23E+00	1.5740			
6	5	2.00E+00	7.31E-01	1.58E+00	7.31E-01		1.5526	37.8220069	37.8266034	0.012

Figure 33.8e) Density variation data in Excel

G. Theory

9. Supersonic flow over a cone can be found theoretically assuming inviscid flow and a sharp semi-infinite cone that extends to infinity in the streamwise direction. The theoretical solution is called Taylor-Maccoll equation and is given by Anderson[1]. This is an ordinary differential equation with one dependent variable. Solutions to this equation is shown in Figure 33.9 from Lassaline[2].

$$\frac{\gamma-1}{2}\left[V_{max}^2 - V_r^2 - \left(\frac{dV_r}{d\theta}\right)^2\right]\left[2V_r + \frac{dV_r}{d\theta}\cot\theta + \frac{d^2V_r}{d\theta^2}\right] - \left(\frac{dV_r}{d\theta}\right)^2\left[V_r + \frac{d^2V_r}{d\theta^2}\right] = 0 \qquad (33.1)$$

where $V_r = f(\theta)$ is the radial component of velocity and V_{max} is the maximum theoretical velocity. The normal component of velocity $V_\theta = \frac{dV_r}{d\theta}$

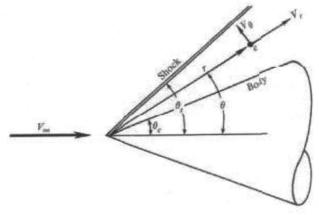

Figure 33.9a) Spherical coordinate system for a cone, from Anderson[1]

885

We can introduce a non-dimensional velocity $V' = \dfrac{V}{V_{max}}$ and equation (1) can be written as

$$\frac{\gamma-1}{2}\left[1 - V'^2_r - \left(\frac{dV'_r}{d\theta}\right)^2\right]\left[2V'_r + \frac{dV'_r}{d\theta}\cot\theta + \frac{d^2V'_r}{d\theta^2}\right] - \left(\frac{dV'_r}{d\theta}\right)^2\left[V'_r + \frac{d^2V'_r}{d\theta^2}\right] = 0 \quad (33.2)$$

It can be shown that the non-dimensional velocity is a function of the Mach number

$$V' = \frac{1}{\sqrt{1 + \frac{2}{(\gamma-1)M^2}}} \quad (33.3)$$

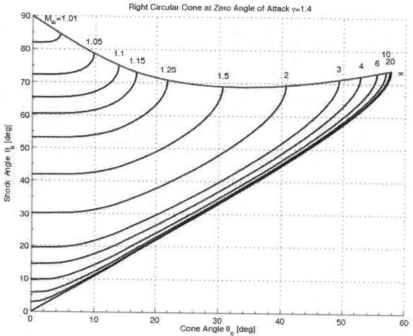

Figure 33.9b) Shock angle versus half cone angle, from Lassaline[2]

We define the Reynolds number Re on the cone as

$$Re = \frac{V_2 L \rho_2}{\mu_2} \quad (33.4)$$

where V_2 is the free stream velocity after the shock wave, L is the slant height of the cone, ρ_2 is the density after the shock wave and μ_2 is the dynamic viscosity after the shock wave. The Mach numbers are defined as

$$M_\infty = \frac{V_\infty}{a_\infty}, \; M_2 = \frac{V_2}{a_2} \quad (33.5)$$

Where ∞ is related to conditions before the shock and 2 is after the shock wave.

H. References

1. Anderson, J.D., "Fundamentals of Aerodynamics", McGraw-Hill Education, 6th Ed., 2011.
2. Lassaline, J.V., "Supersonic Right Circular Cone at Zero Angle of Attack", Ryerson University, 2009.

I. Exercises

33.1 Create the geometry and mesh for your cone angle as shown in Table 33.1. Use the same free stream pressure 100000 Pa and free stream temperature 300 K as shown in this chapter. Use https://devenport.aoe.vt.edu/aoe3114/calc.html
to calculate outlet temperature for Ansys Fluent simulations.

	Mach number	Cone Half-Angle (deg.)	Shock wave angle (deg.) from Ansys Fluent	Shock wave angle (deg.) from Figure 33.9	Percent difference
Student A	1.25	5			
Student B	1.25	10			
Student C	1.25	15			
Student D	1.25	20			
Student E	1.5	10			
Student F	1.5	15			
Student G	1.5	20			
Student H	1.5	25			
Student I	2	10			
Student J	2	20			
Student K	2	30			
Student L	2	35			
Student M	3	10			
Student N	3	20			
Student O	3	30			
Student P	3	40			
Student Q	3	45			

33.2 Create the geometry and mesh for your cone angle as shown in Table 33.1. Use the same free stream pressure 25000 Pa and free stream temperature 100 K as shown in this chapter. Use https://devenport.aoe.vt.edu/aoe3114/calc.html
to calculate outlet temperature for Ansys Fluent simulations.

	Mach number	Cone Half-Angle (deg.)	Shock wave angle (deg.) from Ansys Fluent	Shock wave angle (deg.) from Figure 33.9	Percent difference
Student A	1.25	20			
Student B	1.5	20			
Student C	1.75	20			
Student D	2	20			
Student E	2.25	20			
Student F	2.5	20			
Student G	2.75	20			
Student H	3	20			

Student I	3.25	20			
Student J	3.5	20			
Student K	3.75	20			
Student L	4	20			
Student M	4.25	20			
Student N	4.5	20			

CHAPTER 34. SHOCK DIAMONDS

A. Objectives

- Use Mesh from Ansys Meshing for 2D Axisymmetric Flow Through a Nozzle
- Inserting Boundary Conditions
- Running Turbulent 2D Axisymmetric Ansys Fluent Simulations
- Using Contour Plots for Visualizations

B. Problem Description

In this chapter we will study the flow through a conical nozzle. You will be using the mesh from Ansys Meshing for a nozzle with 40 degrees cone angle.

C. Starting Ansys Workbench and Setting Up Ansys Geometry

1. You can skip to step 3 if you are reading the mesh directly in Ansys Fluent standalone. Otherwise, start by launching Ansys Workbench. Double click on Fluid Flow (Fluent) under Analysis Systems in the Toolbox. Right click on Geometry in Project Schematic and select Properties. Select 2D Analysis Type under Advanced Geometry Options. Right click on Geometry and select New Design Modeler Geometry. Select Units>>Inch from the menu. Select XYPlane in the Tree Outline, right click on XYPlane and select Look at. Roll the mouse wheel backward so that the horizontal scale at the bottom of the graphics window is showing 0.000 to 4.000 (in).

    Select the Sketching tab under the Tree Outline and select Draw>>Polyline. Start 1.5 inch above the origin and draw a 2.25-inch long vertical line upward followed by a 7.5-inch long horizontal line to the left in the second quadrant. Continue drawing a 0.855-inch long vertical line downward followed by a horizontal line to the right and finally an inclined line downward and to the right that ends at the starting point on the vertical axis. Right click and select Closed End.

    Add the dimensions as listed above. Start by selecting Dimensions>>Length/Distance in the Sketching Toolboxes and select the origin followed by clicking on the lower end point of the vertical line on the vertical axis. Set this length to 1.5 inch. Next, select Dimensions>>General and select the vertical line on the vertical axis. Set the vertical length to 2.25 inch. Select the top horizontal line and set the length to 7.5 inch. Select the left vertical line and set the length to 0.855 inch. Select Dimensions>>Angle and set the angle of the inclined line from the vertical to 50 degrees. You will need to first select the vertical axis and then select the inclined line in order to set this angle, see Figure 34.1a).

Figure 34.1a) Nozzle with dimensions

Select Concept>>Surfaces from Sketches from the menu. Select Sketch1 under XYPlane in the Tree Outline and Apply the sketch as Base Object in Details View. Click on Generate, see Figure 34.1b).

Figure 34.1b) Generated surface sketch for nozzle

Next, we will be sketching the mesh region around the nozzle. Select the XYPlane in the Tree Outline and select new sketch to generate a sketch. Select the Sketching tab and Draw>>Rectangle by 3 Points. Start by clicking on the horizontal axis to the left of the origin followed by clicking on the horizontal axis to the right of the origin and finally clicking in the first quadrant to complete the rectangle, see Figure 34.1c).

Figure 34.1c) Sketch for mesh region around nozzle

Select Dimensions>>Horizontal in the Sketching Toolboxes and select the vertical axis and the left vertical edge of the rectangle. Set the horizontal dimension to 7.5 inch. Select the right vertical edge of the rectangle and the vertical axis and set the dimension to 150 inch. Select Dimensions>>Vertical. Select the horizontal axis and the upper horizontal edge of the rectangle and set the dimension to 69-inch. Right click in the graphics window and select Zoom to Fit, see Figure 34.1d).

Figure 34.1d) Dimensions for mesh region around nozzle

Select Concept>>Surfaces from Sketches from the menu. Select Sketch2 under XYPlane in the Tree Outline and Apply the sketch as Base Object in Details View. Select Add Frozen as Operation in Details View. Click on Generate, see Figure 34.1b).

Figure 34.1e) Generated surface sketch for mesh region

Select Create>>Boolean from the menu. Select Subtract Operation from Details View. Select the mesh region in the graphics window and Apply as Target Body in Details View. Right click in the graphics window and select Cursor Mode>>Box Zoom. Zoom in on the nozzle in the graphics window. Right click once again in the graphics window and deselect the Box Zoom by selecting Cursor Mode>>Box Zoom. Select the nozzle and Apply it as Tool Body in Details View. You will need to select the second rectangle as shown in the lower left corner in Figure 34.1f) in order to select the nozzle. Click on Generate. The nozzle has now been subtracted from the mesh region, see Figure 34.1g).

Figure 34.1f) Selection of nozzle as tool body

Figure 34.1g) Nozzle subtracted from mesh region

Select the XYPlane in the Tree Outline and select new sketch to generate a sketch. Select the Sketching tab and Draw>>Line. Draw a vertical line through the mesh region starting at the origin and drawing the line upward on the vertical axis to the upper horizontal edge of the mesh.

Select Concepts>>Lines from Sketches and select the vertical line that you created. Apply the line in Details View and select Generate, see Figure 34.1h).

Figure 34.1h) Completed first vertical line through mesh region

Repeat this step and create a new sketch in the XYPlane with a horizontal line through the mesh region from the left vertical edge of the mesh region to the right vertical edge of the mesh region. The line will be coincident with the upper horizontal edge of the nozzle.

Select Concepts>>Lines from Sketches and select the horizontal line that you created. Apply the line in Details View and select Generate, see Figure 34.1i).

Figure 34.1i) Completed first horizontal line through mesh region

Repeat this step once again and create a new sketch in the XYPlane with a horizontal line through the mesh region from lower right corner of the nozzle to the right vertical edge of the mesh region. The horizontal line will be coincident with the lower right corner of the nozzle. Start by drawing the horizontal line from the lower right corner of the nozzle and draw the line horizontally to the right all the way to the vertical right edge of the mesh region. Select Concepts>>Lines from Sketches and select the horizontal line that you created. Apply the line in Details View and select Generate, see Figure 34.1j).

Figure 34.1j) Completed second horizontal line through mesh region

Repeat this step the last time and create a new sketch in the XYPlane with a vertical line through the mesh region from the middle lower corner of the nozzle to the horizontal axis. Select Concepts>>Lines from Sketches and select the vertical line that you created. Apply the line in Details View and select Generate, see Figure 34.1k).

Figure 34.1k) Completed second vertical line through mesh region

Select Tools>>Projection from the menu and select Edges on Face as Type in Details View. Apply the mesh region as Target in Details View. Control-select a total of 5 Edges for the four lines (Line1 – Line4) in the Tree Outline. This includes two parts of the longer vertical line, one part of the upper horizontal line and the remaining vertical and horizontal lines that only have one part each, see Figure 34.1l). Apply the 5 Edges in Details View and select Generate. You will have six different mesh regions as a result of the projection. Close DesignModeler.

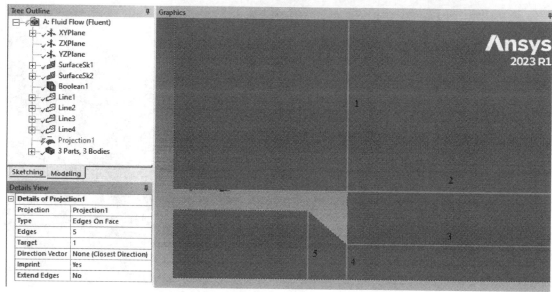

Figure 34.1l) Five edges used for projection

D. Launching Ansys Meshing

2. Double click on Mesh under Fluid Flow (Fluent) in Project Schematic in Ansys
    Workbench. Right click on Mesh in Ansys Meshing Outline and select Insert>>Face
    Meshing. Select the six meshing regions in the graphics window and Apply face as
    Geometry under Scope in Details of "Face Meshing".

    Right click on Mesh in Ansys Meshing Outline and select Insert>>Sizing. Right click in
    the graphics window and select Cursor Mode>>Edge. Select the two horizontal edges as
    shown in Figure 34.2a) and Apply the edges as Geometry under Scope in Details of
    "Edge Sizing". Select Number of Divisions as Type under Definition and set the Number
    of Divisions to 70. Set Capture Curvature to No under Advanced and set Behavior to
    Hard. Select the first Bias Type and set the Bias Factor to 100.

Figure 34.2a) Two horizontal edges used for first sizing

    Right click on Mesh in Ansys Meshing Outline and select Insert>>Sizing. Right click in
    the graphics window and select Cursor Mode>>Edge. Select the four vertical edges as
    shown in Figure 34.2b) and Apply the edges as Geometry under Scope in Details of
    "Edge Sizing". Select Number of Divisions as Type under Definition and set the Number
    of Divisions to 65. Set Capture Curvature to No under Advanced and set Behavior to
    Hard. Select the third Bias Type and set the Bias Factor to 50, see Figure 34.2c)

Figure 34.2b) Four vertical edges used for second sizing

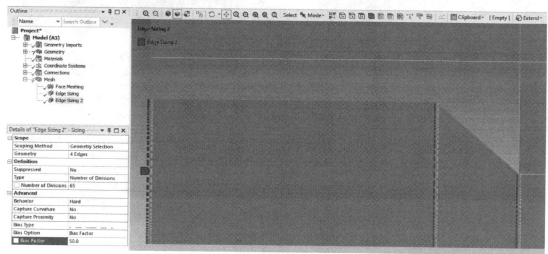

Figure 34.2c) Details of "Edge Sizing" for second sizing

Right click on Mesh in Ansys Meshing Outline and select Insert>>Sizing. Right click in the graphics window and select Cursor Mode>>Edge. Select the two horizontal edges as shown in Figure 34.2d) and Apply the edges as Geometry under Scope in Details of "Edge Sizing". Select Number of Divisions as Type under Definition and set the Number of Divisions to 80. Set Capture Curvature to No under Advanced and set Behavior to Hard. Select the third Bias Type and set the Bias Factor to 10.

Figure 34.2d) Details of "Edge Sizing" for third sizing

Right click on Mesh in Ansys Meshing Outline and select Insert>>Sizing. Right click in the graphics window and select Cursor Mode>>Edge. Select the two horizontal edges as shown in Figure 34.2e) and Apply the edges as Geometry under Scope in Details of "Edge Sizing". Select Number of Divisions as Type under Definition and set the Number of Divisions to 100. Set Capture Curvature to No under Advanced and set Behavior to Hard. Select the first Bias Type and set the Bias Factor to 100, see Figure 34.2f).

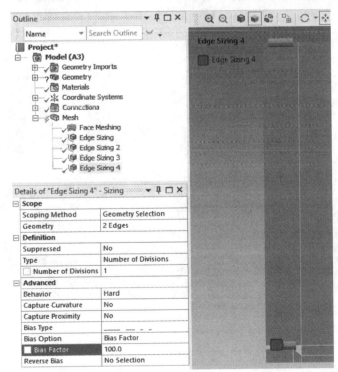

Figure 34.2e) Details of "Edge Sizing" for fourth sizing

Figure 34.2f) Zoomed in details of "Edge Sizing" for fourth sizing

Right click on Mesh in Ansys Meshing Outline and select Insert>>Sizing. Right click in the graphics window and select Cursor Mode>>Edge. Select the four horizontal edges as shown in Figure 34.2g) and Apply the edges as Geometry under Scope in Details of "Edge Sizing". Select Number of Divisions as Type under Definition and set the Number of Divisions to 100. Set Capture Curvature to No under Advanced and set Behavior to Hard. Select the second Bias Type and set the Bias Factor to 1000. Select the upper edge and Apply this edge for Reverse Bias.

Figure 34.2g) Details of "Edge Sizing" for fifth sizing

Right click on Mesh in Ansys Meshing Outline and select Insert>>Sizing. Right click in the graphics window and select Cursor Mode>>Edge. Select the three long vertical edges as shown in Figure 34.2h) and Apply the edges as Geometry under Scope in Details of "Edge Sizing". Select Number of Divisions as Type under Definition and set the Number of Divisions to 100. Set Capture Curvature to No under Advanced and set Behavior to Hard. Select the first Bias Type and set the Bias Factor to 5000. Select the right most edge and Apply this edge for Reverse Bias.

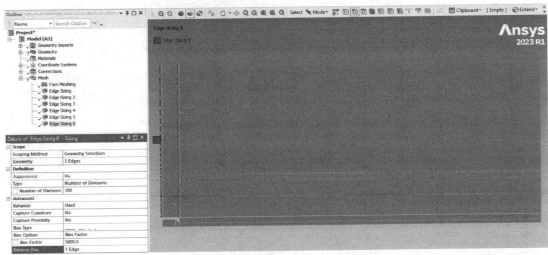

Figure 34.2h) Details of "Edge Sizing" for sixth sizing

Right click on Mesh in Ansys Meshing Outline and select Insert>>Sizing. Right click in the graphics window and select Cursor Mode>>Edge. Select the two vertical edges as shown in Figures 34.2i) and 34.2j) and Apply the edges as Geometry under Scope in Details of "Edge Sizing". Select Number of Divisions as Type under Definition and set the Number of Divisions to 65. Set Capture Curvature to No under Advanced and set Behavior to Hard. Select the third Bias Type and set the Bias Factor to 70. Right click on Mesh in Outline and select Generate Mesh. Open Statistics in Details of Mesh. The mesh has 42750 Elements and 43331 Nodes.

Figure 34.2i) Details of "Edge Sizing" for seventh sizing (left edge)

Details of "Edge Sizing 7" - Sizing

Scope	
Scoping Method	Geometry Selection
Geometry	2 Edges
Definition	
Suppressed	No
Type	Number of Divisions
Number of Divisions	65
Advanced	
Behavior	Hard
Capture Curvature	No
Capture Proximity	No
Bias Type	– —— —— —— –
Bias Option	Bias Factor
Bias Factor	70.0

Figure 34.2j) Details of "Edge Sizing" for seventh sizing (right edge)

Figure 34.2k) Finished mesh

Right click in the graphics window and select Cursor Mode>>Edge. Select the vertical inlet to the nozzle, right click and select Create Named Selection…. Enter the name *inlet* in the Selection Name window and click OK to close the window.

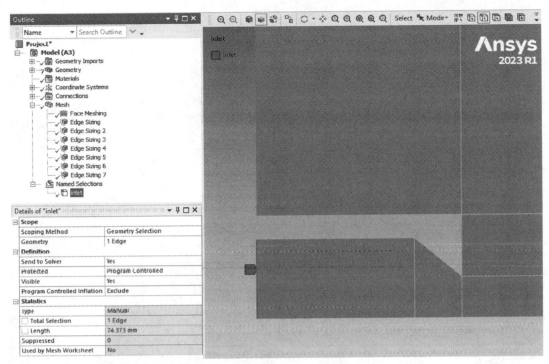

Figure 34.2l) Inlet section

Repeat this step and control select the three lower horizontal edges and enter the name *axis*.

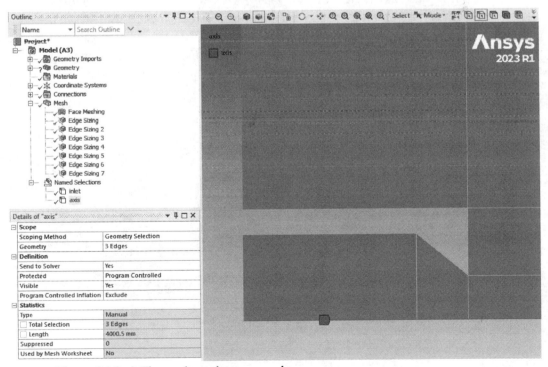

Figure 34.2m) Three edges chosen as axis

Once again, repeat this step and control select the four edges as shown in Figure 34.2n) and enter the name *wall*.

Figure 34.2n) Four edges chosen as wall

Finally, repeat this step and control select the six remaining outer edges as shown in Figure 34.2o) and enter the name *outlet*. Open Geometry under Project in Outline. Control select the two Line Bodies, right click and select Suppress Body. File>>Export>>Mesh>>FLUENT Input File>>Export under the menu and save the mesh file with the name *mesh-40-degrees-nozzle.msh*. Select Close Ansys Meshing. Right click on Mesh under Project Schematic in Ansys Workbench and select Update. Select Files>>Save As… from the menu in Ansys Workbench and save the project with the name *40-degrees-nozzle*.

Figure 34.2o) Six edges chosen as outlet

E. Launching Ansys Fluent

3. You can read the mesh directly to Ansys Fluent by starting Ansys Fluent in standalone mode

    Fluent 2023 R1 that is available in the Ansys folder. Select the Solution tab under Home in Fluent Launcher, see Figure 34.1a). Read the Mesh file *mesh-40-degrees-nozzle.msh*. Check the box for Double Precision under Solver Options in the Ansys Fluent Launcher and set the number of Solver Processes. Select the General Options tab in the menu and check the box for Display Mesh After Reading. Click on Start.

    Alternatively, right click on Mesh in Ansys Workbench Project Schematic and select Update. Double click on Setup under Project Schematic in Ansys Workbench, select Double Precision and set the number of Solver Processes. Select the General Options tab and check the box under Options for Display Mesh After Reading.

    Click on Start. Select Axisymmetric as 2D Space under Solver on the Task Page in Ansys Fluent.

Figure 34.3a) Fluent Launcher

Figure 34.3b) Mesh in Ansys Fluent

4. Double click on Models and Energy (Off) under Setup in the Outline View. Check the Energy Equation box and select OK to close the Energy window. Double click on Models and Viscous (SST k-omega) under Setup in Outline View. Keep the settings as shown in Figure 34.4a). Click OK to close the Viscous Model window. Select the Physics tab in the menu and select Operating Conditions…. Set the Operating Pressure to 0.

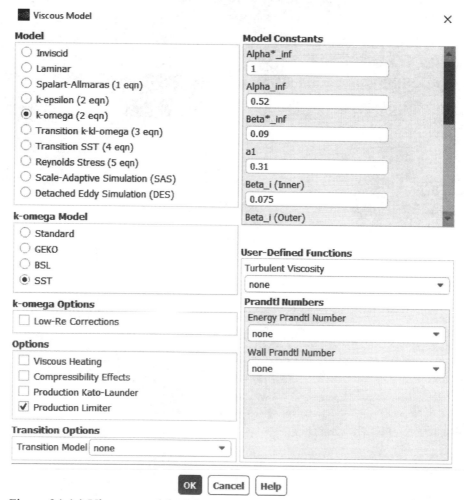

Figure 34.4a) Viscous model settings

Figure 34.4b) Setting the operating pressure

5. Double click on Materials, Fluid and air under Setup in the Outline View. Select *ideal-gas* for Density and select *sutherland* for Viscosity. Select Three Coefficient Method and OK to close the Sutherland Law window. Click on Change/Create and Close the Create/Edit Materials window.

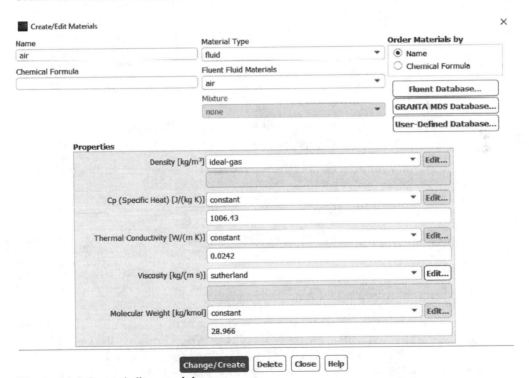

Figure 34.5 Create/edit materials

6. Double click on Boundary Conditions under Setup in the Outline View. Select *inlet* as Zone under Boundary Conditions on the Task Page. Select pressure-inlet as Type on the Task Page. Set the Gage Total Pressure [Pa] to 405411.6 and set Supersonic/Initial Gauge Pressure [Pa] to 101352.9. Select Intensity and Hydraulic Diameter as the Specification Method under Turbulence. Set Turbulence Intensity [%] to 1 and Hydraulic Diameter [m] to 0.147066, see Figure 34.6a). Select Apply and select the Thermal tab and set the temperature to 277.74 K. Select Apply and Close the Pressure Inlet.

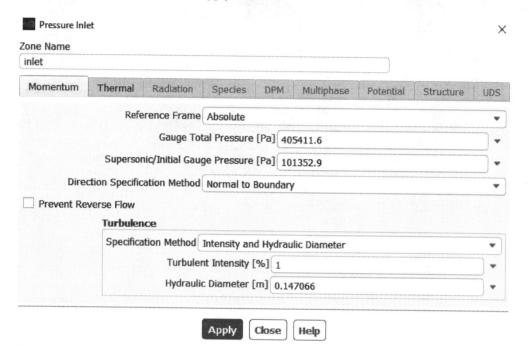

Figure 34.6a) Pressure inlet momentum

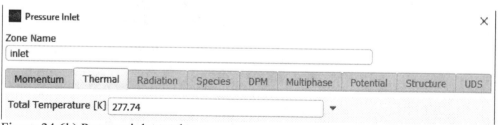

Figure 34.6b) Pressure inlet total temperature

Select *outlet* as Zone under Boundary Conditions on the Task Page. Select pressure-outlet as Type on the Task Page. Select Edit... Set the Gage Pressure [Pa] to 101352.9. Select Intensity and Viscosity Ratio as the Specification Method under Turbulence. Set Turbulence Intensity [%] to 0.5 and Backflow Turbulent Viscosity Ratio to 2, see Figure 34.6c). Select the Thermal tab and set the temperature to 277.74 K. Select Apply and Close the Pressure Inlet.

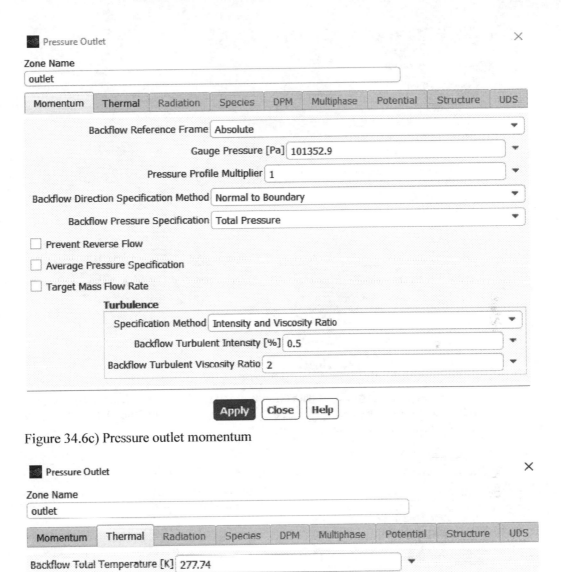

Figure 34.6c) Pressure outlet momentum

Figure 34.6d) Pressure outlet total temperature

Double click on Reference Values under Setup in the Outline. Select Compute from Inlet and select solid-surface_body as Reference Zone.

Figure 34.6c) Reference values

7. Double click on Methods under Solution in the Outline View. Keep the settings as shown in Figure 34.7a). Double click on Controls under Solution in the Outline View. Select the settings as shown in Figure 34.7b).

Figure 34.7a) Solution methods

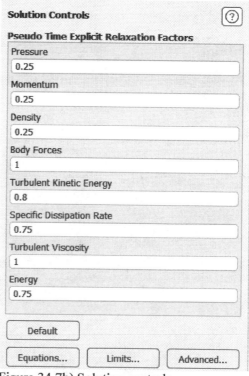

Figure 34.7b) Solution controls

Double click on Monitors and Residual under Solution in the Outline. Set all Residuals to 1e-06.

Figure 34.7c) Residual monitors

F. Initializing the Solution and Running the Calculations

8. Double click on Report Definitions under Solution in the Outline View. Select New>>Flux Report>>Mass Flow Rate…. Select the inlet under Boundaries in the Flux Report Definition window. Check the box for Print to Console Under Create. Enter the Name mass-flow-rate. Select OK to close the window. Close the Report Definition window. Double click on Initialization under Solution in the Outline View. Select Hybrid Initialization and click on Initialize. Select File>>Export>>Case & Data… from the menu. Enter the name *NPR=4-Nozzle40degrees.cas* and select OK. Double click on Run Calculation under Solution in the Outline View. Enter 1000 as Number of Iterations and select Calculate.

Figure 34.8a) Scaled residuals

Figure 34.8b) Mass flow rate

G. Post-Processing

9.   Double click on Graphics and Contours under Results in the Outline View. Select Contours of Pressure and Static Pressure. Deselect all surfaces and select Colormap Options…. Set Font Behavior to Fixed and Font Size to 12. Select Type as float and set Precision to 1 under Number Format. Select Apply and Close the Colormap window. Select Save/Display and Close the Contours window. Select the View tab in the menu and click on Views…. Select *axis* under Mirror Planes, click on Apply and Close the Views window. Select View>>Display>>Headlight>>On from the menu.

Figure 34.9a) Static pressure, Mach number, density and static temperature contours for NPR = 4

Figure 34.9b) Axial and radial velocity contours for NPR = 4

Double click on Plots and XY Plot under Results in the Outline View. Check the box for Node Values under Options and uncheck all the other boxes. Select Velocity and Mach number as Y Axis Function and select Mesh and X-Coordinate as X Axis Function. Select axis under Surfaces.

Select Axes… and select Y Axis in the Axes – Solution XY Plot window. Set Type to float and Precision to 1 under Number Format. Click on Apply. Close the window. Select Save/Plot in the Solution XY Plot window.

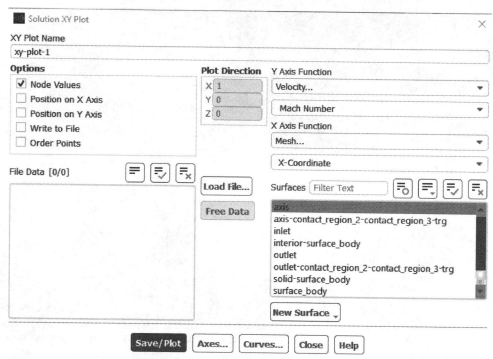

Figure 34.9c) Solution XY plot

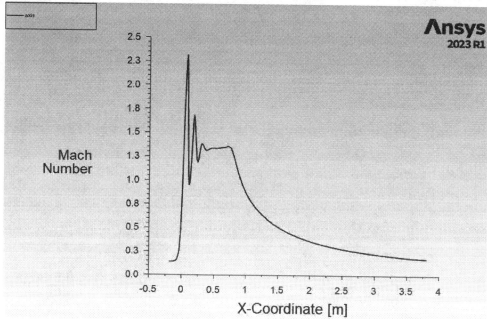

Figure 34.9d) Mach number variation along the axis in Ansys Fluent for NPR = 4

Select File>>Export>>Solution Data… from the menu. Select ASCII as File Type in the Export window. Select Node under Location, Space under Delimiter, axis under Surfaces and Mach number under Quantities. Select Write… and save the file with the name *40-deg-nozzle-axis-Mach-number-variation.*

Open the saved file in Excel. Select Next in steps 1 and 2 and select Finish in step 3. Delete column C. Select columns B and C and Insert>>Charts>>Scatter with Smooth Lines from the menu. Save the file in .xlsx format.

Figure 34.9h) Mach number variation along the axis in Excel for NPR = 4

Select Results>>Surface>>Create>>Line/Rake… from the menu. Set y1 [m] to 0.0381. Enter throat as New Surface Name.

Select Results>>Reports>> Surface Integrals from the menu. Choose Mass Flow Rate under Report Type in the Surface Integrals window. Select *inlet* and *throat* as Surfaces and click on Compute. The throat mass flow rate is shown in the console and the value is $\dot{m}_{Fluent}$ = 4.1092 kg/s. We can now determine the discharge coefficient as shown in equation (6) in the theory section as the mass flow rate found from Ansys Fluent divided by the ideal mass flow rate as shown in equation 8. The value for the ideal mass flow rate $\dot{m}_i$ = 4.4839 and the discharge coefficient is C_d = 0.9164.

Select Results>>Reports>>Surface Integrals from the menu. Select Area-Weighted Average as Report Type in the Surface Integrals window. Select Pressure… and Static Pressure as Field Variable. Select *throat* under Surfaces and click on Compute. The Ansys Fluent Area-Weighted Average pressure at the throat is shown in the Console and is p_{avg}= 257367.68 Pa.

Change the Report Type to Mass-Weighted Average and select Velocity… and Axial Velocity as Field Variable. Click on Compute. The mass-weighted average axial velocity at the throat is U_{ave} = 248.83979 m/s. We can now determine the thrust coefficient as shown in equation (7) in the theory section. The value for the thrust coefficient is C_t = 0.9878.

Right click on Report Definitions under Solution in Outline View and select New>>Surface Report>>Mass Flow Rate…. Enter the Name *mdot* and select *throat* under Surfaces in the Surface Report Definition window. Uncheck the box for Report Plot under Create. Select OK to close the window.

Right click on Report Definitions under Solution in Outline View and select New>>Surface Report>>Facet Average…. Enter the Name *pt* and select *inlet* under Surfaces in the Surface Report Definition window. Select Pressure… and Total Pressure as Field Variable. Uncheck the box for Report Plot under Create. Select OK to close the window.

Right click on Report Definitions under Solution in Outline View and select New>>Surface Report>>Facet Average…. Enter the Name *t0* and select *inlet* under Surfaces in the Surface Report Definition window. Select Temperature… and Total Temperature as Field Variable. Uncheck the box for Report Plot under Create. Select OK to close the window.

Right click on Report Definitions under Solution in Outline View and select New>>Expression…. Enter the Name *cd* and select *mdot* under Report Definitions in the Expression Report Definition window. Add the expression as shown in Figure 34.9i) to define the drag coefficient. You will need to select *pt* and *t0* under Report Definitions as you enter the expression as shown in Figure 34.9i). Check the box for Create Output Parameter. Select OK to close the window.

```
Expression Report Definition                                              ×

Name cd

-{mdot}/(0.8101851852*{pt}*0.0045603673/sqrt(1.4*287*{t0}))    [ Functions    ▼ ]
                                                                [ Variables     ▼ ]
                                                                [ Constants     ▼ ]
                                                                [ Report Definitions ▼ ]
                                                                [ Locations     ▼ ]
```

Figure 34.9i) Expression report definition for drag coefficient C_d

Right click on Report Definitions under Solution in Outline View and select New>>Surface Report>> Area-Weighted Average …. Enter the Name *pinf* and select *outlet* under Surfaces in the Surface Report Definition window. Select Pressure… and Static Pressure as Field Variable. Uncheck the box for Report Plot under Create. Select OK to close the window.

Right click on Report Definitions under Solution in Outline View and select New>>Expression…. Enter the Name *ui* and select Functions>>Mathematical>>sqrt in the Expression Report Definition window. Select *t0* under Report Definitions to add the expression as shown in Figure 34.9j) in order to define *ui*. You will need to select *pinf* and *pt* under Report Definitions as you enter the expression as shown in Figure 34.9j). Uncheck the box for Report Plot under Create. Select OK to close the window.

Expression Report Definition

Name | ui

$$sqrt((2.8*287*\{t0\}/0.4)*(1-(\{pinf\}/\{pt\})^(.4/1.4)))$$

Figure 34.9j) Expression report definition for ideal velocity U_i

Right click on Report Definitions under Solution in Outline View and select New>>Expression…. Enter the Name *fi* in the Expression Report Definition window. Select *mdot* and *ui* under Report Definitions to add the expression as shown in Figure 34.9k) in order to define *fi*. Uncheck the box for Report Plot under Create. Select OK to close the window.

Expression Report Definition

Name | fi

$$-\{mdot\}*\{ui\}$$

Figure 34.9k) Expression report definition for ideal force F_i

Right click on Report Definitions under Solution in Outline View and select New>>Surface Report>> Area …. Enter the Name *at* and select *throat* under Surfaces in the Surface Report Definition window. Uncheck the box for Report Plot under Create. Select OK to close the window.

Right click on Report Definitions under Solution in Outline View and select New>>Surface Report>> Area-Weighted-Average…. Enter the Name *pave* and select *throat* under Surfaces in the Surface Report Definition window. Select Pressure… and Static Pressure as Field Variable. Uncheck the box for Report Plot under Create. Select OK to close the window.

Right click on Report Definitions under Solution in Outline View and select New>>Surface Report>> Mass-Weighted-Average…. Enter the Name *uave* and select *throat* under Surfaces in the Surface Report Definition window. Select Velocity… and Axial Velocity as Field Variable. Uncheck the box for Report Plot under Create. Select OK to close the window.

Right click on Report Definitions under Solution in Outline View and select New>>Expression…. Enter the Name *fj* in the Expression Report Definition window. Select *mdot* , uave, pave, pinf and *at* under Report Definitions to add the expression as shown in Figure 34.9l) in order to define *fj*. Uncheck the box for Report Plot under Create. Select OK to close the window.

Expression Report Definition

Name | fj

$$-\{mdot\}*\{uave\}+(\{pave\}-\{pinf\})*\{at\}$$

Figure 34.9l) Expression report definition for force F_j

Right click on Report Definitions under Solution in Outline View and select New>>Expression.... Enter the Name *ct* and select *fj and fi* under Report Definitions in the Expression Report Definition window. Divide *fj* by *fi* as shown in Figure 34.9m) to define the thrust coefficient. Check the box for Create Output Parameter. Select OK to close the window.

Expression Report Definition

Name ct

fj/fi

Figure 34.9m) Expression report definition for thrust coefficient C_t

You will create the case and scheme files to complete a batch job. Select File>>Write>>Case & Data... from the menu. Save the case and data files in the *Working Directory* folder with the name ***conicalnozzle40deg.cas.***

Open Notebook and create and save the scheme file with the name ***shock-diamond10X.scm*** for the batch job. The scheme file is shown in Figure 34.9n) and is available for download at *sdcpublications.com*. Select File>>Read>>Scheme... from the menu in Ansys Fluent. Open the scheme file to start the batch job. Plotdigitizer is available at https://plotdigitizer.com/app and can be used to extract experimental and Wind-US data from reference 1 for comparison with Ansys Fluent results.

shock-diamond10X.scm - Notepad

File   Edit   Format   View   Help

```
(Do ((x 141894.06 (+ x 20270.58))) ((> x 729740.88))
(begin
      (Ti-menu-load-string (format #f "file read-case conicalnozzle40deg.cas"))
      (Ti-menu-load-string (format #f "def b-c p-i inlet y n ~a n 101352.9 n 277.74 n y n n n y 0.01 0.147066" x))
      (Ti-menu-load-string (format #f "solve initialize hyb-initialization yes"))
      (Ti-menu-load-string (format #f "solve iterate 1000"))
      (Ti-menu-load-string (format #f "dis s-p nozzle_40_degrees_mass_flow_rate_plot_p0=~a.jpg" x))
      (Ti-menu-load-string (format #f "r s-i m-f-r inlet throat () y nozzle_40_degrees_mass_flow_rate_data_p0=~a" x))
      (Ti-menu-load-string (format #f "r s-i a-w-a throat () p y nozzle_40_deg_area_weight_ave_statpressure_p0=~a" x))
      (Ti-menu-load-string (format #f "r s-i m-w-a throat () a-v y nozzle_40_deg_mass_weight_ave_axial_velocity_p0=~a" x))
      (Ti-menu-load-string (format #f "display objects display contour-1"))
      (Ti-menu-load-string (format #f "dis s-p nozzle_40_degrees_static_pressure_contour_plot_p0=~a.jpg" x))
      (Ti-menu-load-string (format #f "display objects display contour-2"))
      (Ti-menu-load-string (format #f "dis s-p nozzle_40_degrees_Mach_number_contour_plot_p0=~a.jpg" x))
      (Ti-menu-load-string (format #f "display objects display contour-3"))
      (Ti-menu-load-string (format #f "dis s-p nozzle_40_degrees_density_contour_plot_p0=~a.jpg" x))
      (Ti-menu-load-string (format #f "display objects display contour-4"))
      (Ti-menu-load-string (format #f "dis s-p nozzle_40_degrees_temperature_contour_plot_p0=~a.jpg" x))
      (Ti-menu-load-string (format #f "display objects display contour-5"))
      (Ti-menu-load-string (format #f "dis s-p nozzle_40_degrees_axial_velocity_contour_plot_p0=~a.jpg" x))
      (Ti-menu-load-string (format #f "display objects display contour-6"))
      (Ti-menu-load-string (format #f "dis s-p nozzle_40_degrees_radial_velocity_contour__plot_p0=~a.jpg" x))
      (Ti-menu-load-string (format #f "display objects display xy-plot-1"))
      (Ti-menu-load-string (format #f "dis s-p nozzle_40_degrees_Machnumber_axis_xy_plot_p0=~a.jpg" x))
      (Ti-menu-load-string (format #f "file export ascii nozzle_40_deg_Machno_axis_data_p0=~a 2 () n mach-number q n" x))
      (Ti-menu-load-string (format #f "plot residuals y y y y y"))
      (Ti-menu-load-string (format #f "dis s-p nozzle_40_degrees_residuals_plot_p0=~a.jpg" x))
))
```

Figure 34.9n) Batch job scheme file for 40-degrees conical nozzle

Figure 34.9o) Nozzle pressure ratio versus drag coefficient for 40-degree conical nozzle

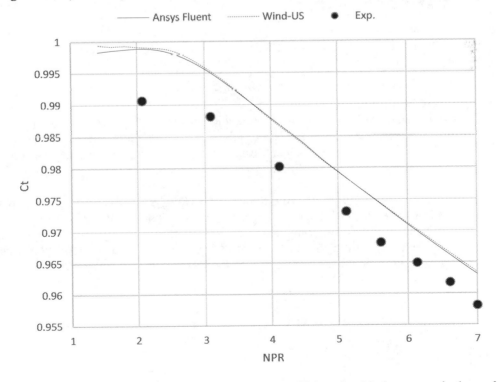

Figure 34.9p) Nozzle pressure ratio versus thrust coefficient for 40-degree conical nozzle

10. Double click on Boundary Conditions under Setup in the Outline View. Select *inlet* as Zone under Boundary Conditions on the Task Page. Select pressure-inlet as Type on the Task Page. Set the Gage Total Pressure [Pa] to 253382.25. Select Apply and Close the Pressure Inlet.

Double click on Initialization under Solution in the Outline View. Select Hybrid Initialization and click on Initialize. Double click on Run Calculation under Solution in the Outline View. Enter 1000 as Number of Iterations and select Calculate.

Select Results>>Surface>>Create>>Iso-surface…. Enter *sonic-line* as New Surface Name. Select Surface of Constant Velocity… and Mach Number. Select *solid-surface_body* from under From Zones and select *interior-surface_body* from under From Surface. Use the slider to slide the Iso-Value and set value to 1.000. Select Create and Close the Iso-Surface window.

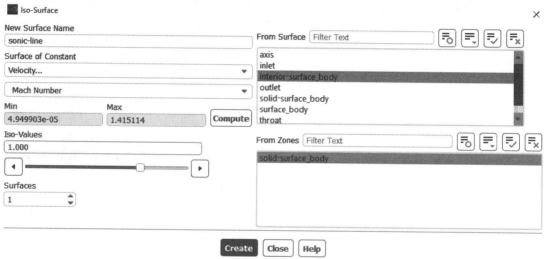

Figure 34.10a) Iso-surface settings

Double click on XY Plot under Plots and Results in the Outline View. Set the XY Plot Name to *sonic-line-mach-number-1* in the Solution XY Plot window. Uncheck Position on X Axis under Options. Select Mesh… and Y-Coordinate as Y Axis Function and select Mesh… and X-Coordinate as X Axis Function.

Select Axis… and uncheck Auto Range under Options for X Axis. Set Minimum to -0.01 and Maximum to 0.07 under Range. Set Type to general and Precision to 3 under Number Format. Select Apply and select Y Axis. Set Type to general and Precision to 3 under Number Format. Select Apply and Close the Axes – Solution XY Plot window. Select sonic-line under Surfaces and click on Save/Plot.

Figure 34.10b) Solution XY Plot window settings

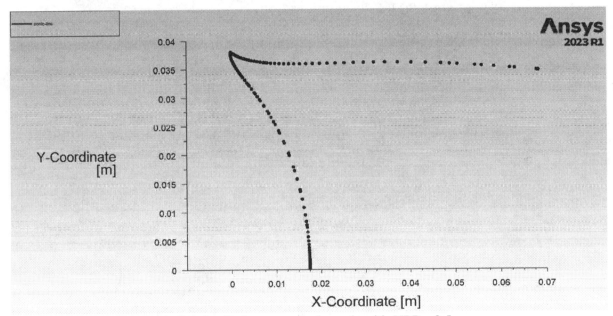

Figure 34.10c) Sonic line for 40-degree conical nozzle with *NPR* = 2.5

Select File>>Export>>Solution Data from the menu. Set the File Type as ASCII. Select Node Location and Space Delimiter. Select *sonic-line* under Surfaces. Select Write… and save the file with the name *sonic-line- coordinates-NPR=2.5.* Close the Export window. Open the file in Excel and delete column A with node numbers.

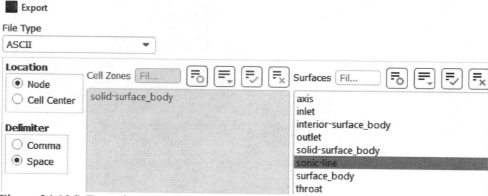

Figure 34.10d) Exporting coordinates for sonic line

Sort by x-coordinate from smallest to largest and right click and clear contents both columns of data for all data above $x = 2e-2$. Copy negative x-coordinate data and corresponding y-coordinate data and paste them in columns D-E. Clear the same data in columns A-B. Sort columns D-E by y-coordinate from smallest to largest.

Copy the remaining data from columns A and B and paste them in columns G and H. Sort columns G-H by y-coordinate from smallest to largest. Enter =SQRT(G2^2+H2^2) in cell I2 and drag this cell down the whole column. Clear content of columns G-I with data that have values in the I column larger than 0.0368. Copy the data in columns D-E and paste these data points at the end of the data in columns G-H.

Enter =G2/0.0762 in cell J2 and =H2/0.0762 in cell K2. Drag the J2 and K2 cells down the whole columns of data. Label the J column as x/D and K column y/D. Plot the data and include reference 1 experimental and Wind-US data for comparison.

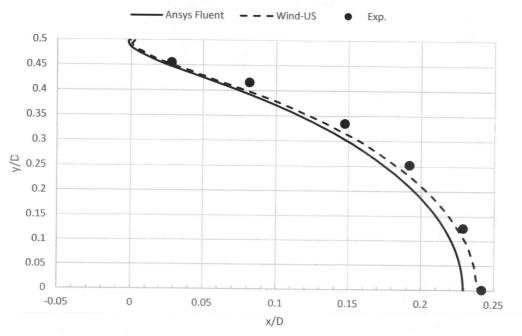

Figure 34.10e) Sonic line for 40-degree conical nozzle with $NPR = 2.5$

H. Theory

11. We define the static throat pressure p_t related to the nozzle pressure ratio NPR and total stagnation pressure p_0 at the inlet

$$p_t = NPR \cdot p_0 \tag{34.1}$$

Based on the NPR value, we can determine the ideal Mach number M_i, ideal temperature T_i, ideal density ρ_i and ideal velocity U_i of the jet

$$M_i = \sqrt{\frac{2}{\gamma-1}\left(NPR^{\frac{\gamma-1}{\gamma}} - 1\right)} \tag{34.2}$$

$$T_i = \frac{T_t}{1+\frac{\gamma-1}{2}M_i^2} \tag{34.3}$$

$$\rho_i = \frac{\rho_t}{\left(1+\frac{\gamma-1}{2}M_i^2\right)^{\frac{1}{\gamma-1}}} \tag{34.4}$$

$$U_i = M_i \cdot \sqrt{\gamma R T_0} \tag{34.5}$$

where ρ_t, T_t are throat density and throat temperature, respectively, T_0 is the stagnation temperature at the inlet, γ is the ratio of specific heats and $R = 287$ J/kgK is the gas constant. We can determine the discharge coefficient and the thrust coefficient for the 2D axisymmetric nozzle as

$$C_d = \frac{2\pi \int_0^{r_t} \rho u r \, dr}{\dot{m}_i} = \frac{\dot{m}_{Fluent}}{\dot{m}_i} \tag{34.6}$$

$$C_t = \frac{2\pi \int_0^{r_t} [\rho u^2 + (p-p_\infty)] r \, dr}{U_i \dot{m}_{Fluent}} = \frac{\dot{m}_{Fluent} U_{ave} + (p_{avg} - p_\infty) A_t}{U_i \dot{m}_{Fluent}} \tag{34.7}$$

where ρ, u and p are local density, axial velocity and pressure. The radius and area at the nozzle exit or throat are r_t and A_t, respectively. The mass flow rate from Ansys Fluent simulation is $\dot{m}_{Fluent}$. The Ansys Fluent area-weighted average pressure at the throat is p_{avg} and p_∞ is ambient pressure. The Ansys Fluent mass-weighted average velocity at the throat is U_{ave}. The ideal isentropic mass flow rate can be determined as

$$\dot{m}_i = \frac{\psi A_t p_0}{\sqrt{\gamma R T_0}} \tag{34.8}$$

where the flow factor ψ depends on the critical pressure p^*

$$\psi = \gamma \left(\frac{2}{\gamma+1}\right)^{\frac{\gamma+1}{2(\gamma-1)}} \qquad \text{if} \qquad p < p^* \tag{34.9}$$

$$\psi = \frac{2\gamma^2 NPR}{\gamma-1}\left[1 - NPR^{\frac{\gamma-1}{\gamma}}\right] \qquad \text{if} \qquad p \geq p^* \tag{34.10}$$

$$p^* = \left(\frac{2}{\gamma+1}\right)^{\frac{\gamma}{\gamma-1}}$$

(34.11)

A_t	throat area	0.004560367	m^2
C_d	discharge coefficient	0.917	
D	Throat diameter	0.0762	m
γ	ratio of specific heats	1.4	
$\dot{m}_i$	ideal mass flow rate	4.4839	kg/s
M_i	ideal Mach number	1.559	
NPR	nozzle pressure ratio	4	
p	local pressure		Pa
p^*	critical pressure	0.52828	Pa
p_0	stagnation pressure at inlet	101352.9	Pa
p_t	throat pressure	405411.6	Pa
p_∞	ambient pressure	101352.9	Pa
R	gas constant	287	J/kgK
r_t	throat radius	0.0381	m
T_i	ideal temperature	186.905	K
T_0	stagnation temperature at inlet	277.74	K
T_t	throat temperature		K
u	local axial velocity		m/s
U_i	ideal axial velocity	427.1851988	m/s
ρ	local density		kg/m^3
ρ_i	ideal density		kg/m^3
ρ_t	throat density		kg/m^3
ψ	flow factor	0.81	

Table 34.1 Nomenclature

I. References

1. Dippold, V.F., "Computational Simulations of Convergent Nozzles for the AIAA 1st Propulsion Aerodynamics Workshop", NASA/TM-2014-218329, 2014.
2. Sasanpuri, B., Kumar, M. and Wirogo, S., "Simulation of flow through Supersonic Cruise Nozzle: A validation study", TFAWS 2011, August 15 – 19, NASA Langley Research Center, 2011.
3. Spotts, N.G., Guzlk, S. and Gao, X., "A CFD Analysis of Compressible Flow through Convergent-Conical Nozzles", AIAA 2013-3734, 2013.
4. Thornock, R.L. and Brown, E.F., "An Experimental Study of Compressible Flow Through Convergent-Conical Nozzles, Including a Comparison with Theoretical Results.", Journal of Basic Engineering, December, 1972.

J. Exercises

34.1 Create the mesh for the 15 degrees conical nozzle and run the simulations as shown in this chapter. Compare with experimental data and Wind-US simulations, see ref. 1.

34.2 Create the mesh for the 25 degrees conical nozzle and run the simulations as shown in this chapter. Compare with experimental data and Wind-US simulations, see ref. 1.

34.3 Create the mesh for the reference conical nozzle and run the simulations as shown in this chapter. Compare with experimental data and Wind-US simulations, see ref. 1.

INDEX

Notes: